FRACTURE MECHANICS TECHNOLOGY APPLIED TO MATERIAL EVALUATION AND STRUCTURE DESIGN

Fracture Mechanics Technology Applied to Material Evaluation and Structure Design

Proceedings of an International Conference on 'Fracture Mechanics Technology Applied to Material Evaluation and Structure Design', held at the University of Melbourne, Melbourne, Australia, August 10–13, 1982

edited by

G. C. SIH
Institute of Fracture and Solid Mechanics,
Lehigh University, Bethlehem, Pennsylvania, U.S.A.

N. E. RYAN and R. JONES
Aeronautical Research Laboratories,
Melbourne, Australia

1983 **MARTINUS NIJHOFF PUBLISHERS**
a member of the KLUWER ACADEMIC PUBLISHERS GROUP
THE HAGUE / BOSTON / LANCASTER

Distributors

for the United States and Canada: Kluwer Boston, Inc., 190 Old Derby Street, Hingham, MA 02043, USA
for all other countries: Kluwer Academic Publishers Group, Distribution Center, P.O.Box 322, 3300 AH Dordrecht, The Netherlands

Library of Congress Cataloging in Publication Data

International Conference on Fracture Mechanics Technology
    Applied to Material Evaluation and Structure Design
    (1982 : University of Melbourne)
    Fracture mechanics technology applied to material
evaluation and structure design.

    1. Fracture mechanics--Congresses. 2. Strength of
materials--Congresses. 3. Structural design--Congresses.
I. Sih, G. C. (George C.) II. Ryan, N. E. III. Jones, R.
IV. Title.
TA409.I444 1982   620.1'126   83-19508
ISBN 90-247-2885-1

ISBN 978-94-009-6916-2   ISBN 978-94-009-6914-8 (eBook)
DOI 10.1007/978-94-009-6914-8

Copyright

CONTENTS

PREFACE

The International Conference on Fracture Mechanics Technology Applied to Material Evaluation and Structure Design was held in Melbourne, Australia, from August 10 to 13, 1982. It was sponsored jointly by the Australian Fracture Group and Institute of Fracture and Solid Mechanics at Lehigh University. Professor G. C. Sih of Lehigh University, Drs. N. E. Ryan and R. Jones of Aeronautical Research Laboratories served as Co-Chairmen. They initiated the organization of this international event to provide an opportunity for the practitioners, engineers and interested individuals to present and discuss recent advances in the evaluation of material and structure damage originating from defects or cracks. Particular emphases were placed on applying the fracture mechanics technology for assessing interactions between material properties, design and operational requirements. It is timely to hold such a Conference in Australia as she embarks on technology extensive industries where safeguarding structures from premature and unexpected failure is essential from both the technical and economical view points.

The application of system-type approach to failure control owes much of its success to fracture mechanics. It is now generally accepted that the discipline, when properly implemented, provides a sound engineering basis for accounting interactions between material properties, design, fabrication, inspection and operational requirements. The approach offers effective solutions for design and maintenance of large-scale energy generation plants, mining machineries, oil exploration and retrieval equipments, land, sea and air transport vehicles. Although much know-how is readily available for application, new design requirements and materials are being constantly introduced that there is always a relentless pressure for technology advancement. The need for a better understanding of the fundamentals of fracture mechanics becomes apparent, if not only to reduce the research results to design methods. A series of general lectures were given to further emphasize the objectives. Over fifty technical papers were presented covering topics such as dislocations, kinetics of crack growth, fracture criteria, fatigue and stress concerned with material testing and structural application. The four-day technical sessions were well-attended by participants from eleven countries: Australia, Canada, France, Holland, India, Japan, Korea, New Zealand, Singapore, UK and USA. Much of the success of the Conference was attributed to the lively discussions after the technical presentations and during the coffee breaks.

The Conference was officially opened by Dr. W. J. McG. Tegart, Secretary of the Department of Science and Technology who presented an overview of the technological and scientific development in Australia. Technology transfer was mentioned as a key element in Conferences of an international character. Fracture mechanics can, no doubt, provide cost-effective solutions to many of the present and future structure-related problems. Equally successful was the social programs organized by the local committee. They were particularly delightful to the foreign partici-

x

pants. A special note of appreciation goes to Mrs. Dawn Ryan who patiently accompanied the ladies to visit shops and historic sites in Melbourne. The individual Committee Members

C. G. Chipperfield
A. R. Ellery
J. R. Griffiths
R. Jones
R. H. Leicester
M. T. Murray
J. C. Ritter
N. E. Ryan
J. F. Williams

of the Australian Fracture Group should be recognized for their cooperative efforts in organizing the Conference. Acknowledgements are also due to the support of The Australian Institute of Metals, Gas and Fuel Corporation of Victoria, ESSO Australia (Sale), Comalco Ltd., Steel Mains, Pty. Ltd., Victorial Branch, Australia Applied Mathematics Division, Aeronautical Research Laboratories, Computer Engineering Application Pty. Ltd., and Lehigh University.

The readers will also notice the excellent typing of the manuscripts by Mrs. Barbara DeLazaro and Constance Weaver. Their efforts contributed towards the speedy publication of this volume.

August, 1982
Melbourne, Australia

G. C. Sih
N. E. Ryan
R. Jones

OPENING ADDRESS

I consider it a signal honour to be invited to deliver the opening address to the International Conference on Fracture Mechanics Technology Applied to Material Evaluation and Structure Design, organized jointly by the Australian Fracture Group and Lehigh University in the U.S.A. When The Organizing Committee issued the invitation, it reminded me that the Australian Fracture Group traces its origins to an initial meeting of fracture mechanics enthusiasts in 1970 at the BHP Melbourne Research Laboratories in Clayton. At that time, I was Research Manager and the Laboratories were still in their formative years with an enthusiastic dialogue and interaction between scientists and engineers on the properties and application of steels. Naturally, the topic of fracture mechanics and its applications was of considerable interest to us and we developed a considerable expertise to which I will refer later.

The Australian Fracture Group has grown in strength and in scope of interests, as you will learn during the Conference, and has sponsored National Tri-Ennial Fracture Mechanics Symposia in 1974, 1977 and 1980. This is its first venture into the co-sponsorship of an International Conference and I must congratulate the organizers on their success in attracting such a distinguished array of speakers from overseas. To these, I give a warm welcome and trust that they will enjoy their stay in Australia and gain some appreciation of the Australian scene in science and technology. Equally, we look forward to learning from them of the latest developments and research in their countries.

For us, this Conference can be regarded as an exercise in technology transfer since, with a population of only 15 million, we can contribute only roughly 1% of the world's output of science and technology and thus must depend heavily on the rest of the world for advances in most fields. However, in a number of fields, because of unique situations in Australia, we have developed a leading role. In the context of this Conference, the increasing loads in the operation of the heavy rail iron ore railroads in the Pilbara district of North-Western Australia have posed unique materials problems which have led to the development of considerable local expertise in rail steel metallurgy and engineering. In such situations, the cost penalty of fractures is extremely high. Thus, a broken rail which causes a derailment can easily lead to an operating loss of $1 million through lost production plus a replacement cost of several hundred thousand dollars for track and rolling stock. Further, the rapid expansion in minerals and coal mining together with extensive exploitation of oil and gas deposits has highlighted the need for cost-effective solutions to materials usage in severe operating environments. Again, the cost penalties of failures are very high and we have developed some unique expertise in Australia. You will have an opportunity to learn of this during the Conference.

In general, however, Australian industry depends on transfer of technology from overseas. Some years ago, my former colleague, Dr. Ward (General Manager,

Research and New Technology of BHP) carried out an analysis of the rate of adoption of new technology in the Australian steel industry. He showed that the lapsed time between first commercial world application and commissioning in Australia had decreased markedly over the years, particularly in the last two decades. This reflected partly the rapid development of the local industrial base with market growth and sophistication as the rate determining factors and partly the improved access to information. In the case of major plant technological introductions, the average lapsed time was about 9 years of which perhaps 4 years are required for design construction and installation leaving about 5 years for recognition and evaluation of overseas innovations. In the case of produce development, the average lapsed time was about 7 years, of which about 4 years can be attributed to recognition and evaluation of overseas innovations and 3 years to implementation.

One might expect that techniques such as fracture mechanics could be more readily transferred but my recent examination of the Australian scene at the formation of the AFG suggests that at least 5 years ensued between application overseas and acceptance in Australia. This is perhaps not surprising since we are all aware of the lead times involved in learning new techniques, building new equipment and carrying out an adequate test program. In that particular case, the catalyst of technology transfer was the Australian steel industry through its own R&D and plant laboratories with the support of the marketing division. However, a significant transfer agent was the Australian Welding Research Association whose Project Panels rapidly took up the concepts and applied them to practical situations such as pipelines and welded structures and developed industry quality acceptance criteria based on fracture mechanics.

AWRA is an excellent example of a multiplier agency for technology transfer since it operates with only five full-time staff, but utilizes the services of 150 voluntary part-time workers in 13 Project Panels in areas of specialized interest. Apart from monitoring and assessing overseas technology, these Panels stimulate research and development in both industrial and government laboratories and in tertiary institutions. Funding is provided by subscriptions from member companies and organizations and a matching grant from the Federal Government.

With the current high rate of development of technology, the efficiency of the technology transfer process within the Australian context compared to that of our principal competitors will be a critical determinant of the competitiveness of Australian industry and thus of its future. However, there are significant barriers to technology transfer to manufacturing industry in Australia. Thus, while technology can be transferred in documented form, a key element in the transfer process is people experienced in the application of the technology to be transferred. These may be academics or R&D workers in government or private laboratories and plants. At the receiving end, it is necessary to have people who are well-informed of their company's requirements and operations and who can interact with potential sources of new technology.

In Australia, manufacturing industry is characterized by a large proportion of companies, particularly but not exclusively small to medium-sized companies which lack the necessary financial and resource capability to mount effective R&D programs and which are dependent on outside sources of technological expertise. For example, firms in the fabricated metal products sector with less than 100 employees account for 93% of the establishments of the sector and provide about half the employment, turnover and value added. Many do not employ technical staff who can evaluate and apply new developments.

This problem has been recognized for a number of years and the development of research associations such as AWRA has been a partial answer. However, much more is needed to be done and the Federal Government responded by forming the Department of Productivity in 1976. This Department initiated a number of innovative programs designed to improve technology transfer to manufacturing industry. In 1981, these programs were moved to the Department of Science and Technology and have recently been reviewed by an expert committee drawn from industry. As a result, some are being phased out, but some are being strengthened.

I propose to briefly describe the latter to indicate the current thrust of Government involvement in technology transfer in Australia. They are the Manufacturing Technology Transfer Program, the Information Technology Program and the Manufacturing Technology Program.

Manufacturing Technology Transfer Program. The major activity has been the establishment of a technical referral network for the metals manufacturing industry. The Department acted as the catalyst to bring together a number of professional and industry groups to define industry needs and provided the financial resources needed to underwrite the proposed program of activities. In order to maintain a strong industry orientation and ensure that Government involvement was at arms length, a non-profit making company, the Technology Transfer Council, was established in October 1979 by the Confederation of Australian Industry and the Metal Trades Industry Association of Australia.

After a period of establishment, the TTC commenced formal operations in June 1980 on a 3 year pilot program. The network currently has eight operational centres sited in the five major mainland capitals staffed by technical liaison officers. The centres are hosted by a variety of Government, academic and industry organizations.

Since its inception, the TTC has developed rapidly. Thus:

1. It has responded to over 1500 technical queries from industry. Of these, about 60% can be dealt with rapidly, about 20% require a visit by a technical liaison officer, and the remaining 20% lead to major consultative activities.

2. It has conducted 25 practical workshops attended by over 1000 industrial personnel.

3. It has organized and conducted a number of group projects involving groups of companies coming together to address a common problem or to evaluate a common technology. A current example is the evaluation of the Toyota diffusion process which is a surface coating technology that could significantly improve the life of tooling. Approximately 60 firms are involved in a project to assess the areas of economic application of this technology.

The review committee strongly supported the continuance of the TTC for a further period and the Government has agreed to continue support.

Information Technology Program. This program was established to encourage the effective application of information system technologies. The Department seeks to provide publicity, education and demonstration of the potential benefits of applying modern information technologies. Apart from the continuing evaluation of technology developments and of their impact on society, the major activity has been the creation of Information Technology Week.

This program is operated in co-operation with Telecom and The Australian Computer Society, and is designed to facilitate a public demonstration of technology and promote a greater awareness of information technology within the community. The Department provides a small secretariat to coordinate the national ITW program to which firms, industry associations and tertiary institutions contribute significant resources.

This year, Information Technology Week will be run over the period August 15 to September 4, depending on the state involved and our overseas colleagues will be able to judge the scope and range of the activities offered from the newspaper advertisements that are already appearing. (In Victoria, ITW will run from August 29 to September 4). Well over a hundred separate activities ranging from small specialist seminars to major exhibitions will take place during the various ITW in the states. In addition to promotion of community awareness of the potential benefits of information technology, ITW has significant benefits for equipment and software suppliers and their potential customers in industry. This program will be continued with Government support.

Manufacturing Technology Program. This program is aimed at developing new manufacturing technology and introducing it to Australian industry. Projects have been organized to foster the introduction and development of new technology in individual firms by providing funding under contract outside the normal provisions of the IR&D Incentives Scheme. Projects are supervised by steering committees comprised of representatives from the Department, the contracting firm and other interested organizations. A significant task of the steering committee is to ensure widespread dissemination of results to industry and the TTC clearly has a role to play in this process.

Projects to date have concentrated on applications of computer-aided design and computer-aided machining and of robotics in manufacturing industry. The review committee recommended a greater degree of industry involvement in project selection and funding and the program is being restructured to take account of these recommendations.

I have chosen to develop the theme of technology transfer in view of my earlier remarks on the relative contribution of Australian science and technology to the world pool of knowledge and the perceived need for the assessment and introduction where applicable of overseas research and development. I believe that International Conferences have a significant role to play in the technology transfer process, both by the material presented and by the relationships developed between the participants. I therefore have great pleasure in declaring open this International Conference on Fracture Mechanics Technology Applied to Material Evaluation and Structure Design and wish you all a successful and producting meeting.

August, 1982
Melbourne, Australia

W. J. McG. Tegart

N. E. Ryan, Co-Chairman, welcomed the participants in
the Opening Session

G. C. Sih, Co-Chairman, remarked on the application of
fracture mechanics technology to engineering problems

A typical technical presentation in Laby Theater

Participants attending a general lecture in Melbourne University

Social hour before the Banquet

From left to right: George Sih, Jennice Sih, Rhys Jones,
Neil Ryan, Dawn Ryan and Bill Smith

Posing for a picture at the Banquet Dinner

A view of Ormond Hall in the evening of August 12, 1982

N. E. Ryan commenting on the success of the Conference

Co-Chairmen N. E. Ryan and G. C. Sih exchange plaques

A post conference fracture mechanics picnic in Melbourne, August 14, 1982

GENERAL LECTURES

MECHANICS OF SUBCRITICAL CRACK GROWTH

G. C. Sih

Lehigh University
Bethlehem, Pennsylvania 18015 USA

INTRODUCTION

Subcritical crack growth is a commonly observed but still not well-understood phenomenon. It is generally identified with the process of slow crack growth in metals subjected to rising or cyclic load. The phenomenon, however, is not exclusively associated with ductile* fracture nor with plastic deformation. Cracks can spread slowly in an elastic stress environment as long as the crack driving force is kept below the critical state. More precisely, it is the combined interaction of loading, geometry and size of specimen, material and environment that determines the crack growth characteristics.

Despite the many attempts [1-5] made to analyze subcritical crack growth, the viewpoints on this subject remain diversified. There are different schools of thought that often appear to be a matter of personal appreciation. The education received, the traditions immersed in and the motive of research can highly affect the degree of this appreciation. The idea of quantifying the different failure modes of measuring the Charpy V-notch energy, the fracture toughness, crack opening displacement, etc., has led to an overwhelming number of parameters. However, it is not at all clear as to how they could be applied to situations other than those tested. While the material testing approach serves a useful purpose in cataloging material behavior under simulated laboratory conditions, it provides little or no useful purpose to a designer who needs to know the allowable load and net section size of structural components. Predictive capability is lacking in the material testing approach.

The initiation and termination of subcritical crack growth cannot be assessed quantitatively by a single parameter. Many of the current approaches involving the arbitrary selection of different and independent failure criteria for the same material damage process have not had success in that the results tend to be highly inconsistent. It should be recognized that global nonlinearity of load and displacement often observed during slow crack growth is fundamentally connected

---

*Ductility as defined by the uniaxial tensile test may not apply to other situations where the character of loading and specimen size are different.

with non-self similar[*] profiles of the crack and permanent distortion of the neighboring material commonly known as plastic deformation. These two effects have to be treated simultaneously and are inherently load history dependent. Should plastic deformation occur along the prospective fracture path, then adjustments must also be made for the change of material resistance as the crack grows. To be remembered is that the determination of material resistance to fracture serves only as a pre-requisite for predicting the global behavior of a system containing flaws or imperfections. Hence, the validity of any fracture toughness parameters[**] can only be tested by its predictive capability. The basic problem of subcritical crack growth lies in determining the available energy to drive the crack which is not always straightforward when energy dissipation due to plasticity is present.

The main objective of this communication is to illustrate how slow crack growth can be characterized in a consistent fashion. This involves the determination of crack initiation, slow growth and termination. The global response of an initially cracked specimen will be predicted from the uniaxial tensile data. Subcritical crack growth accompanied by irreversible plastic deformation is analyzed by assuming that the strain energy density function $(dW/dV)_c^*$ along the crack path attains various threshold values. When yielding prevails ahead of the crack, the energy lost $(dW/dV)_p$ due to plastic deformation is no longer available at the time of macrocrack growth. It must therefore be subtracted from $(dW/dV)_c$ which represents the total area under the nonlinear true stress and strain curve to yield $(dW/dV)_c^*$. The crack tip energy density field is calculated for each load increment by employing the twelve-node isoparametric finite element procedure. On the basis of the general relation $dW/dV = S/r$ where S is the strain energy density factor and r the distance from the crack tip, the condition $dS/da = $ const. was found to yield a straight line relationship between S and half crack length a. The slope of this line changes as the load increments are varied. This information is valuable for determining the influence of loading rate on crack growth.

CRITERION ON CRACK INITIATION AND TERMINATION

The selection of an appropriate crack growth criterion for characterizing ductile fracture has been problematic. There is always the temptation to prematurely fit data with analytical results. Agreements concocted by simple models are frequently found to be case-specific and shed no light on understanding the crack growth process. Consistency in reasoning and application of the mechanics discipline should be observed when choosing failure criteria. It is in this respect that the strain energy density criterion [6,7] has been chosen in this work.

---

[*] This feature has been frequently neglected in the analysis and can lead to gross error in predicting the allowable load on a flawed structural member.

[**] The K_{1c} quantity should not be regarded as a material constant but material *behavioral* parameter that is identified with the phenomenon of the sudden exchange of stored energy with free surface energy. The result is the onset of rapid crack propagation. This is simply because surface creation behavior is governed by the combination of loading, specimen size and geometry even if the metallurgical properties of the material are the same.

Strain Energy Density Function. The energy per unit volume in a continuum element can be computed from

$$\frac{dW}{dV} = \int_{0}^{\varepsilon_{ij}} \sigma_{ij} d\varepsilon_{ij} \tag{1}$$

in which σ_{ij} and ε_{ij} are the stress and strain components in a rectangular Cartesian coordinate system. Equation (1) obviously is valid for any constitutive relations. Without loss of generality, a strain energy density factor S can be defined as

$$\frac{dW}{dV} = \frac{S}{r} \tag{2}$$

Here, r is the distance between the nearest neighbor element and point[*] under investigation. The 1/r singular behavior holds for any material and geometric discontinuities. This fundamental character of the strain energy density criterion is unmatched by other criteria.

The quantity dW/dV can be measured experimentally as the area under the true stress and strain curve. Critical values of $(dW/dV)_c$ for a host of metal alloys can be found in [2,3] and used to predict the failure initiation of local elements in a continuum. The onset of global instability is governed by the critical value of S_c such that $(dW/dV)_c = S_c/r_c$ with r_c being the critical ligament size of the material.

Stationary and Critical Values. In general, dW/dV will vary from point to point in a solid. Its stationary values[**] can be identified with failure by yielding and fracture. The location of dW/dV maximum or $(dW/dV)_{max}$ is associated with failure by excessive change in shape or yielding while dW/dV minimum or $(dW/dV)_{min}$ is connected with failure by excessive change in volume or fracture. Crack initiation is assumed to occur when $(dW/dV)_{min}$ reaches a critical value $(dW/dV)_c$. If the material element ahead of the macrocrack has already been yielded corresponding to state p in Figure 1, then the amount of energy density $(dW/dV)_p$ dissipated due to plastic deformation prior to crack growth must be subtracted from $(dW/dV)_c$. This difference is

$$\left(\frac{dW}{dV}\right)_c^* = \left(\frac{dW}{dV}\right)_c - \left(\frac{dW}{dV}\right)_p \tag{3}$$

---

[*] It normally corresponds to the location of failure initiation such as a crack tip, void, inclusion or any other mechanical imperfections.

[**]To be understood is that there can be many dW/dV maxima and minima in a continuum body. It is the maximum of $(dW/dV)_{max}$ and $(dW/dV)_{min}$ that will first reach their respective thresholds of yielding and fracture.

Referring to Figure 1, $(dW/dV)_c^*$ represents the area p'pcb while $(dW/dV)_c$ and $(dW/dV)_p$ are given by oecb and oepp', respectively. The line pp' being parallel

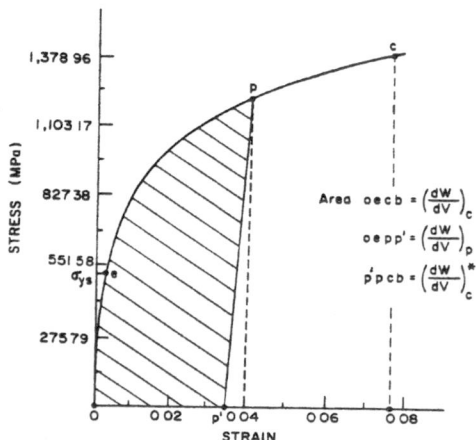

Fig. (1) - True stress and strain response of an elastic-plastic material

to oe is the unloading path. As the macrocrack grows, the point p takes different values between e and c and hence $(dW/dV)_c^*$ or the resistance to crack growth will change according to the load history. When crack growth takes place beyond the yield point p, $(dW/dV)_p$ is zero and $(dW/dV)_c^* = (dW/dV)_c$.

Conditions of Crack Growth. Since crack growth is strictly path dependent, the rate at which it grows depends on the load increment and is assumed to be governed by

$$\left(\frac{dW}{dV}\right)_c^* = \frac{S_1}{r_1} = \frac{S_2}{r_2} = \ldots = \frac{S_j}{r_j} = \ldots = \frac{S_c}{r_c} = const. \tag{4}$$

If the loading increments are such that

$$S_1 < S_2 < \ldots < S_j < \ldots < S_c$$

$$\tag{5}$$

$$r_1 < r_2 < \ldots < r_j < \ldots < r_c$$

then the system will eventually reach global instability when the condition S_c/r_c = const. is satisfied as shown in equation (4). The inequalities in equations (5) are to be reversed if the loading increments lead to crack arrest. The ratio S_c/r_c in equation (4) will then be replaced by S_0/r_0 where S_0 and r_0 correspond to the values at arrest. In the case of Mode I crack propagation, $dr = da$ and the growth rate condition can be expressed simply as

$$\frac{dS}{da} = \text{const.} \tag{6}$$

Fatigue. Equation (6) applies equally well to fatigue crack growth. The critical value $(dW/dV)_c^*$ in equation (3) for monotonic loading has to be modified for cyclic loading such that a material element will no longer fail on the first cycle but energy has to be accumulated after many cycles, say ΔN, before crack growth Δa occurs. Assumed is the critical value

$$\left(\frac{dW}{dV}\right)_c^* = \left(\frac{dW}{dV}\right)_c - \left[\left(\frac{dW}{dV}\right)_o\right]_{r=\Delta a} = \sum_{j=1}^{\Delta N} \left[\Delta\left(\frac{dW}{dV}\right)_j\right]_{r=\Delta a} \tag{7}$$

in which $(dW/dV)_o$ is the energy density corresponding to the mean stress level and $\Delta(dW/dV)_j$ is the energy accumulated during the jth loading cycle[*]. In view of equation (2), the expression in equation (7) may be written as

$$\left[\Delta\left(\frac{dW}{dV}\right)_j\right]_{r=\Delta a} = \frac{1}{\Delta a}\left[(\Delta S)_j\right]_{r=\Delta a} \tag{8}$$

Combining equations (7) and (8), a cumulative damage fatigue crack growth rate relation is obtained [8]:

$$\frac{\Delta a}{\Delta N} = \left[\left(\frac{dW}{dV}\right)_c^*\right]^{-1} \frac{1}{\Delta N} \sum_{j=1}^{\Delta N} \left[(\Delta S)_j\right]_{r=\Delta a} \tag{9}$$

Note that equation (9) contains only a single experimentally determined parameter, i.e., $(dW/dV)_c^*$. With an appropriate constitutive relation of the material, fatigue life curves and actual crack growth paths may be obtained numerically in incremental steps.

ELASTIC-PLASTIC MATERIAL AND STRESS ANALYSIS

Damage accumulation of the material will be modeled by the incremental theory of plasticity such that the successive plastic strain increments are added together algebraically. Yielding of the material is assumed to obey the J_2-flow rule with

---

[*]The symbol Δ has a different meaning than those used in the linear crack growth models. It refers to change in dW/dV over a full cycle.

the approximation that the plastic deformations are incompressible[*]. The uniaxial stress-strain relation is of the form

$$
\varepsilon = \begin{cases}
\dfrac{\sigma}{E}, & \sigma \leq \sigma_{ys} \\[2ex]
\dfrac{1}{E} \left\{ \sigma + \alpha \left[\left(\dfrac{\sigma}{\sigma_{ys}}\right)^{\beta} - 1 \right] \sigma_{ys} \right\}, & \sigma > \sigma_{ys}
\end{cases}
\tag{10}
$$

where σ_{ys} is the yield strength of the material. The hardening coefficient is α
with exponent β. A typical σ versus ε behavior of an alloy steel frequently used
in engineering applications is shown in Figure 1 for $\alpha = 0.02$ and $\beta = 5.0$. The
material properties are:

Poisson's ratio: $\nu = 0.3$

Young's modulus: $E = 2.068 \times 10^5$ MPa $\qquad\qquad$ (11)

Yield strength: $\sigma_{ys} = 517.11$ MPa

while the corresponding fracture properties are:

Critical stress intensity factor: $k_{1c} = 103.52$ MPa$\sqrt{m}$

Critical strain energy density factor: $S_c = 1.349 \times 10^4$ N/m $\qquad$ (12)

Critical strain energy density function: $(dW/dV)_c = 48.624$ MJ/m^3

The quantity k_{1c} in equations (12) differs from the fracture toughness value K_{1c}
by the factor $\sqrt{\pi}$.

The modified PAPST [9] computer program that treats material and geometric
nonlinearities will be employed to find the stresses, strains and strain energy densities for each increment of crack growth. Cubic displacement shape functions are
used for an accurate evaluation of the near field solution by placing the two side
nodes at the 1/9 and 4/9 distance from the corner node at the center tip. Refer
to [9] for details.

CRACK GROWTH UNDER MONOTONIC LOADING

Consider a center cracked specimen 25.40 cm wide by 50.80 cm long as given in
Figure 2. It is loaded uniformly by a stress σ_a whose magnitude increases incrementally while a through crack with initial half length $a_1 = 2.54$ cm starts to
grow at $(dW/dV)_c^* = 35.52$ MJ/m^3. Three different sets of load increments will be
considered from a stress level of $\sigma_a = 266.62$ MPa. They will be referred to as
load type A, B, and C corresponding to small, medium and large load increments,

---

[*]This assumption is inadequate for describing the state of affairs near the crack
tip where the dilatational component of dW/dV is not negligible. The J_2-flow rule
considers only the distortional component of dW/dV.

9

respectively. Since the material elements along the x-axis ahead of the crack in Figure 2 will be loaded to yield, the threshold values of $(dW/dV)_c^*$ in equation (3) will change depending on the crack length relative to specimen width. Their values depend on the distance r from the crack tip and are plotted as a function of crack half length a in Figure 3 for r = 0.064, 0.127 and 0.191 cm.

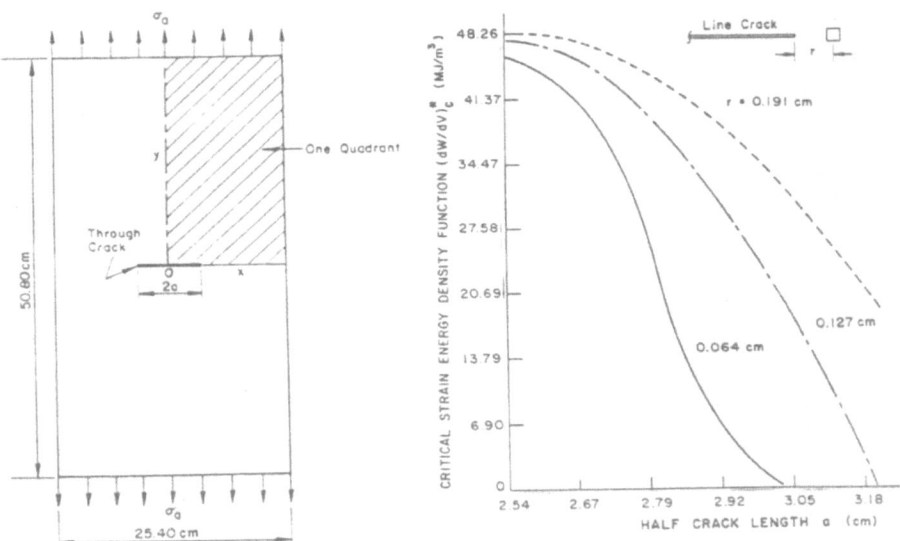

Fig. (2) - A center cracked panel loaded incrementally

Fig. (3) - Critical strain energy density function versus half crack length

The resistance to crack growth is seen to decrease rapidly as the crack grows. This effect becomes more pronounced in the immediate vicinity of the crack tip corresponding to small values of r.

   Load Type A. Six load increments are taken such that the crack is assumed to grow in accordance with equation (4), i.e., $r_1, r_2, ..., r_6$ while $S_1, S_2, ..., S_6$ are computed from the finite element method with S_j/r_j (j = 1,2,...,6) being kept constant. The different values of a_j, $(\sigma_a)_j$ and S_j are shown in Table 1. The strain energy density factors S_j are obtained from the relation $r_j[(dW/dV)_c^*]_j$ where $[(dW/dV)_c^*]_j$ are found from Figure 3 depending on the values of r_j. The results will be discussed together with those obtained for load type B and C.

   Load Type B. By increasing the magnitude of the load increments, the crack is again made to grow from an initial half length of a = 2.54 cm. Figure 4 gives

TABLE 1 - CRACK GROWTH STEPS FOR LOADING TYPE A

Growth Step No.	Half Crack Length a (cm)	Applied Stress σ_a (MPa)	Strain Energy Density Factor S ($\times 10^3$ N/m)
1	2.540	426.61	9.560
2	2.567	439.54	10.367
3	2.596	452.47	11.078
4	2.628	465.40	12.109
5	2.662	491.25	13.802
6	2.700	517.11	15.485

a contour plot of the effective stress

$$\sigma_{eff} = \frac{1}{\sqrt{2}} [(\sigma_1-\sigma_2)^2 + (\sigma_2-\sigma_3)^2 + (\sigma_3-\sigma_1)^2]^{1/2} \tag{13}$$

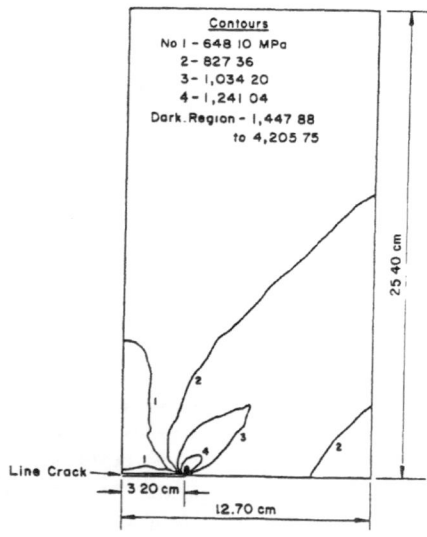

Contours
No I - 648 IO MPa
2 - 827 36
3 - 1,034 20
4 - 1,241 04
Dark.Region - 1,447 88
to 4,205 75

25 40 cm

Line Crack
3 20 cm
12.70 cm

Fig. (4) - Effective stress contours for load type B with a = 2.54 cm and σ_a = 426.62 MPa

for an applied stress of σ_a = 426.62 MPa. For plane strain, $\sigma_3 = \nu(\sigma_1+\sigma_2)$. With reference to the material property data in equations (11), yielding has occurred near contour No. 6. The intensity is seen to increase rapidly as the crack tip is approached. In this case, ten load increments are taken and the crack is grown up to a maximum half length of a = 3.20 cm corresponding to σ_a = 904.94 MPa. Figures 5 and 6 give a graphical account of how the shaded areas S_1 and S_{10} are computed respectively from $r_1[(dW/dV)_c^*]_1$ and $r_{10}[(dW/dV)_c^*]_{10}$. Because of the nature of loading, each increment of crack growth increases tending towards global instability. This can be seen from the results in Table 2. Comparing the results in Tables 1 and 2, it is evident that the crack growth rate process is highly load history dependent.

Load Type C. Further increase on the magnitude of load increments is made such that the crack will grow in even larger increments. It suffices to take five crack growth steps to illustrate the effect of increasing load steps. Refer to the results in Table 3.

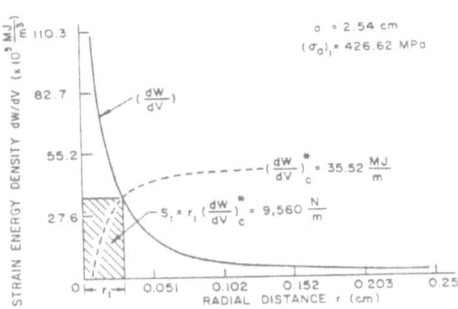

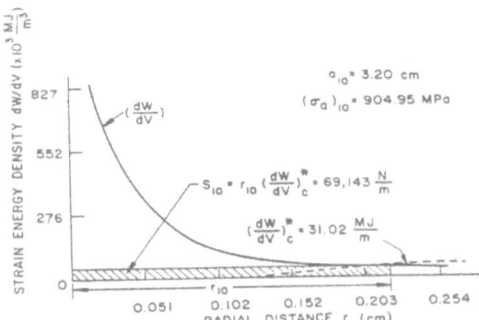

Fig. (5) - Variations of dW/dV with r
for the first increment of
crack growth

Fig. (6) - Variations of dW/dV with r
for the tenth increment of
crack growth

TABLE 2 - CRACK GROWTH DATA FOR LOAD TYPE B

Growth Step No.	Half Crack Length a (cm)	Applied Stress σ_a (MPa)	Strain Energy Density Factor S ($\times 10^3$ N/m)
1	2.540	426.62	9.560
2	2.567	452.47	10.657
3	2.596	491.25	13.436
4	2.632	517.11	15.002
5	2.674	581.74	21.364
6	2.729	646.39	26.843
7	2.803	711.03	32.870
8	2.900	775.67	39.995
9	3.022	840.30	56.352
10	3.200	904.94	69.143

Discussions. It is of fundamental interest to plot S against a for load type A, B and C. The straight line relationship verifies the condition dS/da = const. By increasing the magnitude of load step, the slope of the S versus a line increased accordingly. For a given S_c, the critical crack length a_c tends to decrease as larger load steps are taken. This agrees with experimental findings.

It is also of interest to note that had S been computed from the linear theory of elasticity, the relationship between S and a will no longer be linear. This will be illustrated for load type B. Using the σ_a data in Table 2 together with the equation

TABLE 3 - CRACK GROWTH DATA FOR LOAD TYPE C

Growth Step No.	Half Crack Length a (cm)	Applied Stress σ_a (MPa)	Strain Energy Density Factor S ($\times 10^3$ N/M)
1	2.540	426.61	9.560
2	2.567	465.40	11.773
3	2.599	517.11	15.056
4	2.639	581.75	21.062
5	2.694	646.39	26.395

$$S = \frac{(1+\nu)(1-2\nu)\sigma_a^2 a}{2E} \qquad\qquad (14)$$

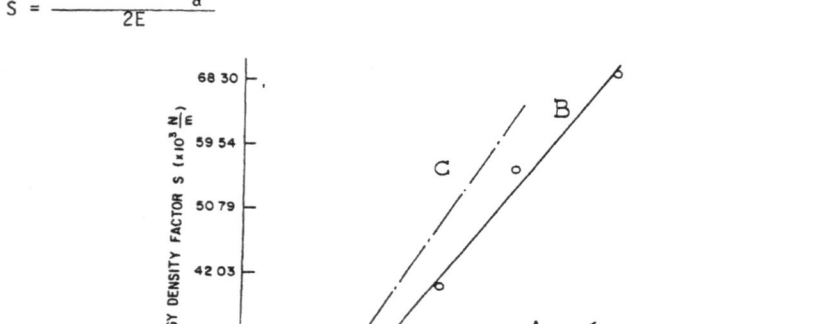

Fig. (7) - Change of strain energy density factor with half crack length

it can be shown that S becomes a nonlinear function of half crack length a as indicated in Figure 8. The deviation from linearity can be seen by comparing the solid curve with the dotted line which corresponds to the initial tangent. In view of the different interpretations of S versus a as shown in Figures 7 and 8, attention should be focused on the physical meaning of the parameters measured rather than the mere agreement between analytical and experimental data.

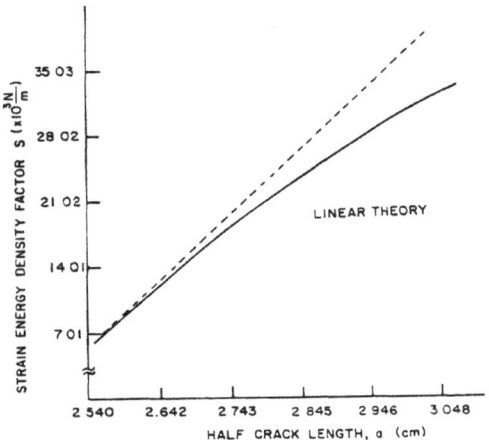

Fig. (8) - Linear elastic strain energy density factor as a function
of half crack length

PATH DEPENDENCY IN FATIGUE

Instead of increasing the load monotonically, its maximum amplitude will be
lowered* and repeated many times. Consider a sinusoidal loading of the form

$$\sigma_a = \frac{1}{2} \sigma_{ys} [1 - 0.6 \cos(2\pi t)]$$ (15)

where t is a dimensionless time parameter. The amount of energy accumulated in
the material ahead of the crack will be computed for each cycle of loading as required by equation (9). The computational requirements can be overwhelming since
it may take many cycles to initiate crack growth.

A conservative assumption will be made by letting the energy accumulation per
cycle to be equal to the amount accumulated during the first load cycle of the
current growth step. Let ΔS be the change in S during the first cycle, then equation (9) may be approximated as

$$\frac{\Delta a}{\Delta N} = [(\frac{dW}{dV})_c^*]^{-1} (\Delta S)_{r=\Delta a}$$ (16)

The critical value $(dW/dV)_c^*$ in fatigue is now given by equation (6). Both the
center and edge cracked panel will be analyzed. The full crack length being 2a

_____

*The maximum applied stress in fatigue is usually taken about fifty percent of
the yield stress.

for the center crack specimen, Figure 2, and a for the edge crack specimen, Figure 9. The material is the same as that used for monotonic loading. Refer to

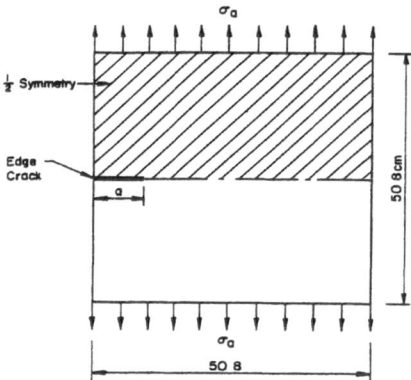

Fig. (9) - Edge crack subjected to cyclic loading

equations (11) and (12) for data on material. Equation (16) is solved numerically for each crack growth increment Δa corresponding to a constant number of elapsed loading cycles, $\Delta N = 100$ cycles*. The nonuniform rate of crack growth will again be guided by the relation S_j/r_j (j = 1,2,...,n) as in the case of monotonic loading.

Fatigue Life and Crack Growth Rates. Starting with the same initial values of a = 2.54 cm, fatigue life predictions are made for both the center and edge crack panels in terms of a versus N plots. Figures 10 show that both curves rise slowly at first and then more rapidly. For the same crack growth range, the center crack exhibited a steeper tangent, more rapid acceleration and shorter life. The predicted fatigue life is 1,261 cycles for the center crack and 1,503 cycles for the edge crack. Again, both specimens are made of the same material subjected to the same loading cycle. The difference in crack growth is solely due to change in the crack geometry.

The crack growth rate $\Delta a/\Delta N$ as a function of a is shown in Figure 11 for both the center and edge crack specimens. The trend of both curves is similar. They change abruptly at first, maintain a slowly rising range and jump at final instability of the specimen. A larger stable acceleration range is seen for the center crack.

---

*The specification of crack growth at every 50 cycles will yield different results as fatigue is strictly a load history dependent process. For more details, refer to the results in [8].

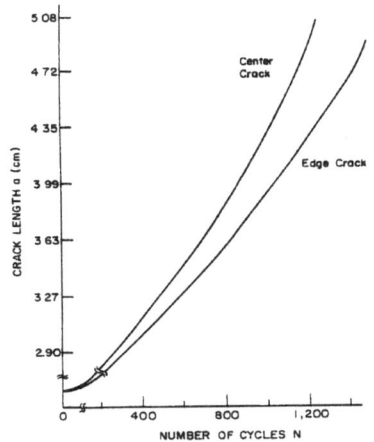

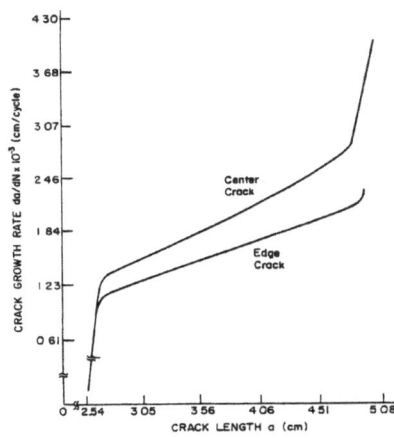

Fig. (10) - Fatigue life predictions on center and edge cracked specimens

Fig. (11) - Crack growth rate curves for center and edge specimens

Comparison with Linear Models. The nonlinear effects included in the accumulated damage model of fatigue crack growth can be best shown by correlating the results to the linear models* using ΔK_1 and ΔS. The most commonly used relation

$$\frac{\Delta a}{\Delta N} = C(\Delta K_1)^n \tag{17}$$

involves the range of the Mode I stress intensity factor and parameters C and n. When the load is not normal to the crack plane and the direction of crack initiation is no longer obvious, the relation

$$\frac{\Delta a}{\Delta N} = B(\Delta S)^m \tag{18}$$

has been used [10]. In both equations (17) and (18), ΔK_1 and ΔS are calculated from the linear theory of elasticity and hence damage accumulation is ignored. The parameters C, n, B and m will be shown to be crack geometry dependent. Based on the $\Delta a/\Delta N$ values in Figure 11, ΔK_1 and ΔS can be calculated to evaluate C, n, B and m for both the center and edge crack specimens. The results are

---
*The symbol Δ in the linear models, equations (17) and (18) designates the range of function evaluated at the maximum and minimum applied stress. This differs from the meaning of Δ in equation (9) as mentioned earlier.

$$C = 1.427 \times 10^{-24}, \; n = 2.454 \quad \text{(Center crack)}$$

$$\text{(19)}$$

$$C = 1.143 \times 10^{-19}, \; n = 1.810 \quad \text{(Edge crack)}$$

A linear least square regression analysis was used to obtain the C and n values. Similarly, the parameters B and m in equation (19) take the values

$$B = 3.061 \times 10^{-10}, \; m = 1.227 \quad \text{(Center crack)}$$

$$\text{(20)}$$

$$B = 6.104 \times 10^{-9}, \; m = 0.905 \quad \text{(Edge crack)}$$

Large deviations are seen for the values of C, n, B and m due to change in crack geometry. These parameters cannot be regarded as material constants. Care should be exercised when using the linear models given by equations (17) and (18). Large errors can result if they are used indiscriminantly without a knowledge of their range of limitation.

CONCLUSIONS

Subcritical crack growth is inherently a load history dependent process. Appropriate stress and failure analysis must be performed for each increment of load in order to have an accurate description of the crack growth characteristics. Both monotonic and cyclic loadings can be treated by the strain energy density criterion in a consistent fashion. The nonlinear nature of crack growth is significant and provides guidelines on the range of applicability of the linear models currently used by the practitioners.

Regardless of the nature of loading and crack geometry, the nonlinear crack growth model presented in this communication contains only two parameters $(dW/dV)_c^*$ and S_c. The former accounts for initiation and slow growth[*] and latter for the onset of rapid crack propagation. The critical ligament size of the material is obtainable from the relation $r_c = S_c/(dW/dV)_c^*$. The critical value S_c can be measured in the same way as K_{1c} in the ASTM standard plane strain toughness test. The procedure for determining $(dW/dV)_c^*$ is also standard as it involves the measurement of area under the true stress strain curve obtained from a uniaxial tensile test. Adjustments, however, should be made for change in loading rates and specimen sizes. Another important contribution of this work is the verification of the condition dS/da = const. Large load steps on cracked structural members are undesirable as they yield smaller critical crack length. This intuitively obvious result is presented analytically for the first time and is a useful piece of information for the designer.

---

[*] The only difference between initiation and slow growth is that failure takes place at different distances ahead of the crack.

The assumption that the crack growth direction corresponds to the maximum of $(dW/dV)_{min}$ is also verified for the case even when material separation occurs in the yielded portion of the material. Figure 12 shows a plot of dW/dV versus the polar angle θ referenced from the line of expected crack extension. The curve corresponds to σ_a = 904.94 MPa and r = 0.191 cm for the center crack specimen in Figure 2. The maximum of $(dW/dV)_{min}$ indeed occurred at θ = 0° which is the line of symmetry under Mode I fracture. The numerical values of dW/dV are not accurate as θ approaches the free crack surface (θ = 180°) near which the interface of two adjacent finite elements are located.

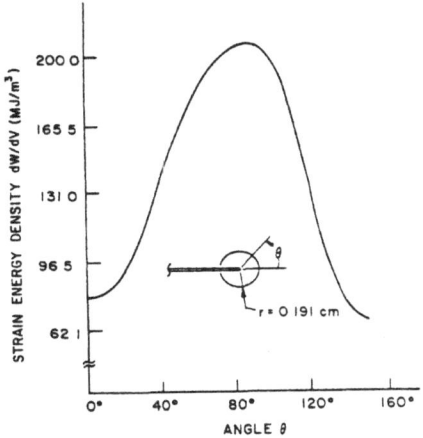

Fig. (12) - Angular variation of strain energy density function at r = 0.191 cm and σ_a = 904.94 MPa for the center crack

All of the basic assumptions of the strain energy density criterion are valid for the fatigue of mixed mode crack growth and three-dimensional crack configurations. The value $(dW/dV)_c^*$ in fatigue corresponds to the energy accumulated within the hysteresis loops. What remains to be improved is the continuum theory of plasticity. In particular, the J_2-flow rule may not be adequate for describing the local stress and strain fields.

ACKNOWLEDGEMENT

This research work was performed in the Institute of Fracture and Solid Mechanics at Lehigh University and sponsored by the Office of Naval Research under Contract No. N00014-76-C-0094. A special note of appreciation is extended to Dr. Yappa Rajapakse for his interest in this work.

REFERENCES

[1] Sih, G. C., "Fracture Toughness Concept", American Society of Testing Materials, STP 605, pp. 3-15, 1976.

[2] Gillemot, L. F., "Criterion of crack initiation and spreading", J. of Engineering Fracture Mechanics, Vol. 8, pp. 239-253, 1976.

[3] Absorbed Specific Energy and/or Strain Energy Density Criterion, edited by G. C. Sih, E. Czoboly and F. Gillemot, Akadémiai Kiadó, Budapest, 1982.

[4] Post-Yield Fracture Mechanics, edited by D. G. H. Latzko, Applied Science Publishers, Ltd., London, 1979.

[5] Turner, C. E., "A review of elasto-plastic fracture design methods and suggestions for a related hierarchy of procedures to suit various structural uses", Fracture Mechanics Technology Applied to Material Evaluation and Structure Design, edited by G. C. Sih, R. Jones and N. Ryan, Martinus Nijhoff Publishers, The Netherlands, pp. 39 - 64 , 1983.

[6] Sih, G. C., "A special theory of crack propagation", Methods of Analysis and Solutions of Crack Problems, edited by G. C. Sih, Noordhoff International Publishers, The Netherlands, pp. XXI-XLV, 1973.

[7] Sih, G. C., "Experimental fracture mechanics: strain energy density criterion", Mechanics of Fracture, Vol. VII: Experimental Evaluation of Stress Concentration and Intensity Factors, edited by G. C. Sih, Martinus Nijhoff Publishers, The Netherlands, pp. XVII-LVI, 1981.

[8] Sih, G. C. and Moyer, E. T., Jr., "Path dependent nature of fatigue crack growth", Journal of Engin. Fracture Mechanics, Vol. 17, No. 3, pp. 269-280, 1983.

[9] Gifford, L. N. and Hilton, P. D., "Preliminary documentation of PAPST nonlinear fracture and stress analysis by finite elements", David W. Taylor Naval Ship Research and Development Center, Bethesda, Maryland, January 1981.

[10] Sih, G. C. and Barthelemy, B. M., "Mixed mode fatigue crack growth predictions", Int. J. of Fract. Mech., Vol. 13, No. 3, pp. 439-451, 1980.

DAMAGE TOLERANCE THROUGH THE CONTROL OF MICROSTRUCTURE

J. F. Knott

University of Cambridge
Cambridge, England

ABSTRACT

The paper explores some aspects of the application of fracture mechanics to the assessment of material "quality", as affected by alloy-type, processing, heat-treatment and embrittlement. The plane-strain fracture toughness, critical J-integral value, and fatigue-crack growth-rate are used as illustrative parameters in quality-assessment, and contrasts are made, in a quantitative manner, between the effects of changes in "toughness" or in initial defect distribution. Examples chosen to demonstrate the approach are: effects of material selection and heat-treatment on the critical defect size in materials of high strength/weight; effects of temper-embrittlement in steel on critical defect size, and through this, choice of steelmaking route; fatigue-crack growth in aluminium-bronze castings as a function of inherent defect distribution; a speculative investigation of processing variables for powder-formed super-alloys. In all these examples, the reliability of non-destructive examination is considered as a critical feature of the assessment.

INTRODUCTION

During the last fifteen years, application of fracture mechanics to the engineering design of structural components has greatly improved the engineer's ability to produce designs which are both efficient and safe. These features are of increasing importance in the fields of transportation and energy-production, where scarcity of fuel demands higher efficiencies, generally involving higher static and dynamic stresses, often applied in environments which can affect crack growth-rates through mechanisms of corrosion, embrittlement or creep. As society also becomes more safety-conscious, these more stringent service conditions must be sustained with minimal risk of failure. There are conventional engineering design codes which relate the lifetime of a structure to the stresses or strains that it experiences, but it is often the case that these codes are, implicitly, if not explicitly, founded on fracture mechanics concepts, because the structures contain populations of defects. Additionally, if individual defects are detected during the periodic inspection of a structure, the decision which must be taken, on whether to scrap the part; to remove the defect and carry out a repair procedure, or to allow the defective part to continue in operation; can only be made rigorously, following a fracture mechanics analysis.

In addition to the ways in which fracture mechanics affects concepts of engineering design, it also has important consequences with respect to the evaluation of engineering alloys. Firstly, measurements of fracture-toughness or crack growth-rate in standard testpieces may be related to different alloy compositions and heat-treatments, so that a data-bank of materials' properties is generated, to enable the engineer to select a material whose resistance to fracture and crack-growth is specified in a quantitative manner. In these tests, it is essential that the mechanical state (crack length, thickness, triaxiality, extent of yielding) pertaining in the testpiece is a fair representation of that in service. If it is more severe, the engineering consequences may be over-conservative (inefficient) design or use of unduly expensive material. If it is less severe, there is a risk of premature failure in service.

The need for more efficiency in design develops a requirement for materials of higher (static) strength/weight ratio. This trend has led to catastrophic buckling problems in thin-walled, high-strength, box-girder sections, of which the Westgate Bridge disaster was a particularly tragic example. It can also generate an increasingly severe risk of fracture, because a material's fracture toughness usually decreases with increase in yield stress, yet the higher yield-strength materials are chosen specifically to bear higher service stresses. The critical defect size is therefore decreased on two counts, placing more onerous demands on the reliability and precision of processing, fabrication and non-destructive-testing. It will be argued later in this paper that there may be a case, in some applications for <u>reducing</u> the conventional safety-factor (based on prevention of yielding) in order to utilise material of higher toughness. Designs employing high static stresses may also be subjected, either to higher dynamic stress-amplitudes, or to stress-amplitudes at higher levels of mean-stress.

Standard fracture-toughness and crack growth-rate tests measure properties of material which is implicitly assumed to be a homogeneous continuum. This is not unreasonable if the length of the fatigue pre-crack is much greater than that of any inherent defect which may be present in the material, but it is important to note that, in practical engineering alloys, the processing and fabrication routes can produce defect populations which, in the absence of a single, long, pre-crack, dominate the fracture behaviour or lifetime of the fabricated part. The variety of cracks which can be produced during welding is well-recognised, but defects such as shrinkage-cavities, porosity, graphite flakes or silicon needles may be present in castings, silicate or sulphide inclusions may be deformed into thin sheets following the directions of flow in rolled or forged products, hard oxide inclusions may be present in powder-formed components, and cracking may be produced during heat-treatment. Such defects may not always show large effects on monotonic fracture toughness, but they are often of major significance with respect to the fatigue life of a component, because their presence obviates the need for a long "initiation" period. The larger the size of an "inherent" defect, the fewer the number of cycles required to increase its size to the critical value, and the shorter the lifetime that results.

In the future, the application of fracture mechanics to material evaluation may not only encompass the quantitative measurement of properties but could also be employed as a quantitative, diagnostic tool to assess (from the viewpoint of the user, not the producer) the virtues of competing processing and fabrication routes with respect to the defect populations that are generated, and the effects of these on service performance. If these assessments can be made by a combination of analysis and a relatively small number of experimental checks on test specimens, rather than by having to accumulate experience on full-scale service components, money and effort will be saved, and safety to the operator will be enhanced.

FRACTURE TOUGHNESS AND DAMAGE ACCUMULATION

In the following, it is assumed that a structural component is designed in accordance with conventional codes, in which attention is paid to elastic deflections and instability, but in which the main failure mode is taken as <u>general yield</u> or <u>plastic collapse</u>. The static design stress, σ_{des}, is related to the material's yield stress, σ_Y, or 0.2% proof stress, $\sigma_{0.2}$, by means of a "safety factor", F, as in equation (1).

$$\sigma_{des} = \sigma_Y/F \tag{1}$$

The safety-factor is introduced to try to take account of any extra, unanticipated loading that may occur during erection or operation and varies with application, degree of redundancy of the structure and "integrity" of the material. Typically, F takes values of the order of 1.5-2.0 for wrought steels in applications such as boilers, bridges or pressure-vessels, a value of 4 for castings [1] and 5-10 for steel-rope used in suspension cables or lift-hoists. The design philosophy permits local yielding around stress-concentrators and it is possible for this local yield to operate fracture mechanisms at a stress below the design stress, if the stress-concentrators are sufficiently severe. A catastrophic fracture of this type is a "brittle" fracture, regardless of the micromechanism of failure. Non-catastrophic, but usually accelerating, crack growth at low stress is referred to as "sub-critical" crack growth.

Under conditions which may be described as "linear elastic" (LEFM) or "small-scale yielding" (SSY), the strength of the stress field close to a crack tip is characterised by the "<u>stress-intensity factor</u>", K, which is related to the applied stress, σ_{app}, the crack length, a, and the width of the body, W, through the expression:

$$K = \sigma_{app} Y(a/W) \cdot Q(a/c) \tag{2}$$

where $Y(a/W)$ is a geometrical (compliance) function for through-thickness cracks and $Q(a/c)$ is a factor to take account of elliptical or semi-elliptical cracks (of surface length 2c). Initially, taking $Q(a/c) = 1$, and knowing $Y_1(a/W)$ for a given testpiece geometry, it is possible to measure the fracture stress, $\sigma_{app} = \sigma_F$, and use equation (2) to derive the materials' fracture toughness, K_{crit}. If plane-strain conditions hold, K_{crit} is written as K_{Ic}. Then, for a service application, in which σ_{app} is the design stress, σ_{des}, it is possible, if the appropriate compliance function $Y_2(a/W)$ is known, to calculate the critical defect size, a_{crit}, necessary to cause catastrophic failure. With suitable scaling-down to provide a safety margin, a_{crit} can be used as an NDT inspection limit. Often, the service structure can be treated as infinite or semi-infinite and equation (2) then becomes:

$$K = \alpha\sigma_{app}(\pi a)^{1/2} \tag{3}$$

for through-thickness cracks, where $\alpha = 1.12$ for an edge crack of length a, and α = unity for a central crack of length 2a, normal to σ_{app}. For a "penny-shaped" central crack of radius, a:

$$K = 2\pi^{-1/2} \sigma_{app} a^{1/2}$$

$$(4)$$

In lower strength steels at normal service temperatures, fracture does not occur under linear-elastic conditions in testpieces and either a critical value of the J-integral, J_{crit}, or a critical value of the crack-(tip)-opening-displacement (COD). δ_{crit} may be used to characterise initiation (J_i, δ_i) or the slope of the ductile crack-growth "resistance" curve (dJ/da, $d\delta/da$). Values of J and δ are related by the expression:

$$J = M\sigma_Y \delta$$

$$(5)$$

where M takes values in the range 1-3, depending on the testpiece geometry, the material's work-hardening exponent, and the precise definition of σ_Y (which may be the lower yield stress, σ_Y, the 0.2% proof stress, $\sigma_{0.2}$, the "flow stress", $\bar{\sigma}$ = $0.5(\sigma_Y+\sigma_U)$ where σ_U is the U.T.S. or a scaling parameter, σ_0, assuming a "power-law hardening" type of stress-strain curve). If the service application can be treated as that of "small-scale yielding" in an infinite body, use may be made of the expressions:

$$K^2(1-\nu^2)/E = J = M\sigma_Y \delta$$

$$(6)$$

to calculate critical values of K and hence of critical crack length from critical values of J or δ (ν is Poisson's ratio, E is Young's modulus). If this cannot be done, direct calibrations between applied load, or pressure, and J, or δ, must be available for the structure.

Common modes of sub-critical crack growth are fatigue, creep and stress corrosion. In fatigue, the rate of crack advance per cycle, da/dN, is related to the alternating stress intensity, $\Delta K = (K_{max}-K_{min})$ where K_{max} and K_{min} are the maximum and minimum values in the cycle, through the expression [2]

$$da/dN = A\Delta K^m$$

$$(7)$$

where A and m are constants. In detail, a graph of $\log(da/dN)$ versus $\log\Delta K$ exhibits three regions, of which only the central region is linear, in accordance with equation (7). At low stress-intensities, a "threshold", ΔK_{th}, exists and sharp increases in growth-rate are observed as K_{max} approaches K_{Ic}. A more extended form of equation (7) becomes [3]:

$$da/dN = (A'/\sigma_Y E)(\Delta K-\Delta K_{th})^2 \{1 + \Delta K/(K_{Ic}-K_{max})\}$$

$$(8)$$

where the threshold ΔK_{th} varies with mean stress, expressed through the "R-ratio", R = (K_{min}/K_{max}) as:

$$\Delta K_{th} = \{(1-R)/(1+R)\}^{1/2} \Delta K_o$$

$$(9)$$

Over limited ranges of growth rate, both stress corrosion and creep crack growth rate, (da/dt) can be expressed as polynomials in monotonic stress intensity, K, and a more general creep "damage" parameter, defined as "loss-of-continuity", has been expressed as a polynomial in stress [4].

The lifetime of a component is calculated by integrating an expression such as equation (7) or (8) between limits a_o, representing the initial size of any defect present, and a_f, the size at which the component "fails". Failure may be determined by the onset of fast fracture or plastic collapse, so that a_f, in principle, depends on the applied stress level. In practice, the failure life can be estimated by assuming that a_f is constant, because a growing crack accelerates so rapidly as failure is approached that (small) variations in a_f are insignificant. Consider a generalised crack "growth-rate", da/dx, where the "period" variable x may refer to time, t, or number of cycles, N, and assume that this is related to stress, σ, (either monotonic, σ_{app}, or alternating, $\Delta\sigma_{app}$) and crack length, a, through the expression:

$$da/dx = B\sigma^p a^q \qquad (10)$$

where B, p and q are constants. The term a^q could be replaced by a polynomial in compliance function $\{Y(a/W)\}^q$ is required. Separation of variables gives:

$$a^{-q}da = B\sigma^p dx \qquad (11)$$

which, integrated between limits on crack length, a_o and a_f, corresponding to limits on period x = o, x = x_f (failure) gives:

$$C\sigma^p x_f = a_o^{-q+1} - a_f^{-q+1} \qquad (12)$$

If a_o and a_f can be taken as constants, the form $\sigma^p x_f$ = constant results. In creep, where x = time, $x_f = t_f$ and Norton's Law, $\sigma_{app}^p t_f$ = constant, is obtained. In fatigue, $\sigma = \Delta\sigma_{app}$, $x_f = N_f$ and, from equation (7), p = m, because ΔK is linear in $\Delta\sigma_{app}$. Here, it is conventional to take logarithms to obtain:

$$LogN_f = -mlog\Delta\sigma_{app} + constant \qquad (13)$$

which is the "S-N" curve derived from a crack-growth basis. This is not always admissible in pure materials, because an "initiation" period is required to generate incipient cracks, but in many commercial alloys as well as in structures, there are often sufficient initial inherent defects for the approach to be justified. Incorporation of a threshold, ΔK_{th}, value - see equations (8) and (9) - produces a "fatigue limit" to the S-N curve. Practical examples of this integration will be discussed later. It is interesting to note that the general integration does not require the assumption of LEFM as such, because p and q in equation (10) are independent powers.

Variable amplitude damage may be treated in a similar general manner. Following equations (11) and (12), the failure period $x_{f(c)}$ at any constant stress σ_c is given by:

$$C\sigma_c^p x_{f(c)} = I_{of} \qquad (14)$$

where I_{of} represents the integral between crack lengths a_o and a_f (constants of integration being included in C). Now consider a loading sequence in which blocks of stress levels σ_i (i = 1,2,3,4,...etc.) are applied for periods x_i. During the first block, at stress σ_1, the crack grows from a_o to a_1 and the period taken, x_1, is given by $(I_{o1}/C\sigma_1{}^P)$ where I_{o1} is the integral of equation (11) between limits a_o and a_1. From equation (14), the life, $x_{f(1)}$, at stress σ_1 is $(I_{of}/C\sigma_1{}^P)$, so that $x_1/x_{f(1)} = I_{o1}/I_{of}$. At stress σ_2, the crack grows from a_1 to a_2 in a period $x_2 = (I_{12}/C\sigma_2{}^P)$ and $x_2/x_{f(2)} = I_{12}/I_{of}$. In general, $x_i/x_{f(i)} = I_{(i-1),i}/I_{of}$ and, writing this as a summation,

$$\Sigma x_i/x_{f(i)} = I_{o1}/I_{of} + I_{12}/I_{of} + I_{23}/I_{of} +$$
$$= (1/I_{of})(I_{o1} + I_{12} + I_{23} + ...) \tag{15}$$

By the definition of integrals, $I_{o1} + I_{12} = I_{o2}$; $I_{12} + I_{23} = I_{o3}$ etc., so that the R.H.S. of equation (14) becomes $(I_{of}/I_{of}) = $ unity. Hence,

$$\Sigma x_i/x_{f(i)} = 1 \tag{16}$$

If x is taken to represent time, in a creep test for example, equation (16) becomes the life-fraction rule: if x is the number of cycles in a fatigue test, equation (16) is the Palmgren-Miner relationship.

Note that these damage accumulation expressions again do not require "classical" LEFM formulation but do require that the growth-rate expression holds at all stress levels and that a_f is sensibly constant. This places restrictions on the approach and a number of further factors, such as retardations following overloads in fatigue, can seriously affect the linear summation assumption. In particular, attempts to combine creep "damage" and fatigue "damage", under conditions of high-temperature cyclic loading, by means of linear summations of "life-fractions", are seldom based on clear physical principles.

In the preceding discussion, it has been assumed that the material contains a single defect of length a_o. This is usually the case in a standard testpiece, but, in a component in service, there will often be a distribution of defect lengths and orientations arising from processing and fabrication. In some brittle materials, such as sintered, polycrystalline ceramics, the distribution may be derivable from a knowledge of the grain sizes of sieved particles, combined with the degree of compaction and sintering time, since these factors determine the size of intergranular pores. More usually, the distribution cannot be determined in this way and recourse is made to Weibull (or other probabilistic) analysis of the fracture stresses of a number of test samples, taken from a given batch of material. Although the fracture stress distribution reflects that of the defects, the relationship between the distributions is complicated, because defect size, shape and orientation are all variables.

In structural metals, dye-penetrant or magnetic-particle inspection (MPI) may be used to detect and size surface cracks. A sensitivity of 0.25 mm for surface length may be assumed, but, for wrought steel, only about 65% accuracy in detection and 35% accuracy in sizing cracks of surface length 2.5 mm has been quoted [1]. For buried, crack-like defects, ultrasonics detection is better than X-radiography, and has a limiting sensitivity of about 2 mm diameter for a ("square-on") penny-shaped flaw in moderately thick-section alloy steel. Complete reliability of detection of buried flaws is difficult to establish experimentally but the diameter of 6 mm quoted for surface flaws [1] is probably an underestimate. In very thick pressure-vessel plates, even 25 mm long cracks may be detected with only 90% accuracy. In the following sections, examples are chosen to illustrate effects of toughness and initial defect population on a material's resistance to fracture, and the sensitivity of NDT is seen to be a vital factor in these assessments.

STRENGTH AND TOUGHNESS IN ULTRA-HIGH-STRENGTH ALLOYS

Conventional engineering codes, equation (1), expresses the design stress σ_{des}, as a fraction of the yield (or proof) strength, to prevent failure by general yield. A typical code for structures might specify that $\sigma_{des} = \sigma_Y/1.5$. To save weight in applications such as aircraft undercarriages or rocket-motor cases, there is incentive to use steel of the highest possible yield strength, consistent with fabricability. A choice might lie between various grades of maraging steel or, as a cheaper substitute, the steel 300M, which is based on 4340 (0.4C, 1.8Ni, 0.8Cr, 0.7Mn, 0.25Mo) but with 2%Si added to modify the tempering characteristics. Table 1 shows calculated critical defect sizes for 300M tempered at 320°C and 450°C and

TABLE 1 - CRITICAL DEFECT SIZES IN ULTRA-HIGH STRENGTH STEELS

Material	Yield Stress MNm^{-2}	K_{Ic} MNm$^{-3/2}$		Critical Defect Size, mm @ $\sigma_{des}=$		
				σ_Y 1.5	1600 MNm^{-2}	1267 MNm^{-2}
G150 Marageing	2400	34	E	0.1	0.1	0.2
			C	0.3	0.3	0.5
			P	0.7	0.7	1.1
G125 Marageing	1900	76	E	0.9	0.6	0.9
			C	2.3	1.4	2.3
			P	5.7	3.6	5.7
G110 Marageing	1700	100	E	2.0	1.0	1.6
			C	5.0	2.5	4.0
			P	12.3	6.2	10.1
300M T320°C	1910	40	E	0.3	0.2	0.3
			C	0.6	0.4	0.6
			P	1.6	1.0	1.6
300M T450°C	1510	50	E	0.6	-	0.4
			C	1.6	-	1.0
			P	4.0	-	2.5

E = edge crack   C = central crack
P = penny-shaped crack   (see text)

Data courtesy of Dr B. Wiltshire and Dr J.E. King and ref. [5].

for three grades of maraging steel: G150, G125, G110, (strengths 330 ksi, 280 ksi and 240 ksi respectively). The calculations are made for edge (E), central (C) and penny-shaped (P) cracks, using equations (3) and (4) with σ_{app} values equal to

σ_{des} ($\sigma_Y/1.5$), to 1600 MNm^{-2}, or to 1267 MNm^{-2}, (the design stresses for G150 and G125 respectively). The critical defect size is "a" for an edge crack, "2a" for a central crack and "2a" (diameter) for a penny-shaped, buried crack.

From Table 1, it can be seen that use of the strongest maraging steel, G150, with the conventional safety factor leads to an edge crack length of 0.1 mm and a penny-shaped crack diameter of 0.7 mm. The detectability by MPI of the surface length of a crack 0.1 mm deep depends on the aspect ratio, and unless c/a >> 1, the calculation itself is strongly dependent on the Q(a/c) factor in equation (2). Full reliability of detection of a crack of surface length even equal to 2 mm (c/a = 10) cannot necessarily be guaranteed, however [1], and the figure of 0.7 mm for the diameter of a buried crack is below the sensitivity of ultrasonic or radiographic techniques. If the "safety factor", F, in equation (1) were reduced, the same stress, of 1600 MNm^{-2}, could be sustained in G125 with an increase in defect size to 0.6 mm (edge crack) or 3.6 mm (penny-shaped crack). The reliability of detecting cracks of this size is very much improved, albeit at the expense of a high ratio of $\sigma_{des}/\sigma_Y = 0.84$. The manufacturing processes associated with these ultra-high strength steels are such that high-quality materials and well-controlled construction procedures are employed throughout, so that the likelihood of obtaining unanticipated stresses, which would lead to premature yielding, is relatively small. There is, therefore, a case for reducing the safety factor (based on yield) to improve safety with respect to fast fracture. Use of G110 would increase the critical defect size to 1 mm (edge crack) or 6.2 mm (penny-shaped crack), but the associated ratio of $\sigma_{des}/\sigma_Y = 0.94$ is so high that the most stringent precautions would need to be taken during manufacture to avoid failure by yield.

The same principle applies for slightly lower-strength, tougher materials. The final column in Table 1 shows critical defect sizes for a design stress of 1267 MNm^{-2}, which equates to a safety factor of 1.5 applied to the yield stress of G125 (or 300M tempered at 320°C). Use of G110 increases the critical defect size from 0.9 mm to 1.6 mm (edge crack) or from 5.7 to 10.1 mm (penny-shaped crack) with an increase in the ratio of σ_{des}/σ_Y to 0.745. The table also enables a quantitative assessment to be made of the value of using tougher, but more expensive, maraging steel, compared with low-alloy steel (300M). At a yield strength of 1900 MNm^{-2} and a design stress of 1267 MNm^{-2}, the critical defect size in G125 is 0.9 mm (edge crack) or 5.7 mm (penny-shaped) crack. These figures are reduced to 0.3 mm or 1.6 mm respectively in 300M and the latter figure is very much on the borderline of, if not below, the sensitivity limit for ultrasonic detection. Some advantage could possibly be gained by using 300M tempered at 450°C. Here, the critical defect size at a stress of 1267 MNm^{-2} is 0.4 mm (edge crack) or 2.5 mm (penny-shaped crack). In this temper, the ratio of design stress to yield stress is 0.84.

The figure of 2.5 mm is close to the ultrasonic NDT limit whereas the equivalent value of 5.7 mm in G125 is significantly above the limit. The economic analysis which has to be made with respect to choice of material must balance prime material costs, which would favour 300M, with the cost of more rigorous control of manufacture and inspection.

An alternative strategy for high strength/weight (more strictly, for high specific load-bearing capacity) applications is to use high-strength alloys of density lower than that of steel (approximately 7900 Kgm^{-3}). Recent developments of the 7000 series (Al-Zn-Mg-Cu) aluminium alloys [6] may combine a yield strength of 500 MNm^{-2} with a K_{Ic} value of 36 $MNm^{-3/2}$ and a density of some 2800 Kgm^{-3}, which is a weight advantage of X 2.82. For uniaxial tensile loading, it is possible to contemplate increasing the cross-sectional area by this factor without any weight penalty, so that the load which produces a stress of 1267 MNm^{-2} in steel is associated with a stress of 450 MNm^{-2} in the aluminium alloy. The ratio of σ_{des}/σ_Y is 0.9, but the values of critical defect size are substantially increased, to 1.6 mm (edge crack) or 10 mm (penny-shaped crack). Similarly, high-strength titanium alloys, such as Ti318 (Ti-6Al-4V), may combine a density of approximately 4500 Kgm^{-3} with a yield strength of 960 MNm^{-2} and a fracture toughness of 92 $MNm^{-3/2}$, although this value is affected strongly by testpiece orientation. The stress of 1267 MNm^{-2} is now reduced to an effective 721 MNm^{-2}, which gives critical defect sizes of 4.1 mm (edge crack) or 25.6 mm (penny-shaped crack). These increases in critical defect size are extremely attractive, but use of aluminium alloys or titanium alloys in high strength/weight applications is restricted by a number of factors, including prime material cost, the compactness of the component, the magnitude of elastic deflections, and sensitivity to stress-corrosion.

In steels, there are practical limits to the use of lower yield strength materials, in order to utilise their higher fracture toughness values. Failure by yielding must be prevented and high ratios of σ_{des}/σ_Y can be tolerated only if reproducible, high-quality construction techniques are employed. Some aerospace applications already make use of σ_{des}/σ_Y ratios greater than 0.67 in high yield-strength alloys, which not only limits the room for manoeuvre in terms of reducing the yield stress, but also implies extremely small critical defect sizes. A further point is that many components are subjected to a "proof test" or "overstressing" procedure before they enter service. In the proof test, the loading may be increased typically to 0.8 σ_Y, i.e., a 20% increase over the normal service stress of 0.67 σ_Y. If the 20% figure is maintained, it is clear that service stresses of greater than 0.83 σ_Y cannot be contemplated, because they would produce general yielding in the proof test. Some processes, such as the autofrettaging of gun-barrels deliberately generate extensive yielding, or even general yielding, to produce compressive residual stresses in service, but such yielding would more usually be an embarrassment for conventional structural components.

The proof test may not only improve subsequent service performance, but is also a useful check on the maximum initial size of defect present in the component. From equations (3) and (4), the defect size is inversely proportional to the square of the applied stress, so that an increase in stress of 20% reduces the figures in Table 1 by a factor of 1.44. If the component survives the proof test, it cannot have contained a defect of a size which would produce failure at 0.8 σ_Y. The proof test is then used as an NDT technique, and, in high-strength materials, its sensitivity with respect to small defects is better than that of ultrasonics or radiography. From Table 1, for example, if a component made in G150 maraging steel survives a proof test at 0.8 σ_Y, it cannot contain any penny-shaped cracks greater than 0.7/1.44 = 0.5 mm in diameter. This size is well below the ultrasonic sensitivity limit. The corollary of this use of the proof test as a sensitive NDT technique is that, if a new processing route is being explored, the possibility of failure by fast fracture in the proof test cannot be ignored.

QUENCHED-AND-TEMPERED LOW-ALLOY STEELS

The previous discussion has been based on LEFM calculations in ultra-high-strength material, but similar trends may be observed in J_{Ic} results, recently obtained by Slatcher for a quenched-and-tempered low-alloy steel (0.34C, 2.7Ni, 0.7Cr, 0.55Mo) in air-melted (A) and vacuum-melted (V) conditions. Calculated values of critical defect size (edge cracks and penny-shaped cracks) are given in Table 2 for design stresses of $\sigma_Y/1.5$, 933 MNm^{-2} (1400/1.5) and 800 MNm^{-2}

TABLE 2 - CRITICAL DEFECT SIZES IN Q.T. LOW-ALLOY STEEL

| Steel Condition | Yield Stress MNm^{-2} | J_{Ic} kJm^{-2} | | Critical defect size, mm @ $\sigma_{des}=$ | | |
				$\frac{\sigma_Y}{1.5}$	933 MNm^{-2}	800 MNm^{-2}
	1000	70	E	9.1	4.7	6.4
			P	56.3	29.1	39.6
Air-melted (A)	1200	36	E	3.3	2.4	3.3
			P	20.4	14.9	20.4
	1400	29	E	1.9	1.9	2.7
			P	11.8	11.8	16.7
	1000	135	E	17.6	9.1	12.3
			P	108.9	56.3	76.1
Vacuum-melted (V)	1200	50	E	4.5	3.4	4.5
			P	27.9	21.0	27.9
	1400	26	E	1.7	1.7	2.4
			P	10.5	10.5	14.9

Data courtesy S. Slatcher

(1200/1.5). As for the higher strength materials, the critical defect size can be increased by utilising material of lower yield strength with design stresses which are higher fractions of the yield stress. If proof testing is to be employed, the figures in Table 2 should be divided by 1.44. The design stress ratios of 933/1200 = 0.78 and 800/1000 = 0.8 will be increased to 0.94 and 0.96 respectively in the (20% overstress) proof test, so that the margin below general yield is small.

It is interesting to use Slatcher's results to examine the effects of processing (air-melting or vacuum-melting) on toughness. In both conditions, the inclusion content is small and the spacing between inclusions is approximately 20 μm. The S content in air-melted material is 0.012 and that in vacuum-melted material is 0.009. From Table 2, it is clear that the change in inclusion distribution produced by vacuum melting has a large effect on J_{Ic} in steel tempered to a yield strength of 1000 MNm^{-2} (70 kJm^{-2} increased to 135 kJm^{-2}) but has negligible effect in steel of 1400 MNm^{-2} yield strength (26 vs 29 kJm^{-2}). This lack of effect can perhaps be explained on the basis of a model [7] which calculates the size of plastic zone associated with a given crack-opening displacement and compares this with the inclusion spacing. Using equivalent finite-element analysis to establish a relationship between plastic zone size and COD, we may write:

$$R_{IY} = 0.32 \ (E/\sigma_Y)\delta \tag{17}$$

where R_{IY} is the extent of the plastic zone. If δ is controlled by the initiation of voids on tempered carbides [8], a displacement of approximately 0.4 μm might be contemplated (a strain of unity acting on particles of diameter 0.4 μm) so that $R_{IY} \simeq 19$ μm for $\sigma_Y = 1400$ MNm^{-2} and $R_{IY} \simeq 27$ μm for $\sigma_Y = 1000$ MNm^{-2}. These values straddle the material's inclusion spacing and the argument would be that the zone size associated with initiation in the higher strength material would in general not involve inclusions, because initiation is controlled by the carbides. However, the value of δ_i calculated from a J_{Ic} value of 28 kJm^{-2} (corresponding to an average for 1400 MNm^{-2} yield strength material, Table 2) is 10 μm, so that R_{IY} is 0.48 mm. It may be that the important feature is the "process-zone" size associated with initiation and that this can be calculated from the "microstructural component" of δ_i, as above. The practical importance of this lack of effect of vacuum-melting is that there would appear to be a limit to the extent to which the inclusion content need be reduced (inter-inclusion spacing increased) in high-strength steel, because further reductions would not produce any increase in toughness.

The monotonic toughness of quenched-and-tempered steels is also affected strongly by trace impurity elements, such as P, Sn and Sb, which may produce intergranular embrittlement. Dramatic changes in toughness are produced by selection of material and heat-treatment and the need for materials selection can be assessed in terms of critical defect size, to attempt to quantify the value judgments. In (U.K.) practice, five options may be recognised, assuming that the alloy steel is made in an electric furnace, to minimise loss of alloying element and to control sulphur. The economics of the electric furnace are improved by the use of metal scrap, which if not sorted carefully, may contain Sn (from tin-plate) Sn and Sb (from white-metal bearings) or P (from high-P low-grade steel scrap). The options available to meet design requirements may be listed as follows:

1. "run-of-the-mill" production is sufficient,

2. following property-tests, the best x% of normal production is sufficient (and there are less stringent markets for the remaining production),

3. careful sorting of scrap is required to make premium grades,

4. steel must be made from impurity-free ores,

5. the part must be re-designed, because even the best obtainable properties are insufficient.

The option followed is likely to depend on the application (stress level) and it is possible to illustrate this by making use of equations (3) and (4) for two toughness levels: K_{Ic} = 120 MNm$^{-3/2}$ (unembrittled), K_{Ic} = 40 MNm$^{-3/2}$ (embrittled); and two stress levels: 500 MNm^{-2} (yield strength 750 MNm^{-2}) and 800 MNm^{-2} (yield strength 1200 MNm^{-2}). The critical defect sizes (diameters of penny-shaped buried defects) become: 90 mm and 35 mm for the unembrittled condition; 10 mm and 4 mm for the embrittled condition. The corresponding critical lengths of edge cracks are 14.6 mm and 5.7 mm; 1.6 mm and 0.6 mm. All sizes should be decreased by a further factor of x1.44 to guard against failure in a (20% overstress) proof test. Although every application will possess its own idiosyncrasies, the general conclusion to be drawn from the above might be that for the high-strength application, such demands would be placed on the NDT inspection in embrittled material (4/1.44 = 2.8 mm for buried cracks: 0.6/1.44 = 0.42 mm for surface cracks) that option 4 should be followed, because inspection limits of 35/1.44 = 24 mm and 5.7/1.44 = 4 mm are relatively easy to guarantee. In the lower-strength application, there would be more incentive to follow option 2, because even a mean toughness of 80 MNm$^{-3/2}$ would give a critical (buried) defect size of 30/1.44 = 21 mm. A problem with heavy forgings is, however, that of banding: regions of the Hinkley Point steam-turbine disc [9] were segregated and had a toughness of 40 MNm$^{-3/2}$ whereas other regions of the same disc had a toughness of 120 MNm$^{-3/2}$. The probability of any such effect clearly features in the choice of an appropriate option.

The preceding discussion has concentrated on the balance of design stress and toughness under monotonic loading. The following section considers the implications of changes in K_{Ic} on life, using fatigue-crack growth to illustrate general effects on all sub-critical growth processes, and then examines the importance of a material's inherent defect population in applications where there is no single, large crack to dominate behaviour.

EFFECTS OF TOUGHNESS AND DEFECTS ON SUBCRITICAL CRACK GROWTH

The general relationships between crack growth-rate per cycle (da/dN) and alternating stress intensity, ΔK, have already been given in equations (7) to (9) but for illustration, it is convenient to use a simple form, which holds for ferritic steels:

$$da/dN = 10^{-11} \Delta K^3 \qquad (18)$$

where da/dN is in m/cycle and ΔK is in MNm$^{-3/2}$. For applications such as turbines or aircraft engines, the fatigue spectrum combines a large "on-off" cycle (up to full design stress and back to zero) every shift or every flight, combined with high-frequency, low-amplitude cycles, arising from out-of-balance vibrations, small pressure fluctuations, the demands of manoeuvring, etc. In the absence of interaction-effects, the total life may be estimated from the Palmgren-Miner relationship, see equation (16).

Consider the "on-off" component, as applied to edge cracks, so that K = 1.12 $(\sigma_{des}-o)(\pi a)^{1/2}$. Substitution and rearrangement of equation (18) then gives:

$$a^{-3/2}da = 7.82 \times 10^{-11} \sigma_{des}{}^3 dN \qquad (19)$$

which integrates to

$$3.91 \times 10^{-11} \; \sigma_{des} \; N_f = a_o^{-1/2} - a_f^{-1/2} \tag{20}$$

where σ_{des} is in MNm^{-2} and a_o and a_f are in m. For a steam turbine, operating one shift per day for 80% of a design life of 40 years, N_f is of order 1.15×10^4 cycles. A turbine disc in an aero-engine would have a shorter design life in years, but probably more cycles (take-off/landing) per day. For an edge crack in a steam turbine assuming that σ_{des} = 500 MNm^{-2} as before, we have:

$$4.89 \times 10^{-3} \; N_f = a_o^{-1/2} - a_f^{-1/2} \tag{21}$$

For a surface crack in tough material (K_{Ic} = 120 $MNm^{-3/2}$), a detection system capable of setting a_o as 0.2 mm (2×10^{-4} m) might be possible; from the previous section, a_f = 14.6 mm. Hence, $a_o^{-1/2}$ = 70.7, $a_f^{-1/2}$ = 8.28, N_f = 1.3×10^4 cycles, which is just greater than the expected life, provided that the high frequency, low-amplitude cycles are insignificant. The initial value of ΔK, for $\Delta\sigma = \sigma_{des}$ = 500 is approximately 14 $MNm^{-3/2}$ for a crack of length 0.2 mm, i.e., significantly above the threshold, ΔK_{th}.

Assuming equation (20) and the values of a_o and a_f above, together with a "vibration" stress even of 10% of the "on-off" stress (i.e., $\Delta\sigma$ = 50 MNm^{-2}) and a threshold value, ΔK_{th} = 7.5 $MNm^{-3/2}$, the crack has to grow to a length of 5.7 mm by "on-off" stresses before the "vibration" stress produces a stress intensity which exceeds the threshold value. This requires 1.18×10^4 cycles, so that the contribution from "vibration" stresses is, indeed, small.

It is interesting to rework the figures for a penny-shaped, buried crack. Now, $K = 2\pi^{-1/2} \; \sigma_{des} \; a^{1/2}$, equation (4), and equation (19) becomes:

$$a^{-3/2}da = 1.44 \times 10^{-11} \; \sigma_{des}^3 dN \tag{22}$$

which integrates to

$$7.2 \times 10^{-12} \; \sigma_{des}^3 N_f = a_o^{-1/2} - a_f^{-1/2} \tag{23}$$

or, with σ_{des} = 500 MNm^{-2}

$$9 \times 10^{-4} \; N_f = a_o^{-1/2} - a_f^{-1/2} \tag{24}$$

For tough material (K_{Ic} = 120 $MNm^{-3/2}$), a_o may be set at the limit for detection of a buried crack, i.e., 2 mm (radius $a_o = 10^{-3}$ m) and a_f, as in the previous section, at 45 mm (diameter 90 mm). Hence, $a_o^{-1/2}$ = 31.6, $a_f^{-1/2}$ = 4.7, N_f = 3×10^4 cycles. It is important to note the sensitivity of this result to the value of a_o and the relative insensitivity to a_f. If K_{Ic} were lowered to 40 $MNm^{-3/2}$, so that a_f = 5 mm, $a_f^{-1/2}$ = 14.1 and N_f = 1.9×10^4 cycles, but if a_o were increased to 5 mm, even with a_f at 45 mm, $a_o^{-1/2}$ = 10, $a_f^{-1/2}$ = 3.3, N_f is decreases to 1×10^4 cycles, i.e., to less than the required design life. This is why the inherent defect population of a material and the reliability of detecting defects by NDT is so crucial to lifing procedures.

The effect of increasing the design stress to 800 MNm^{-2} for a buried crack is to alter equation (23) to

$$3.7 \times 10^{-3} \ N_f = a_o^{-1/2} - a_f^{-1/2} \tag{25}$$

where a_o remains at the NDT inspection limit ($2a_o = 2$ mm) and a_f is decreased to 17.5 mm; $a_f^{-1/2} = 7.56$. The life, N_f, is decreased to 6.5×10^3 cycles. The effect of embrittlement is to reduce a_f even more: $a_f = 2$ mm, $N_f = 2.5 \times 10^3$ cycles, even assuming that NDT is completely reliable in detecting a 2 mm defect.

For the brittle condition in the high yield strength material, it is interesting to contrast the NDT approach and the proof test approach for the assessment of a_o. In estimates of N_f above, it has been assumed that all buried cracks greater than 2 mm in diameter would have been detected and removed before the component went into service. This is clearly a major assumption and, in practice, one which cannot be guaranteed in thick sections. (In thick-walled pressure-vessels, even 20 mm diameter defects may be detected with only some 90% confidence). If a rather brittle material has survived a (20% overstress) proof test, however, it is possible to set an upper limit to a_o with high confidence. For K_{Ic} = 40 MNm$^{-3/2}$ and $\sigma_{app} = 1.2 \ \sigma_{des} = 960$ MNm^{-2}, the maximum diameter of a penny-shaped buried crack which can be present in material which has survived a proof test is 2.73 mm. Given this value ($a_o = 1.365$ mm) and $a_f = 2$ mm, the fatigue life, N_f is reduced to 1.3×10^3 cycles (about half that calculated using $2a_o$ = 2 mm). The proof test approach allows the upper limit to a_o to be determined with high accuracy, but it must be appreciated that if the material processing were such that a proportion of components contained defects greater than 2.73 mm in diameter, fast fracture could occur in the proof test.

These illustrative examples emphasise the importance of initial defect size on fatigue life. If the quality of the material were such that it contained no defects of diameter greater than 0.2 mm, the fatigue life would be increased to 2×10^4 cycles even in the brittle condition at high design stress. The traditional method of assessing such material is to establish a composition and a processing route and then to carry out a large number of tests on nominally smooth testpieces to obtain S-N data.

Recent work by Taylor has shown how this general illustrative analysis is applicable to an aluminium-bronze casting, which contained a distribution of casting defects, mainly in the size range 0.1-0.5 mm, but containing a significant population of size greater than 0.5 mm. Figure 1 shows the conventional fatigue-crack growth-rate curve for the material and Figure 2 illustrates the way in which this is divided into linear segments to facilitate the integration, and to take the effect of the threshold into account. The result of the integration, assuming different initial defect sizes, is shown in Figure 3, together with experimental S-N data for nominally smooth specimens (which, in fact, contain casting pores). Agreement with experiment is quite satisfactory, but it should be noted that de-

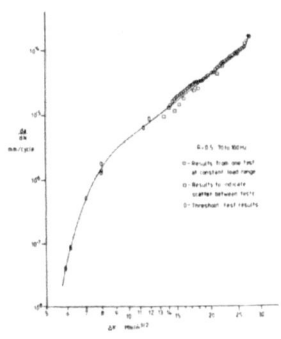

Fig. (1) - Fatigue-crack growth-rate
curve for aluminium bronze
casting

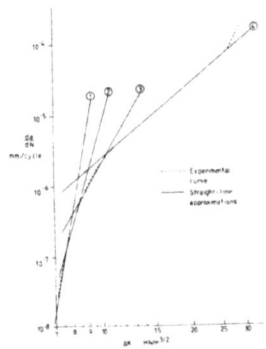

Fig. (2) - Linearization of curve in
Figure 1, to assist inte-
gration

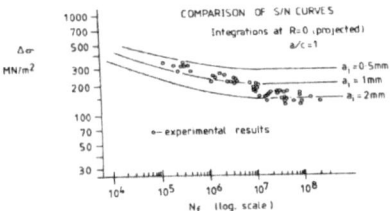

Fig. (3) - Comparison of fatigue-crack growth-rate predictions with S/N curve

fects smaller than approximately 0.35 mm in size grow anomalously rapidly [10].
This did not affect the results in Figures 1-3, because a sufficient population
of larger defects was present in the specimens, but the validity of integration
from the da/dN curve for cracks smaller than 0.5 mm in other, higher strength
materials has not yet been proved rigorously. It is, however, likely that the
size below which the simple LEFM approach cannot be applied decreases with increase in yield stress. A major advantage of the integration method with respect
to the application to the aluminium-bronze is that this material is used for large
"one-off" castings for marine propellers. For these, there is no opportunity to
test individual components or even large-scale models. Once the validity of the
integration has been established, it can be used to predict behaviour in service

from growth-rate data obtained on small specimens. Initial calculations are encouraging.

The method has potential application to a number of other situations. One of particular interest might be an assessment of the use of powder-formed components in nickel-base superalloys. Here, the alloy is atomised by pouring it as a molten stream through argon, and the metal powder is then sieved, hot-isostatically-pressed (HIP) and forged. One possible problem is that small pieces of refractory oxide may be scoured from the lining of the tundish through which the metal is poured and these may act as inherent defects to reduce fatigue life. The maximum size of such a defect is controlled by the mesh size in the sieve and a fracture mechanics calculation may be used to calculate the effect of initial defect size (mesh size) on the fatigue life.

As an example, consider the manufacture of a superalloy turbine disc from powder-formed material, to last 3300 cycles in service. Data from Coles [11] based on the U.S. Federal Aviation Agency (FAA) recommendations, suggests a distribution in the schematic field lives of a "3300 cycle" disc as having a mean of 2.3×10^4 cycles and "mean minus three standard deviations" of 10^4 cycles. The FAA requires the latter figure to be divided by three, to give the 3300 cycle life in service. The powder-formed alloy AP1 (0.03C, 15Cr, 17Co, 5Mo, 4Al, 3.5Ti, bal Ni) has a yield strength of approximately 1050 MNm^{-2} [12,13] and so a stress cycle of 0-700 MNm^{-2} might be assumed for "on-off" loading. Fatigue-crack growth-rate data on a similar alloy [14] tested in air at 600°C and R = 0.1 conform to the expression:

$$da/dN = 4 \times 10^{-12} \, \Delta K^{3.3} \tag{26}$$

for $\Delta K > 10$ MNm$^{-3/2}$. The threshold, ΔK_{th} is approximately 7 MNm$^{-3/2}$, ($\Delta K_o \simeq 7.8$ MNm$^{-3/2}$ using equation (9)). These data show accelerating growth at $\Delta K \simeq 70$ MNm$^{-3/2}$, and an underestimate of K_{Ic} would therefore be 70/(1-0.1) $\simeq 80$ MNm$^{-3/2}$. In the calculation below, K_{Ic} will be taken as 100 MNm$^{-3/2}$.

Equation (25) may be integrated to give:

$$9.48 \times 10^{-3} \, N_f = a_o^{-0.65} - a_f^{-0.65} \tag{27}$$

Using equation (4) with σ_{app} = 700 MNm^{-2}, K_{Ic} = 100 MNm$^{-3/2}$, a_f is calculated as 16 mm (diameter $2a_f$ = 32 mm). The life is taken as 2.3×10^4 cycles, and equation (26) may then be evaluated to give a_o = 0.23 mm (diameter = 0.46 mm). At an applied stress of 700 MNm^{-2}, this size of initial defect produces an initial ΔK of 12 MNm$^{-3/2}$, which is above the threshold ($\Delta K_o \simeq 7.8$ MNm$^{-3/2}$) and corresponds to the linear range of the (da/dN) curve. Additionally, Brown and Hicks [14] have shown that cracks of size 0.5 mm follow the same (da/dN) curve as that for long cracks. If the defect is regarded as a very thin penny-shaped object, the diagonal spacing of a square mesh must be less than 0.46 mm, to prevent the defect slipping through, i.e., the conventional mesh spacing must be smaller than 0.325 mm (approximately 1/80 inch or 80 mesh). If the defects are allowed to be elliptical in cross-section, with an aspect ratio of 2:1, the mesh size should be decreased to approximately 150 mesh (and small changes must be made to the figures above). Modest changes in K_{Ic} do not affect the lifetime significantly, because $a_o^{-0.65}$, with a_o = 0.23 mm, is 232, whereas $a_f^{-0.65}$, with a_f = 16 mm, is only 14.7. If a_f = 10 mm, $a_f^{-0.65}$ = 20; if a_f = 22 mm, $a_f^{-0.65}$ = 12; so that a change of 37.5% in a_f (approximately 19% change in K_{Ic}) leads to only approximately 1.6% change in N_f.

Applications such as turbine discs provide great incentive for the use of design stresses which are high fractions of the yield stress and it is instructive to re-work the figures for an applied stress level of 840 MNm^{-2} (0.8 of the yield stress). The LHS of equation (27) becomes 1.73×10^{-2} N_f rather than 9.48×10^{-3} N_f. With a_o = 0.23 mm, a_f = 16 mm, as above, N_f is reduced to 1.4×10^4 cycles, i.e., to 61% of the required design life. To obtain the required life of 2.3×10^4 cycles, a_o must be reduced to 0.095 mm: allowing for a 2:1 aspect ratio and the possibility of slipping diagonally through a square mesh, a mesh spacing of 0.067 mm (380 mesh) would be necessary. Test data are quoted for nominally-smooth testpieces of API for a stress of 1050 MNm^{-2} at 600°C [12]. Although LEFM should strictly not be used for such a high stress level, the conclusion from integration of the crack growth rate is that if a_o = 0.23 mm, N_f = 6830 cycles: if a_o = 0.095 mm, N_f = 11,010 cycles. The S/N data give the endurance as 7765 and 12,730 cycles for a stress of 1050 MNm^{-2}, so that the general agreement is remarkably close, bearing in mind the improper use of LEFM and the lack of knowledge of defect sizes and distribution in the S/N specimens.

The figures above have been used as part of a hypothetical study to illustrate the general principle of the relationship between fatigue life and initial defect distribution and the conclusions on defect size should be regarded as comparative, rather than absolute. One serious problem is that the published (da/dN) curves have been obtained in air, whereas buried cracks grow in conditions approximating to those in vacuum. The effect of this change of environment on crack growth rate could be significant. Two further points are that the calculations have been made only for "on-off" loading cycles, so that vibration stresses are not included; and that, in practice, probabilities must be assigned, firstly to the aspect ratio of any given oxide defect, and, secondly, to the statistics of such a particle passing through the sieve (i.e., some particles will approach the mesh parallel to a side of the square, some parallel to the diagonal, and many in other orientations). The figures have been calculated for a mean life of 2.3×10^4 cycles, but whether three standard deviations below this figure would correspond to 10^4 cycles, as in the data presented by Coles [11], is a matter of conjecture.

CONCLUSIONS

The aim of this paper has been to indicate ways in which fracture mechanics could be used to provide quantitative assessments of the effects that processing and manufacturing routes may have on a material's resistance to fast fracture or to sub-critical crack growth, illustrated here by fatigue crack growth. Clearly, detailed predictions must sometimes make use of yielding fracture mechanics and incorporate other mechanisms of crack growth, but useful and interesting conclusions can often be obtained from simple analysis. In the present paper, comments have been made on effects of alloy selection, heat-treatment and impurity content on monotonic fracture toughness and critical defect size in high-strength materials, but the judgments which are made as a result of the calculations depend critically on the sensitivity and reliability of NDT. The calculations have been extended to estimate fatigue lives under cyclic loading by integration of crack growth rates and the importance of the original defect distribution has been emphasised. In many cases, this must be assessed by NDT, although, in some cases, useful information may be gained from a proof test. Sample calculations, dealing with powder-formed alloys, demonstrate the use of fracture mechanics in producing recommendations concerned directly with process variables. The approach is intended to serve as a model for the more general development of quantitative relationships between a material's "damage tolerance" and the control of its microstructure through alloy chemistry, impurity content, processing route and heat-treatment. It is such relationships that will represent the application of fracture mechanics technology to material evaluation.

ACKNOWLEDGEMENTS

Thanks are due to Professor R. W. K. Honeycombe F.R.S. for provision of research facilities and to Dr. J. E. King, Mr. S. Slatcher, Dr. D. Taylor and Dr. B. Wiltshire for helpful discussions and for permission to use previously unpublished results.

REFERENCES

[1] Jackson, W. J. and Wright, J. C., "Fracture toughness approach to steel castings quality assurance", Metals Technology, 11, p. 425, September 1977.

[2] Paris, P. C. and Erdogan, F., "A critical analysis of crack propagation laws", J. Basic Engineering, Trans. ASME, Series D, 85, p. 528, 1963.

[3] McEvily, A. J., "Current aspects of fatigue", Metal Science, 11, p. 274, 1977.

[4] Rabotnov, Yu. N., "Creep problems in structural members", North-Holland Publishing Co., London, 1969.

[5] Marrison, T., "Characteristics of maraging steels", The Metallurgist, 8, p. 80, 1976.

[6] Hunsicker, H. Y., "Development of Al-Zn-Mg-Cu alloys for aircraft", in Proc. Rosenhain Conference "Contribution of Physical Metallurgy to Engineering Practice", Royal Society, p. 245, 1975.

[7] Smith, E., Cook, T. S. and Rau, C. A., "Flow localization and the fracture toughness of high strength materials", Proc. 4th Intl. Cong. on Fracture, Waterloo, ed., D. M. R. Taplin, Pergamon/Univ. Waterloo Press, 1, p. 215, 1977.

[8] Knott, J. F., "Micromechanisms of fibrous crack extension in engineering alloys", Metal Science, 14, p. 327, 1980.

[9] Mogford, I. L., "The analysis of catastrophic failures", Proc. Conf. on "The Practical Implications of Fracture Mechanisms", Institution of Metallurgists, 2, 10, 604-73-Y, p. 63, 1973.

[10] Taylor, D. and Knott, J. F., "Growth of fatigue cracks from casting defects in nickel-aluminium bronze", Metals Technology, May 1982.

[11] Coles, A., "Material considerations for gas turbine engines", Proc. 3rd Intl. Conf. on Mechanical Behaviour of Materials, Cambridge, ed. K. J. Miller and R. F. Smith, Pergamon, 1, p. 3, 1980.

[12] "Superalloys for Powder Atomisation", Henry Wiggin and Co., Ltd., publication 3712, 1978.

[13] King, J. E., "The effects of grain size and microstructure on threshold values and near-threshold crack growth in a powder-formed Ni-base superalloy", Metal Science, July 1982.

[14] Brown, C. W. and Hicks, M. A., "Fatigue growth of surface cracks in nickel-based superalloys", Int. J. Fatigue, p. 73, April 1982.

A REVIEW OF ELASTO-PLASTIC FRACTURE DESIGN METHODS AND SUGGESTIONS FOR A RELATED
HIERARCHY OF PROCEDURES TO SUIT VARIOUS STRUCTURAL USES

C. E. Turner

Imperial College
London SW7 2BX, England

ABSTRACT

An approximate J-based design or assessment procedure, for the avoidance of
fracture, broadly comparable to the better known COD and R-6 methods, has been
proposed recently. Further developments and comparisons with these other methods
are described here, together with related simplified approaches suitable for preliminary use or for projects requiring less stringent assessment. To recognize
that a number of engineering approximations are made and that the estimation method is not rigorous, and also to distinguish from other J-based proposals, it is
referred to as the EnJ method. The procedure includes treatment of shallow and
deep notch cases in two and three dimensions, thermal, residual and biaxial loading, and preferred J-based test methods to allow for use of either an initiation
toughness or a post-initiation (R-curve) value. The simplified method offers
rules applicable where the load in the cracked ligament does not exceed 0.8 of
the fully plastic load. For yet more routine applications selection of material
by Charpy value alone is still quite feasible, thus presenting the designer with
three levels of treatment to be used according to the complexities and risks of
particular applications.

INTRODUCTION

Several concepts exist for assessing the significance of defects in the elasto-plastic regime. In practice, three inter-related contributions are necessary:
a) estimation of applied severity of loading; b) test procedures for measurement
of material property; c) recognition of the status of both defect and structure.
Most of the concepts concentrate on the estimation of an applied severity and
virtually all can be regarded as functions which interpolate between the well-
established theories of linear elastic fracture mechanics (LEFM) and plastic collapse. Several such theories of elasto-plastic fracture mechanics (EPFM) were reviewed [1] are are illustrated, Figure 1. However fruitful these may have been
in establishing present understanding, only two concepts are further discussed
here, crack opening displacement (COD) and the J-contour integral (J), since these
seem to embrace all the others. The background to COD and J was also reviewed [1]
and has, of course, been described in several books such as [2-4]. Four COD or J
based proposals for use in design have emerged. The first is the well-known COD
design curve [5] now transformed [6] into what is probably the most complete pro-

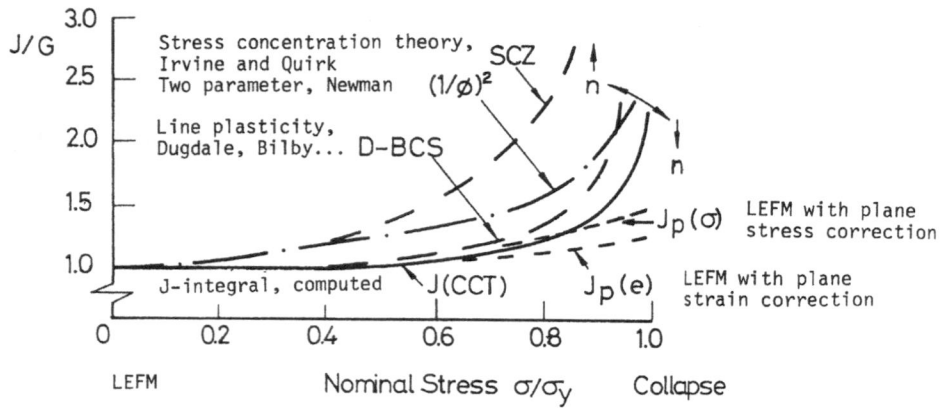

Fig. (1) - Comparison of several elasto-plastic fracture theories

cedure, encompassing all three aspects A, B, C, above. The second procedure often seen as a modification of the so-called two-criteria concept [7] is now called the R-6 method [8,9], and despite original associations with the Dugdale-BCS line plasticity model for COD, it is now generally regarded as an estimate of $\sqrt{G/J}$ (i.e., of LEFM severity/elastic-plastic severity) as a function of load, which must itself be limited by plastic collapse. The third procedure was outlined in a preliminary stage [1] as a J-design curve. It has recently been developed in much more detail [10,11] to cover a variety of circumstances as wide as the COD or R-6 methods, and is here called the "engineering J-method" (EnJ) to acknowledge an intended approximate engineering use of J and to distinguish it from the other J-based methods. The most recent proposal is the EPRI J-based method [12] which is aimed primarily at pressure vessel problems.

The first object of this paper is to compare these four methods in order to show their similarities and differences. The second and main object is to use the essential unity that emerges, despite many differences of detail, to propose a related hierarchy with three levels of complexity to suit the need of various users. It is accepted that it may be better for some purpose to retain or further develop quite specific and perhaps ad hoc design procedures relating to a clearly defined usage. Nevertheless, the development of several "general purpose" fracture safe assessment methods seems to indicate a continuing need for methods not closely object-related, or at least applicable to objects for which more routine or ad hoc methods have not yet been developed. The three levels of complexity proposed here are: i) selection of material to avoid brittle fracture; for structural steels this method would not normally use fracture mechanics concepts; ii) an extension of LEFM methods applicable for regular use and casual abuse of structures up to about 0.8 of net section yield; iii) fully elastic-plastic J-based methods for circumstances where net-section yield is to be exceeded, perhaps during fabrication or in gross overload or emergency, where catastrophic fracture must be avoided but subsequent re-use probably implies re-building.

THE COD, R-6, ENGINEERING J AND EPRI METHODS

In LEFM, $J \equiv G$ and it may sometimes be convenient to express J in K-like units by using

$$K^2 = E'G \text{ for LEFM}; \quad K_J^2 = E'J \text{ for EPFM} \tag{1a}$$

where E' is the effective modulus, E for plane stress, $E/(1-\nu^2)$ for plane strain. As plasticity becomes more extensive for a given configuration, then, as seen in Figure 1, J exceeds G and of course K_J exceeds K. The excess is fairly small in the near LEFM regime but quite large as plastic collapse is approached.

It must also be recognized that the COD method [6] is aimed at assessing an acceptable size of defect, and contains a deliberate factor known to be at least two-fold (in relation to G or crack size) in the LEFM regime, whereas the other methods do not. The relationships between COD and J can be expressed as

$$J = m\sigma_y\delta \tag{1b}$$

where σ_y is the uniaxial yield stress.

If m were indeed a constant then COD and J concepts would be identical. Insofar as m differs with geometry, extent of plasticity and degree of work hardening then the concepts are not identical. In practice, it seems [1] $1 < m < 3$ and often $m \approx 2$. Thus, provided calculations are not made which imply a change in value of m (e.g., from component data to test piece data) difference between a consistent use of J or consistent use of COD may be quite small but the factor of safety must be eliminated if a direction comparison of like with like is to be made. This factor is strictly 2m in the LEFM regime but not explicitly known in the plastic regime where the basis of the COD curve is experimental.

CHOICE OF AXES

The four methods COD, R-6, EnJ and EPRI are illustrated, Figure 2, together with the governing equations (2a-d) that define the estimation procedures. The COD curve is presented both in its current form from [6] and also in its original form [5] for ease of comparison with the other methods. That difference only involves the scales of axes with no change in equations. Also shown is a COD-type curve with a factor of safety of 2 eliminated for purposes of more direct comparison here. It must be emphasized that such a curve is not part of the standard usage [6]. J-related scales, with m=1 and m=2, are also shown to aid direct comparison. The other main differences in the comparison is that COD uses an effective crack size, $\bar{a}$, which is related to the actual crack size, a, after the estimation procedure has been used so that the non-dimensional COD axis [5] of ϕ is related to the J axis by

$$\phi = JE/2\pi m\sigma_y^2\bar{a} \tag{3}$$

Notionally, $\bar{a}$ and a are related to the LEFM shape factor Y by

$$\pi\bar{a} = Y^2 a \tag{4}$$

where Y is defined by $K = Y\sigma\sqrt{a}$, in which case

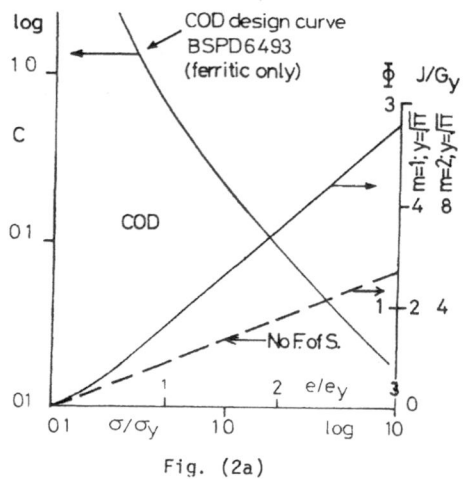

Fig. (2a)

$\phi = (e/e_y)^2$ for $e/e_y \leq 0.5$ (i)

$\phi = (e/e_y) - 0.25$ for $e/e_y \geq 0.5$ (ii)

$C = 1/2\pi\phi$ in BSPD 6493

(2a)

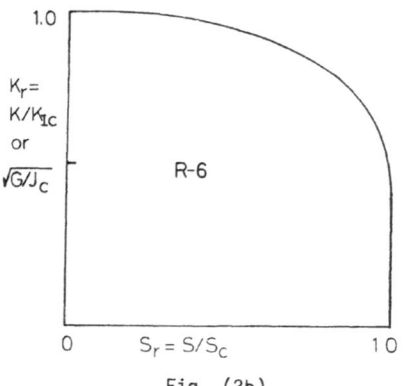

Fig. (2b)

$$K_r = S_r[(8/\pi^2)\{\ln \sec(\pi S_r/2)\}]^{-1/2}$$

(2b)

Fig. (2) - The several elasto-plastic design curves

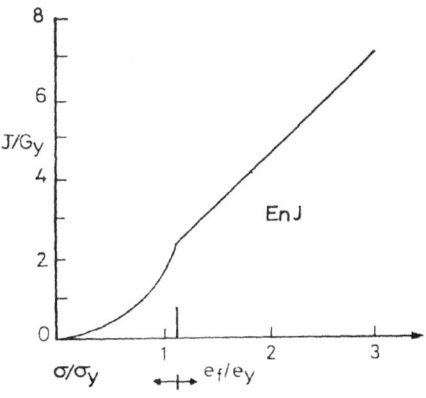

Fig. (2c)

$$J/G_y = (\sigma/\sigma_y)^2\{1+0.5(\sigma/\sigma_y)^2\} \text{ for } \sigma/\sigma_y \leq 1.2 \qquad (i)$$
$$J/G_y = 2.5\{(e_f/e_y) - 0.2\} \text{ for } e_f/e_y \geq 1.2 \qquad (ii) \qquad (2c)$$
$$\text{If } (W/b)(\sigma/\sigma_y) \geq 1.0 \text{ or } b/W \leq 0.5 \quad \text{check for collapse } (iii)$$

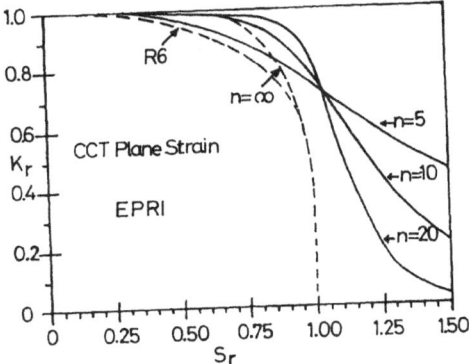

Fig. (2d)

Fig. (2) - continued

$$2m\phi = JE/Y^2\sigma_y a = J/G_y \tag{5}$$

where G_y is the value of G at $\sigma = \sigma_y$, but in [6], the conventional values of Y are not always used, notably in respect of part through thickness cracks subjected to bending, so that equation (4) is not necessarily satisfied if [6] is followed precisely.

If, for a first comparison, all the caveats are swept aside, then as seen in Figure 2, the COD and EnJ estimation curves are recognizably similar, being presentations of a non-dimensionalized term J/G_y (or ϕ as in equation (5)) versus strain ratio e/e_y in the body. For many cases where [6] does indeed use equation (4), then the ordinate of the COD or EnJ diagrams differ only by the "factor of safety" of $2m$, already discussed. The actual formulae are not quite identical, COD being based on experiment, EnJ on computations, but both have a parabolic form near LEFM and a linear relationship for strains above yield. EnJ can similarly be related directly to R-6, since if the ordinate just described as J/G_y were multiplied by $(\sigma_y/\sigma)^2$ it would become J/G which when inverted and square-rooted is the ordinate of R-6 and indeed, the EPRI diagram. If J is restricted to J_i, or with equation (1), K_j to K_{1c}, then the standard R-6 ordinate of $K_r = K/K_{1c}$ is obtained. The further main difference between COD and EnJ on the one hand and R-6 and EPRI diagram on the other, is that the abscissa of the former is strain based, whilst the latter is stress based. In fact, in the near LEFM regime, both the former accept $e/e_y \simeq \sigma/\sigma_y$ so the only point of real substance is that the COD and EnJ methods state that plastic collapse shall be estimated separately as a matter of normal design procedure not explicitly related to fracture, whereas the R-6 and EPRI diagrams imply that collapse, however defined, is an integral part of the fracture diagram. If the EnJ diagram is scaled and inverted, as just described, it is clearly identifiable as the "hump" common to R-6 and EPRI plus the "tail" which features in the EPRI diagram, although the abscissa above yield differs in being some nominal strain for EnJ, nonexistent beyond "nominal collapse" for R-6 and "work hardened collapse load" for EPRI.

This quite identifiable relationship between all the diagrams, most clearly seen by relating the EnJ diagram to each in turn, cannot, however, be extended so clearly to either the treatment of secondary stresses (residual, thermal, local at stress concentrations, biaxial and so on) or to the selection of an appropriate value of fracture toughness or to the status of defect and structure, much of which tend to be matters of subjective judgment not clearly specified in the various procedures as at present defined. These differences in treatment are summarized here but can be followed more explicitly in worked examples, or in the case studies described in detail in the reference already cited [13].

A. Deep Notches

The direct effect of notch depth in the LEFM regime is accommodated by use of the shape factor. However, the major difference between deep notch (a/W large) and shallow notch (a/W small) in the context of plasticity, arises if the ligament size and the effect of work hardening is such that (for tension)

$$\sigma_{f1} b > \sigma_y W \quad \text{(inequality 1)} \tag{6}$$

When this is so the ligament can support a load in excess of that to cause first yield of the gross section, so that there is some measure of uniformity of strain over the whole body, despite the obvious concentration at the crack. Where equation (6) is not met, all displacement subsequent to ligament yield must be concentrated into the crack in some idealized COD type behaviour. In the limit of no work hardening (or a long plateau of yield), this effect occurs for all depths of notch after net section yield. The R-6 method recognizes this behaviour by the adoption of the ln sec term which itself arises from the quasi-elastic Dugdale model where plastic strain is always very localized, and the method does not permit usage above some particular definition of net section collapse. The EPRI computations extend the regime to above "first collapse" by the inclusion of work hardening. The COD and EnJ methods adopt strain for the abscissa and use equations, experimental in the former, computed in the latter, based mainly on cases for which equation (6) is satisfied, i.e., there is some uniformity of deformation above yield. The effect of localization of strain in that regime is mitigated in COD [6] by limiting the strain to $e/e_y < 2$ unless particular estimation is made and by the inclusion of a factor of safety as already discussed. In the EnJ method, the concept of an augmented value of strain the "cracked body structural strain" (cbse) was introduced [10] whereby the effective strain is taken as

$$e_f/e_y = \{W/b \text{ (or } B/b)\}(\sigma/\sigma_y) \tag{7}$$

if the strain controlled regime above net section yield is to be entered. A further refinement of e_f is to apply the augmentation only to the plastic component of strain. For many problems, these complications are not necessary because the requirements of conventional design will restrict conditions to below net ligament yield so that in the simplified EnJ method proposed later no distinction is made between deep and shallow notches, other than by use of the appropriate shape factor.

B. Stress Concentrations

For a crack in a region of stress concentration, using COD, the abscissa is entered at $e/e_y = k_t\sigma/\sigma_y$ where k_t is the conventional elastic stress concentration factor (SCF). R-6 allows for stress concentrations via LEFM, which affects the ordinate, K_r, but not the abscissa, S_r, since stress concentration does not affect plastic collapse. The EPRI method gives explicit estimates of J for the case of a nozzle on a flat plate representative of a pressure vessel problem but as yet gives no estimate for other more general cases. In the EnJ method [11], the abscissa is entered at $e_f/e_y = (k_t\sigma/\sigma_y)^2$ following re-analysis of the original results [14] using the Neuber relationship, $k_\sigma k_e = k_t^2$, where k_σ is the plastic stress and k_e the plastic strain concentration factors respectively. One of the modifications leading to the simplified EnJ diagram suggested here is to treat stress concentration cases via the ordinate, since as shown by the original analysis [14] the estimate of J made from LEFM, together with a plastic zone correction factor was adequate up to about 0.86 net section yield, and equation (2ci) incorporates what is in effect a plastic zone correction for plane stress, which will be adequate for plane stress and conservative for plane strain.

C. Residual Stresses

If the value of residual stress is not known it must be assessed, often as σ_y for weldments with no stress relief, or as zero for machined or heat treated components, but some value other than zero, e.g., $\sigma_y/4$ may be thought by some to be more representative of the stress relieved state. These values and the distribution through the thickness are not closely specified in the descriptions cited for the various methods, although a pattern which reduces rapidly below the surface is implied in [15] and often used elsewhere. If values σ_r and distributions are known or implied, COD then enters the abscissa at $e/e_y = (\sigma_m + \sigma_r)/\sigma_y$, where σ_m is the mechanical stress term. A combined stress concentration with full yield level residual stress σ_y would be entered at $e/e_y = (k_t\sigma/\sigma_y)+1$. The distribution of residual stress is not directly required but translation from $\bar{a}$ to a would usually imply that the through thickness distribution of mechanical and residual stress is the same. R-6 again allows for residual stress on the ordinate and in the latest revision amends the shape of the diagram by an empirical procedure to allow for the fact that the ln sec plasticity term relates to the mechanical component of stress only. The distribution of residual stress must be assumed in order to evaluate the K term for the residual stresses. The EPRI method [12] does not as yet cater for residual stress explicitly. In EnJ, residual stress also affects the ordinate. A first proposal [16] simply added a value of unity to the ordinate J/G_y to allow for yield level residual stresses. In later developments [10,11], a J for residual stress is used, implying values and distributions either known or assumed. In the LEFM regime, K(mech) and K(resid) are added. In the plastic regime J(mech) and J(resid) are added where each term is separately estimated for the EnJ design curve for the known value of stress. For intermediate cases

$$J = \{(J^{\alpha}_{mech}) + (J^{\alpha}_{resid})\}^{1/\alpha}$$

(8a)

where

$$\alpha = 0.5 + \{(\sigma_m + \sigma_r)/2\sigma_y\} \leq 1$$

(8b)

This is just an ad hoc interpolation between $\alpha = 1/2$ for low stress (so that $J^{1/2}$ or K is added) and for high stress $(\sigma_m + \sigma_r \rightarrow \sigma_y)$, $\alpha = 1$ so that J values are added. It is appreciated that strictly J is not rigorous for residual stresses but by analogy with COD and R-6 methods some approximate usage must be feasible. It is argued [11] that if overall reaction stresses are induced, e.g., by welding in a preheated or undersized patch, then the problem should be treated as a separate step in the analysis rather than as a subsequent problem of combined residual and mechanical loading, since the final severity depends upon the sequence of events, for which an overall simple rule would often be unreasonably conservative.

BIAXIAL STRESS OR STRAINS

In strict LEFM biaxiality has no effect on values of K induced for a given applied stress, σ. The actual value of σ may of course be affected if the system is not statically determinate. The main effect of a biaxial system is to modify the load at which full plasticity occurs and thus the value of load at which J increases rapidly above G. This could be reflected in R-6 by a different value of the collapse load, S_c, against which the abscissa is normalized. Considering only tension or compression loading for simplicity, if the transverse stress, σ_t, is given by

$$\sigma_t = \beta\sigma \tag{9}$$

$$\sigma_n = \sigma W/b \tag{10}$$

$$\sigma_t = \gamma\sigma_n = (\beta W/b)\sigma \tag{11}$$

then the fully plastic load is

$$Q_{fp} = Bb\sigma_{ny} \tag{12}$$

where the net section stress σ_n and the transverse stress σ_t combine to cause yield at σ_{ny}. If the ligament is taken to be in plane stress, use of von Mises criteria gives

$$Q_{fp} = Bb\sigma_y/(1-\gamma+\gamma^2)^{1/2} \tag{13}$$

If the ligament were to be in plane strain, Q_{fp} might be appreciably greater for $\beta \to +1$ but the plane stress values are used here for conservatism. The main effect is thus on the value of load at which either collapse is assumed to be the limiting factor, e.g., in R-6, or deformation controlled behaviour ensues as in the post-yield regime of EnJ. It is possible to envisage biaxial displacements. If these occur due to mechanical cause, the true loads induced must be evaluated since the effect of (say) transverse compressive load is to augment J (for a given σ) because the fully plastic load is reduced, whereas transverse compressive displacement (at fixed axial displacement) reduces σ and hence J by the Poisson ratio effect. The most likely case of biaxial displacement is thermal straining, as discussed below.

In the simplified EnJ method, which is restricted to below full plasticity, it is suggested that if the transverse stress is tensile, it is neglected on the argument that full plasticity for a ligament in plane stress is not affected. If the transverse stress is compressive, then equation (13) can be used, or yet more simply, using the Tresca shear stress criterion

$$Q_{fp} = Bb\sigma/(1-\gamma) \quad (\gamma\text{-ive only}) \tag{14}$$

THERMAL STRAINS

If a body is restrained in both x and y directions and cooled by an amount ΔT, the induced stress in each direction is

$$\text{plane stress (thin sheet)} \quad \sigma = E\alpha\Delta T/(1-\nu) \tag{15a}$$

$$\text{plane strain (thick body)} \quad \sigma = E\alpha\Delta T/(1-2\nu) \tag{15b}$$

Cooling of the surface of a thick body probably induces stresses in between these values, although the surface is strictly in plane stress. The estimation equations must therefore be entered at the appropriate value of stress, not $\sigma = E\alpha\Delta T$, which is appropriate if constrained in the axial direction only. Although thermal stress cannot itself cause plastic collapse, the magnitude of ΔT may readily cause strains greater than e_y. If these occur in an uncracked body, or of course,

if the strains remain elastic, the distribution is uniform and if the body subsequently cracks, the stresses are partly relaxed. If, however, the body is cracked
before cooling, then any excess of thermal contraction over and above e_y must be
met by mechanical strain above yield. The uniform value of mechanical strain is
limited by the load that can be transmitted by yield of the ligament so that localization of deformation must occur with any excess of strain focused into the crack
tip as a true COD type behaviour. For the simplified EnJ diagram, it is proposed
that thermal stress, unless compressive, must be restricted so that

$$\Delta T \leq \sigma_y(1-2\nu)/E\alpha \qquad\qquad\qquad (16)$$

COMPARISON OF THE METHODS

The main conclusions of [13] were that five matters accounted for the main
differences between the original calculations. These matters were: (1) assessment
of acceptable defect size by COD, but critical size by R-6 and EnJ (and also EPRI)
methods: to compare like with like a factor of two was removed from the COD equations; (2) calculation of a defect size relevant to fracture only, or the inclusion of the risk of collapse as an inherent part of the fracture calculation (as
in R-6). To compare like with like, approximate estimates of collapse were made
in association with all methods; (3) distribution of residual stresses: for the
present purpose a distribution that died away rapidly with thickness was agreed.
In fact, further differences over choice of the LEFM Y factor occurred for part
through cracks where no exact solution exists, so that for a strict comparison of
methods, agreed values of Y were used for both mechanical and residual stress pat-
terns; (4) use of initial yield stress σ_y or flow stress $\sigma_{fl} = (\sigma_y+\sigma_u)/2$; COD and
EnJ used the former, R-6 used the latter. Many small differences resulted and re-
mained unresolved, including different aspects of using upper or lower bound values
of strength. Lower bound gives conservative estimate of collapse relevant if the
component is load controlled, but an unconservative estimate of J if the component
is displacement controlled (as in the above yield regime of EnJ); (5) value of
toughness: strictly, R-6 uses K, COD uses δ_c, EnJ uses a value of J, which may be
either J_i (at initiation), J_{1c} (standard test [17]), or J_r (after stable growth).
The material data was in fact J-based and made consistent with K and COD via equations (1) and (2). Particularly for the COD method this raises the uncertainty
over m, equation (1b), and whether it is the same for the test piece and the structure, and whether the amount of slow growth permitted in finding the J_c data given
was the same as implied in a standard COD test [18]; no doubt it was not.

In [13], no direct attempt was made to allow for slow stable growth, the exercise being based on a given value of toughness, with no discussion of its physical
meaning. The exercise was therefore a comparison of estimation procedures only.
The question of what definition of toughness to use is discussed further later on
in the paper.

A THREE LEVEL RESPONSE TO FRACTURE ASSESSMENT

Aside from the uncertainties in the theories used, the main problem is recognition of the level of assurance of safety required in a structure. With tough
material, a structure is limited by the plastic collapse load. If that limit is
to be reached and the actual act of collapse not truncated by rapid fracture, then
local tearing or cracking type fracture must be avoided at the high levels of plas-

tic straining which may occur during plastic collapse. Such deformation would normally make the structure unserviceable so at this level avoidance of catastrophic fracture is a safeguard under accident conditions. It may also be desirable to use such studies as guides to absolute limits of behaviour which are then avoided by a known margin. For materials that experience a transition in the micro-mode of fracture, notably the ferritic structural steels, both ductile tearing and brittle cleavage fracture must be avoided. The only absolute assurance of the latter is to use materials sufficiently far onto the ductile upper shelf regime. The definition of that "upper shelf" must, for absolute assurance, be rigorous in respect of size loading rate and material degradation. These effects are shown schematically, Figure 3. Such an upper shelf is likely to be well

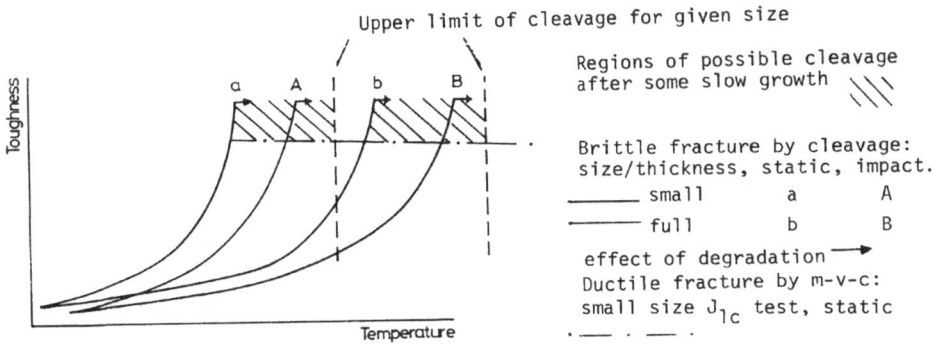

Fig. (3) - Effect of temperature, size and strain rate on the toughness of steel (schematic)

above the onset of Charpy V upper shelf and to be rigorous would imply valid dynamic K_{1c} testing. For most practical purposes, such a level of assurance is uneconomic and it is in the relaxation of various severities that requirements for one application may differ from another.

MATERIAL SELECTION BY CHARPY V TEST

A number of proposals have been made, such as [19,20] based mainly on experience of use of particular grades of steel and conditions of service. The basis of a rational scheme exists in that G or J can be related to work done, w, via the equations of the form

$$G \text{ (or } J) = \eta w/Bb \qquad (17)$$

where η is a geometric factor. For a three-point bend test, $S/W = 4$, $a/W \geq 0.4$, $\eta \simeq 2$ for both elastic and plastic behaviour. For the Charpy geometry of $a/W = 0.2$, the elastic term is rather less than 2. Thus, neglecting the difference between the sharp fatigue crack in a fracture mechanics type test and the V-notch in the Charpy test, and also the difference between static and dynamic yield strengths and ductilities, there is a formal link between K_{1c}^2 and C_v, so that in principle, setting a Charpy energy value would correspond to setting a desired K_{1c} value. The further theoretical link exists on yield stress of steel so that the difference in static or dynamic yield for K_{1c} and Charpy tests respectively can be accommodated by an equivalent change in testing temperature. As far as the writer knows, these ideal links have not been fully explored and empirical relations are used. Those suggested [21], based on low alloy pressure vessel steels, when translated to S.1 units ($MNm^{-3/2}$ for K_{1c}; Joules for C_v; MN/m^2 for σ_y) are

Brittle: $\quad K_{1c} = 14.5\, C_v^{1/2}$ $\hfill$ (18a)

Ductile: $\quad (K_{1c}/\sigma_y)^2 = 0.68\, \{(C_v/\sigma_y) - 0.01\}$ $\hfill$ (18b)

For a strength of about $300\ MN/m^2$, these two equations give similar relationships. The particular Charpy values proposed for material selection [19] are summarized, Appendix 1. The requirement there for $C_v = 41J$ translates to $K_{1c} \simeq 94\ MN/m^2$, using equation (18a), at a temperature that depends on the circumstances. A fully fracture mechanics approach would specify the temperature of interest and assess the K_{1c} (or C_v) value to suit the stress level and defect sizes of interest. A comparison of the methods could no doubt be taken further but the immediate point is that several simple non-fracture mechanics procedures exist if the fracture assessment rests on material selection alone and does not enter into the significance of defects.

A SIMPLIFIED ESTIMATION PROCEDURE FOR NORMAL USE (CASE ONE)

An objection sometimes raised is that the fracture design curves are either too complicated or not accurate enough. It seems impossible to resolve these two conflicting statements within one method. The present proposal is to simplify the EnJ curve still further by limiting its use to normal design and safety assessment for regular use, i.e., Case One, rather than to exceptional conditions of overload or fault that imply a severity probably putting the structure out of use, Case Two. The basis of the argument is that LEFM, when corrected for size of plastic zone, provides an adequate estimate of J up to an applied stress of about 0.8 or 0.82 of net section collapse, as indeed shown somewhat schematically, Figure 1, in a rather different context. If the conditions for regular use, casual abuse and continued re-use are identical with restricting the load to 0.8 of that to cause a full plasticity of the ligament, then an LEFM based method should be adequate. Thus, although in COD, the linear regime covers residual stress, stress concentrates and above yield behaviour generally, and in EnJ it is used for stress concentration and above yield loads, in fact only the latter need be so treated. Using EnJ, residual stresses are already treated on the ordinate, and with one small reservation, stress concentration cases can be similarly dealt with, as indeed they are in R-6. The assessment of J/G_y at a stress concentration in terms of the nominal stress σ rather than the local stress or strain was demonstrated [14] and the departure from that method in favour of entry on the abscissa at the true local strain was unnecessary for use up to 0.8 ligament yield. This could arise if the adjoining ligament yields, in Case 1 explicitly precluded, or if the crack were very small in relation to the size of plastic zone at the stress concentration, but the absolute value of K is then low because of the small crack size, and relevant to fatigue rather than fracture. Thus, an estimate of J based on LEFM with plastic zone correction seems perfectly feasible for stress concentration cases, as well as the existing use for primary loading up to $0.8\sigma_y$ and for residual stresses, as at present [10, 11]. The LEFM abscissa of σ/σ_y is therefore retained provided the load Q is less than $0.8Q_{fp}$ where Q_{fp} (or the corresponding moment M_{fp}) is the load (or moment) for full plasticity of the ligament. In terms of net section (elastic) stress σ_n,

$$\sigma_{nt} = \sigma W/b = Q/Bb < 0.8Q_{fp}/Bb, \text{ i.e., } 0.8\sigma_y \tag{19a}$$

for tension, or

$$\sigma_{nb} = \sigma(W/b)^2 = 6M/Bb^2 < 0.8(6M_{fp}/Bb^2), \text{ i.e., } 0.8(1.5)\sigma_y \tag{19b}$$

for bending.

The factor of 1.5 in bending is the plastic shape factor (for a rectangular cross-section) which recognizes the difference between first yield and fully plastic moment while the notch constraint factor is neglected. For combined bending and tension, the limiting stresses should strictly be worked out from plastic collapse theory but consistent with the simple methods intended here, it is suggested that the abscissa of σ/σ_y be used up to a stress level such that

$$\sigma_{nt} + (\sigma_{nb}/1.5) < 0.8\sigma_y \tag{20}$$

For biaxial loading, the net ligament equivalent or shear stress would be restricted to $0.8\sigma_y$ noting that for thermal cases, the true biaxial stress rather than the nominal value $E\alpha T$ must be used if biaxial conditions do indeed exist. The estimate of J, for loads up to $0.8Q_{fp}$, is then

$$J/G_y = (\sigma/\sigma_y)^2(1+0.5(Q/Q_{fp})^2) \tag{21}$$

The reasoning for this particular approximation and some more detailed comments on its application with high loading, and certain other problems are given in Appendix 2. It is noted here that the terms differ very little from R-6, and when restricted as here, to 0.8 of full yield, imply that J does not differ from G by more than 32%, so that estimates of crack size might be in error by that amount, or estimates of stress level by 18% if LEFM alone were employed. This value of J/G_y is judged conservative for plane strain. While such errors are not negligible, other uncertainties as already discussed in relation to [13] and yet to be discussed in relation to toughness, are likely to be more significant.

AN ESTIMATION METHOD FOR EXTENSIVE YIELD (CASE TWO)

If use above 0.8 of full ligament yield arises, by intent to avoid catastrophe or by the unnoticed development of a deep crack, then recourse to the existing EnJ proposal [10,11] seems appropriate. In principle, the COD curve may be relevant but some restrictions exist if $e/e_y > 2$ as is likely in this regime, and the various differences already discussed of continuity with the implications of LEFM remain. The R-6 method is inherently restricted to $S/S_c < 1$, i.e., to ligament collapse, again with the uncertainties of what the particular usage of S_c implies; in short, the usage may not be sufficiently conservative for normal design level but too conservative for estimates of extreme design requirements. The EPRI method inherently extends above net yield, where work hardening exists. The EnJ proposal accepts use beyond ligament yield primarily because elastic paths remain in parallel with the ligament at the crack. It is noticeable that the COD, R-6 and EnJ methods make no direct reference to the degree of work-hardening, whereas in the EPRI method, various values of the work-hardening exponent n are one of the main variables. This was rationalized [11] in terms of the extent of plastic strain in the ligament being a more important controlling factor than the local strain hardening at the crack tip, so that provided the estimate of an effective yield stress reflects the collapse behaviour more exact modelling of the work hardening is not vital. The full EnJ method therefore uses estimates of J/G_y from equations (2a,b), provided the checks on collapse from equation (2c) show that physical collapse is not being approached, so that ligament yield would be exceeded, if at all, in a deformation controlled sense. The cbse term, equation (7), is W/b for a through crack or B/b for a part-through crack. The former may well be for a two-dimensional situation which, if load controlled, will be protected against collapse only by the effects of work hardening. The latter may well be protected by the elastic material providing parallel load paths so that the ligament is really deformation controlled and equation (2b) relevant. The cbse term thus fulfills two requirements. It augments the strain for displacement controlled ligament yield via equation (2b) and notifies of the risk of plastic collapse via equation (2c). As discussed under the simplified EnJ method, use of equation (2b) for stress concentration cases, entered at $k/e_y = (k_t \sigma/\sigma_y)^2$ seems necessary only when $\sigma/\sigma_y > 0.8$, or if the crack size is much less than the size of plastic zone at the region of concentration. Residual stresses are treated via equations (8a,b). Biaxial stress may have a major effect on the collapse load and thus affects the use of equation (2a) via equations (13) and (14). Thermal strains, if in excess of equation (16), require augmentation by the cbse term since as already noted, localization of the deformation can occur in the loading of an already cracked body. The further features that distinguish the full EnJ procedure from the simplified one relate to the selection of appropriate values of toughness as discussed in the following.

CHOICE OF TOUGHNESS VALUE

The selection of a toughness value and the method used to determine it are not entirely divorced from the estimation procedure used. Three aspects are discussed: i) whether cleavage can occur; ii) whether a post-initiation (R-curve) value can be used; iii) whether toughness of parent or degraded material shall be used. All methods use a high constraint configuration such as deep-notch three-point bend or compact tension, accepting that other circumstances with less constraint may give a benefit in toughness. It is possible that yet higher constraint with an implication of reduced toughness is possible, e.g., in biaxial bending [22], but for general engineering purposes, no further allowance is here made. Directly impulsive or explosive loading is excluded from the present discussion, although normal structural dynamic loads or local high rates from "pop-in" are not.

Ideally, R-6 requires the use of K_{Ic}. If a valid K_{Ic} cannot be found, full thickness tests are required, interpreted by an equivalent energy type analysis with certain modification not here detailed. The COD test also requires a full thickness test. Both methods are seeking to ensure that if cleavage can occur in the structure, it will be found in the test. Where the R-6 method adopts a test invalid by LEFM standards, a maximum load toughness is accepted. The COD method [18] accepts one of four measures of critical toughness, δ_i, δ_c, δ_u, δ_m, according to whether a true initiation, value δ_i, can be determined, whether or not pop-in or cleavage behaviour occurs on a rising load diagram, or whether a "roundhouse" type diagram with a maximum load is obtained after some crack growth. No explicit instability analysis is used [6] but it is generally argued that a post-initiation toughness value can be accepted if the ligament of the component is greater than the ligament of the test piece, and if the fracture test has been conducted with a compliant machine or system that will tend to maintain the load. The EnJ procedure [11] also requires the use of a full thickness test to ensure the risk of cleavage is guarded against, though interpreted in terms of J. One such procedure was proposed [23], although others might be devised. Where there is no risk of cleavage, a K_J value obtained from the recently standardized J-test would be acceptable for EnJ. Such a test would normally satisfy [17]

$$B > b; \ b > 25J/\sigma_{f1} \tag{22}$$

which gives an adequate degree of plane strain to give J_{1c} substantially independent of thickness for a ductile micro-mode of failure. Experimental evidence on whether such small pieces give sufficient constraint to cause cleavage, if cleavage is in fact possible, is not conclusive, hence the above preferences for full thickness pieces. At the present stage of development, the EPRI method is formulated primarily as an estimation procedure and toughness data for valid K_{1c} tests within the transition range of steels, or from K_J tests on the "upper shelf" or if cleavage cannot occur, is implied. The differences, so far, lie in the insistence of R-6, COD and EnJ on full thickness tests, if the risk of cleavage is not known to be non-existent, and in the case of R-6 and COD, the use of full thickness even in a ductile situation.

Related questions are whether a static or impact test, and whether parent or degraded material should be used to determine the risk of cleavage. A static test for full thickness, is commonly agreed to be adequate for most structures where explosive loading is not a risk to be considered. Some values of applied strain rate for various structures were given [24] and it was seen that live loading of large components does not of itself usually produce the rate of straining associated with impact tests, no doubt because of inertia of the structure. The risk of dynamic loading from a "pop-in" from degraded material remains, however. The Charpy test, though dynamic, does not combine the full effects of strain rate, notch acuity and full size and thus responds to some arbitrary combination of these three effects, plausible for some applications but not separately quantifiable or interpretable. Where deterioration may occur, of the bulk material, (by nuclear irradiation) or of a continuous region, (e.g., HAZ) along which a crack may propagate, then degraded toughness is relevant. Where a local degraded region is "surrounded" by un-degraded bulk material, the question is less clear cut, the answer depending on the size of the degraded region and the ratio of degraded to un-degraded toughness, including dynamic effects in the latter (bulk) properties. No true solution is known so that ad hoc rules must be used to suit various situations. The R-6 method seems not to refer to degraded material so that static tests on parent material are implied. The COD method requires tests on the appropriate part of a weld region. Small sized COD tests or Charpy V tests are accepted, the latter, though itself dynamic, being correlated to a K_{1c} value.

Such toughness data for particular regions within a weld are usually used together with a known defect or close knowledge of the type of defect likely to be found in a particular welding procedure. Thus, although in principle, it is possible to set down relationships between weld defect, stress level and required toughness, it is not easy to propose values realistic to a wide range of circumstances, and neither R-6 nor COD attempts to do so. Nevertheless, even though weld toughness surveys may be a necessary part of more elaborate safety studies, it seems inconsistent with the simple method sought here, over and above the specification of welding procedures known to be satisfactory for comparable circumstances. Some rather tentative suggestions are made, Appendix 2, to complement the basic assessment of toughness for the parent material. The intention is to ensure that the weld/HAZ toughness is adequate firstly, for either as-welded (un-stress relieved) use with negligible mechanical load, or for stress relieved use with mechanical load; secondly, for un-stress relieved use with mechanical load, and thirdly, to ensure parent material is adequate if a small crack were to "pop-in" because the weld/HAZ toughness was not adequate. Either these datum values of toughness or that found by the EnJ estimation procedure would then have to be met, whichever is the greater.

The remaining question is whether the toughness value measured should relate to initiation or some post-initiation event. A schematic representation of toughness versus temperature for steel, was shown, Figure 3. Note the upper shelf is defined here by the initiation of ductile tearing. This is likely to be more severe than behaviour in the Charpy V test because of the small size of the latter. It follows, however, that tests conducted below the upper shelf regime so defined will fail by cleavage with no prior slow growth. For tests just above the "cross over" of micro-modes, cleavage can occur after some slow growth and it is in this regime that uncertainty exists on how thickness dependent the "cross over" point is, and how sensitive it is to strain rate effects if "pop-in", or dynamic loading does indeed occur. Above some yet slightly higher temperatures, presumably cleavage cannot occur at all. Provided due care is taken to allow for scatter of data, it seems that a toughness measured at either pop-in, final (cleavage) fracture or at onset of maximum load (as specified in the COD method) is satisfactory, even though beyond initiation, with the restriction to full thickness data if cleavage is possible. If, therefore, slow stable growth is permitted, either above the "cross over" point or with materials which are always ductile in micro-mode, it is necessary to guard against unstable ductile tearing. This can be done by making a full instability analysis using the concept

$$dJ/da \text{ (applied)} < (dJ/da) \text{ material} \tag{23a}$$

or

$$T \text{ (applied)} < T \text{ (material)} \tag{23b}$$

where $T = (E/\sigma_y^2)(dJ/da)$, the so-called tearing modulus [25]. An alternative approach by balance of energy rates has been proposed [1,26]

$$I < (b/\eta)(dJ_r/da) \tag{24}$$

where I is the energy release rate in the elastic-plastic case, $G<I<J$, and η relates J or the work rate $\partial w/\partial a$ to work, w, equation (17).

For small amounts of growth defined [27] by

$$\omega = (b/J)(dJ/da) \gg 1 \tag{25}$$

the two methods are identical in concept since I is related to T by

$$T = IE\eta/b\sigma_y^2 \tag{26}$$

It should be noted that both T and I depend upon the compliance of the structure, so that, unlike brittle fracture, unstable ductile tearing is machine or structure dependent. An approximate estimate of T can be made by differentiating the estimation expressions for J, such as equation (2a). An upper bound value is given for constant load, Q, when, following the separation of the variable of crack geometry and loading

$$T = (E/\sigma_y^2)(J/G)(\partial G/\partial a)|_Q \tag{27}$$

The material property, dJ/da (mat) is obtained from an R-curve expressing J versus crack growth Δa. For large amounts of crack growth, the T concept is restricted both by the reduction in ω as the slope of the R-curve decreases and by a restriction to $\Delta a/b < 0.06$ in order to maintain a J derived stress field at the crack tip. The energy balance method is not explicitly restricted in this way since the term dJ_r/da (mat) is defined in a way that attempts to be independent of the precise crack-tip stress state, although it is not proven whether or not this is so.

In determining the material resistance curves, there is still doubt on whether or not it is a property independent of geometry, and prudence would dictate the use of a lower bound R-curve, probably found using side-grooved pieces (to eliminate shear lip that may not exist for growth of a part through crack), in a high constraint configuration (such as deep notch bending) with growth in the relevant direction if anisotropic effects exist. Still further consideration must be given to whether, if determined on small pieces, there is a risk of change to cleavage, and whether time dependent effects are important if the load is maintained, particularly if the net section is above yield. Although all these complications are relevant to use of a fracture safe procedure, be it R-6, COD, EnJ, EPRI or other, if taken above net section yield as may happen in the Case 2 design problem of gross overload, or an accident condition, an analysis less than a full examination of instability was proposed [28].

If a post-initiation value of toughness, J_c, selected above initiation from an R-curve determined on a test piece (with suitable provision for lower bound data as mentioned above) is to be used in a structure, then

$$J \text{ (applied to structure)} < J_c \tag{28}$$

and

$$(b/\eta) \text{ structure} > (b/\eta) \text{ test piece} \tag{29}$$

where approximate allowance is made for the compliance of the structure [11] by using

$$\eta(\text{structure}) = \eta(\text{component})(1+\Phi) \tag{30}$$

where Φ is the ratio of compliances

$$\Phi = \phi(\text{structure})/\phi(\text{uncracked component}) \tag{31}$$

Estimating the necessary structural compliances either here, or for use in the T-analysis is not always easy, although in particular experimental or case studies, it seems feasible.

For Case 1 design for normal use, for which the simplified EnJ method is proposed, it seems inconsistent to call for toughness test methods that are too demanding. Either determination of an R-curve of toughness versus crack growth, or detailed surveys across weldments may seem unrealistic to the user. Both these and other more specialized methods remain open of course to the Case 2 user intent on conducting the most searching study. To avoid determining an R-curve or making a precise measure of initiation toughness, it is suggested that toughness J_m be accepted at onset of maximum load in a full thickness J test. A standard full thickness J test does not yet exist, although one method was outlined [23]. Where there is no risk of cleavage, the full thickness requirements can be relaxed and the standard (small sized) J test used if preferred to determine J_{1c} [17], although that is a multi-test piece method. If a maximum load toughness, J_m, is used, a restriction is made, Appendix 2, to guard against the event that the unknown amount of slow growth Δa_m, will also occur in the component so that if the stress level is already near to 0.80_{fp} based on original ligaments b_c, then $0.8Q_{fp}$ might be exceeded based on $(b_c - \Delta a_m)$. Both the risk of collapse and fracture under the enhanced value of J, equation (24) is being entered at a higher value of Q/Q_{fp}, must be guarded against. The restriction suggested is that if $0.7Q_{fp} < Q < 0.80_{fp}$, then the required level of toughness and size of ligament be restricted such that

$$J < 0.75\ J_m \text{ and } b(\text{component}) > 2b(\text{test piece}) \tag{32}$$

These steps are seen to follow the spirit of the present use of maximum load COD, often coupled with the requirement that b(comp) > b(test). No formal justification of that procedure is known. The present proposal is slightly more conservative but has some basis of justification, albeit not rigorous.

The question of whether detailed surveys across a weldment to determine possible regions of low toughness are necessary has no general answer. The COD procedure [6] calls for such a survey during procedural tests. Some structural standards, recently revised, have called for certain toughness values based on some assumed standard defect unlikely to be exceeded in specified circumstances of welding. That line of reasoning leads to the suggestion, Appendix 2, that for Case 1 use certain toughness levels are called for, which once demonstrated by a controlled procedure, may then be assumed, or perhaps checked for only the weld metal deposited, without detailed surveys for local regions of low toughness. Tentative suggestions for suitable levels of toughness are made, Appendix 2, but the question of what types of weldments and materials can be so treated, and whether the suggested levels would vary with weld detail clearly needs further discussion. The concept, though not the technical detail, could perhaps be likened to the well-known existing classi-

fication of welded joints for the purposes of fatigue strength. Thus, selection
of toughness itself falls into three broad categories: parent material assessed
by simple methods such as Charpy; an estimate of fracture toughness based on various simplifications such as maximum load; full determination of fracture toughness
in both value (with or without R-curve) and location. These three levels are
roughly matched to the three levels of procedures of specifying Charpy energy, a
simplified EnJ or the full EnJ estimation methods.

CONCLUSIONS

1. A three level treatment of fracture safe assessment or design has been
outlined, ranging from material selection only, to an essentially LEFM based use
of fracture mechanics and a full elasto-plastic fracture mechanics treatment.

2. Reliance on materials selection is often adequate for less demanding usage
and several elasto-plastic methods have been introduced already. The main attention
here has been to the development of an intermediate level method, based primarily
on LEFM, and to inter-relate the levels of treatment so that use can be made of
any of them in a flexible way. The procedure for estimating J, denoted the EnJ
method, is shown graphically, Figure 4, together with the bands of data for various
configurations.

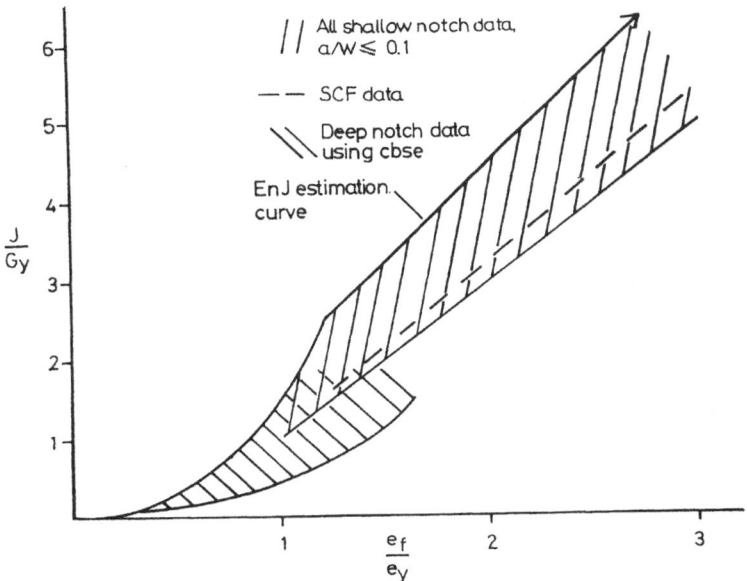

Fig. (4) - The EnJ estimation curve

3. The simplified EnJ method limits the net ligament state to not more than 0.8 of the fully plastic load, itself neglecting work hardening and constraint for reasons both of a margin below collapse and for the simplicity of applying an LEFM based analysis. In the full EnJ assessment, overloads beyond yield are considered. In both procedures rules, empirical in detail but based on rational argument, are suggested for residual stress, thermal stress, stress concentration, biaxial stress and so on.

4. Appropriate measures of toughness are discussed, variously limited to initiation, J_i, an R-curve value, J_r, or maximum load value J_m, according to circumstance.

5. The development of a full thickness J test is seen as highly desirable.

APPENDIX 1 - SELECTION OF STRUCTURAL STEEL TO AVOID BRITTLE FRACTURE

A method of selecting a suitable grade of C-Mn structural steel to avoid brittle fracture of welded components in general engineering service not below -20°C was given [19]. The method defines an energy level C_v that must be attained in the Charpy V test at a test temperature T°C. The energy C_v is defined according to the yield stress of the steel.

Yield Stress, σ_y		Required Energy, C_v	
Up to 16 ton/in.	245 MN/m²	20 ft.lb.	27J
16-20	240-310	25	34
20-23	310-350	30	41
23-25	350-380	35	48

The test temperature T depends upon yield stress and other factors as follows:

1. take minimum design service temperature plus 30°C

2. add design adjustments

   a. if stress concentrations are minimized, good workmanship and inspection ensured, add 10°C

   b. if welds are stress relieved, add up to 30°C

3. add or subtract thickness adjustment

Strength Level	Up to 245 MN/m²			Up to 380 MN/m²			
Thickness in.	3	2	1 1/2	1 1/4	1	3/4	1/2
mm.	76	51	38	32	25	19	13
Adjustment °C	-30	-25	-20	-15	-10	0	+10

4. subtract stress level adjustment

Maximum Design Stress	150 MN/m² (10 ton/in.²)	230 MN/m² (15 ton/in.²)
Adjustment °C	-10	-15

5. subtract usage adjustments

   a. if consequence of failure is serious, deduct 10°C, if very serious, deduct 20°C

   b. if loads are cyclic, deduct 10°C, if impact, deduct 20°C

APPENDIX 2 - THE SIMPLIFIED EnJ METHOD

A. Justification.

In the full EnJ proposal protection against collapse is given by equation (2ciii) directing attention to a risk, which is then assessed. If there is no risk despite high ligament stress, for example because of elastic material in parallel, then the use of equation (2cii) guards against deformation controlled fracture under extreme overload. In the simplified method, a conservative but simple estimate of full ligament plasticity, equations (19) and (20) are used. This avoids the need for detailed assessment of collapse and also assumes that full plasticity is not reached, as a desiderata in its own right. Little is known of shakedown or growth at a crack under a few repetitions of load [29] so avoidance of full plasticity of the ligament by some margin (i.e., $Q < 0.8\ Q_{fp}$) is seen as a guard where the load Q is itself a maximum "normal overload" rather than a regular in-service level. It may be reached "occasionally" and the component retained in service. Clearly, the combination of possibilities being guarded against is very wide and the choice rather arbitrary.

A further guard against the combination of high sustained load and further slow growth is contained in Note 2 below.

The factor of 0.8, rather than some other, is dictated largely by the desire to use LEFM based procedures. The choice of plastic zone correction in equation (2ci) is based on a recognition of the plane stress correction factor as being comfortably conservative for plane strain. The modification to equation (21) is also seen as conservative for all the cases likely to be met, particularly while Q_{fp} is itself an underestimate of true collapse. Correlations with the computed data for J are closer if the shape factor Y is related to the original crack size, not altered to some effective crack size, $(a+r_p)$, except for the cases (such as stress concentration) where Y reduces as the crack length increases, and even the $(a+r_p)$ correction is an over-estimate.

The simplified EnJ procedure is summarized below.

Procedures

1. Restrict load to $0.8\ Q_{fp}$ where Q_{fp} is the fully plastic load of the net ligament based on yield stress and net section, (increased in bending by the plasticity factor) with no allowance for constraint or work hardening.

2. Estimate J from

$$J/G_y = JE/Y^2\sigma_y^2 a = (\sigma/\sigma_y)^2\ \{1 + 0.5(Q/Q_{fp})^2\} \tag{A2-1}$$

which implies LEFM plus a (plane stress) plastic zone correction factor, where Y is the LEFM shape factor based on initial crack size.

3. Relate the value of J to an acceptable toughness, J_c, meeting the least of either the parent material, degraded if so be, in the bulk or along a crack path (e.g., by ageing, radiation, HAZ, etc.) using the test methods outlined below or the following datum values of toughness (see Note 1).

a. $K_J > 0.85\sigma_y\sqrt{B}$ for unstress relieved weld or HAZ with negligible load or stress-relieved weld/HAZ with full mechanical load;

b. $K_J > 1.6\sigma_y\sqrt{B}$ for unstress-relieved weld/HAZ carrying mechanical load up to $0.8 \, Q_{fp}$;

c. $K_J > 1.6\sigma_y\sqrt{B}$ for parent material with weld/HAZ not meeting a or b.

4. Test Methods

a. If cleavage is possible, use static full thickness bend or high constraint pieces at the lowest relevant temperature to determine J_c. A maximum load toughness implying some small amount of growth is acceptable, provided $Q < 0.7Q_{fp}$ and b (component) > 2b (test piece) (see Note 2).

b. If cleavage is not possible, either use full thickness test with restriction as above, or use standard (small size) J_{1c} test [17] to determine toughness at or near initiation*.

Note 1: These values are derived by considering a submerged defect, diameter $B/2$, with J estimated for either yield level residual stress or a mechanical stress $0.8\sigma_y$ (Case i) or both (Case ii). For Case iii, a crack might jump from such a defect if the toughness specified (i) or (ii) is not reached. It is supposed this would present the parent plate with a through thickness dynamic crack $2a = B$ for which $\sigma = 0.8\sigma_y$ and $K_{1d} = 0.67 \, K_{1c}$ has been assumed. As an example, if σ_y is taken as 360 MN/m^2 (units in MN, not N), B = 50 mm (units, m) then for i) $K_{1c} > 72$ MNm$^{-3/2}$ ($C_v = 25J$); ii) $K_{1c} > 130$ MN/m$^{-3/2}$ ($C_v \simeq 80J$); iii) K_{1c} 130 MNm$^{-3/2}$ ($C_v = 80J$).

Note 2: The basis of this estimation is an approximate assessment of the relationship between the unknown amount of slow growth, Δa_m, up to maximum load, Q_m and displacement q_m in a deep notch three-point bend test, where J_m is measured and the corresponding change in deflection, ΔQ, subsequent to start of growth, i.e., where growth started at $(q_m - \Delta q)$ at a value of toughness, J_i. The relationship used is

$$2\Delta a/b = (\Delta q/q)\{1-(1/\alpha)\} \quad \text{(Note } \Delta q \text{ is negative)} \qquad \text{(A2-2)}$$

where the work done to maximum load, $w_m = \alpha Q_m q_m$ (NB. $\alpha = 1/(n+1)$ for the power law hardening case). The change in deflection, Δq, is related again approximately to change in work done and hence in J through the ratio J_i/J_m and n value, which are not of course explicitly known. Reasonable bounds are used. The same growth, Δa_m, if met

---

*There may yet prove to be materials for which the particular details of crack blunting line and crack growth restrictions [17] are not appropriate.

in the component, should not increase the severity of the ligament state from $0.8Q_{fp}$ to more than $0.9Q_{fp}$ both to avoid the risk of collapse and to limit the increase of J/G_y (from equation (21)) to 1.35 fold. The former restriction is met in broad terms if b (component) > 2b (testpiece), (although the factor 2 depends strictly on the ratio J_i/J_m, n and N where 1<N<2 as the stress system passes from tension to bending). If the load indeed approaches $0.8Q_{fp}$, then the further restriction $J<J_m/1.35$ (i.e., the acceptable toughness should be restricted to about 0.75 J_m) is needed to avoid excessive increase in J. It is believed, without proof, that the general restriction to $0.8Q_{fp}$ will also avoid any possibility of crack growth under sustained load.

REFERENCES

[1] Turner, C. E., "Methods for post-yield fracture safety assessment", Post-Yield Fracture Mechanics, D. G. H. Latzko, ed., App. Science Pub., Chapter 2, 1979.

[2] Knott, J. F., Fundamentals of Fracture Mechanics, Pub. Butterworths, 1973.

[3] A General Introduction to Fracture Mechanics, J. of Strain Monograph, Pub. MEP, 1978.

[4] Rice, J. R., Fracture: An Advanced Treatise, H. Liebowitz, ed., Acad. Press, 3, Chapter 2, 1968.

[5] Burdekin, F. M. and Dawes, M. G., Practical Use of Linear Elastic and Yielding Fracture Mechanics with Particular Reference to Pressure Vessels, Inst. Mechn. Engs., London, pp. 28-37, 1971.

[6] Guidance on Some Methods for the Derivation of Acceptance Levels for Defects in Fusion Welded Joints, British Standards Inst., PD6493, 1980.

[7] Dowling, A. R. and Townley, C. H. A., "Effect of defects on structural failure: a two-criteria approach", Int. J. Pres. Ves. & Piping, 3, pp. 77-107, 1975.

[8] Harrison, R. P., Loosemore, K. and Milne, J., Assessment of the Integrity of Structures Containing Defects, CEGB Report R/H/R-6, 1976 and supplements, 1979, 1981.

[9] Milne, I., "Failure analysis in the presence of ductile crack growth", Mats. Sci. & Engng., 39 (1), pp. 65-80, 1979.

[10] Turner, C. E., "Further developments of a J-based design curve and its relationship to other procedures", Presented 2nd Int. Symp. on Elasto-Plastic Fracture Mechanics, Philadelphia, 1981. To be published ASTM.

[11] Turner, C. E., "The J-estimation curve, R-curve and tearing resistance concepts leading to a proposal for a J-based design curve against fracture", Fitness for Purpose Validation of Welded Constructions, Welding Institute, London, Paper 17, November 1981.

[12] Shih, K. F., German, M. D. and Kumar, K., "An engineering approach for examining crack growth and stability in flawed structures", Int. J. Pres. Ves. & Piping, 9, pp. 159-196, 1981.

[13] Burdekin, F. M. et al, "Comparison of COD, R-6 and J-contour integral methods of defect assessment, modified to give critical flaw sizes", see [11].

[14] Sumpter, J. D. G. and Turner, C. E., "Fracture analysis in areas of high nominal strain", Proc. 2nd Int. Conf. Pres. Ves. Techn., ASME (New York), 2, pp. 1095-1104, 1973.

[15] ASME Boiler & Pressure Vessel Code, Section III, ASME, New York.

[16] Advances in Elasto-Plastic Fracture Mechanics, L. H. Larsson, ed., Pub. Appl. Sciences, pp. 301-318, 1979.

[17] A Standard Test for J_{1c}: a Measure of Fracture Toughness, ASTM E813-8, ASTM, Philadelphia, 1981.

[18] "Methods for crack opening displacement (COD) testing", BS5762, Br. Stand. Test, 1974.

[19] Brittle Fracture in Steel Structures, G. M. Boyd, ed., Butterworths, 1970.

[20] Charleux, J., "Selection of steel quantities for welds of structural elements", Journal of Society of Naval Architects & Marine Engineering, Calgary, 1981 (to be published).

[21] Sailors, R. H. and Corten, H. T., "Relationship between material fracture toughness using fracture mechanics and transition temperature tests", ASTM STP 514, pp. 164-191, 1972.

[22] Ziebs, J. F. et al, "The influence of stress state on fracture toughness", Int. J. Pres. Ves. & Piping, 8, pp. 131-142, 1980.

[23] Sumpter, J. D. G. and Turner, C. E., "Method for the laboratory determination of J_c", Cracks and Fracture, ASTM STP 601, pp. 3-18, 1976.

[24] Christopher, P. et al, "Influence of strain rate and temperature on the fracture and tensile properties of several metallic materials", Dynamic Fracture Toughness, WI/ASM, London, 1976.

[25] Paris, P. C. et al, "Instability of the tearing mode of elastic-plastic crack growth", ASTM STP 668, pp. 5-36, 1979.

[26] Turner, C. E., "Description of stable and unstable crack growth in the elastic-plastic regime in terms of J_r resistance curves", Fracture Mechanics, ASTM STP 677, pp. 614-628, 1979.

[27] Hutchinson, J. and Paris, P. C., "The theory of stability analysis of J-controlled crack growth", Elastic-Plastic Fracture, ASTM STP 668, pp. 37-64, 1979.

[28] Turner, C. E., "A J-based engineering usage of fracture mechanics", Advances in Fracture Research, ICF-5, D. Francois, ed., et al, Pergamon Press, pp. 1167-1189, 1981.

[29] Landes, J. D. and McCabe, D. E., "Local history effects on the J_R curve", see [10].

THE FATIGUE STRENGTH OF WELDED TUBULAR STEEL OFFSHORE STRUCTURES

K. J. Marsh

National Engineering Laboratory
East Kilbride, Glasgow, United Kingdom

ABSTRACT

In many fields of mechanical engineering, the design and development process for components and structures incorporates a prototype testing phase. Such fatigue testing under service loading conditions has been increasingly assessed by simulation testing in the structural fatigue test laboratory. In the field of offshore structures, such prototype testing is not possible; the designer must get it right at the design stage. It is therefore particularly important that he has adequate fatigue design data, including data from tests on realistic welded tubular steel test pieces, simulating welded joints in offshore steel jacket platforms.

In the United Kingdom, it was concluded that the available information on the behavior of welded tubular joints in marine environments was insufficient to enable fatigue lives to be calculated reliably. Hence, in 1975, approval was given by the UK Department of Energy to a large programme which became known as the UK Offshore Steels Research Project (UKOSRP). This was a very wide-ranging project, but concentrated on fatigue tests on realistic weldments, under representative variable amplitude loading, in a seawater environment. In particular, a large number of tests were carried out on welded tubular steel test pieces; this enormously expanded the available data base and resulted in proposals to modify the recommended design S/N curves applicable to welded offshore structures. The first phase of the project is virtually complete; a second phase, supported largely by major oil companies, is now proposed. The present paper will attempt to give an overview of the project, stressing where fracture mechanics approaches have enabled correlation between the results.

INTRODUCTION

In many fields of mechanical engineering, certainly in ground vehicles and aircraft structures, there is an increasing tendency for the design and development process to incorporate a prototype testing phase. Thus, when the designer has arrived at a preliminary design, whether it be a vehicle, a crane, an engine or an aircraft, a prototype is built and testing and development are carried out to evaluate and improve this prototype. In recent years, such development fatigue testing, simulating service loading conditions, has been increasingly undertaken in the structural testing laboratory using servo-hydraulic testing equipment.

Amongst others, my own Establishment, the UK National Engineering Laboratory (NEL), has been very active in this area [1,2] as can be seen from the examples of service loading fatigue testing on a cross-country vehicle, Figure 1, and on a 30 metre long mobile crane jib, Figure 2.

Fig. (1) - Service-simulation fatigue test on a cross-country vehicle

However, in the field of offshore structures, this prototype testing phase is, of course, not possible; the designer must get it right at the design stage. The penalty for failure can indeed by catastrophic, as the recent "Alexander L. Keilland" disaster, Figure 3, reminds us. (This cannot be attributed, of course, to a simple design fault. Numerous factors contributed to the failure). It is therefore particularly important that the designer does have adequate design data to work with, and that this data base is in a form that he can use. Although much useful information can be obtained from fatigue tests on simple materials test pieces, and fracture mechanics based methods may be extremely useful to correlate results or extrapolate to different situations, this approach must be balanced by data from tests on more realistic specimens such as welded steel test pieces. While fatigue design guidance information for offshore structures had existed in the past [3], there was, in particular, a great lack of fatigue data on tubular welded joints, and those available had been derived mainly from model-scale test pieces.

In 1973, a review [4] of the fatigue and fracture behaviour of the complex welded steel structures required for the exploitation of North Sea gas and oil

Fig. (2) - Service loading fatigue test on a complete crane jib

gave, as a major conclusion, that the available information on the behaviour of welded tubular steel joints in marine environments was insufficient to enable fatigue lives to be reliably calculated. After a period of preliminary corrosive environment experiments and detailed programme definition, approval was given in 1975 by the Department of Energy to a large programme which became known as the UK Offshore Steels Research Project (UKOSRP). This was a very wide-ranging project covering stress analysis methods, fracture toughness studies and basic fatigue studies but, in particular, a major effort on tubular welded joint fatigue testing.

This paper describes the scope of UKOSRP, gives some indication of the results and their implications and outlines probable future developments.

THE UKOSRP PROGRAMME

The bulk of the work was carried out in a number of government-affiliated establishments, including NEL. (Separately and in parallel to this project there has, of course, been extensive activity on the subject taking place in UK universities). The project was divided into four inter-related parts, as follows, although this paper will concentrate on the latter two parts.

Part 1 consists of stress analysis studies of a number of different types of tubular joint, typical of those found in offshore structures, namely T-joints,

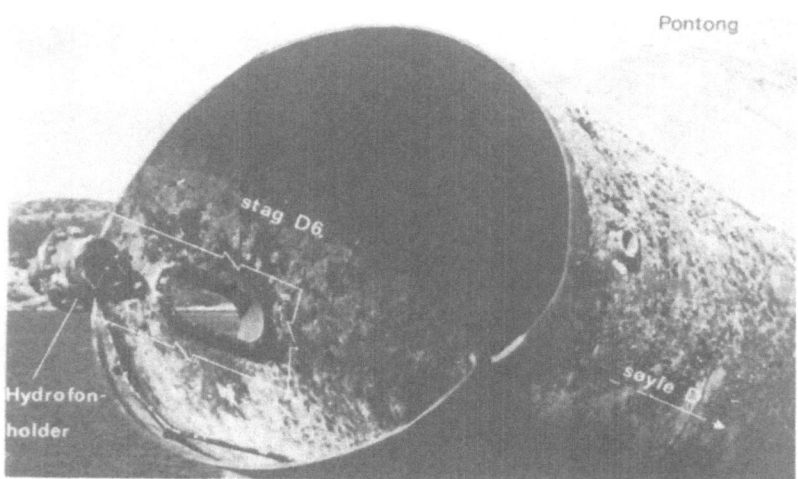

Fig. (3) - The fractured leg of the "Alexander L. Keilland"

K-joints and KT-joints of various brace/chord diameter ratios. Various analytical and experimental methods were used to predict the stress distributions in such joints including finite-element analysis, photoelastic model studies and strain-gauged acrylic tubular models. The results were compared with those from strain-gauge measurements on the tubular steel test pieces in the fatigue testing program.

Part 2 consisted of fracture toughness studies of weldments of various grades of steel, providing information on the effects of temperature, plate thickness and post-weld heat treatment. Some 600 specimen tests were involved including Charpy impact tests, COD tests and wide plate tests.

Part 3, entitled Basic Fatigue Testing, included both crack propagation tests on thick plate specimens and tests on simple weldments. The crack propagation tests were carried out on specimens having through-thickness cracks and those having surface cracks. A wide range of parameters was investigated with tests in air and artificial seawater including the effects of stress ratio (from R = -1 to R = 0.85), of temperature, frequency and oxygen concentration in seawater, of free corrosion, intermittent immersion and various degrees of cathodic protection. The results of these tests and indeed all tests in the UKOSRP Phase I program have been published at various stages [5-8] and specifically summarized at the Institution of Civil Engineers' Conference [9]. However, it may be noted in passing that at high stress ratios crack growth rates in free corrosion seawater conditions were increased by a factor of six over those in air. (This work was carried out at the UK Atomic Energy Authority establishments at Harwell and Springfields).

The welded joint tests in this part of the program (numbering some 550 tests) were carried out at NEL and at the Welding Institute on test pieces as shown in Figure 4, i.e., transverse load-carrying fillet welded specimens and longitudinal

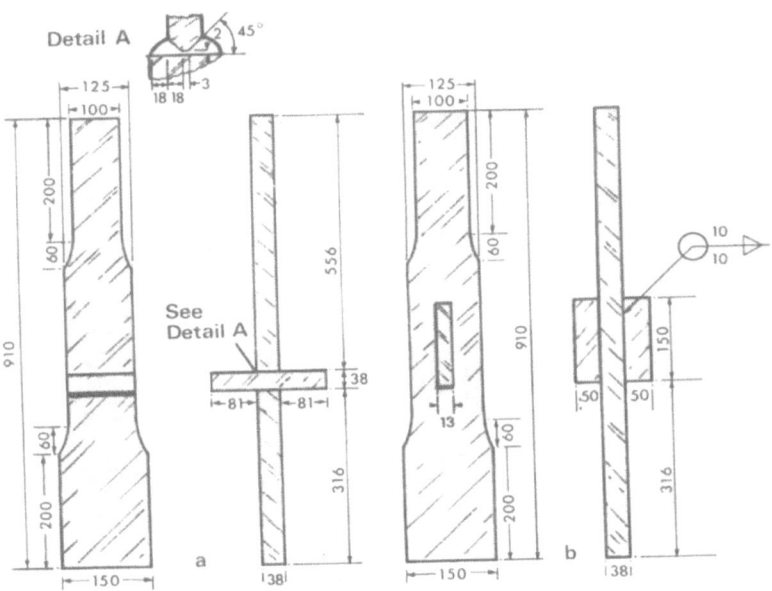

Fig. (4) - Cruciform and longitudinal welded plate joints

non-load-carrying fillet welded specimens of 38 mm thick steel plate. Tests have been carried out in axial loading and in bending, in air and in an artificial sea-water corrosive environment, at various stress ratios, to investigate the effects of various factors, such as partial penetration welding, weld-toe grinding, ca-thodic protection in seawater tests and cumulative damage prediction methods for various variable amplitude load histories. Figure 5 shows a typical fatigue test rig for cantilever bending of cruciform joints in an artificial seawater environ-ment; this is a 3-station rig using micro-processor controlled servo-hydraulic loading actuators. Detailed results from such tests have been extensively pub-lished as mentioned, but in very general terms the presence of a seawater environ-ment (free corrosion) reduces lives by only a factor of about 1.6, both in con-stant amplitude and random loading tests. (Figure 6 shows results for constant amplitude tests on cruciform joints). Cathodic protection improves the lives, particularly in zero mean load tests.

The random loading tests in this program used a non-stationary narrow-band random loading history [10] intended to be representative of wave loading on North Sea structures. It consisted (Figure 7) of four different levels of stationary NBR loading arranged in rising and falling sequence, with an overall block length of 100,000 cycles. Typical results are those in Figure 8, for in air tests at zero

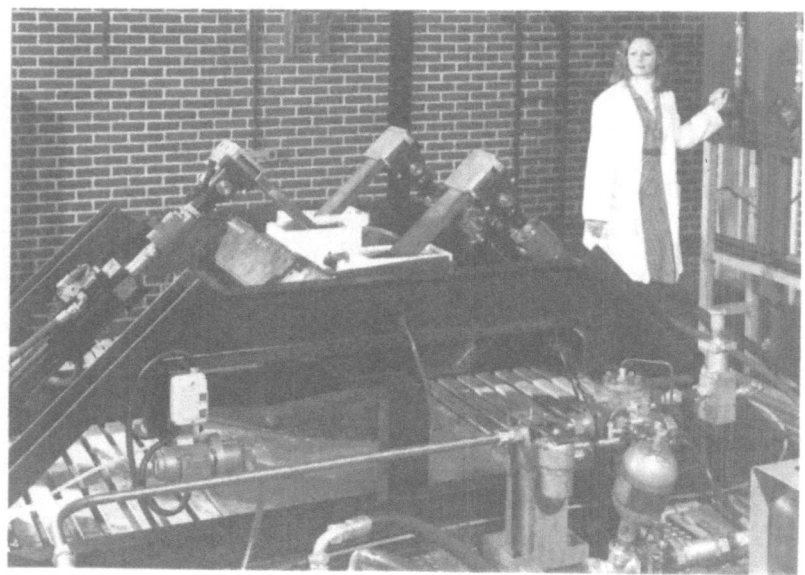

Fig. (5) - Three-station cantilever bending fatigue test rig for cruciform
joints with artificial seawater environment

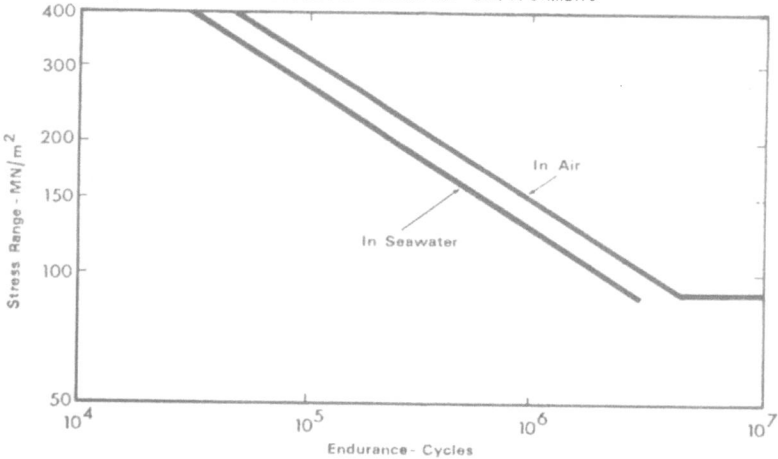

Fig. (6) - The effect of an artificial seawater environment on the
constant-amplitude bending fatigue strength of cruciform
joints

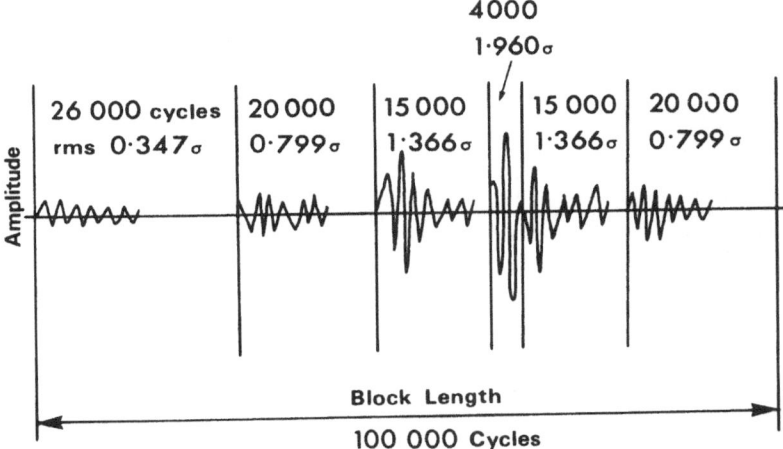

Fig. (7) - Non-stationary narrow-band random loading history

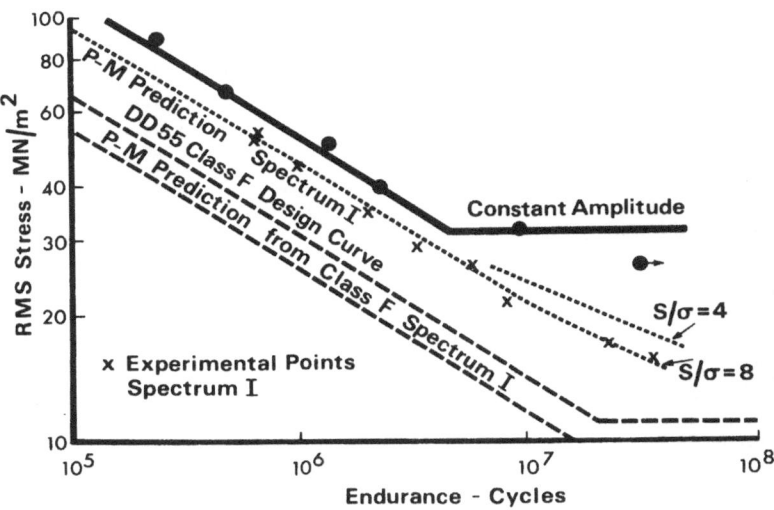

Fig. (8) - Random-loading tests on cruciform joints in air, zero mean load

mean load on cruciform joints, which show that the Palmgren-Miner summation predicts lives under random loading adequately (although somewhat unconservatively at low truncation levels and low stresses) and that the UK Class F Design Curves are adequately conservative for these joints.

However, although much useful information was gained from the tests on welded plate specimens, the program was perhaps dominated by those, in Part 4, on tubular welded joints, which totalled some 200 tests. The test pieces vary from model-sized joints of 170 mm chord diameter to the largest, intended to represent offshore welding practice realistically, having chord members of 1.8 m diameter, 75 mm thick, Figure 9. There were three different geometries of tubular joint in the program. All joints involve a chord member, usually of larger diameter, and one, two or three (usually smaller) brace members, the joints being described respectively as T-joints, K-joints and KT-joints. Only T-joints were made in the largest sizes. Three different modes of loading were investigated: axial loading in the brace member (with the chord free to bend and bulge), in-plane bending of the brace member with respect to the chord, and out-of-plane bending.

Tests were carried out either in purpose-built test rigs or using existing fatigue machines, but invariably using servo-hydraulic loading actuators. For example, Figure 10 shows an in-plane bending test on a 0.45 m chord diameter T-joint. Figure 11 shows an axial-load test on a 0.45 m brace/0.9 m chord T-joint in a 2.5 MN capacity fatigue machine and Figure 12 a similar test in a test rig (in this case on a cast steel node). Finally, Figure 13 shows an in-plane bending test on a double T-joint of 1.8 m chord diameter. In this test, two loading actuators act out of phase between the ends of the two chord members joined by the brace.

Much information has been derived from the fatigue tests on these tubular joints. Extensive strain gauging has enabled detailed investigation of stress distributions around tubular welded intersections and correlation with analytical methods (Part 1). The tests have allowed the evaluation of various devices for the detection and monitoring of crack growth in realistic situations. Stress/endurance data have been derived for lives to the first detectable crack, to the first through-thickness crack and to the termination of the test (this usually involved cracking around some 180°, i.e., effectively complete failure). Hence, in addition to S/N data, valuable information on the rate and modes of crack propagation around various tubular welded intersections was obtained (Figure 14).

However, perhaps the major result arising from this part of the project was the demonstration of a marked thickness effect in as-welded tubular joints. Figure 15 shows the general trend of a lowering of the S/N lines as the chord diameter increases from 0.17 m to 0.9 m, corresponding to thickness increasing from 6 m to 32 mm. It can be seen that the scatter band for experimental results on the thickest joints crosses the AWS Design Line at long endurances and that the slope of the experimental lines increases with increasing size. Further analysis of all the available results led to the preparation of recommendations for modifying the Fatigue Section of the Guidance Notes for Offshore Structures [11] which are now being implemented.

FRACTURE MECHANICS METHODS

The original design of the UKOSRP program, involving a number of related parts, assumed that wherever possible data from one part would be correlated with data

Fig. (9) - Double T tubular welded joint of 1.8 m chord diameter

Fig. (10) - In-plane bending fatigue test on a 0.45 m chord diameter T-joint

Fig. (11) - Axial-load test on 0.45 m brace/0.9 m chord T-joint in 2.5 MN
capacity fatigue machine

from other parts and that fracture mechanics methods would necessarily be the
basis of any such correlation. In general terms, this is almost axiomatic but
some more specific examples may be noted here.

In recent years, Pook has extensively investigated [12] the prediction of
the behavior of welded joints by fracture mechanics analysis. In a recent paper
[13], he applies his analysis to the random loading test results from the cruciform welded joints in the UKOSRP program and couples this with an interesting
fractographic analysis, measuring crack growth rates from markings on the fracture faces of the welded specimens, as in Figure 16. It was shown that at medium
crack depths, residual stress and clipping ratio have little effect on crack growth
rates, which can be predicted fairly accurately from constant amplitude data using
linear summation. Crack growth rates in seawater with cathodic protection are
significantly slower than in air. Crack growth rates for shallow (less than 1½ mm)
cracks are strongly influenced by residual stresses and events in this region dominate the overall life.

Fig. (12) - Axial-load test on 0.45 m brace/0.9 m chord cast steel T-joint

Applying fracture mechanics to the prediction of the lives of tubular welded joints is a somewhat more difficult problem. In an early report [14], Tomkins and Scott, considering the case of axial brace-loading in a simple T-joint, proposed that only surface crack propagation around the joint intersection needed to be considered. This led to a relatively simple expression for the stress intensity factor along the growing crack and enabled the calculation of stress/endurance curves for various combinations of parameters. In this way, the effect of initial defect size and material crack growth behavior was evaluated, the choice of the former being particularly critical.

In parallel to UKOSRP, extensive testing of tubular welded joints has been carried out by Dover and his associates. As another example of the application of fracture mechanics analysis, his papers [15,16] on random loading fatigue crack growth in T-joints under in-plane bending can be cited. He defines the stress intensity factor for these joints as $K = Y(S) \cdot Y(\sigma) \cdot \sigma\sqrt{\pi a}$ where the symbols have

76

Fig. (13) - In-plane bending test on double T-joint of 1.8 m chord diameter

Fig. (14) - Fatigue crack in 1.8 m diameter chord tubular joint

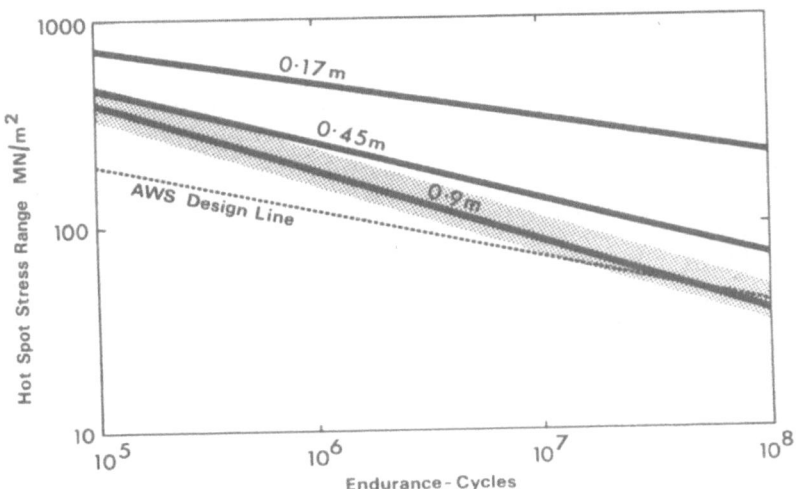

Fig. (15) - S/N curves for tubular joints of various sizes

Fig. (16) - Fracture surface of cruciform welded joint

the usual meanings, $Y(S)$ is a factor dependent on crack shape and path and $Y(\sigma)$ is dependent on the applied forces, the joint geometry and the weld metal geometry. This approach has had some success in predicting crack growth rates and lives for these joints.

In general, the difficulty in these approaches is to be able to generalize from what works for one geometry to what applies to a very different joint geometry where there may be very different bending and tensile components of stress, and hence very different crack growth modes or shapes. Because of these difficulties, at this stage fracture mechanics approaches must still be backed up by extensive S/N data for tubular welded joints.

FUTURE WORK

It can be seen from the previous Sections that considerable progress has been made in recent years in our understanding of fatigue in offshore structures. A very good summary of the whole European program and the problems of predicting the life of offshore structures is given by Schütz [17] in the 1981 Paris Conference. Certain general conclusions emerge from the work so far, although the extent of their generality is not always clear. For example, the Palmgren-Miner rule is adequate for fatigue life prediction, particularly if a damage sum of 0.5 is used. The effect of seawater corrosion on fatigue life is quite small (a factor of less than 2 on life) under realistic stress sequences. Nevertheless, further work is desirable to clarify some of the problem areas. Because of this a second phase of the UK program is about to commence. The new program (UKOSRP -II), of some 5 years duration, is funded by a consortium of oil companies (75 percent) and other offshore-oriented parties, which demonstrates the importance attached to the program.

The technical content of the program results from extensive discussion with researchers and with the industry. The work in UKOSRP-I has provided basic S/N curves for tubular joints in air up to a thickness of 75 mm and a correction for the effect of seawater deduced from welded plate tests. The new program will provide modifications to the basic S/N curves as affected by:

1. Increased plate thickness, in both tubular and plate joints

2. Seawater corrosion

    a. tests on tubular joints in air and seawater at a standard level of cathodic protection

    b. tests on welded plate specimens at varying R ratios and levels of cathodic protection

    c. the effect of protective coatings in above-water zones

    d. effect of temperature

3. Post weld heat treatment and its relation to applied stress ratio

4. Weld improvement techniques, weld profiling, toe grinding, hammer peening

5. Joint geometry

    a. ring stiffening in T-joints

    b. overlapping brace K-joints with low stress concentration factors

6. Variable amplitude loading, including broad-band spectra

Hence, it can be seen that the major effort of the last few years in providing design data is being continued within UKOSRP-II and it is almost inevitable that parallel work will also occur in other European countries and in the UK universities.

In conclusion, the exploitation of oil reserves in the severe conditions of the North Sea involves a relatively new technological approach compared with, for example, sophisticated industries like the aircraft industry. Nevertheless, it is one where fatigue is a major problem. In the last few years, an intensive effort has been made to improve the design data base in this field and to identify the remaining problem areas. Fracture mechanics methods have played a major part as a framework for correlating experimental data.

REFERENCES

[1] Marsh, K. J., "Full-scale testing - an aid to the designer", J. Soc. Environmental Engrs., 13(4), pp. 15-22, 1974.

[2] Marsh, K. J., "The new strong floor structural testing building at NEL", Int. J. Fatigue, 1(1), pp. 3-6, 1979.

[3] Department of Energy. Offshore installations: guidance on design and construction. London: HMSO, 1974.

[4] Hicks, J. G., "A study of material and structural problems in offshore installations", Report No. 3384/1/73. Abington, Cambridge: The Welding Institute, 1973.

[5] Proc. European Offshore Steels Research Seminar, Cambridge, November 22-29, 1978, Abington, Cambridge: The Welding Institute, 1980.

[6] Proc. 2nd Int. Symp. on Integrity of Offshore Structures (Ed. Faulkner, D., Cowling, M. J. and Frieze, P. A.), London: Applied Science Publishers, 1981.

[7] Proc. Int. Conf. on Steel in Marine Structures. Paris, October 1981. Luxembourg: European Coal & Steel Community, 1981.

[8] BOSS 1979, Proc. 2nd Int. Conf. on Behaviour of Offshore Structures, London, 1979. Cranfield: BHRA Fluid Engineering, 1979.

[9] Proc. Conf. on Fatigue in Offshore Structural Steels, Instn. of Civil Engrs., London, February 24-25, 1981. London: Thomas Telford Ltd., 1981.

[10] Pook, L. P., "An approach to practical load histories for fatigue testing relevant to offshore structures", J. Soc. Env. Engrs., 17(1), pp. 22-23, 25-28, 31-35, 1978.

[11] Department of Energy. Background to proposed new fatigue design rules for steel welded joints in offshore structures, Report of Department of Energy, "Guidance Notes", Revision Drafting Panel, London: Department of Energy, May 1981.

[12] Pook, L. P., "Fracture mechanics analysis of the fatigue behaviour of welded joints", Weld Res Int., 4(3), pp. 1-24, 1974.

[13] Pook, L. P., "Fatigue crack growth in cruciform welded joints under non-stationary narrow-band random loading", In: Residual Stress Effects in Fatigue, ASTM STP 776, pp. 97-114, 1982.

[14] Tomkins, B. and Scott, P. M., "An analysis of the fatigue endurance of tubular T-joints by linear elastic fracture mechanics", Interim Technical Report UKOSRP 4/01, Department of Energy, 1977.

[15] Hibberd, R. D. and Dover, W. D., "Random load fatigue crack growth in T-joints", Proc. 9th Offshore Technology Conf., Paper OTC 2853, Houston: Offshore Technology Conference, 1977.

[16] Dover, W. D., "Fatigue fracture mechanics analysis of offshore structures", Int. J. Fatigue, 3(2), pp. 52-60, 1981.

[17] Schütz, W., "Procedures for the prediction of fatigue life of tubular joints", Proc. Int. Conf. on Steel in Marine Structures, Paris, October 1981, Special and Plenary Sessions, pp. 254-308. Luxembourg: European Coal & Steel Community, 1981.

ACKNOWLEDGEMENT

This paper is presented by permission of the Director, National Engineering Laboratory, Department of Industry, UK. It is Crown Copyright.

RECENT STUDIES ON BRITTLE CRACK PROPAGATION AND ARREST IN JAPAN

T. Kanazawa

University of Tokyo
Chiyoda-ku, Tokyo, 101 Japan

S. Machida

University of Tokyo
Bunkyo-ku, Tokyo, 113 Japan

and

H. Yajima

Mitsubishi Heavy Industries, Ltd.
Fukahori-machi, Nagasaki, 851-03 Japan

ABSTRACT

This report reviews studies made in Japan recently on predicting the brittle crack propagation and arrest behaviors using the dynamic or simplified dynamic technique. Also reviewed in the last chapter of this report is an analysis, though statical in technique employed, made of an upper deck plating brittle failure experienced by a 10379 G.T. bulk carrier.

INTRODUCTION

In Japan, the statical analyses [1,2] used to be employed, in characterizing brittle crack behaviors in steel plates, to relate the crack motions in a small plate of several hundred millimeters in width to those in a larger specimen simulating an actual structure.

Recently, there has been a strong demand for the safety and the structural integrity of liquified gas carriers, nuclear vessels, storage tanks, etc., in response to the gradual increase in number of such sophisticated structures. Nowadays, in characterizing brittle cracks, a more rational and precise approach [3-10] considering the dynamic effect of the phenomenon has become available and popular. Until now, however, almost all the dynamic analyses conducted were limited to the characterization of the material resistance using the standard shape test specimen. Very few trials have been done to predict the crack behavior in the real steel structure because of the troublesomeness of the dynamic calculations.

This report reviews studies made in Japan recently on predicting the brittle crack propagation and arrest behaviors using the dynamic or simplified dynamic technique. Also reviewed in the last chapter of this report is an analysis, though statical in technique employed, made of an upper deck plating brittle failure experienced by a 10379 G.T. bulk carrier [11].

SIMPLIFIED DYNAMIC CALCULATION FOR PREDICTING BRITTLE CRACK PROPAGATION AND ARREST IN SHIP FRAME STRUCTURE

In this study, firstly the dynamic fracture toughness of the common ship hull steel was obtained on the basis of the dynamic estimation. Secondly, the experiment and the forecasting calculation were made of the brittle crack propagation in the uniform plate with the stiffener. And the simplified static estimation as a substitute for the iterative dynamic calculations was confirmed to be promising in predicting the fluctuation of the brittle crack velocity and the crack arrest [9].

A. Brittle Crack Propagation Test

The test specimens were prepared from 20 mm thick NK (Nippon Kaiji Kyokai) standard KAS class ship steel plates. The specimens consisted of two large size double tension test pieces of 1600 mm in width (named SI-1 and SI-2 specimen), and of the two stiffened pieces of the same size (named BS-1 and BS-2 specimen) as shown in Figure 1.

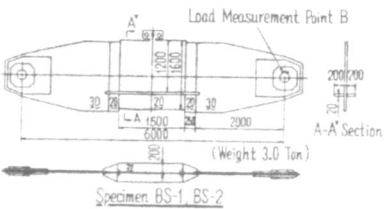

Fig. (1) - Specimen configuration

Crack detecting gauges were attached to measure the crack velocity by the time intervals at which they are cut. And a strain gauge was attached at the edge of the tab plate to pick up the transient load changes.

The specimen was first cooled down to the prescribed temperature distribution over the width and was loaded to 12 kgf/mm^2, and the brittle crack was made to initiate at the tip of the notch and run into the main plate by the sub-tension load. The temperature distributions along the crack path are shown in Figure 2.

As shown in Figure 2, the brittle crack was arrested at 785 mm from the crack initiation point in the specimen SI-1, and the specimen SI-2 was severed. In the specimen BS-2, the brittle crack ran through the intersection, cutting the stiffener, and was arrested at the location 170 mm away from the stiffener.

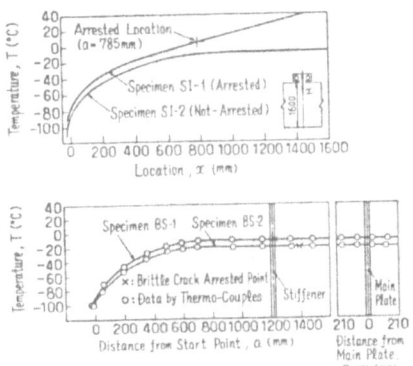

Fig. (2) - Temperature distribution

B. Brittle Crack Behavior Prediction

The transient movement of the specimen and the test rig during the crack propagation was investigated using the simple mass-spring model.

The test rig and the specimens were, considering the symmetry, idealized as the mass-spring system as shown in Figure 3. In this model, the crack advance represented by the decrease in the crack length dependent spring constant k_{SC} [12].

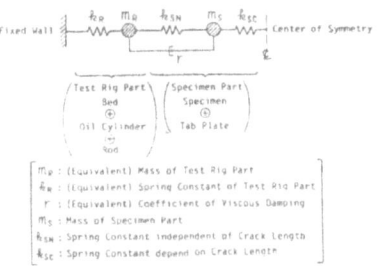

Fig. (3) - Idealized mass-spring system

The behavior of the mass-spring system can be revealed by numerically solving the dual equations of motion for this system. Once the motions of specimen part m_S and test rig part m_R are solved, the load acting at the end of specimen P_{end}

84

and the energy components for both parts can be calculated.

From the conservation of energy and assuming that the dissipated energy except the work-done by viscous damping in the energy dissipated in fractured surface layer, the dynamic fracture toughness K_D is defined by analogy with static case as expressed by equation (1).

$$K_D = \{\frac{E}{(1-\nu^2)} \cdot \frac{dD}{da}\}^{1/2}$$ (1)

The numerical constants m_R, k_R, r governed by test rig should not be altered as long as the same testing machine is used, but cannot be determined easily in the idealized model. Therefore, conducting the iterative calculations with arbitrary values, an attempt was made to find out the best set of the constants that matches the experimentally obtained applied load fluctuation characteristics during the crack propagation. As the results, the adopted set of values was found to give the best agreement with the experimental results as shown in Figure 4.

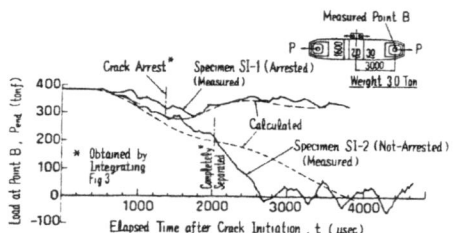

Fig. (4) - Transient load fluctuation

Using those numerical constants for the test rig, the dissipated energy was calculated for the specimen SI-1 and SI-2. The changes in the dynamic fracture toughness K_D were obtained as shown in Figure 5. In general, the dynamic

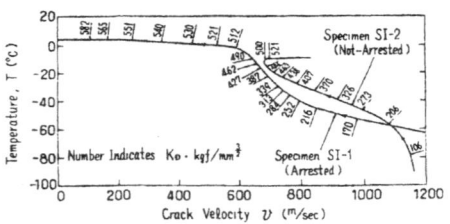

Fig. (5) - Calculated changes in K_D

fracture toughness K_D depends on the crack velocity v and temperature T. Assuming that K_D is expressed by a relatively simple function, it is empirically determined as shown in Figure 6. The above procedure requires the incremental dynamic crack

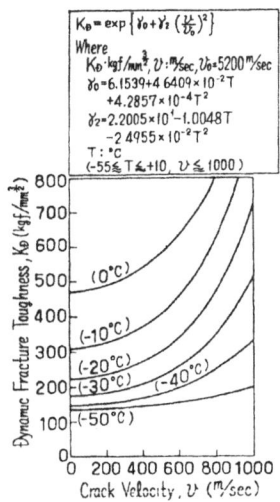

Fig. (6) - K_D versus v and temperature

analysis. This orthodox approach would take great amount of time and cost for calculations.

    Therefore, instead of the iterative dynamic analysis, a substitute method that is a substantially static approach was employed to estimate the relevant parameters and the dynamic stress intensity factor $K^{(D)}$. This simplified approach is based on Freund's solutions [13] about the stationary crack propagation in infinite plate, and is intended to be expediently extended to nonstationary problem. By Freund, it was revealed that under the uniform remote stress acting on the half infinite length crack, the dynamic stress intensity factor $K^{(D)}$ and the dynamic energy release rate $G^{(D)}$ are given by the following equations.

$$K^{(D)} = k_{(v)} K^{(S)} \tag{2}$$

$$G^{(D)} = \frac{(1-\nu^2)}{E} A_{(v)} \{K^{(D)}\}^2 \tag{3}$$

where $K^{(S)}$: static stress intensity factor; $k_{(v)}$ and $A_{(v)}$: function of crack velocity.

The energy conservation criterion for the propagating crack is the equality of the dynamic energy release rate $G^{(D)}$ and the dissipated energy rate dD/da. Consequently, the necessary condition during the crack propagation is given by equation (4).

$$K_D = \{A_{(v)}\}^{1/2} k_{(v)} K^{(S)} \tag{4}$$

The right side value in equation (4) can be approximated by simplified equation as follows:

$$K_D = \sqrt{1 - \frac{v}{v_R}} \cdot K^{(S)} \tag{5}$$

v_R is the Rayleigh wave velocity ($v_R \simeq 0.92 \sqrt{\frac{G}{\rho}}$, when $\nu = 0.25$).

In the following equation (6), which is modified from equation (5), the left side of the equation depends only on temperature and crack velocity but not on the crack length, and the right side of the equation depends only on crack length.

$$K_D / \sqrt{1 - \frac{v}{v_R}} = K^{(S)} \tag{6}$$

Namely, neither side of equation (6) needs to be incrementally calculated or dynamically estimated. This provides a simple and promising tool to predict the brittle crack behaviors. This procedure was used for the specimens SI-1 and SI-2, and the results are shown in Figure 7, in which ten percent scatter of K_D is allowed. Except immediately before the specimen severance it seems that the estimated velocities are in good agreement with the velocities measured.

Next, the prediction trials as to the stiffened specimens were done by the same procedure. K_D is a function of the static stress intensity factor $K^{(S)}$. The K value estimation about a crack in a stiffened plate is not easy, but could be approximately evaluated by some manipulations. Using relevantly evaluated static stress intensity factor for a crack in stiffened plate [14,15], the predicted crack velocity variance with crack advance is shown in Figure 8, (specimen BS-1).

As it is said that there is a lower limit of a crack velocity for a brittle crack, it was assumed that the crack should immediately stop if the velocity decrease under 200 m/sec. Considering this effect, the crack that stopped at the stiffener intersection in the specimen BS-1 and the crack that ran through the intersection in the BS-2 can be well predicted by the calculation.

C. Some Considerations and Conclusion

The equations (2) and (3) by Freund were derived from the assumption of the stationary crack propagation in the infinite plate. These equations were extended to the non-stationary problem in this study. The main difference is that there is no effect of stress waves reflected back from the boundaries in former

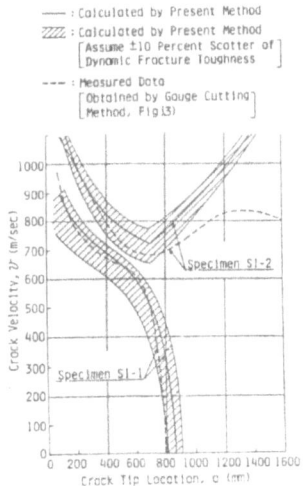

—— : Calculated by Present Method
▨▨ : Calculated by Present Method [Assume ±10 Percent Scatter of Dynamic Fracture Toughness]
--- : Measured Data [Obtained by Gauge Cutting Method, Fig.13)]

Fig. (7) - Calculated crack velocity

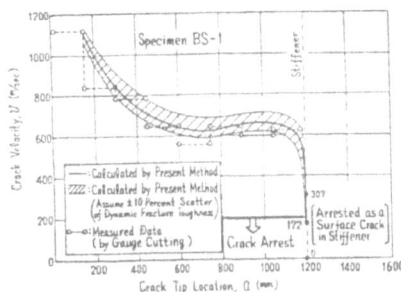

Fig. (8) - Comparison between calculated and measured crack velocity

case. Another point is whether the load boundaries make the additional work or not during the crack propagation, namely, whether the additional displacements occur or not at the locations where the loads are applied.

In the present experiments, it was estimated that the load boundary condition can never affect the crack behavior for the crack length less than 1000 mm and so that the predicted behaviors do not contain any considerable errors due to the simplifications in the procedures.

EVALUATION OF DUPLEX TYPE CRACK ARRESTER BY DYNAMIC TECHNIQUE

From fracture dynamics analysis on high-velocity crack propagation observed in double tension test and DCB test on the ship mild steels, it has been revealed that the crack propagation and arrest characteristics could be defined with a fair degree of accuracy using the dynamic fracture toughness [3-6,8].

On the basis of the conclusion obtained from these studies, the crack arrest performance of the duplex type crack arrester was reevaluated by dynamical technique based on the results of the duplex type double tension test [7].

A. Brittle Crack Propagation Test

The materials used are 20 mm thick ordinary ship plate KAS as the crack starter material and ship plate with higher toughness KEN as the crack arrester material.

First of all, the 500 mm wide standard double tension test was performed on each of these two materials. The specimen configuration, and attachment positions of strain gauges and crack detecting gauges are shown in Figure 9. The

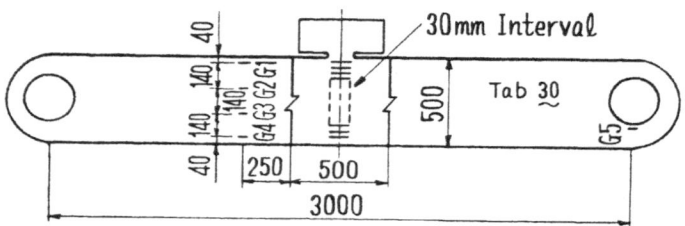

Fig. (9) - Specimen configuration and gauge positions for A and E series

change in strain that took place with the crack propagation was measured as well as crack velocity. Both series of specimens were of temperature gradient type. The tests were performed under various applied stresses.

The duplex type double tension test specimen which is called AE series was prepared by joining the KAS and KEN steel plates by electron-beam welding 230 mm from the entrance into the crack propagation part. The specimen configuration is shown in Figure 10. To reduce the residual stress, the test specimen was preloaded close to the yield point. The test temperature was -10°C maintained uniformly in most of crack propagation part. An example of change of crack velocity is shown in Figure 11.

It is obvious that beyond the weld seam, the crack velocity decreased abruptly by about 200~300 m/sec in every instance. The crack running into the arrester behaved in three different manners. The crack under 10 kgf/mm^2 applied stress was arrested immediately after running into the arrester and it was arrested after running about 40 mm through the arrester under 12 kgf/mm^2. On the other hand, the cracks under 14 and 16 kgf/mm^2 applied stress ran through the whole width of the arrester respectively.

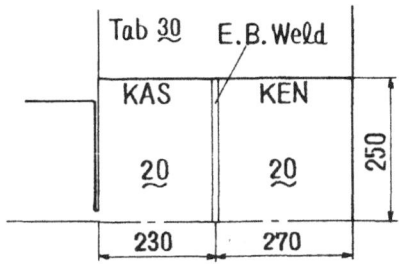

Fig. (10) - Specimen configuration
for AE series

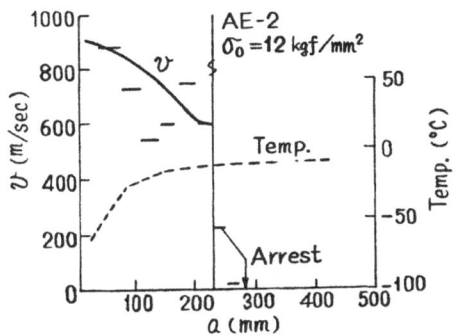

Fig. (11) - Variation of crack velocity
(AE series)

B. Analysis and Consideration

Assuming the test specimen to be a two-dimensional elastic body, the equations of motion were numerically solved by finite difference technique using the measured crack velocity and strain distribution on the prescribed boundary as the time dependent boundary conditions. Then variation of energy components of the system can be calculated.

As the dissipated energy D is mostly the energy dissipated in the fractured surface layers, dynamic fracture toughness K_D is defined from the analogy with the static case.

K_D was calculated for each of the A series, E series and AE series. An example of K_D and crack velocity plotted against the crack length for the duplex specimen is shown in Figure 12. For reference, the static stress intensity factor $\sigma_0\sqrt{2B\tan(\pi a/2B)}$ and temperature distribution are also indicated. The crack was arrested by the arrester in AE-1, under the uniform temperature condition. Although the crack propagated about 40 mm after running into the arrester in AE-2, K_D increased once the crack ran into the arrester as shown in Figure 12. In case of AE-4 and AE-8, the crack ran through the arrester due to high applied stress (16 kgf/mm^2 and 18 kgf/mm^2, respectively).

In Figure 13, K_D obtained by the basic E series double tension test are correlated using temperature and crack velocity as parameter for the KEN steel.

Figure 14 shows K_D versus crack velocity relations at -10°C obtained from gradient-temperature type double tension test for A series and E series, and AE series. Despite the difference in the above two testing methods, unique relations between K_D and crack velocity are defined. The horizontal arrows in the figure mean the levels of dynamic K_D values corresponding to available energy at the in-

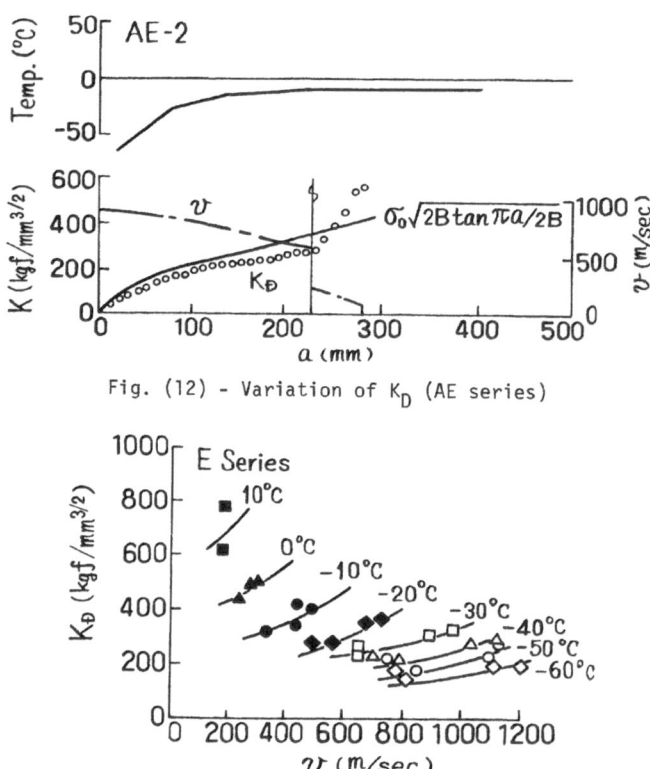

Fig. (12) - Variation of K_D (AE series)

Fig. (13) - K_D of KEN as a function of crack velocity

stant that a crack in starter plate reach the arrester plate.

In AE-4 and AE-8 tests, cracks from the starter part were abruptly slowed down at their entrance into arrester but they ran through whole width of arrester plate with increasing speed following the K_D curve in upper right direction. In AE-2 test, a crack from the starter part was abruptly slowed down at its entrance and arrester and it propagated with very low speed until it was arrested in the arrester plate. The crack in the arrester plate showed a fracture appearance with some degree of ductility accompanied by shear lips and quite different from usual brittle crack. K_D value for this crack defines the low speed region of K_D curve for KEN as shown in Figure 14 where K_D increases as crack speed decreases, and it has a minimum value at the crack speed of about 200 m/sec. In AE-1 test, the level of dynamic K value at the entrance of arrester plate was less than minimum value

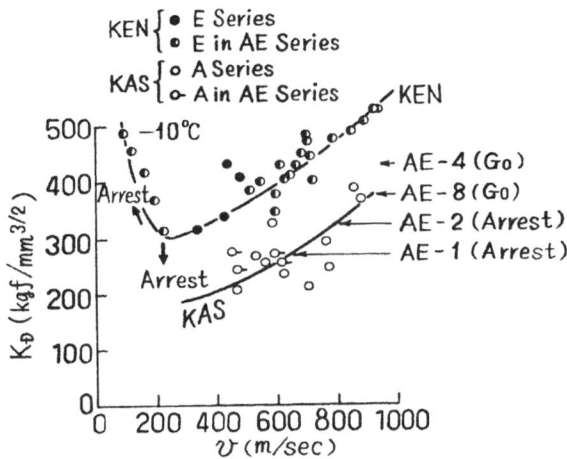

Fig. (14) - K_D versus crack velocity at -10°C for KAS and KEN
obtained from conventional type and duplex type
double tension tests

of K_D of the arrester material, and the crack could not get into arrester plate
and was arrested. The above three different crack propagation behaviors are schematically shown in Figure 15 in terms of the characteristics of K_D curves. A crack

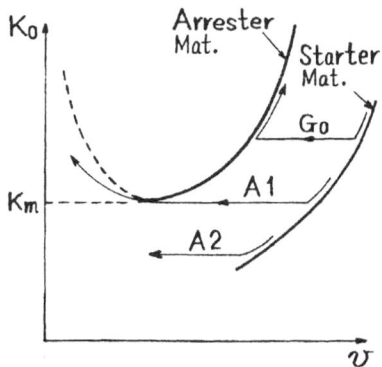

Fig. (15) - Schematic illustration for three different behaviors
observed in duplex type double tension test (AE series)

running into arrester material with available dynamic K_D value close to the mini-mum value of its K_D follows the lower speed part of K_D curve (A1 in Figure 15) showing fracture appearance with some extent of ductility which is different from the brittle appearance observed in higher crack velocity region.

Applying the dynamic fracture toughness curve to the double tension test performed, the brittle crack propagation and arrest behaviors were simulated by previously mentioned simplified approximation based on Freund's equation. The calculated variation of crack velocity for AE series are compared well with the experimental observations as shown in Figure 16. This is probably because the

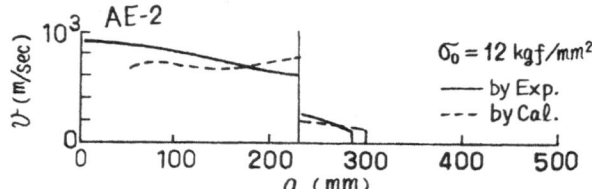

Fig. (16) - Comparison between measured and calculated crack velocity (AE series)

test specimens used were long enough and the effects of stress wave reflected from the end remained not so pronounced until the crack was arrested.

EVALUATION OF BRITTLE CRACK PROPAGATION AND ARREST BEHAVIORS IN 9% Ni STEEL BY DYNAMIC TECHNIQUE

In this study, extra-wide 9% Ni steel duplex double tension test specimens were used under the cryogenic LNG temperature, with analysis performed using the statical and dynamical techniques, to evaluate the large brittle crack propagation and arrest characteristics in the 9% Ni steel [10].

A. Brittle Crack Propagation Test

The test specimens were prepared from quenched and tempered 27 mm thick 9% Ni steel plate. The test specimen was 3 m in width and of duplex type double tension test, and its configuration is indicated in Figure 17.

The test temperature was at the LNG level (-162°C). The crack velocity and the change in strain distribution on a prescribed boundary were dynamically measured and the dynamic change in applied load was also measured by the strain gauge on the tab plate.

In No. 1 specimen, the crack widely ramified as can be seen from Figure 18. Two of the ramified cracks reached to but did not run into the 9% Ni steel plate. In this particular instance, however, it appears likely that the ramifica-tion served to considerably weaken the momentum of crack propagation.

Figure 19 shows the crack propagation path observed in No. 3 specimen. The crack ran straight across the starter part of embrittled 9% Ni steel without

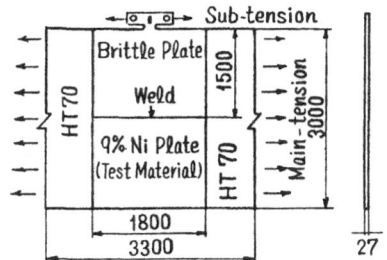

Fig. (17) - Specimen configuration of
the duplex type double
tension test

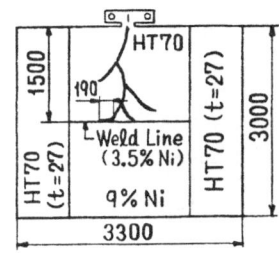

Fig. (18) - Crack path of the
No. 1 specimen

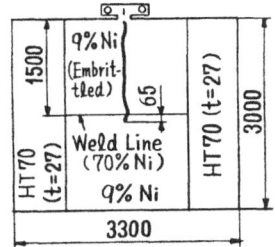

Fig. (19) - Crack path of the No. 3 specimen

any ramification and, after crossing the 70% Ni weld line, penetrated 60 mm into
the 9% Ni steel plate. The crack running into the 9% Ni steel plate was all ductile except for slight brittleness in the vicinity of the mid-plate thickness at
the weld bond, suggesting that the crack as brittle crack was arrested at the weld.
It appeared that the weld metal also contributed greatly to arresting the crack
propagation.

B. Analysis and Consideration

To analyze the experimental results, an approximation was attempted by
static analysis that took into account the effects of load decreasing with crack
propagation, by applying the compliance method [17]. Figure 20 shows the model
used in the calculation. The specimen is treated as a spring-supported rectangular plate of uniform displacement. The static stress intensity factor $K^{(S)}$ value
of the crack can be expressed as follows.

$$K^{(S)} = F(\frac{a}{b}) \cdot \sigma\sqrt{\pi a} = F(\frac{a}{b}) \cdot \frac{P}{2bt} \sqrt{\pi a} \tag{7}$$

94

(a) No Load

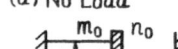
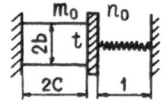

m_0 : Compliance of the specimen
n_0 : Compliance of the test rig
t : Thickness of the specimen

(b) Loaded

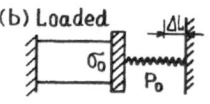

$\tilde{\sigma}_0$: Mean stress of the specimen
P_0 : Load

(c) Cracked under the load

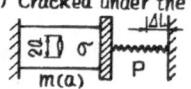

a : Half of the crack length
P : Dropped load under the crack
m : Compliance of the cracked specimen
$\tilde{\sigma}$: Mean gross stress

Fig. (20) - Schematic representation for calculating K value
by compliance method

From the relationship between the strain energy release rate and $K^{(S)}$ value, the compliance m(a) of the specimen as a function of crack length 2a is given by

$$m(a) = m_0 + \frac{\pi}{E \cdot t} \int_0^{a/b} \{F(x)\}^2 x dx \qquad (8)$$

The decrease in load is also expressed as follows.

$$\frac{P}{P_0} = \frac{m_0 + n_0}{m_0 + n_0 + \frac{\pi}{Et} \int_0^{a/b} \{F(x)\}^2 x dx} \qquad (9)$$

where m_0 and n_0 are compliances of the uncracked specimen and the test rig, respectively. Compliance n_0 was estimated from relevant measurement of relative displacement of the test rig. Then, the $K^{(S)}$ value that takes into account statically the decrease in test rig load can be given by

$$K^{(S)} = \sigma_0 \sqrt{\pi a} \cdot \frac{P}{P_0} \cdot F(x) \qquad (10)$$

Since the pulling tab plate is high in rigidity, it is assumed that the uniform displacement is assured at the boundary of specimen. Therefore, F(x) is numerically calculated using Ishida's solution [18].

Figure 21 shows the static load drop for No. 3 specimen calculated in this manner compared with the dynamically measured load drop.

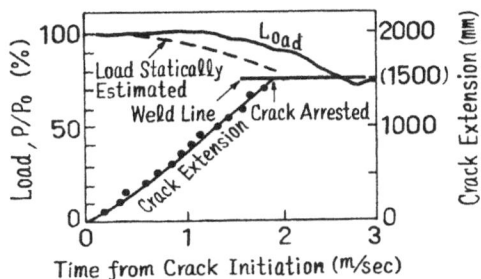

Fig. (21) - Load drop and crack extension in the No. 3 specimen

From equation (10), it is possible to easily determine the progressive change in $K^{(S)}$ value that takes place with crack propagation if only the initial stress σ_o is known. The results of calculation performed for No. 3 specimen are shown in Figure 22. The $K^{(S)}$ value at the time of crack arrest indicates considerable differences from the values under constant displacement condition and constant load condition.

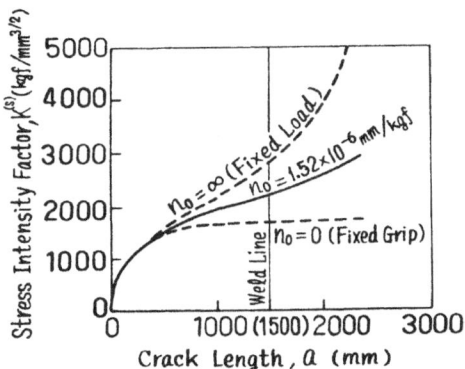

Fig. (22) - $K^{(S)}$ value in the No. 3 specimen by the static model

Next, the dynamical analysis [3-6] was attempted. Assuming the test specimen to be a two-dimensional elastic body, the equations of motion were numerically solved by finite difference technique using the experimentally obtained time dependent boundary conditions in the same way as stated in the preceding chapter.

Figure 23 shows the dynamic stress intensity factors K_D calculated for No. 1 specimen. The results of static analysis are also indicated in this figure.

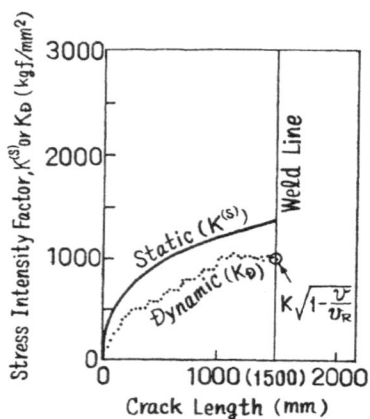

Fig. (23) - Dynamic stress intensity factor during the crack propagation in the No. 1 specimen

K_D value for No. 1 specimen in which the crack propagation velocity runs high is considerably smaller than the static stress intensity factor, being about 1000 kgf/mm$^{3/2}$ immediately before the crack runs into 9% Ni steel plate.

The dynamic stress intensity factor becomes smaller than the static stress intensity factor by dint of kinetic energy. Although how small the former will become than the latter depends on the crack propagation velocity, determining the dynamic stress intensity factor requires not only complex calculation but also dynamic data measurement for calculation input. A simplified approximation was made by combining the static and dynamic techniques, using Freund's equations as mentioned before.

Figure 23 shows also the result obtained by this method with the open circle, which is in fairly good agreement with K_D values obtained by dynamic analysis. Therefore, only by measuring the compliance of the test rig including the tab plate, the brittle crack propagation and arrest in duplex type double tension test can easily be evaluated by dynamic analysis. In case of the present experiments, it can be concluded that the toughness of 9% Ni steel tested is greater than the K_D value estimated at the crack arrest point.

ANALYSIS OF SHIP HULL DAMAGE BY STATICAL TECHNIQUE

A static analysis performed on an upper deck plating brittle failure which
was experienced by a 10379 G.T. bulk carrier [11]. While the ship was under way,
a crack was produced in the top side tank upper deck plating near midship section.
The crack rapidly grew and, after propagating 1250 mm through the upper deck plating and severing the shear strake, ran 2590 mm through the side shell plating before arresting as shown in Figure 24. A fatigue crack, about 400 mm in length,
was located at the place of crack initiation.

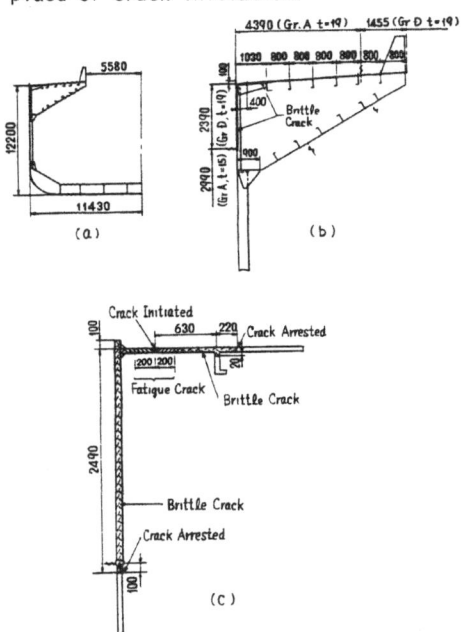

Fig. (24) - Damage plan of brittle failure in bulk carrier

The atmospheric temperature and seawater temperature were 1°C and 5°C, respectively. The ship was in ballast condition, and the estimated stress in the
upper deck plating was 5.2 ± 2.5 kgf/mm².

K_{CA} value at the temperature of failure was estimated from the V-notch Charpy
value of test specimen removed from the failed upper deck plating. On the other
hand, K value at the crack tip was estimated by static analysis applying K_{eff}'s
concept [2].

The results of these analysis are indicated in Figure 25. It appears that
how the brittle crack was arrested can be explained fairly well by these calculations.

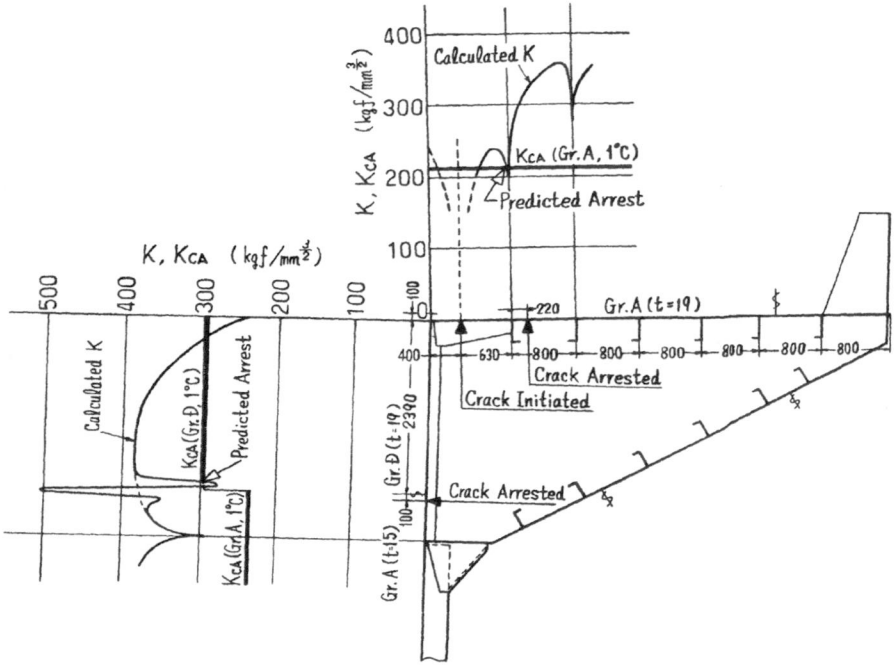

K value is obtained in statical calculation and modified by K_{eff} concept in higher K value region.

K_{CA} values are estimated as the lowest toughness from the damaged plates V-notch Charpy tests.

Fig. (25) - Calculation on brittle crack arrest in bulk carrier

CONCLUDING REMARKS

Dynamic analysis or simplified dynamic analysis recently performed in Japan to predict the brittle crack propagation and arrest behaviors have been reviewed in this report. Also, an analysis of brittle crack propagation and arrest experienced by a bulk carrier, though using a modified statical technique, has been reviewed.

Welded structures used in low-temperature conditions, such as LNG carriers, LPG carriers, pipelines, pressure vessels, etc., are on the increase. These structures are exposed to the danger of brittle failure even under normal service conditions, and their brittle failures generally are catastrophic. Thus, in response to the demands for assurance of safety against the brittle crack developing into a catastrophic failure, the usefulness of the crack arrester has come to attract renewed attention in the general move towards the fail safe design.

It is expected that the demands for establishing or refining the brittle crack propagation and arrest prediction technique will increase more and more henceforth.

REFERENCES

[1] Kihara, H., Kanazawa, T., Ikeda, K. et al, "Effectiveness of crack arrester (1st Report)", S.N.A.J., Vol. 122, 12, 1967, (2nd Report), S.N.A.J., Vol. 124, 12, 1968.

[2] Machida, S. and Aoki, M., "Some basic considerations on crack arresters for welded steel structures (the 7th Report)", S.N.A.J., Vol. 131, 6, 1972, Kanazawa, T., et al, "Study on brittle crack arrester", Selected Paper J. of S.N.A.J., Vol. 11, 1973.

[3] Kanazawa, T., Machida, S. and Teramoto, T., "Study on fast fracture and crack arrest - the 1st Report", S.N.A.J., Vol. 141, 6, 1977.

[4] Kanazawa, T., Machida, S. and Niimura, Y., "Study on fast fracture and crack arrest - the 2nd Report", S.N.A.J., Vol. 142, 12, 1977.

[5] Kanazawa, T., Machida, S., Niimura, Y. and Yoshinari, H., "Study on fast fracture and crack arrest - the 3rd Report", S.N.A.J., Vol. 144, 12, 1978.

[6] Teramoti, T., Machida, S. and Kanazawa, T., "Study on fast fracture and crack arrest - the 4th Report", S.N.A.J., Vol. 146, 12, 1979.

[7] Kanazawa, T., Machida, S. and Teramoto, T., et al, "Study on fast fracture and crack arrest - the 5th Report", S.N.A.J., Vol. 150, 12, 1981.

[8] Kanazawa, T., Machida, S. and Yoshinari, H., "Evaluation of brittle fracture propagation arrest characteristics by DCB testing", S.N.A.J., Vol. 146, 12, 1979.

[9] Kanazawa, T., Machida, S., Yajima, H. and Kawano, H., "Study on brittle crack propagation and arrest behavior in plates and beams structure (1st Report)", S.N.A.J., Vol. 149, 6, 1981.

[10] Machida, S., Kawaguchi, Y. and Tsukamoto, M., "An evaluation of the crack arrestability of 9% Ni steel plate to an extremely long brittle crack", S.N.A.J., Vol. 150, 12, 1981.

[11] Yajima, H. and Kawano, H., "Analysis of ship hull damage", Commission No. 1, Welding Research Committee, S.N.A.J., 1-550-82, 5, 1982.

[12] Tada, H., The Stress Analysis of Cracks Handbook, DEL Research Corp., 1973.

[13] Freund, L. B., "Crack propagation in an elastic solid subjected to general loading-I", J. Mech. Phys. Solids, Vol. 20, 1972.

[14] Ishida, M., "Analysis of stress intensity factors for plates containing random array of cracks", Bulletin of Japan Soc. Mech. Engrs., Vol. 13, 1970.

[15] Kanazawa, T., Machida, S. and Doi, H., "Some basic consideration on crack arresters - the 6th Report", S.N.A.J., Vol. 124, 12, 1968.

[16] The 101 Research Committee of The Japan Shipbuilding Research Association, Report No. 78, 1968.

[17] Kanazawa, T., Machida, S., Yajima, H. and Aoki, M., "Effects of specimen size and loading condition on the brittle fracture propagation arrest characteristics", S.N.A.J., Vol. 130, pp. 343-351, 12, 1971.

[18] Ishida, M., "Effect of width and length on stress intensity factors of internally cracked plates under various boundary conditions", Int. J. Fract. Mech., Vol. 7, p. 301, 1971.

SECTION I
MICROSTRUCTURE

RELATIONSHIPS BETWEEN MICROSTRUCTURE AND FRACTURE TOUGHNESS

J. G. Anderson and R. H. Edwards

Australian Iron and Steel Pty. Ltd.
Port Kembla, New South Wales, Australia

ABSTRACT

Although quantitative relationships between microstructure and Charpy impact toughness are well established, similar relationships between microstructure and elasto-plastic properties such as COD, have not yet been determined and at best remain qualitative. Attempts have been made to improve these relationships by studying the effects of specific microstructural features such as grain size, dislocation density, dispersed phases and inclusion shape on the fracture toughness of steels. This paper presents the findings of several of these studies.

INTRODUCTION

Relationships between the microstructure of steels and their toughness properties, can to some extent, allow quantitative predictions to be made, for example, grain size of mild steel has been correlated direct with its stress intensity factor at low temperatures [1]. However, relationships between microstructures and elasto-plastic properties such as COD remain qualitative, this paper reviews some progress towards improving those relationships.

EXPERIMENTAL

All fracture toughness testing was carried out on fatigue cracked three point loaded bend test pieces in accordance with BS5762 [2]. However, in the case of crane rails and mill rolls, it was not practicable to use test pieces of full thickness so the largest size capable of being tested on the available equipment was used, namely 60 x 120 mm in cross section.

Fracture surfaces were examined by scanning electron microscopy (SEM) whereas microstructures at the position of crack initiation on COD test pieces were studied by optical and transmission electron microscopy (TEM).

RESULTS AND DISCUSSION

A. Influence of Grain Size

An important factor affecting both yield strength and impact properties in commercial steels is grain size, since excluding other factors, strength and toughness are both linearly related to the square root of the average grain diameter [3]. In modern commercial steels, grain sizes vary generally between about 10-40 μm and the expected increase in yield strength and decrease in impact transition temperature corresponding to a maximum grain size reduction in this range would be 70 MPa and 40°C, respectively.

A simple example illustrating the influence of grain size was provided by the testing of the failed top and bottom chords of an overhead crane which had been in service close to 30 years. Both chords were of steel to AS 1:1956 [4] although they differed in thickness. The average grain sizes for the top and bottom chords were 90 μm and 180 μm, respectively and Charpy V notch values of 19J and 9J and COD values of 0.5 mm and 0.31 mm were obtained at 0°C.

Another case is that of an 11 mm thick submerged arc weld containing acicular ferrite of average plate width and length of 1 and 8 microns respectively, Figure 1, which normalizing converted to equiaxed polygonal ferrite of average grain diameter 20 microns, Figure 2. The COD of the as welded material was 0.20 mm

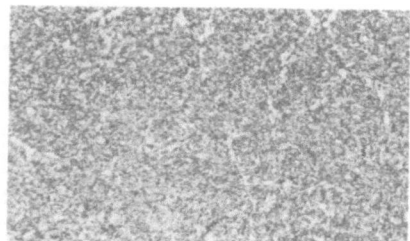   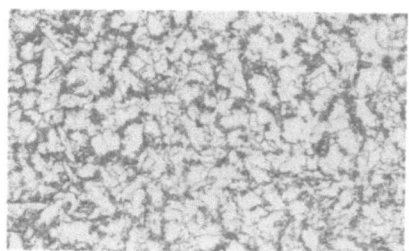

Fig. (1) - Micrograph showing acicular   Fig. (2) - Micrograph showing polygonal
      ferrite microstructure in   ferrite and martensite in
      submerged arc weld. As   submerged arc weld after
      welded condition. X200   normalizing. X200

at -30°C as compared with 0.15 mm in the normalized material and this difference was maintained over most of the transition temperature range. This example shows that in the acicular ferrite microstructure, the anticipated adverse contribution of the higher dislocation density to fracture toughness (see section on dislocation density) was more than compensated for by the significant reduction in grain size accompanying formation of very small acicular ferrite plates in preference to polygonal ferrite grains.

However, in spite of the undisputed effect of grain size on toughness, its influence may be overshadowed by other factors as was illustrated by fracture toughness on 192 Kg/m crane rails. Both standard and high yield strength rails were tested at 0°C and cleavage fracture, with no sign of prior ductile crack growth, occurred in all test pieces. The standard rail had a much coarser grain size than the high yield strength rail but the K_{1c} values were 46 and 32 MPa$\sqrt{m}$ respectively whilst the yield strengths were 507 and 409 MPa. In this case, the effect of the high yield strength in restricting plastic flow at the crack tip and so reducing toughness was greater than the beneficial effect of the finer grain size.

B. Dislocation Density

It is well established [5] that as the dislocation density of a metal increases, tensile strength also increases, the strengthening increment being proportional to the square root of the dislocation density. However, strengthening by a mechanism of this type is invariably accompanied by an upward shift in transition temperature the penalty averaging 0.2°C/MPa [6].

Similar deteriorations in toughness with increased numbers of dislocations have been measured during COD testing of ERW seam welds in 273 mm dia., 6.4 mm wall thickness pipes to API LX46 specification. Specimens were tested in the as welded, post-weld heat treated and normalized conditions, respective mean COD values being 0.10, 0.13 and 0.49 mm. Transmission electron microscopy of as welded and post-weld heat treated test pieces revealed the ferrite component of the microstructures to be heavily dislocated, Figure 3. Differences between these two welds were minimal.

Normalized welds consisted of lightly dislocated polygonal ferrite, Figure 4, and islands of pearlite. It was therefore evident that fracture toughness had been markedly increased by significantly reducing the dislocation density.

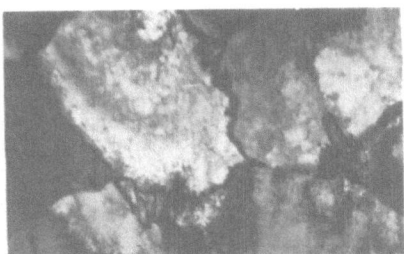

Fig. (3) - Electron micrograph showing heavily dislocated polygonal ferrite in ERW weld - as welded condition. X10,000

Fig. (4) - Electron micrograph showing lightly dislocated polygonal ferrite in laboratory normalized ERW weld. X10,000

It should be noted here that toughness is influenced not only by dislocation density but also by the ease of movement of these dislocations through the crystal lattice under the influence of an applied stress. During COD testing of 864 mm dia. seam welded pipes at various positions removed from the weld centerline, a zone of relatively low toughness was noted approximately 25 mm from the center line of the weld. Microstructural examinations of high (0.38 mm COD at -60°C, 15.2 mm from weld centerline) and low (0.10 mm COD at -60°C, 27.4 mm from weld centerline) toughness zones, by optical microscopy, revealed similar microstructures. However, TEM showed the dislocations in the high COD test pieces to be randomly oriented whereas in low COD test pieces they were clearly aligned, Figure 5. Baird and MacKenzie [7] reported such alignments to result from inability of the dislocations to leave their slip planes as a result of a "locking" mechanism. In the pipe specimens, it seemed probable that strain ageing by nitrogen was responsible for the aligned dislocation arrays. At about 25 mm from the center line of the weld, strain and thermal conditions were apparently favourable for

106

strain ageing and the resultant confinement of the dislocations to their slip planes gave decreased COD values in this zone.

Fig. (5) - Electron micrograph showing aligned dislocations in the low toughness zone of a seam welded pipe. X50,000

C. Effect of Dispersed Phases

An increasing demand for higher strength and finer grained steels has been met by microalloying with elements such as Nb, V and Ti which have a tendency to precipitate as small uniformly distributed carbides, nitrides or carbonitrides. The influence of these precipitates on steel toughness is strongly dependent on particle size and nature of the interface between precipitate and matrix. Coherent dispersions increase strength substantially, generally at the expense of toughness. On the other hand, incoherent dispersions provide little strengthening as a direct result of precipitation but by inhibiting grain growth at normal hot rolling temperatures ensure a fine ferrite grain size and significant improvement in transition temperature.

The laboratory normalizing of the seam weld of 6.4 mm thick pipe referred to previously resulted in an incoherent precipitate of niobium carbonitride, Figure 6, and so the improvement in toughness observed must be attributed both to the nature of this precipitate and to the lower dislocation density.

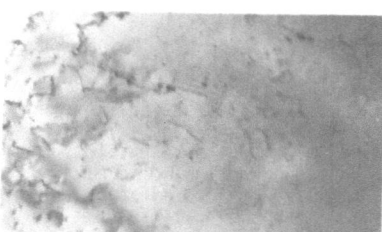

Fig. (6) - Electron micrograph showing dispersion of incoherent niobium carbonitrides in laboratory normalized ERW weld. X13,000

A related effect was observed with the fracture toughness testing of slabbing mill rolls. Test pieces were longitudinal with respect to the rolls, they were 60 x 120 mm in section and were taken close to the roll surface with the notch orientated so that it would propagate in a direction corresponding to the radial direction of the roll. The two rolls were of similar composition and tensile strength but at 50°C, roll A had a COD value of .078 mm compared with .027 mm from roll B. The superior toughness of roll A was confirmed by Charpy tests at 200°C and 300°C although Charpy tests at 50°C hardly differentiated between the rolls. Microexamination showed that roll A was almost completely spheroidized, the carbides thus existing as a coarse globular dispersion, whilst the structure of roll B consisted of lamellar pearlite with only minor spheriodization.

D. Non-Metallic Inclusions

The effects of inclusions on toughness are mostly due to their shape, particularly in the case of plate-like inclusions such as type II MnS which causes splitting of the metal under stress so that a test piece behaves like a number of separate thinner test pieces resulting in a low upper shelf COD. The testing of two 14 mm thick plates of API LX65 at AIS, the plates being of similar analysis except for Ce, showed that the omission of Ce reduced the upper shelf COD at room temperature from 0.35 mm to 0.10 mm. The photograph in Figure 7 shows how the area of slow crack growth, and consequently the COD, is reduced by the occurrence of splits.

Fig. (7) - Macrophotograph of COD test piece fracture face illustrating restriction of stable crack growth by splits.   X5

If insufficient Ce is added to fully modify the shape of the type II MnS inclusions, then partial modification results in partial improvement in toughness. This was shown by COD tests on three 11 mm thick plates of API LX60 grade, similar in composition and tensile properties except that the Ce contents were 0, .014% and .021%. COD values at 20°C were .074 mm, .243 mm and .476 mm, respectively.

CONCLUSIONS

The experimental results demonstrate that fracture toughness is strongly dependent on microstructure. Grain size reduction was an important factor since it promoted improved toughness despite increased strength. This was particularly exemplified by observations that in acicular ferrite microstructures, the anticipated adverse contribution of a higher dislocation density to fracture toughness was more than compensated for by the significant reduction in grain size accompanying formation of very small acicular ferrite plates in preference to polygonal ferrite grains.

The importance of dislocation density was clearly shown, especially in welds, and the same work provides an example of the increase in toughness related to growth of a coherent precipitate to the incoherent state. The beneficial effect on toughness of spheroidizing a lamellar pearlitic structure has been shown and two cases of improved toughness due to sulphide shape modification have been monitored.

Although an attempt was made to isolate individual metallurgical factors and examine their contributions to fracture toughness, it must be borne in mind that most of the variables are interrelated and therefore care must be exercised when interpreting fracture toughness data in terms of microstructure.

REFERENCES

[1] Curry, D. A. and Knott, J. F., Metal Science, 10, 1, p. 1, 1976.

[2] BS5762: 1979 Methods for Crack Opening Displacement (COD) Testing.

[3] Irvine, K. R., Iron and Steel, 31, February 1971.

[4] ASI: 1956 Structural Steel and Rolled Steel Sections.

[5] Morrison, W. B., Scandinavian Journal of Metallurgy, 9, 2, p. 83, 1980.

[6] Gladman, T., Dulieu, D. and McIvor, I., Microalloying 75, Union Carbide Corp., New York, p. 136, 1977.

[7] Baird, J. D. and MacKenzie, C. R., J.I.S.I., 202, p. 427, 1964.

THE INFLUENCES OF DUAL PHASE MICROSTRUCTURES ON CRACK PROPAGATION BY MEAN STRESS

T. Ishihara and Y. Kowata

Ishikawajima-Harima Heavy Industries Company, Ltd.
3-5 Mukodai-cho, Tanashi-shi, Tokyo, Japan

INTRODUCTION

The dependence of fatigue crack propagation rate, da/dN, on alternating stress intensity factor, $\Delta K = K_{max} - K_{min}$, is generally expressed by the relationship da/dN $= C \Delta K^m$ where C and m are constants [1,2]. But, it is said that other factors, such as microstructure, temperature, mean stress and material properties, can markedly influence fatigue crack growth [3,4]. Particularly the effect of microstructure is important. Although many investigations have been made into such effects, they have not been enough. Formerly, Richards and Lindley [5] and Ishihara [6] have suggested that the general effect of mean stress and microstructure on the rates of fatigue crack propagation may be due to modes of "static" or monotonic fracture which can occur in addition to characteristic striation growth. And, it is also said that segments of fracture in microstructure are promoted by an increase in mean stress, and lead to accelerated fatigue crack growth rates.

Through appropriate heat treating procedures, two unique types of microstructures were successfully developed in low carbon steel, which is characterized by martensite encapsulating islands of ferrite (R material) and K material having ferrite encapsulated islands in the martensite. The strength, toughness, ductility, etc., of such structures have been studied and the behavior related to microstructural aspects.

On the other hand, it has been shown that the mechanical properties of these microstructures were associated with the interaction effects of these dual phases rather than the individual properties of martensite and ferrite. And it has been recognized in R material that the fracture occurs in a brittle manner accompanying cleavage cracking of ferrite grains.

However, crack propagation to fatigue crack growth in these unique microstructures has not been explored in any detail, and because of the interesting morphological relationships which can be developed between the phases, it appeared worthwhile to explore the fatigue crack propagation in these two microstructures under various mean stresses where microstructural influence is known to be of importance.

The characteristics of behavior of crack propagation in these dual phase microstructures under various mean stresses might be expected to show a strong dependence on the microstructures.

110

So, in the present paper, attention is given to the crack growth in these two dual phase microstructures. However, questions arise as to how the various mean stresses at the R and K material influence the crack propagation.

EXPERIMENTAL PROCEDURE AND SPECIMENS

A low carbon steel as shown in Table 1 was chosen to be investigated. After certain heat treatments, test pieces were machined to a length of 156 x 10^{-3}m,

TABLE 1 - CHEMICAL COMPOSITIONS

	C	Si	Mn	S	P	Cu	Ni	Cr
S15C	0.15	0.28	0.51	0.006	0.018	0.13	0.04	0.15

width of 39 x 10^{-3}m, and thickness of 10 x 10^{-3}m as shown in Figure 1 [7]. In

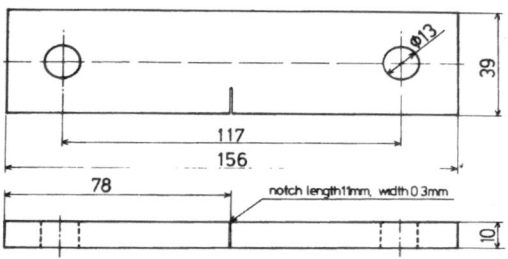

Fig. (1) - Geometries of specimen for fatigue crack growth study

order to develop the desired two phase microstructures, which are K microstructure having ferrite encapsulated islands of martensite and R microstructure having martensite encapsulated islands of ferrite as shown in Figure 4, heat treating procedures were applied as shown in Figures 2 and 3. Quantitative metallographic

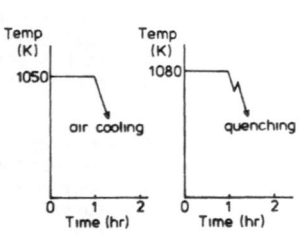

Fig. (2) - Process of the heat treatment for material R

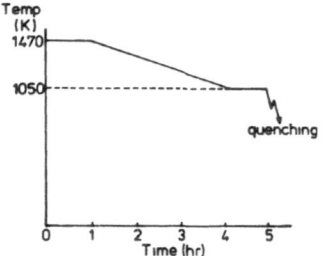

Fig. (3) - Process of the heat treatment for material K

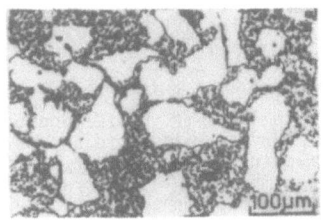

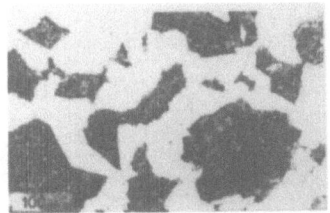

(a) Material R   (b) Material K

Fig. (4) - Microstructures

techniques were employed to determine the ferrite grain size, the volume fraction of martensite, the size of the martensitic structure, and the degree of connectivity of the martensitic structure as shown in Table 2. The fatigue tests were car-

TABLE 2 - MECHANICAL PROPERTIES AND METALLURGICAL PARAMETERS

		Martensitic Structure					
Materials	Ferrite grain size (μ)	Volume fraction (%)	Hardness (Hv 20g)	$\overline{0}$ (%)	Reduction in area (%)	Yield stress $\sigma_{0.2}$ (MPa)	Tensile strength (MPa)
R	49	49	496	94	17	559	990
K	51	45	497	(57)	25	471	775

ried out in a hydraulic testing machine at mean stress, σ_m, of 26 MPa, 53 MPa, 79 MPa and the stress range, $\Delta\sigma$, was 49 MPa with sinusoidal wave in laboratory air. The cyclic frequency was 10 Hz. Twenty electron microscope pictures at the center on the crack surface were taken every 10μ toward the direction of cracking.

The following equation (8) was used to calculate stress intensity range, ΔK (= $K_{max} - K_{min}$).

$$\Delta K = \Delta\sigma\sqrt{\pi a} \ F\left(\frac{a}{W}\right)$$

$$F\left(\frac{a}{W}\right) = \sqrt{\frac{2W}{\pi a}} \cdot \tan\frac{\pi a}{2W} \times \frac{0.75 + 2.02(a/W) + 0.37\{1 - \sin(\pi a/2W)\}^3}{\cos(\pi a/2W)}$$

where a is the crack length and W is the width of test piece.

Therefore, in the case of crack propagation by ductile cracking caused by accumulating strains, the crack propagation is not always accelerated. On the contrary, in the case of R material which easily incurs the static fracture, microfracture mode will be occupied by the static mode.

EXPERIMENTAL RESULTS AND DISCUSSIONS

A. Influence of Mean Stress on Crack Propagation

It is reported that the relationship of da/dN versus ΔK is not always expressed as da/dN = $C(\Delta K)^m$ by Paris et al, but the non-linear and difference of C and m occur because fracture mechanism is varied by the microstructure and ΔK level. Also, it is said that the change of m value, fracture mechanism and fracture percent caused by varied fracture mechanism are very important to the crack propagation [9]. The purpose of this study is to reveal the dependency of crack propagation on mean stress is R and K material.

The microstructure of R material is shown in Figure 4(a) and it is expected to incur cleavage facet from the beginning for this reason; it is expected that most of the microfracture surfaces will be occupied by the static mode and, even if mean stress is alternated, the fracture mode will not transfer from static mode to another.

On the contrary, K material shown in Figure 4(b) will not show the static mode from the beginning and its fracture mode transfers gradually from slipping to static mode according to mean stress level because this structure does not consist of martensite networks, so its dependency on mean stress will be remarkable. We guess that the difference of such a change as mentioned will occur because of alternation of mean stress.

Now, stress range, $\Delta\sigma$, is set constant and mean stress, σ_m, is alternated from two to three times. For the crack propagation curve as shown in Figure 5, we could obtain the relationship of da/dN versus ΔK as shown in Figures 6, 7 and 8.

It is shown by Figures 6, 7 and 8 that crack propagation speed of R material is higher than that of K material. Figure 9 shows each crack propagation curve of R material. At a glance, each curve seems to be a little different. But, if Figure 9 is compared with Figure 10, the differences of the speed at σ_m = 26 MPa 53 MPa and 79 MPa of R are little as shown in Figure 9. Whereas that of K as shown in Figure 10 has a remarkable difference and the crack propagation increases depending on the increase in mean stress, σ_m.

It is widely known that investigation of the alternation at microfracture behavior depending on stress intensity factor, ΔK, is effective for research of the factors influencing fatigue crack propagation. On the fracture surfaces, striations, cleavage facets, quasi facets and dimples are observed. Figures 12-17 show fracture percent of various fracture modes at each mean stress in R and K material.

Now, we would like to investigate from Figures 12-17 the reason why the differences of crack propagation speed occur in only K material. As a whole, each fracture surface percent depends on ΔK level of R and K materials. And, if we observe in detail, fracture surface percents of striation in R and K are in inverse ratio to crack intensity factor, ΔK. On the contrary, fracture surface percents of dimples, cleavage facet and quasi cleavage facet increase depending on ΔK. We investigated the changes of each fracture surface percent of K corresponding

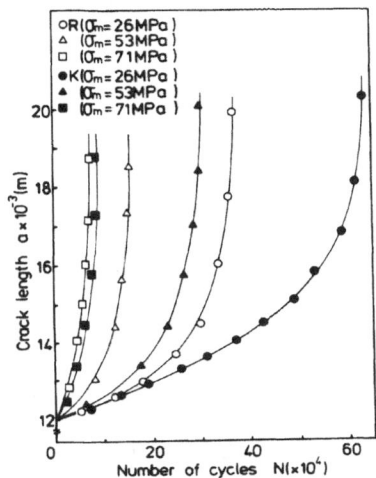

Fig. (5) - Relationship between crack
length and number of cycles

Fig. (6) - Relationship between crack
propagation speed and stress
intensity factor at mean
stress 26 MPa

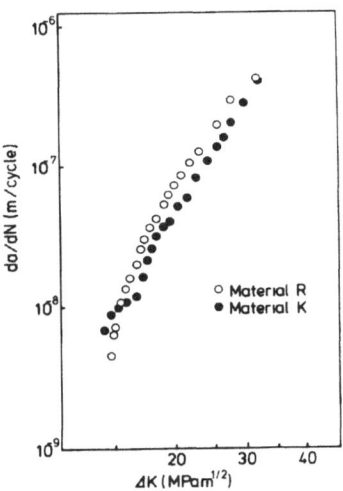

Fig. (7) - Relationship between crack propagation speed and stress
intensity factor at mean stress 53 MPa

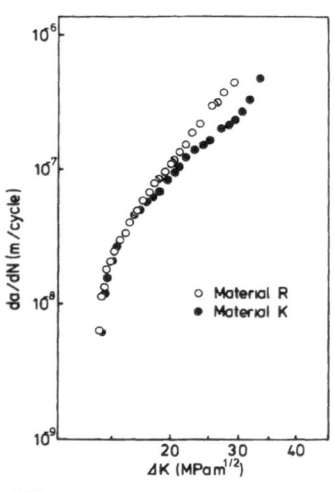

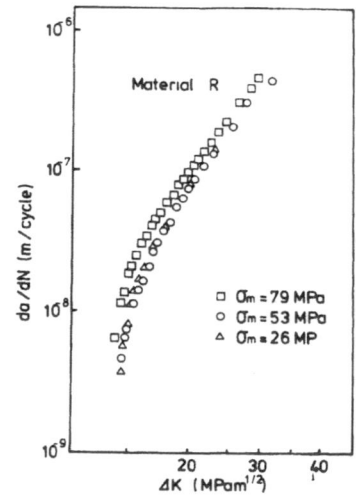

Fig. (8) - Relationship between crack propagation speed and stress intensity factor at mean stress 79 MPa

Fig. (9) - Relationship between crack propagation speed and stress intensity factor

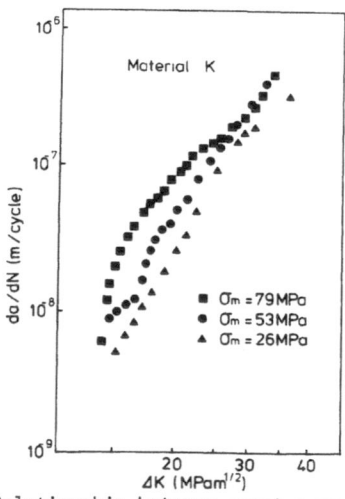

Fig. (10) - Relationship between crack propagation speed and stress intensity factor

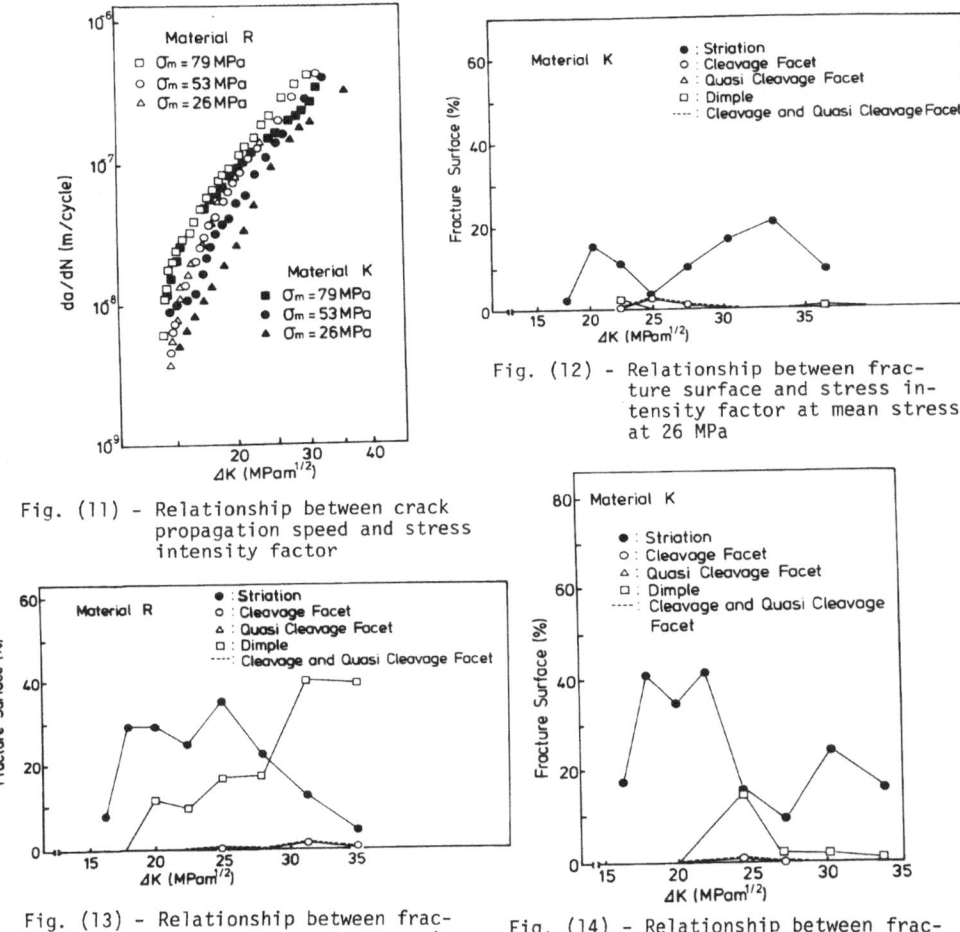

Fig. (11) - Relationship between crack propagation speed and stress intensity factor

Fig. (12) - Relationship between fracture surface and stress intensity factor at mean stress at 26 MPa

Fig. (13) - Relationship between fracture surface and stress intensity factor at mean stress at 26 MPa

Fig. (14) - Relationship between fracture surface and stress intensity factor at mean stress 53 MPa

116

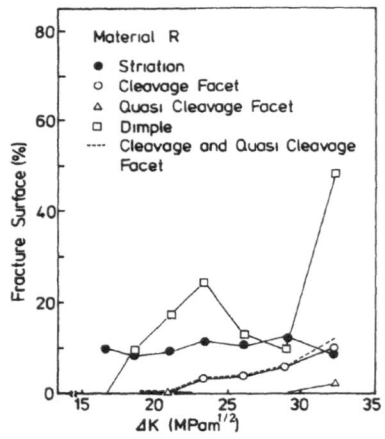

Fig. (15) - Relationship between fracture surface and stress intensity factor at mean stress
53 MPa

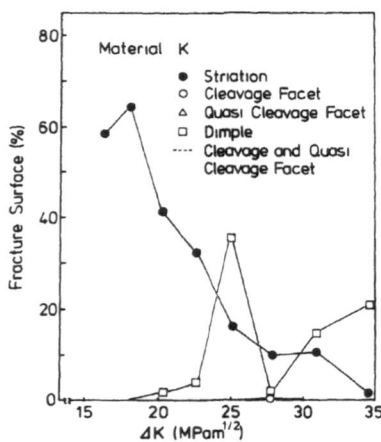

Fig. (16) - Relationship between fracture surface and stress intensity factor at mean stress
79 MPa

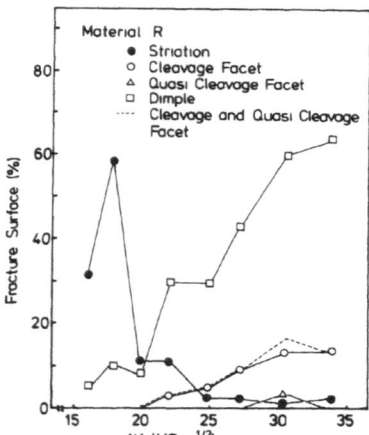

Fig. (17) - Relationship between fracture surface and stress intensity factor
at mean stress 79 MPa

with Figures 6-8. In case of σ_m = 26 MPa, fracture surfaces of cleavage facet,
quasi cleavage facet and dimple of the area below ΔK = 20 MPa are not observed,
so the crack propagation of this stage is 2b stage consisting of striations. The
change from 2b stage to 2c depending on the increase in σ_m occurs at lower ΔK

level. That is to say, it changes from slipping fracture mode to static depending on the increase in σ_m. Change from 2b stage to 2c depending on σ_m causes acceleration of crack propagation and it makes a difference of crack propagation speed depending on σ_m of K. The crack propagation of R is rapid, even if σ_m is low, and the fracture shows static fracture mode mainly consisting of dimples and cleavage etc. This phenomenon continues with the increase in σ_m. The reason why there is almost no difference of crack propagation between each σ_m of R is that static fracture mode occurs from lower σ_m level to rather higher level.

From these facts, we can recognize that the influence by σ_m depending on microstructure for fatigue crack propagation of R material and that of K propagate depending on σ_m. We investigated from the standpoint of microstructural aspect.

Generally, it is said that fatigue crack propagation is accelerated according to the change from striation mode to static as intergranular fracture and cleavage cracking. Dr. Horibe pointed out that the above does not mean absolute.

B. Main Acceleration Factor for Crack Propagation

Every crack propagation speed of R material exceeds that of K material in case of mean stress = 26 MPa, 53 MPa and 79 MPa as shown in Figures 6, 7 and 8. It is well-known that fatigue crack propagation is accelerated by the static fracture modes. It will be interesting to investigate the most effective factor of acceleration for crack propagation by cleavage facets, dimples, etc.

For this purpose, the difference of fracture percent at the same crack intensity factor level of R and K material is determined as the x coordinate axis and the difference of crack propagation speed as the Y coordinate by use of the results of fractography as shown in Figures 12-17. As a result, we could not find any relation between the difference of fracture percent and the difference of crack propagation speed at the dimples as shown in Figure 18. The difference of cleavage facets is determined as the x coordinate axis. So, it becomes clear that the difference of cleavage facets accelerates the crack propagation depending on mean stress. And its tendency is remarkable at high values of mean stress. Next, we paid attention to the form of ferrite cleavage facets in R and K material, and the kind of microfracture behavior. Also, it is important to know their parts at the acceleration of the crack propagation by the form of microfracture behavior. For this purpose, we observed in detail each microfracture surface to investigate the dependency of fracture surface on the crack propagation speed by the form of ferrite cleavage facet behavior. As a result, cleavage facets are observed on the fracture surfaces of R and K material as shown in Figure 20. From our observation, it becomes clear that cleavage facets on R material occur continuously in the neighboring ferrite grains as shown in the left of Figure 20 but, on the contrary, those of K material occur isolated in the microstructures as shown in the right of Figure 20. And, from the increase in continuous cleavage facets on R material as shown in Figure 21 and 23 which has the largest difference of crack propagation speed as shown in Figure 6, etc., the number of cleavage facets shows the tendency of dependence on acceleration of crack propagation. At the 79 MPa, showing the highest value of mean stress, the number of continuous cleavage facets at the high stress intensity range shows a great increase as shown in Figures 21 and 23, which means the tendency of dependence on the difference at R and K material. On the contrary, continuous and isolated cleavage facets on K material hardly occur at

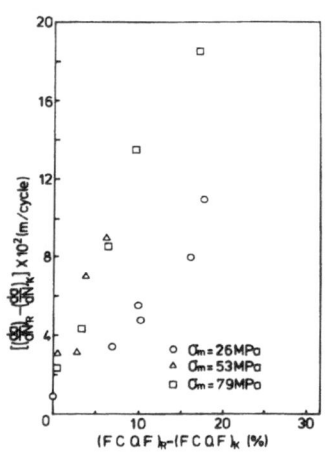

Fig. (18) - Relationship between difference of cleavage surface percent and crack propagation speed from material R to K

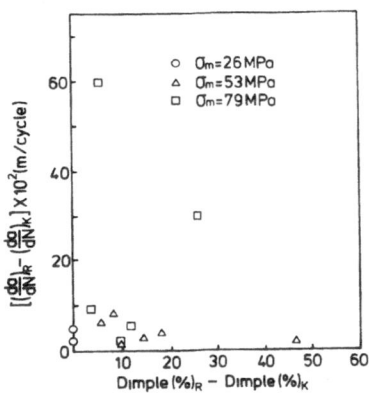

Fig. (19) - Relationship between difference of dimple fracture surface percent and crack propagation speed from material R to K

Direction of Fatigue Crack Propagation

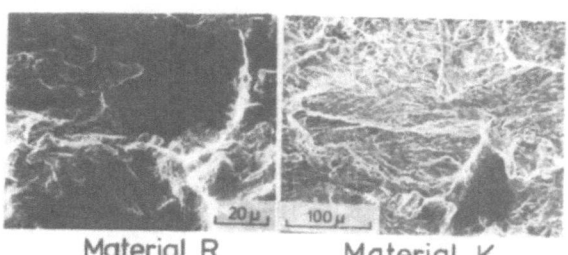

Material R   Material K

Fig. (20) - Difference in cleavage appearance between R and K materials

79 MPa of mean stress as well as at 26 MPa as shown in Figures 21-24. The part of isolated cleavage number on R material is very little to accelerate fatigue crack propagation. For this reason, the influence of dual phase microstructures on crack propagation becomes clear by the difference of microstructural form on the fracture surfaces.

CONCLUSIONS

1. In the case of a material in which cleavage facet does not easily occur at its crack tip, fatigue crack propagation depends on mean stress and also its speed accelerates depending on the change from slipping fracture mode to static.

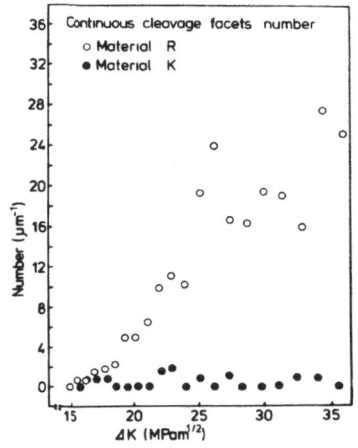

Fig. (21) - Relationship between continuous cleavage facet numbers and stress intensity
factor at mean stress
26 MPa

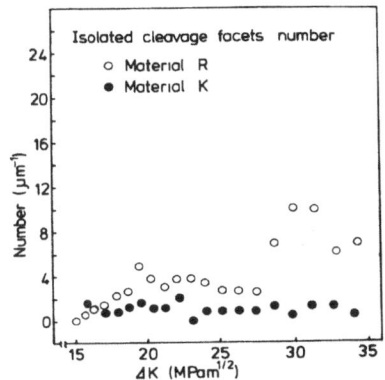

Fig. (22) - Relationship between isolated cleavage facet num-
bers and stress intensity
factor at mean stress
26 MPa

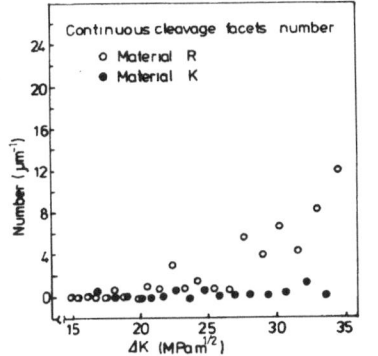

Fig. (23) - Relationship between continuous cleavage facet numbers and stress intensity
factor at mean stress
79 MPa

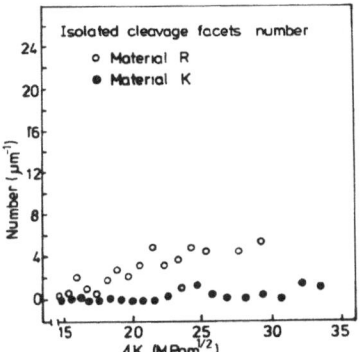

Fig. (24) - Relationship between isolated cleavage facet num-
bers and stress intensity
factor at mean stress
79 MPa

2.   In the case of a material in which cleavage facet easily occurs, the crack propagation depends on static fracture mode from the beginning, and change of propagation speed which depends on mean stress is little because the change of microfracture mode is very small.

3.   The main factor to accelerate fatigue crack propagation depends on the number of continuous cleavage facets which occur at adjoining matrix grains.

ACKNOWLEDGEMENT

The support of Professor S. Yoshida of the University of Tokyo in carrying out this study is appreciated.

REFERENCES

[1]   Paris, P. C., ASME Paper, 62-MET-3, 1962.

[2]   Paris, P. C. and Erdogan, F., Trans. ASME, Ser. D, 85, p. 528, 1963.

[3]   Tomoda, Y., Kuroki, G. and Tamura, I., J. Iron Steel Inst., Japan, 61, p. 107, 1975.

[4]   Hayden, H. W. and Floreen, S., Met. Trans., 1, p. 1955, 1970.

[5]   Richards, C. E. and Lindley, T. C., Engng. Fract. Mech., 4, p. 951, 1972.

[6]   Ishihara, T., JSME, 46-410, p. 1026, 1980.

[7]   Crooker, T. W., ASTM STP 600, p. 209, 1976.

[8]   Tada, H., Stress Intensity Factor Handbook, 2.10, 1972.

[9]   Asami, M., 8, 1979.

A NEW METHOD FOR OBSERVING FATIGUE MICROCRACK PROPAGATION

S. Nunomura, Y. Higo, I. Tsubono and H. Hayakawa

Tokyo Institute of Technology
Nagatsuta, Midori-ku, Yokohama, 227, Japan

ABSTRACT

The influence of material microstructure on fatigue crack growth must be understood before improvement on the fatigue strength of metals could be made. Although observed fatigue striations are informative, they are confined to a grain because the striations are not continuous at the grain boundary. Special loading cycles were computerized in order to obtain information on striations that continue over the grains. Fractography results are obtained for aluminum alloys with modulated spacing striations. Micro-morphology of the crack front and the interaction with the grain boundary and inclusion are discussed.

INTRODUCTION

Macroscopic behavior of fatigue crack growth has been studied by relation of the crack length of the number of loading cycles and the nature of loading. It is generally divided into three stages that vary with the applied stress and material properties. In the second stage, well-defined crack front arrest lines are visible which are commonly known as the "fatigue striations". Despite the number of microscopic fatigue crack growth mechanisms proposed in connection with fatigue striations, the influence of the microstructure on fatigue due to grain boundaries, second phase particles and voids is still not well understood. Since striations appear only under appropriate combination of applied stress and orientation of the crystallographic planes, they are often observed only in part and are discontinuous at the grain boundaries. Such information is limited to within a grain. The morphology of the crack border is no longer clear because it requires the continuous segments of many grains. Macro crack growth rate $d\ell/dN$ in fatigue is often said to agree with that at the microscopic scale level, i.e., the striation spacing, the observed difference is not negligible even if experimental errors are considered. Hence, there is the need to clarify the effect of material microstructure on fatigue crack growth.

McMillan and Hertzberg [1] developed a semi-analytical method of fatigue crack propagation with a periodical overload during the test and found that one striation corresponded to one load cycle while crack growth occurred on the load rising half cycle. By numbering the fatigue striations on the fracture surface, a better definition of fatigue crack front can be made in terms of grain boundaries, inclusions, etc. This work is concerned with a computerized method of generating striations which will be numbered systematically.

MODULATION PATTERN METHOD

The simple periodic cyclic overload method cannot identify the simultaneous generation of striations in the individual grains nor can it count the number of loading cycles spent to advance the crack front through grain boundaries and striations that may not always be visible within a grain. The present method makes use of an electric signal for generating combination patterns from a mini-computer system, Mitsubish MELCOM 70/40. It consists of two pulses in accordance with a certain regulation. Figure 1 shows a part of the modulated load spectrum of a constant K_{mean} test. With this reference signal, fatigue tests are carried

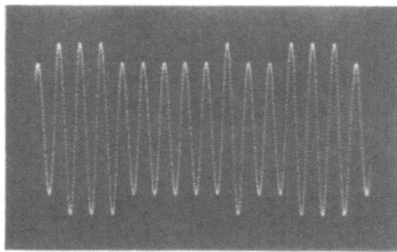

Fig. (1) - Modulation loading spectra for the fatigue test.
Pattern format is 03-S5, 01-S2-03-S(6),---

out by a servo-hydraulic fatigue testing machine under constant K control. The fatigue fracture surfaces show a combination of narrow and wide striation spaces corresponding to the standard and overload cycles, respectively, Figure 2.

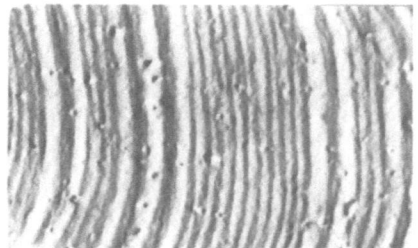

Fig. (2) - An example of modulated fatigue striations ($\Delta K = 12$ MPa$\sqrt{m}$).
Pattern format is 01-S3-02-S10, 01-S3---

Regarding the mean load, the modulation patterns are classified into three; constant K_{mean}, constant K_{min} and constant stress ratio, R. The ratio of overload used was 130% in all tests and it corresponds to about 200% striation spacing increase. Using the symbols "S" and "O" to represent the standard load and the overload cycles, respectively, the modulation pattern used was of the form: O-1,S-i, O-j,S-k (i=1~5, j=1~9 and k=1~10) with a period of 6,525. Simple binary system expresses longer period for a given number of observing striations than the above mentioned system but it was not obvious on the fractography. The period of 6,525 corresponds to the crack length of 0.1 ~ 1 mm in the stable fatigue crack growth range and was enough to identify the number of cycles for all striations of the fracture surface referring to the macroscopic fatigue crack length data. Three real time programs A, B and C were run simultaneously in the modulation pattern fatigue test. Recording and displaying of the data in the fatigue test were controlled by program A, and crack length measurement, load calculation for the constant stress intensity range and designation of the magnitude and phase of wave were accomplished by program B. The analog sine wave signal generated in program C consists of more than 100 samples and is very stable because program C has priority to programs A and B.

MATERIAL AND PROCEDURES

Most of the fatigue tests were performed on compact tension specimens (width W = 25 mm; thickness B = 6 mm) corresponding to the longitudinally transverse orientation (L-T). Some bending specimens (W = 6 mm; B = 10 mm) referred to the orientation (L-S) were also tested to study the influence of grain flow. The material used was 6 mm thick, 7075-T6 aluminum alloy plate which is known to exhibit fatigue striations. Shape and size of grains of 7075-T6 Al alloy on the fatigue crack propagating plane is shown in Figure 3 where the rolling direction was from left to right. The crack propagates from left to right in the compact tension

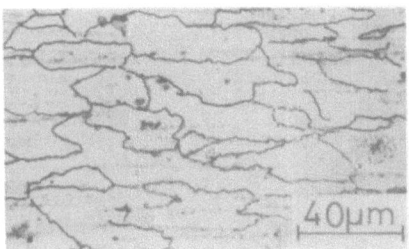

Fig. (3) - Shape and size of the grains on crack propagating plane.
7075-T6. Grain boundaries are emphasized

specimen, and from lower to upper side in the notched bending specimen. The length of the crack front coinciding with grain boundaries is probably longer in the notched bending specimen than in the compact tension specimen. Rows of inclusions were seen in the rolling direction regardless of the grain.

The tests were performed in laboratory air at 20°C and a frequency of 40 Hz. Reference and load spectra were monitored by an oscilloscope and were recorded in a computer memory (40 MB). Output of load cell in the modulation pattern test includes both standard and overload signals. The load is controlled occasionally during the standard wave is repeating for ten cycles. Macroscopic crack length was sensed by the electric potential drop method, calculated and recorded in the computer.

Since the depth of the fatigue striation was less than one hundredth of its spacing, scanning electron microscopy was unsuitable for the fractography observation in this experiment. Two stage chrome-carbon replica of the fracture surfaces was observed by a 200KV electron transmission microscope. Spherical particles of 0.312 μm diameter was set on the first stage replica and metallic chrome was spattered in an angle of 30° with the replica plane in the direction of macroscopic

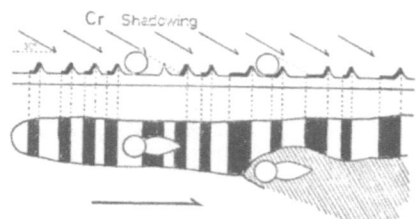

Fig. (4) - Two stage chrome-carbon replica. The relations
in the striation, image and the shadow of spherical particles

crack propagation. The shade of the particles was used to indicate the direction and local inclination of the facets; on the base plane, the length of bright part should be 1.4 times of the particle diameter. If the shade is longer than it, the observed facet tilts up in the crack growth direction. The number, n, of load cycles of a striation can be found either from a reference table or by computer processing. Most of the tests on compact tension specimens were performed with a con stant stress intensity range of 10 or 12 MPa√m (standard load), but the bending tests were done in the constant load range. Many high magnification TEM photograp were assembled. Figure 5 shows an assembled photograph which consists of 154 pictures with 4400 magnification. Figure 6 illustrates the profiles of the crack fronts in Figure 5. Solid lines connected with dotted lines indicate the striatio whose pattern format are labelled in the illustration. The crack line fronts were almost straight with grain size scale unevenness.

FRACTOGRAPHY AND DISCUSSION

Clear evidences of hesitation or acceleration of the fatigue crack propagation were observed only on a few grain boundaries and inclusions. There is stria tion invisible zones at the grain boundary frequently, but the apparent crack propagation rate is same as in the striation zone, Figure 7. It is not clear whether crack propagates continuously in this zone or not. In Figure 8, crack spent a few cycles at the diagonal line (arrow) and appears to hesitate around th grain boundary. This could be explained by crack branching such that it changes the direction of growth after a few cycles.

There are three types of interaction between the inclusion and the striation, Figure 9; the inclusion resists crack propagation (A), assists crack propagation (B) [3] and inclusion is broken before it meets with the crack front where seconda cracking is generated and joined with main crack that continues to propagate (C) | Inclusions in Figure 8 could be classified as type C but they were broken before the crack front has met with the inclusion. Clear reverse propagation of type C : far has not been observed. In Figure 10, the modulation pattern in the upper hal of the figure has the format 01-S1-03-S7 (n = 232~243), and lower half the format 01-S1-04-S6 (n = 332~343). That means crack in the upper half progagated about 100 cycles before main crack arrived due to breaking of the inclusions but the se ondary crack did not propagate in the reverse direction. This is the only observe

Fig. (5) - Assembled TEM photograph consisting with 154 pictures of x 4400.
Modulated striation are shown in about 30% of the field. ΔK
= 12 MPa√m. CT specimen

126

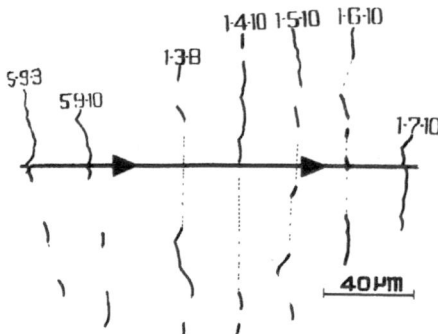

Fig. (6) - Profiles of progressing fatigue crack front. They are almost
perpendicular to the macroscopic crack propagating direction

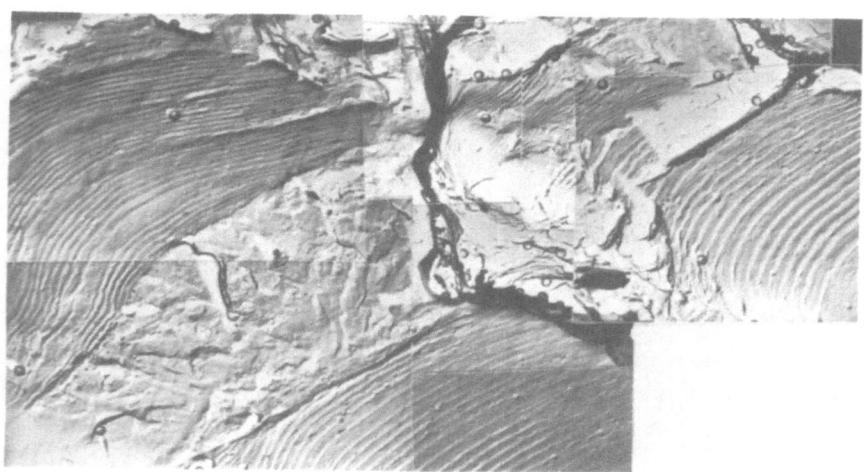

Fig. (7) - Striation invisible zone at the grain boundary. Apparent crack
propagation rate in this zone are same as the striationed zone.
(Notched bending specimen)

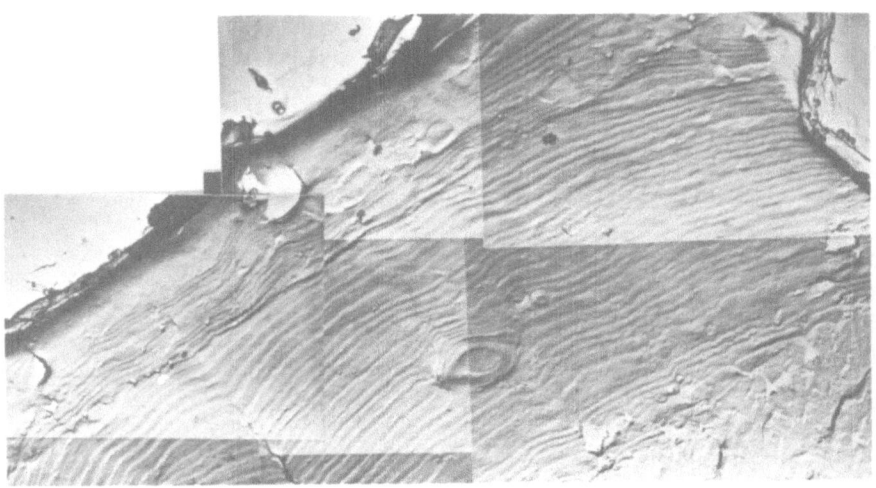

Fig. (8) - Crack seems to change the propagating plane (arrow) near the grain boundary. Preceded striations were shown around the broken inclusion (center). Crack propagated from lower right to upper left

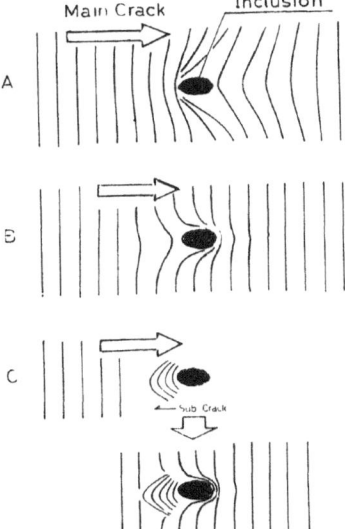

Fig. (9) - Interaction models between the inclusion and the striation

128

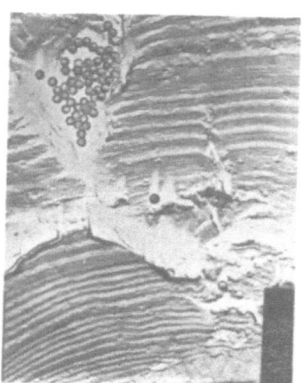

Fig. (10) - Preceded striations were observed around the broken inclusions in a notched bending specimen. The tire truck patterns were found in the lower right corner.

tion where striation preceded for more than 30 cycles. There are many tire track patterns at the lower right corner which are regarded as the result of the hard broken inclusion. Type A interaction were often observed. Inclusions in Figure 11

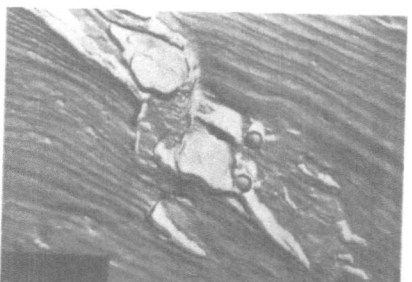

Fig. (11) - Delayed striation due to inclusions

resisted to the crack propagation. In a grain, the striations are convex in the propagating direction and grain boundaries orientated in that direction apparently tend to resist crack propagation, Figure 5.

According to the crystallographic analysis [5], the striation could not bend an angle of more than 30° in a crystal plane, but those curved more than 45° were often observed. Though the striations curved an angle of more than 60° in Figure 12, it could be a result of surface unevenness which can be accounted for by the length and direction of the shadow of the particles set on replica.

Observed modulation patterns in any grain showed good reproducibility. There was no clear trace of the fatigue crack retardation caused by 30% overloading. When

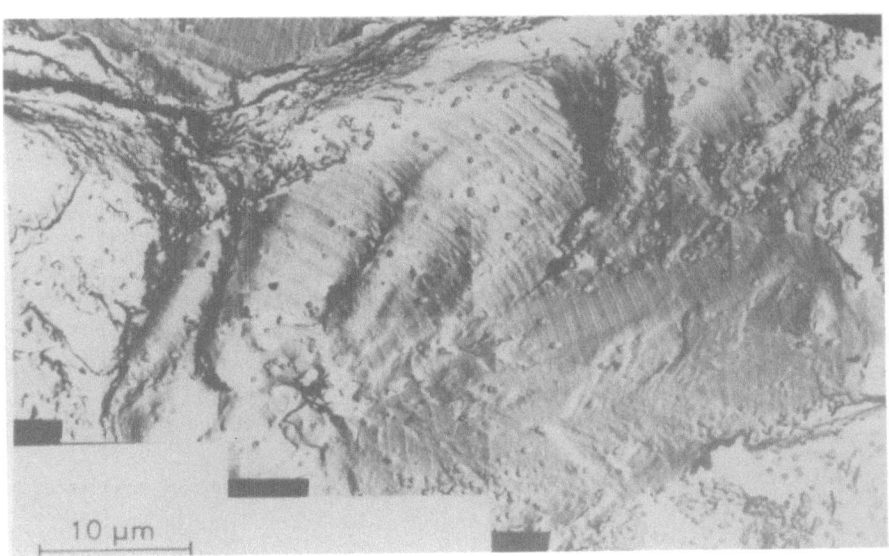

Fig. (12) - Striations look like to trace circular arcs, but they lie
over a few facets which tilt each other. Delay in upper
left facet seems to cause by difference in crystallographic
orientations. Crack propagates from left to right. CT
specimen. $\Delta K = 12$ MPa$\sqrt{m}$

the standard load cycles were less than 10 cycles, then no retardation was expected.
In 2124-T351 Al alloy, 30% overload caused retardation for more than 1000 cycles
after overloading with delayed retardation phenomena [6]. Using a few sets of the
numbers of the standard and the overload cycles N_{st}, N_{ov}, and the length of crack

propagated in a grain L, the crack propagation rate for the standard load cycles,

$(\frac{d\ell}{dN})_{st}$, and overload cycles $(\frac{d\ell}{dN})_{ov}$ were calculated by linear regression of equation
(1).

$$L = (d\ell/dN)_{st}N_{st} + (d\ell/dN)_{ov}N_{ov} \tag{1}$$

At $\Delta K = 12$ MPa$\sqrt{m}$, $(d\ell/dN)_{st} = 120$ nm and $(d\ell/dN)_{ov} = 258$ nm respectively and m
$= 2.9$ (constant K_{mean} test). The exponent m agrees well with common test data,
but the propagating rates were about a quarter of them. Furthermore, the fatigue
life estimated using common test data for a notched bending specimen (bending moment range $= 5.2$ Nm, B $= 10$ mm, W $= 6$ mm and l $= 1$ mm) was less than 5000, but the
experimental fatigue life of this specimen was 18,650 cycles. Therefore, the retardation should be considered in the modulation pattern analysis.

CONCLUSION

It has been shown that fatigue crack propagation tests under program loads have yielded a better understanding of the micro behavior of fatigue cracking. The main findings are as follows:

1. The crack front line was almost straight accompanied with grain scale un-evenness.

2. Clear evidences of hesitation or acceleration of the fatigue crack propagation were observed only on a few grain boundaries and inclusions.

3. Grain boundary seems to prevent crack propagation; a few crack fronts hesitate near the grain boundary and the striations were convex in the direction of crack propagation.

4. Inclusions did accelerate or resist the crack propagation depending on their mechanical properties.

5. Crack propagation caused direction reversal due to secondary cracking ahead was not observed.

6. Retardation of fatigue crack propagation was observed for 30% overloaded program load.

REFERENCES

[1] McMillan, J. C. and Pelloux, R. M., ASTM STP 415, p. 505, 1967.

[2] Hertzberg, R. W., ASTM STP 415, p. 202, 1967.

[3] Kitagawa, H. and Oterazawa, R., Fractography, p. 79.

[4] Kobayashi, K. et al, Tetsu to Hagane, 64, p. 1072, 1978.

[5] Nunomura, S. and Fukui, Y., Jr. JIM., 4, p. 405, 1977.

[6] Robin, C. and Pelloux, R. M., Mat. Sci. Eng., 44, p. 115, 1980.

AN EXPERIMENTAL STUDY ON THE FRACTURE BEHAVIOR OF LAMINATED STEEL COMPOSITES

Y.-H. Yum and C.-K. Shin

Seoul National University
Sinlim-Dong, Kwanack-Ku, Seoul 151, Korea

ABSTRACT

Laminated steel composites are made from spring steel and mild steel by roll-ing process. Experiments for fatigue and impact test are carried out and the fracture behavior of laminated steel composites are compared with that of homo-geneous steel. Also, fracture models of laminate composites are analyzed, finite element method and the computed fracture stress is compared with experimental re-sults.

INTRODUCTION

Recent development on the science and technology has brought significant in-novation into various engineering fields and especially studies on composites are closed up in engineering materials. The development of the homogeneous materials is limited and they could not have enough merits required to satisfy such condi-tions as anti-corrosion, heat resistance and high strength, etc. So many engineers [1-5] are interested in laminated composite materials to compensate some of the deficiencies of homogeneous materials.

The laminated composites not only have good merits as clad material or heat resistance material but also have good merits in strength, fatigue or impact. But, it is very difficult to obtain adequate information for laminated steel composites.

In this study, laminated steel composites are bonded with spring steel and mild steel by rolling mill, and specimens are provided the arrester type crack for fatigue and impact bending. Investigation is carried out on fracture behaviors on crack propagation rate, stress intensity factor, delamination process of lami-nated layer, impact energy-time curve under low temperature and also elastic stress distributions at neighborhood of crack are computed by finite element method.

Fracture behavior and characteristics of laminated steel composites are com-pared with those of an homogeneous steel. The results show that laminated compos-ites displayed significant merits in fatigue and impact load.

EXPERIMENTAL EQUIPMENT AND TESTING METHOD

A. The Specimen

1. Laminated Composite Specimen

The specimens in this study are homogeneous materials such as spring steel (SUP 9), mild steel (SS41) and some laminated steel composites. The laminated composites are composed of three layered (3 SMS), five layered (5 SMS) and seven layered steel (7 SMS).

Chemical composition of specimen is shown in Table 1. The specimens were made by rolling process at 1200°C and the specimens for the fatigue test

TABLE 1 - CHEMICAL COMPOSITION

Material	Composition (%)						
	C	Si	Mn	P	S	Cr	Cr
Spring Steel	0.56	0.29	0.90	0.025	0.04	0.72	
Mild Steel	0.21	0.32	0.45	0.021	0.007		
Copper							99.96

were quenched at 850°C and tempered at 550°C. And those for the impact test were annealed at 850°C. Their mechanical properties are listed in Table 2.

TABLE 2 - MECHANICAL PROPERTIES

Material	Tensile Strength (kg/mm²)	Yield Strength (kg/mm²)	Elongation (%)	Rockwell Hardness (HRC)
Impact Test	38.9 98.9	26.1 65.9	30.4 19.1	40.3 58.6
Fatigue Test	65.2 132.5	49 117	18.2 8	

Figure 1 shows shape of fatigue specimen and Type A is for fatigue life and Type B and types C for crack propagation. Table 3 is cross section of laminated layer for fatigue and impact specimen.

2. Specimen of F.E.M. Model

Figure 2 shows the shape of specimen for finite element method and Table 4 shows the shape and symbol of finite element method computation.

B. Experimental Equipment

The fatigue test was carried out by Schenk's type bending fatigue tester (20 kg-m) and the impact test was carried out by some equipment such as Olsen im-

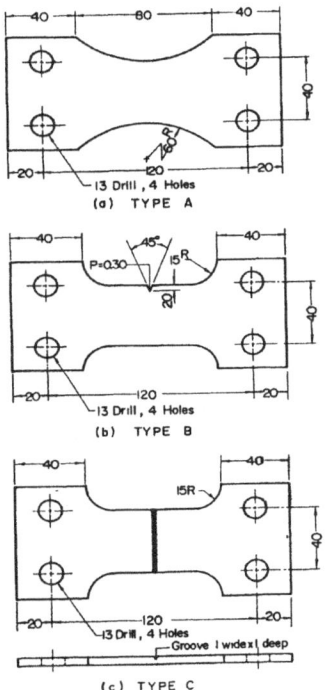

Fig. (1) - Shapes and dimensions of fatigue specimens

pact testing machine, DPE-6E type strain amplifier, storage type Oscilloscope, low temperature apparatus (20°C-180°C) and optical microscope.

C. The Testing Method

1. Fatigue Test

The specimens were tested with the assumption that they were only pure bending stress. The length of the fatigue crack was measured with an optical microscope having a micro-comparator with the precision 0.01 mm. Repeating this procedure, we could measure the total length of the fatigue crack.

2. Impact Test

Impact energy and impact load-time curves are recorded by strain gage method to obtain dynamic fracture behaviors. During impact, although the speci-

TABLE 3 - CROSS SECTION OF SPECIMEN

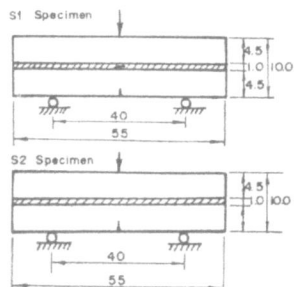

| Material | Fatigue | Impact Specimen | |
(Symbol)	SPecimen	Notched (10 x 10 x 55.6)	Unnotched (10 x 5 x 55.6)
SUP 9	58	10	50
SS 41			
3 SMS		3.5 3.0 3.5	2.0 1.0 2.0
5 SMS	1.6 8 1.0 8 1.6	2.4 1.0 1.0 1.0 2.3	1.2 0.7 1.2 0.7 1.2
7 SMS	1.2 6 .8 6.8 .6 1.2	2.5 7 8 7 8 7 1.8	0.9 0.5 8.5 8.5 0.5 0.9

Legend : ▨ Spring Steel, SUP 9   ▭ Mild Steel, SS 41

S1 Specimen
4.5
1.0 10.0
4.5

40
55

S2 Specimen
4.5
1.0 10.0

40
55

Fig. (2) - Shapes of specimen for F.E.M. calculation

mens are failed by the elastic or plastic deformation due to the impact load, dynamometer [6] of Figure 3 displays always the elastic deformation. The impact
test was carried out at six states such as 18°C, 0°C, -40°C, -80°C, -120°C and
-160°C. Figure 4 shows impact testing assembly of low temperature.

3.   An Analysis for Fracture Model by using F.E.M.

In this paper, the Chiles Program [7] is used, which was proposed in
1973 and can calculate plane strain, displacement and strain, displacement and
stress intensity factor for an elastic fracture mechanics.

Difference between computed results and experimental result is within
5% in case of very small plastic region.

TABLE 4 - SHAPES AND SYMBOLS OF USED SPECIMENS

▨ : Cu
□ : SPS 5

Sort of Crack / Crack Length / Material / Shape Symbol		2 mm	3.5 mm	5 mm	6 mm	2 mm	3.5 mm	5 mm	6 mm
			S 1				S 2		
"A"	Symbol	SI-1A	SI-2A	SI-3A	SI-4A	S2-1A	S2-2A	S2-3A	S2-4A
S Cu S	Shape	▭	▭	▭	▭	▭	▭	▭	▭
"B"	Symbol	SI-1B	SI-2B	SI-3B	SI-4B	S2-1B	S2-2B	S2-3B	S2-4B
Cu S Cu	Shape	▭	▭	▭	▭	▭	▭	▭	▭
"C"	Symbol	SI-1C	SI-2C	SI-3C	SI-4C	S2-1C	S2-2C	S2-3C	S2-4C
SPS 5	Shape	▭	▭	▭	▭	▭	▭	▭	▭
S 3		S3-1	▭	S3-2	▭	S3-3	▭	S3-4	▭

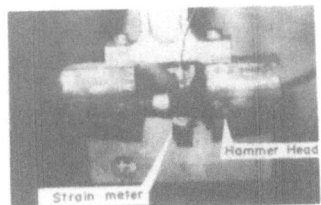

Fig. (3) - Dynamometer

Fig. (4) - Impact testing assembly of low temperature

It is considered how fracture behaviors observed in experiment for laminated composites are influenced by shape, composition and stress distribution at crack tip and crack propagating into specimen as shown in Figure 2, Table 4.

EXPERIMENTAL RESULTS AND DISCUSSION

A. Fatigue Fracture Behavior

1. Relationship Between Crack Propagation Rate da/dN and Stress Intensity Factor K

Due to little knowledge about the crack propagation law for laminated composites, we adopt conveniently the stress intensity factor K satisfying

$$K = 1.005 \ \sigma\sqrt{a} \qquad (1)$$

which was proposed by Bowie [8] for the plane bending problem for one side notched homogeneous materials.

Plotting the experimental data in log-log coordinates with assumption that crack growth rate da/dN is a function of stress intensity factor K, their relationships are plotted as straight lines as shown in Figure 5 and can be represented in the form of equation (2) below, suggested by Paris

136

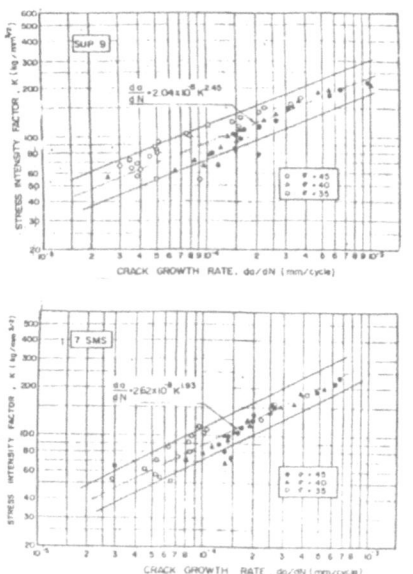

Fig. (5) - Crack growth rate as a function of K

$$da/dN = CK^m \tag{2}$$

where C is material constant and m is exponential constant of K.

$$da/dN = 2.05 \times 10^{-8} K^{2.45} \text{ (SUP9)}$$
$$da/dN = 3.50 \times 10^{-8} K^{1.86} \text{ (5SMS)} \tag{3}$$
$$da/dN = 2.62 \times 10^{-8} K^{1.93} \text{ (7SMS)}$$

Considering equation (3), it can be observed that crack growth rate of laminated composites is lower than that of spring steel SUP9.

This phenomenon can be explained as follows: Crack propagation is dominated by the plastic stress field in the neighborhood of crack front and each layer surface prevents crack from propagation, though laminated composites have lower yield strength than that of spring steel SUP9. Therefore, crack propagation speed decreases and the development of yield in the crack front delays. This tendency coincides with previous report.

2. Fatigue Limit Curve

Fatigue test was carried out with specimens as shown in Figure 1 (A type) and results are represented by fatigue stress-cycle curves in Figure 6.

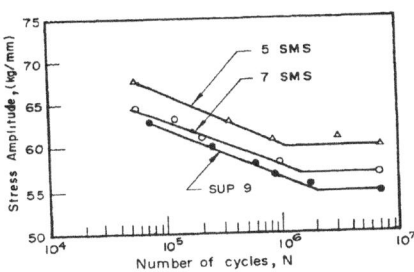

Fig. (6) - Fatigue stress-cycle curve

Fatigue limits for specimens of 5SMS, 7SMS and SUP9 are obtained 60 kg/mm^2, 57 kg/mm^2 and 54.9 kg/mm^2, respectively, according to fatigue stress-cycle curves. Thus, laminated composites have a little higher fatigue limit than those of homogeneous materials.

3. Fatigue Crack Propagation for Arrester Type Crack

Figure 7 is shown a model of interior crack propagation for arrester type 5SMS specimen in Figure 1 (C type). This was observed by optical microscope

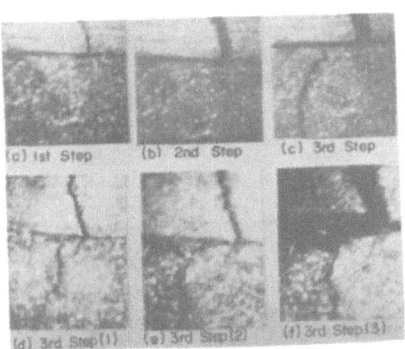

Fig. (7) - Fatigue crack propagation

under the plane bending stress 25 kg/mm^2. The illustrating process of crack propagation through the interface of layers is shown in Figures 7 and 8.

As shown in Figure 7 and Figure 8, the process of crack propagation from specimen surface along the vertical direction for 5SMS can be divided as

138

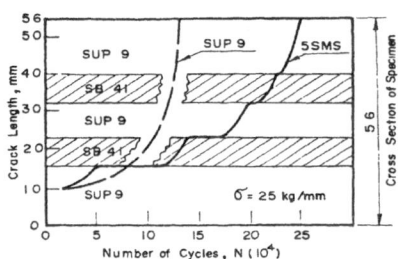

Fig. (8) - Crack length versus cycle curves for 5SMS

follows:

        First, a crack occurs in the notch root and propagates toward the
next layer surface;

        Second, delamination takes place in the layer surface between spring
steel SUP9 and mild steel SS41;

        Third, the crack tip comes to a layer surface;

        Fourth, a new crack occurs on any point of next layer surface.

        It can be observed that fracture propagated and occurred in the direc-
tion of thickness through the four steps above.

        From these results, if the delamination takes place just before crack
tip reaches the layer surface as a crack propagates along the vertical direction,
then energy for crack propagation increases because of the delamination of layers
and the removal of stress concentration; hence, a fatigue life increases.

    B.   Impact Fracture Behavior at Low Temperature

        1. Impact Energy at Low Temperature

        For a temperature region between 18°C and -160°C, the impact energy of
notched and unnotched specimens is measured and shown in Figure 9. In Figure
9(a) for notched specimens, the impact energy of homogeneous specimens and laminated composites is nearly the same value of 3.67 kg-m at room temperature.

        As temperature lowers, the impact energy of homogeneous mild steel
SS41 and spring steel SUP9 decreases abruptly between 10°C and 0°C and it reduces
to 1.2 kg-m at -40°C. But the impact energy of 7SMS is relatively large value of
3 kg-m at -40°C and it decreases abruptly between -40°C and -160°C and it reduces
to 1 kg-m about -120°C.

        In Figure 9(b) for unnotched specimens, the impact energy is 28 kg-m
at 18°C, which is much larger than those of the notched specimens. As temperature

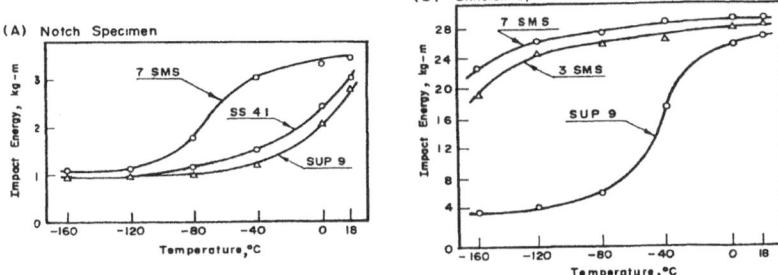

Fig. (9) - Impact energy-temperature curves for notch and unnotched specimens

lowers, while impact energy of spring steel SUP9 decreases between 18°C and -60°C, that of 3 SMS and 7 SMS does until -140°C.

Impact energy of laminated composites is influenced by temperature, material components and bonding force. In general, if delamination occurs in the case of arrester type, impact energy tends to increase.

2. Microscopic Fracture Behavior under the Impact Loading

The fracture behaviors under the impact loading velocity 2.5 m/sec were observed in microscopic fractography. The crack propagation mode of specimen is shown in Figure 10. It is shown that the crack occurs at the notch root

Fig. (10) - Microscopic impact bending fracture mode of 5SMS, temperature; -40°C velocity; 2.2 m/sec

140

and it does not propagate continuously into the interior of specimen. The layer surface is delaminated before the crack tip arrives at layer surface of the specimen. The crack propagation mode in Figure 11 is shown that the phenomenon of de-

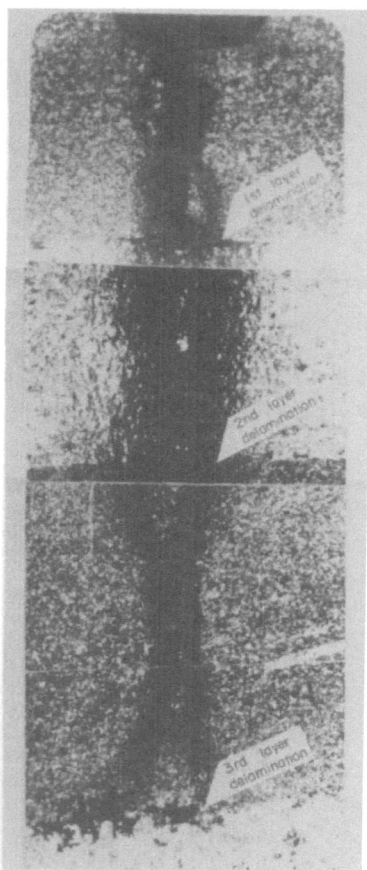

Fig. (11) - Microscopic impact bending fracture mode of 5SMS temperature; -40°C, velocity; 2.5 m/sec

lamination occurred on first and second layers, and delamination crack slightly on third layers before a crack arrives at a layer surface.

The process of crack propagation as shown in schematic Figure 12 can be explained as follows: The first step is crack initiation at the notch root and crack propagation, the second step is a new crack initiation at a layer sur-

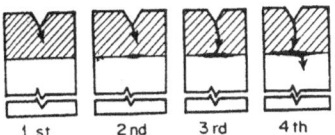

1 st   2 nd   3 rd   4 th

Fig. (12) - Illustration of crack propagation mode of laminated impact specimen

face and delamination of a layer surface, the third step is conjunction between a crack created at the notch root and delaminated layer and the fourth step is a new crack initiation at a delaminated layer. A delaminated layer is fractured as preceding above 1-4 steps. But it is predicted that a new crack initiation at layer surface is more difficult than at the notched layer because of smoothness of delaminated layer surface and that impact energy for laminated composites increases due to large crack initiation energy containing delamination energy.

It is observed through microscopic structure that ferrite layer of highly lower strength than that of parent metal was produced and void, impurity and inclusion were accumulated on the vicinity of layer surface, as shown in Figures 10 and 11. Delamination crack was observed at weak ferrite layer.

3. Impact Load-Time Curve

Charpy's impact specimens of spring steel SUP9 and laminated composite 5SMS at -40°C were tested at impact hammer velocity of 4.4 m/s and the relationship between load and fracture time was obtained through the storage type oscilloscope as shown in Figure 13.

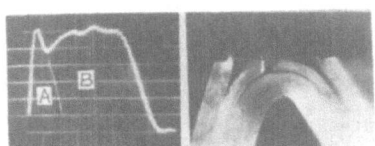

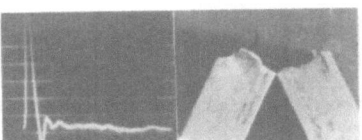

E = 2.63 kg-m, deflection = 2.49 mm   E = 17.13 kg-m, deflection = 15.19 mm

(a) Fracture of homogeneous material   (b) Delaminated fracture of 2 layers

Fig. (13) - Shape of failed notched impact specimen and instrumented impact load-time curves under -40°C

Impact energy was determined by indicating scale and impact load measure by measuring the magnitude of vertical axis scale of oscilloscope. Calibration of impact load was obtained by static test and the latter was roughly calculated by the multiplication of the length of impact time, which appeared on horizontal axis of oscilloscope, and average velocity of impact hammer in fracture.

Fracture energy in Figure 13(b) due to fracture of first layer is represented in the region of A, and immovability of crack due to delamination of layers, new crack initiation and added mode of energy by propagation are represented in the region of B.

Consequently, if the layer is delaminated before crack tip arrives at a layer surface as shown in Figure 13(b) such as laminated composite steel, specimen of arrester type has more impact energy, 17.13 kg·m, due to delamination and unnotched effect of the crack tip, while impact energy in Figure 13(a) is 2.63 kg·m.

It is recognized that due to delamination of layer surface and unnotched effect of a layer surface laminated steel composites has larger impact energy than homogeneous spring steel. Figure 13(a) is represented the applied impact load-time curve of oscilloscope photograph in fracture not to be delaminated.

STRESS ANALYSIS AT CRACK TIP BY USING F.E.M.

A. Stress Distribution at Crack Tip

We want to investigate the fracture behavior for laminated composites, S1 and S2 as shown in Table 3 and to calculate the stress distribution by using F.E. M. in case of arrester type crack. Because it is difficult to compute and analyze for the crack propagation continuously, the crack was made at positions that the distance from the lower side of specimen is 2 mm, 3.5 mm, 5 mm and 6 mm and the concentrated load of 200 kg, 140 kg, 80 kg and 40 kg was applied individually so that stress distribution was calculated by Chiles F.E.M. program.

1. S1 Specimen

Stress distribution at the neighborhood of crack tip and midpoint of specimen for S1 specimen in represented in Table 5 and stress distribution for ScuS is shown in Figure 14 when a crack propagates to vertical direction. Considering fracture behavior during crack propagation, in the case that the length of crack is 2 mm, longitudinal stress is higher than normal stress in whole region.

Therefore, tearing crack only occurs. As the length of crack increases to 3.5 mm, normal stress was higher than longitudinal stress for S1-2A and S1-2C specimens. When the crack tip comes to midpoint of specimen, normal stress is higher than longitudinal stress at front of crack tip for S1-3C specimens. Hence, it is easy to be delaminated. But, in the case of S1-3B specimen, longitudinal stress was higher than normal stress. Especially normal stress for specimen A whose upper and lower material is brittle was higher than that for B and C specimens and specimen A was more easily delaminated.

TABLE 5 - STRESS DISTRIBUTION OF S1 SPECIMEN

Specimen	Load (Kg)	Sort of Stress (Kg/mm²)	Horizontal Distance (mm)	Vertical Distance (mm)	A	B	C
S1-1	200	Longitudinal Stress	0,25	2,25	129,8	123,6	152,3
			0,25	5,25	8,7	7,4	9,72
		Normal Stress	0,25	2,25	107,8	104,4	105,1
			0,24	5,25	2,6	2,0	1,3
S1-2	140	Longitudinal Stress	0,25	3,75	125,4	117,4	122,8
			0,25	5,25	33,8	39,4	37,9
		Normal Stress	0,25	3,75	113,1	115,0	114,9
			0,25	5,25	39,7	38,2	39,7
S1-3	80	Longitudinal Stress	0,25	5,125	131,7	198,9	164,5
			0,25	5,375	150,3	184,0	173,7
		Normal Stress	0,25	5,125	152,0	194,7	173,9
			0,25	5,375	130,0	139,1	137,6
S1-4	50	Longitudinal Stress	0,125	5,125	2,5	2,5	
			0,125	6,125	138,9	140,9	
		Normal Stress	0,125	5,125	27,4	18,3	
			0,125	6,125	153,8	158,8	

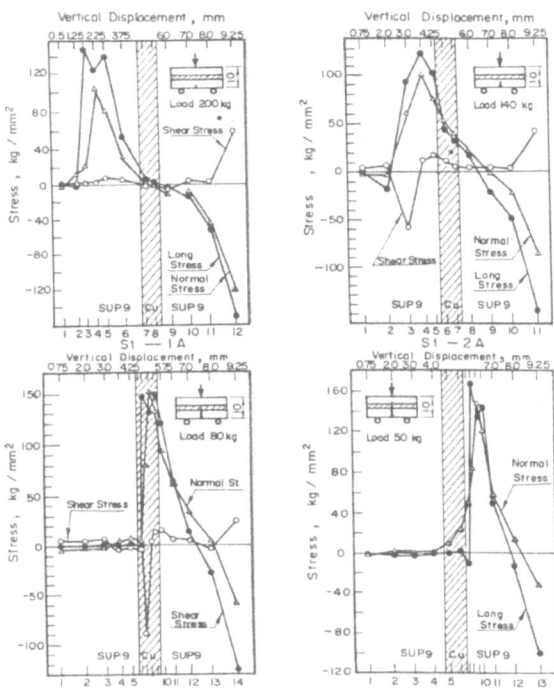

Fig. (14) - Stress distribution along the vertical crack tip direction for S1 specimen

2. S2 Specimen

Stress distribution at neighborhood of crack tip and midpoint of specimen for S2 specimen is represented in Table 6 and stress distribution is

TABLE 6 - STRESS DISTRIBUTION OF S2 SPECIMEN

Specimen	Load (Kg)	Sort of Stress (Kg/mm²)	Horizontal Distance (mm)	Vertical Distance (mm)	A	B	C
S2-1	200	Longitudinal Stress	0,25	2,25	131,1	125,0	128,3
			1,125	5,125	10,8	22,5	16.4
		Normal Stress	0,25	2,25	108,0	104,6	106,2
			1,125	5,125	8,5	13,1	-0,2
S2-2	140	Longitudinal Stress	0,25	3,75	139,2	126,9	129,4
			1,125	5,125	50,3	93,2	70,1
		Normal Stress	0,25	3,75	104,6	109,6	117,9
			1,125	5,125	74,9	98,7	85,4
S2-3	80	Longitudinal Stress	0,25	5,125	163,2	262,1	200,1
			1,125	4,875	25,7	123,1	91,3
			1,125	5,125	15,8	201,6	154,4
		Normal Stress	0,25	5,125	-0,2	3,29	2,59
			1,125	4,875	11,3	131,2	117,7
			1,125	5,125	7,9	101,5	93,9
S2-4	50	Longitudinal Stress	0,25	5,125	0,0003	0,0003	
			1,125	4,875	29,4	39,2	
			1,125	5,125	44,3	61,2	
			1,125	5,875	66,1	60,4	
			1,125	6,250	72,6	67,4	
		Longitudinal Stress	0,25	5,125	0,0001	0,0002	
			1,125	4,875	51,5	56,2	
			1,125	5,125	42,7	50,2	
			1,125	5,875	105,5	10,9	
			1,125	6,250	3,96	4,96	

shown in Figure 15 when a crack proceeds to vertical direction. Considering fracture behavior during crack propagating, in case of S2-1 specimen whose crack length was 2 mm, longitudinal stress at front of crack tip was higher than normal stress. Hence, tearing crack occurred. But in case of 3.5 mm crack length, longitudinal stress as higher than normal stress at neighborhood of crack tip for S2-2A and S2-2B specimens and tearing crack propagated at position that the vertical distance from the lower side of specimen is 5.125 mm, normal stress for S2-2A and S2-2B specimen was higher than longitudinal stress and they were easily delaminated at midpoint of specimen.

In case of 5 mm crack length longitudinal stress at front of crack tip was higher than normal stress and tearing crack occurred and was propagated, but normal stress at rear of crack tip, vertical distance 4.875 mm, was higher than longitudinal stress and delamination occurred. It is considered that delamination length is limited and stopped because normal stress in horizontal direction at crack tip decreased quickly. In case of 6.0 mm crack length, normal stress at midpoint of specimen was larger than longitudinal stress and small delamination occurred, but at front of crack tip, longitudinal stress was higher than normal stress and tearing crack could be propagated.

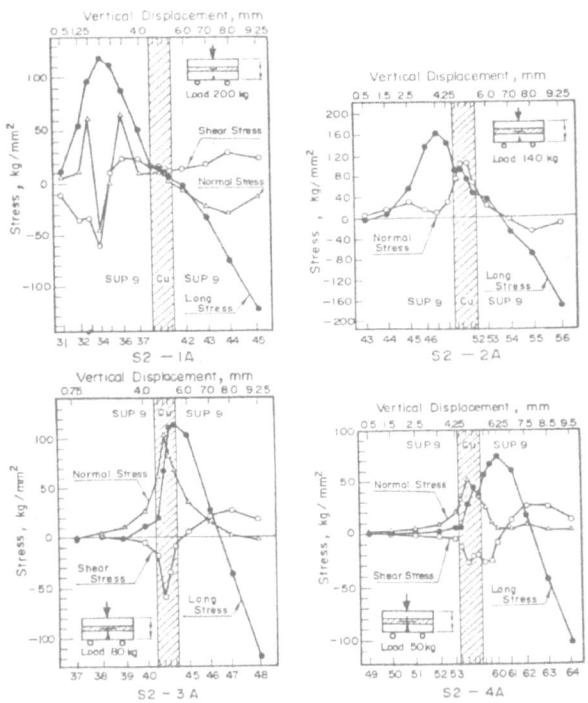

Fig. (15) - Stress distribution along the vertical direction through the
horizontal crack tip for S2 specimen

As the results of specimens of S1 and S2, when arrester type crack
occurs in the lower side of specimen and propagates into interior specimen, layer
of S2 is more easily delaminated than layer of S1, but the delaminated horizontal
crack concentrates stress and has a bad effect on fracture property. In order to
prevent a crack propagating, it is more effective to let delamination occur just
before a crack tip arrives at a layer surface.

B.   The Comparison of Experimental Results and Calculated Results by Finite
Element Method

The fracture behavior of an arrester type for laminated composites are
studied by fatigue test, impact test and F.E.M. simulation of the stress distribution at crack tip.

To summarize the result above; laminated steel composites bonded through the process of hot rolling produces a ferrite layer surface. When an arrester type crack tip propagates vertically to layer surface, delamination occurs at a weak ferrite layer before a crack tip arrives at a layer surface as shown in Figures 10 and 11.

The reason is obvious from the F.E.M. simulation as shown in Table 5 and Table 6, Figures 14 and 15, which says that, when a specimen is subjected to the center concentrated load, sufficient stress to delaminate is generated there before a crack reaches at a layer surface, since normal stress at the near front region at a crack tip is larger than the longitudinal stress and the interior laminated composites is weak.

In this case, required energy for a new crack to initiate is bigger after delamination than before the delamination because, with energy absorbed, the phenomenon of stress concentration at crack tip is vanished by new surface due to layer delamination as shown in Figure 13(b) (impact load-time curve).

CONCLUSIONS

Results from the fracture test and the calculated results by using F.E.M. for laminated composites are as follows:

1. In Paris' relationship between fatigue crack propagation rate and stress intensity factor, $da/dN = CK^m$, the obtained value m of laminated composites (m = 1.86 - 1.93) is lower than that of homogeneous steel (m = 2.45).

2. In spite of higher tensile strength of homogeneous spring steel, laminated steel composites have higher failure limit and impact energy than homogeneous spring steel.

3. By analyzing the fracture behavior, it was found that the crack propagation is delayed in laminated composites by delamination process of interior layers in comparison to homogeneous steels.

4. In impact test for a temperature region between 20°C and -160°C, the impact energy for homogeneous material SUP9 and SS41 decreases abruptly between 20°C and -20°C, but the impact energy is relatively higher value at -40°C for notched arrester type laminated composites and it is even higher at -120°C for unnotched laminated composites.

5. High fracture energy for laminated steel composites results in delamination energy which is required in order to delaminate a ferrite layer surface produced in hot rolling, when an arrester type crack propagates into a specimen, and results in a new crack initiation energy on delaminated layer surface.

6. The reason for delamination is obvious from the results of the F.E.M. calculations. When an arrester type crack occurs in the lower side of specimen and propagates into laminated composites, that is, as a crack tip arrives at a layer surface, normal stress is higher than longitudinal stress. Moreover, if once the interior crack initiates, then it is easily delaminated.

REFERENCES

[1] Chen, E. P. and Sih, G. C., "Interfacial delamination of a layered composite under anti-plane strain", J. Composite Materials, Vol. 5, p. 12, 1971.

[2] Embury, J. D., Petch, N. J., Wraith, A. E. and Wright, W. S., "The fracture of mild steel laminates", Trans. Met. Soc. AIME, 239, p. 114, 1967.

[3] Devine, T. M., Floreen, S. and Hayden, H. W., "Fracture mechanics in maraging steel-iron laminates", Eng. Fracture Mechanics, Vol. 6, p. 315, 1974.

[4] Whitney, J. M. and Nuismer, R. J., "Stress fracture criterion for laminated composites containing stress concentration", J. Composite Materials, Vol. 8, pp. 255-265, 1974.

[5] Pagano, N. J., "On the calculation of interlaminar normal stress in composite laminate", J. Composite Materials, Vol. 8, p. 65, 1974.

[6] Yum, Y. H., "Study on the mechanical properties of unnotch and notch charpy impact specimen", Iron and Steel Japan, 51, p. 2056, 1965.

[7] Benzley, S. E. and Beisinger, Z. E., "A finite element computer program that calculates the intensities of linear elastic singularity CHILES manual", 1973.

[8] Bowie, O. L., "Rectangular tensile sheet with symmetric edge cracks", Journal of Applied Mechanics, 31, pp. 208-212, 1964.

[9] Paris, P. C. and Erdogan, F., "A critical analysis of crack propagation laws", ASME, Trans. Ser. D, Vol. 85, p. 528, 1963.

INFLUENCE OF MICROSTRUCTURES ON CRACK PROPAGATION IN STAINLESS STEEL WELD METAL

G. Venkataraman, S. Arunagiri and A. Srinivasulu

Bharat Heavy Electricals Limited
Tiruchirapalli 620014, India

ABSTRACT

In certain welded fabrications, austenitic stainless steel E309 Nb is over-layed on ferritic steel SA 515 Grade 70 as a buffer layer for subsequent overlay with E347 grade filler material. This was subjected to Stress Relief Heat Treat-ment (SRHT) cycle with holding at 600°C for 12 hours. E309 Nb weld metal showed rapid crack initiation and propagation in 180° bend tests. To study the influ-ence of microstructure on crack initiation and propagation, bend deformation stage was used in SEM. The metallographically polished and etched specimens were fa-tigue precracked and bent in the deformation stage. The crack propagation was preferential along delta ferrite and austenite boundaries, even though the path was longer. Whereas in 347 weld, random crack propagation with appreciable crack tip blunting was seen. However, these two materials without SRHT showed random crack propagation with reference to microstructure and crack tip blunting was simi-lar. X ray diffraction studies revealed appreciable transformation of delta fer-rite to secondary phases in the case of 309 Nb. Detailed thin foil studies in Transmission Electron Microscope revealed selective needle-like precipitation of secondary phases along the interface of delta and gamma phases of 309 Nb. This weakened the interfacial strength. This was the reason for easy crack propagation in 309 Nb after SRHT. Whereas such precipitates were not observed in 347 weld after SRHT and hence it was resisting crack propagation by crack tip blunting.

INTRODUCTION

It is well-known that the fracture process in metallic materials is sensitive to the microstructure which is influenced by factors such as composition and ther-mal-mechanical processings. For the material technologist aiming at the improved fracture resistance of metallic material, a techno-economical and faster method of material development combining both the aspects, i.e., fracture study and micro-structural response is very much essential. An independent study of fracture me-chanics using bulk samples and a parallel study of microstructure using separate samples will demand elaborate working schedule for establishing the mechanisms of fracture in materials having multiple phases with complex shape and distribution, for example, stainless steel weld metal [1]. Earlier work combining fracture process and microstructure by using SEM deformation stage, proved very useful in understanding and controlling the fracture process in pressure vessel steels [2-4]. A typical in-depth study of fracture process as related to microstructures in stainless steel weld metal 347 and 309 Nb will be discussed in this paper.

MATERIALS AND INITIAL TESTS

Four batches of austenitic electrodes, type 309 Nb ϕ 3.25 mm and 4 mm and types 347 ϕ 3.25 mm and ϕ 4 mm were subjected to consumable qualification test in accordance with SFA 5.4 of ASME Section II Part C. Base metal used for welding was carbon steel conforming to SA 515 Gr 70. The parameters used for welding with these electrodes are given in Table 1.

TABLE 1 - WELDING PARAMETERS

S. No.	Electrode	Dia. mm	Volt	Amp	Preheat °C (max)	Interpass Temp °C (max)
1	309 Nb	3.25	26-28	90-100	60	120
2	309 Nb	4.0	26-28	110-130	60	120
3	347	3.25	28	90-110	60	120
4	347	4.0	28	120-140	60	120

The test coupons were subjected to a temperature of 600 ± 10°C for 12 hrs., the rate of heating and cooling above 300°C being 50°C/hr. max. After simulated heat treatment, nondestructive tests were carried out to ensure that the test plates could be used for mechanical tests and they were found to be satisfactory. The following tests were carried out: chemical analysis, tensile test of undiluted weld metal, transverse side bend test and delta ferrite measurements. The results of chemical analysis and other test data are given in Tables 2 and 3 respectively.

TABLE 2 - RESULTS OF CHEMICAL ANALYSIS

Electrode	Dia. mm	C	Mn	Si	S	P	Cr	Ni	Nb	Co
309 Nb	3.25	0.04	0.92	0.51	0.006	0.01	23.20	11.90	0.33	<0.05
309 Nb	4.0	0.04	0.94	0.52	0.008	0.01	23.25	11.70	0.32	<0.05
347	3.25	0.04	0.99	0.50	0.007	0.01	19.80	10.10	0.47	<0.05
347	4.0	0.04	0.91	0.52	0.006	0.01	19.90	9.50	0.32	<0.05

The data for 309 Nb before heat treatment was furnished for ϕ 4 mm electrode only as this was similar to ϕ 3.25 mm electrode. For 347 weld deposit, the mechanical tests before heat treatment were not conducted since the properties were satisfactory even after heat treatment. In the case of type Nb ϕ 4 mm electrode test plate, the side bend specimens were broken into two pieces after bending to an angle of 30°. In other case of failure, severe openings initiated at the fusion boundary were observed. The all weld tensile test also showed poor ductility. The delta ferrite measurements were made using Magne Gauge instrument calibrated according to AWS 4.2.

Charpy impact tests at room temperature were carried out using specimens with 5 mm U notch in the simulated heat treated condition. From the test coupons, impact specimens were prepared for type 347 and 309 Nb electrodes. The impact energy values obtained were 35.3, 35.3, 39.2 Joules and 2.0, 3.9, 2.9 Joules respectively for types 347 and 309 Nb electrodes.

TABLE 3 - PROPERTIES OF WELD DEPOSITS

S. No.	Electrode	Condition	Y.S Mpa	T.S Mpa	%El	%RA	Side Bend 4t, 180	Ferrite Number
1	309 Nb φ3.25 mm	After HT	662.9	749.2	10.4	6.0	Failed	6.6
2	309 Nb φ4.0 mm	Before HT After HT	495.7 651.2	623.8 718.8	33.0 13.3	53.0 17.0	OK Failed	14.1 6.0
3	347 φ3.25 mm	Before HT After HT	- 488.5	- 644.3	- 32.0	- 45.0	- OK	9.6 7.0
4	347 φ4.0 mm	Before HT After HT	- 488.5	- 650.1	- 28.0	- 58.0	- OK	9.9 8.5

SEM DEFORMATION STUDIES

With the objective of tracing out the effect of microstructure on crack initiation and propagation, metallographically polished and etched bend specimens with notch were used. The test specimens as shown in Figure 1 with 28 x 9 x 3 mm size with 0.8 mm deep central Charpy notch were metallographically polished on 28 x 3 mm face and electrolytically etched in 10% oxalic acid in water. Trial samples were grid etched symmetrically over the notch. For evaluating the depth of crack propagation with time in fatigue precracking and also for estimating the biaxial strain at the root of the crack after deformation by bending, Figure 2.

Fig. (1) - SEM bending stage for metallographic samples

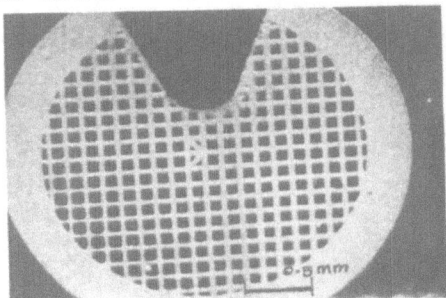

Fig. (2) - 200 mesh TEM grid etched near notch

The impression was permanent and well defined. In this new approach, the surface of the sample was coated with KODAK photoresist and dried in darkness. A 200 mesh grid was placed over the notch and ultraviolet light from 150W Xenon arc lamp source was focussed over the grid and exposed for 5 minutues using Leitz metallux microscope. The sample was developed in trichloroethylene and electrolytically etched. It was soaked in chloroform for 15 minutes to remove the resist completely. At this stage, the square grid pattern was seen clearly. The fatigue precracking of such small samples were done using specially developed fixtures, Figure 3, which were fixed to Avery bending fatigue machine. The effective K values have been calculated using plastic zone correction [5,6]. The effective ΔK was around 15 MPa√m during fatigue precracking. The elastic stress intensity field at the root of the crack after elastic bending in SEM deformation stage was symmetrical near the crack as seen in polarized light, Figure 4. The SEM three-point bending stage for

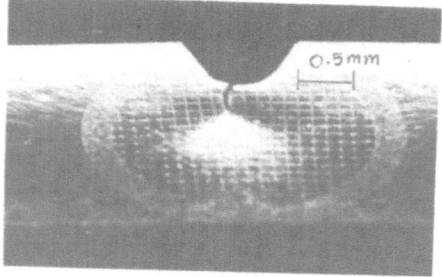

Fig. (3) - Miniature grips for fatigue precracking

Fig. (4) - Symmetrical bending stress field seen in polarized light

these notched samples can produce a load up to 450 Kg by screw advance and strain up to 30% for unnotched samples. Specimens similar to the actual test were used for calibration of load versus load point displacement in a tensile testing machine. From this data, the actual load acting on the sample can be estimated after knowing the displacement of load point. In these experiments, a fatigue precrack of 1 mm was introduced for comparative evaluation, Figure 5. The depth of crack was uniform as shown by earlier experiments done under similar conditions [4].

Fig. (5) - Fatigue precrack starting from center of notch

In 309 Nb material after SRHT, the crack initiation took place at the interface boundaries of austenite and interphase delta network as shown in Figures 6, 7 and 8. The elasto plastic region below the crack developed localized cracking

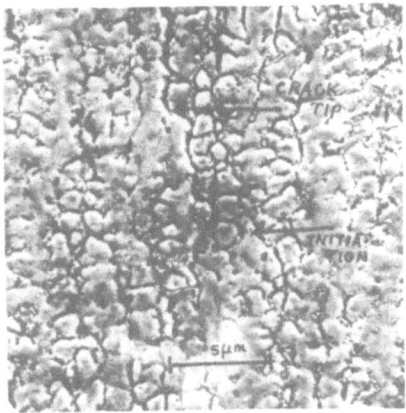

Fig. (6) - Crack initiation in 309 Nb weld after SRHT (continued 7,8)

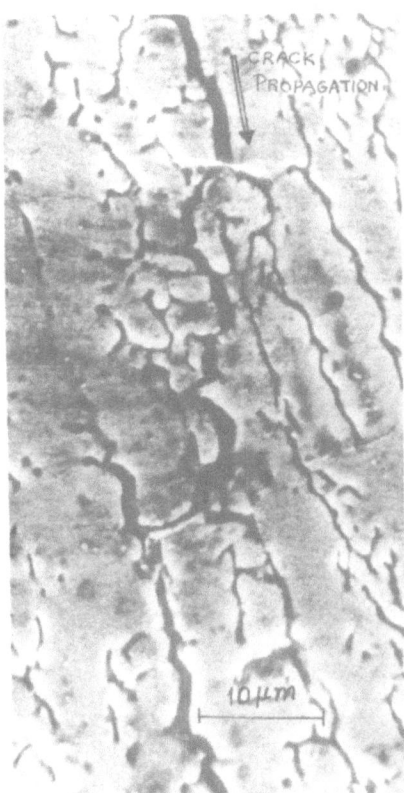

Fig. (7) - Preferential crack propagation along delta phase

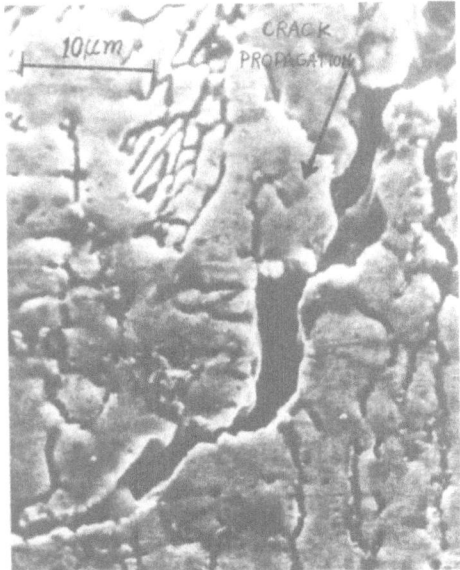

Fig. (8) - Longer crack path along delta after bending

154

along preferential paths located at delta ferrite boundary. The crack propagation took even a longer path along delta ferrite when intercepted by large area of austenite phase which resists crack propagation, Figure 8. Whereas in 347 weld material after SRHT, the crack path was rather short and the crack went through austenite matrix and delta ferrite as shown in Figure 9. The region below crack tip

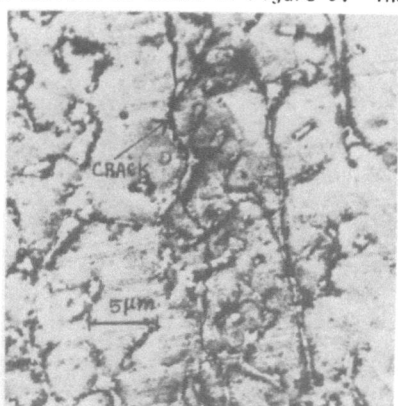

Fig. (9) - Random fatigue crack propagation in 347 weld after SRHT
(continued 10, 11)

showed localized plastic deformation, Figures 10 and 11, as a contrast to 309 Nb which had localized crack initiation without any noticeable deformation, Figure 6.

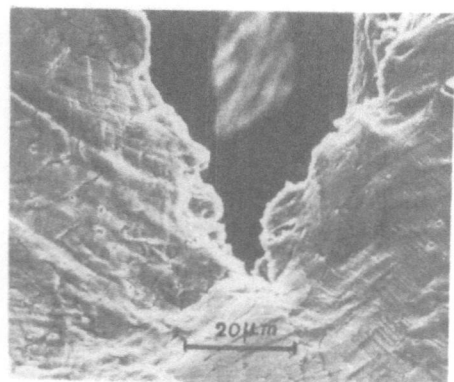

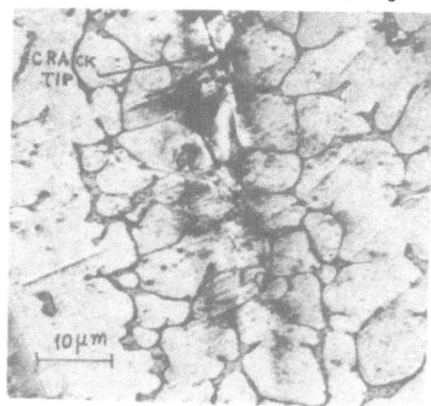

Fig. (10) - Ductile crack tip and
microscopic slip

Fig. (11) - Ahead of crack tip showing
localized slip

Even though delta ferrite had a continuous network in 347 weld, no embrittlement is introduced as seen in Figures 10 and 11.

Both these materials without SRHT were observed in deformation stage and their fracture behaviors were similar to those shown in Figures 9 and 10.

X-RAY DIFFRACTION STUDIES

The secondary phases were extracted by electrolytic process [7] using 10% Hcl in methanol at 2-4°C. A constant potential of 1.5V was set for 4 hours between the specimen and the inconel cathode. The residues were collected by ultrasonic cleaning and repeated twice. Using Debye Scherrer X-ray diffraction camera and Cr K alpha radiation powder, patterns were obtained with these residues for both 347 and 309 Nb after SRHT. 309 Nb material showed majority of $M_{23}C_6$ carbides and some sigma phase. The delta ferrite line became very weak compared to that without SRHT. Thus, considerable decomposition of delta ferrite to secondary phases was noticed. On the other hand, 347 weld after SRHT showed weak lines due to $M_{23}C_6$ and relatively stronger delta ferrite line. The delta ferrite line in the SRHT condition remained relatively stronger indicating better stability of this phase.

TEM OF THIN FOILS

A large number of thin foils were prepared from 3.05 diameter discs by a new double anode electrolytic jet method using 7.5% perchloric acid in methanol and examined in Philips EM 400 Transmission Electron Microscope at 120 KV. 309 Nb material after SRHT showed dense and fine precipitation at the delta ferrite boundaries, Figure 12. In many cases, plate-like precipitates arranged along delta ferrite boundary were observed. Whereas 347 material after SRHT did not show boundary precipitates as seen in Figure 13. However, fine NbC precipitates were

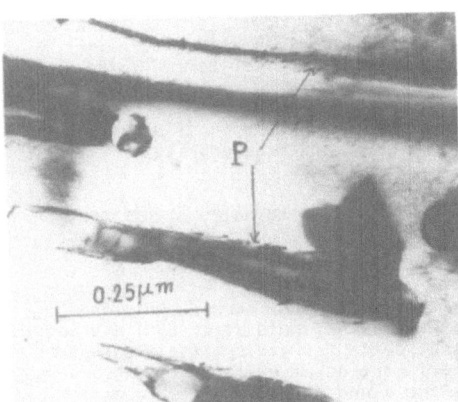

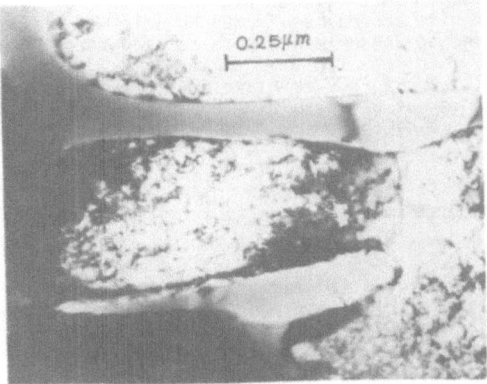

Fig. (12) - TEM of foil from 309 Nb after SRHT; dense precipitates (P) along delta ferrite boundary

Fig. (13) - 347 weld after SRHT showing delta ferrite with negligible precipitates

156

observed in austenite matrix. Because of their extremely fine size, the precipitates could not be identified in TEM. 309 Nb material without SRHT did not show
precipitates at delta ferrite boundary, Figure 14. However, extremely fine NbC

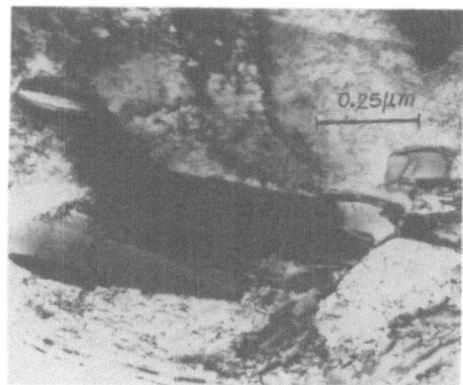

Fig. (14) - 309 Nb without SRHT showing delta ferrite without precipitates

precipitates were seen in the matrix. The boundary precipitates as in Figure 12
will weaken the interface of delta and austenite leading to easy crack initiation
and propagation [1].

CONCLUSIONS

1. SEM deformation studies on metallographically polished bend test specimens
showed that fatigue precrack took a preferential path along delta ferrite and austenite boundaries in 309 Nb weld metal after SRHT. Whereas 347 weld after SRHT
showed random crack path cutting through austenite and delta ferrite.

2. In 309 Nb after SRHT, crack initiation took place ahead of the root of the
crack at delta ferrite boundary whereas 347 weld showed microscopic slip pattern
ahead of the crack. The network delta ferrite did not introduce embrittlement in
347 weld.

3. In 309 Nb after SRHT, X-ray diffraction of residues revealed significant
transformation of delta ferrite to secondary phases such as $M_{23}C_6$ and sigma phase.
Whereas 347 weld after SRHT showed less transformation of delta ferrite.

4. TEM studies of foils from 309 Nb weld metal after SRHT showed dense and
fine precipitates along delta ferrite boundary whereas 347 weld after SRHT did not
show boundary precipitates. It is concluded that delta ferrite by itself cannot
promote crack propagation whereas the dense precipitation at delta ferrite boundary
will promote easy crack initiation and propagation along delta ferrite boundaries.

5. In the as-welded condition, 309 Nb and 347 behaved in a similar fashion in
crack propagation studies. This was attributed to the relatively precipitate free
condition of delta ferrite especially at the boundaries as seen in TEM of foils.

ACKNOWLEDGEMENTS

The authors thank Bharat Heavy Electricals Limited for permitting them to present this paper in International Conference on Fracture Mechanics Technology, Melbourne, Australia, August 10-13, 1982. They thank R. R. C. Kalpakkam and I. I. T. Delhi for sparing TEM and SEM facilities. Thanks are due to the staff of Central Laboratory for their help and cooperation in this work.

REFERENCES

[1] Lai, J. K. and Haigh, J. R., "Delta ferrite transformation in weld metal", Welding Res. Suppl., Welding J., 58, p. 1S, 1979.

[2] Venkataraman, G., Arunagiri, S. and Theobald, S. L., "Effect of plate segregation on weld quality", Weld. Res. Abroad, WRC-USA, 25, p. 46, 1979.

[3] Venkataraman, G., Thyagarajan, V. and Srinivasulu, A., "Embrittlement of SA 515 Gr 70, fusion welded pipes", Proc. of Int. Conf. on Welding in the 80's - WIC, Calgary, Canada, Pergamon Press, Canada, p. 69, November 1980.

[4] Venkataraman, G., Thyagarajan, V. and Srinivasulu, A., "The influence of microstructure and biaxial stress on crack propagation in pressure vessel steels", Proc. of Fifth Int. Conf. on Fracture, Cannes, France, Session 1e, Pergamon Press, UK, April 1981.

[5] Pelloux, R. M., "Fracture mechanics and SEM failure analysis", Proc. of IITRI Chicago 7th Annual SEM Symp., Part IV, p. 851, 1974.

[6] Hertzberg, R. W., "Deformation and fracture mechanics of engineering materials", John Wiley & Sons, Inc., New York, Chapter 8, p. 255, 1976.

[7] Spruiell, J. E., Fett, W. E. and Lundin, C. D., "Delta ferrite stability at elevated temperatures in stainless steel weld metal", 56, p. 289S, 1977.

SECTION II
STRESS AND FAILURE ANALYSIS

NUMERICAL ANALYSIS FOR NOTCHES OF ARBITRARY NOTCH ANGLE

R. H. Leicester and P. F. Walsh

CSIRO Division of Building Research
Melbourne, Victoria, Australia

ABSTRACT

A method is described for the numerical evaluation of the stress intensity factor for structural elements containing notches with arbitrary notch angles. The method is applicable to the state of plane stress for arbitrarily shaped elements subjected to arbitrary loading conditions.

The numerical analysis is based on coupling of conventional finite element analysis program with the singular stress fields that occur in the vicinity of a notch root. Equations are presented for singular stress fields in orthotropic materials, together with a proposed definition of the associated stress intensity factors. The computation method is illustrated by a numerical example.

INTRODUCTION

The stress intensity factor is a useful parameter for estimating the fracture strength of a structural element. However, the available literature on stress intensity factors is concerned almost entirely with sharp cracks. For notches of arbitrary notch angle, there exists a simple method which may be used to evaluate the stress intensity factors at the roots of notches, contained in arbitrarily shaped bodies subjected to a state of plane stress. Unfortunately, there is no convenient single source of information that gives all the necessary details for applying this method. Accordingly, this paper seeks to remedy that deficiency. The following is based largely on two earlier papers by Leicester [1] and Walsh [4].

The method of analysis makes use of the fact that elastic stress fields within the vicinity of notch roots contain one or two eigenfields that have singularities at the notch roots. Because of the infinite stresses associated with the singularities, the eigenfields dominate the stresses near the notch root. Obviously, conventional finite element methods cannot model the infinite stresses of a singularity. However, provided a sufficiently fine mesh of elements is used, the eigenfields will be correctly modelled except for a small area around the notch root. This area will be denoted the "distortion field".

A schematic illustration of the stress fields associated with a notched element is given in Figure 1. In addition to the eigenfield and distortion field mentioned earlier, there is also a "nominal stress field" which may be taken as

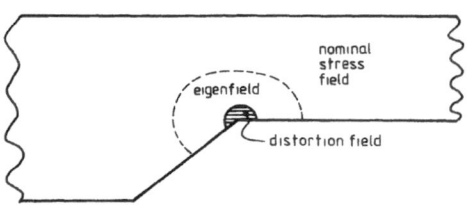

Fig. (1) - Schematic illustration of stress fields

the stress field that would exist without the concentration effects of a notch. The method of numerical analysis is based on the fact that while the intensity of the eigenfield is not affected by the existence of a sufficiently small distortion field, the magnitude of stresses and displacements in the distortion field are directly proportional to the intensity of the eigenfield. Thus, the stress intensity factor at a notch root of a stressed element may be evaluated by comparing the displacements of a distortion field when located in an eigenfield of unit value with those obtained when located in the stressed structural element.

The following illustrates the numerical analysis procedure by application to a notched orthotropic element. The necessary equations for the analysis of isotropic material may be derived in the same way as for orthotropic materials, and are very similar to the stress functions given by Williams [7].

EIGENFIELDS IN ORTHOTROPIC MATERIALS

A. Equations of Elasticity

The detailed derivation of these equations has been given elsewhere [1]. The co-ordinate system and notation for stresses to be used are shown in Figure 2.

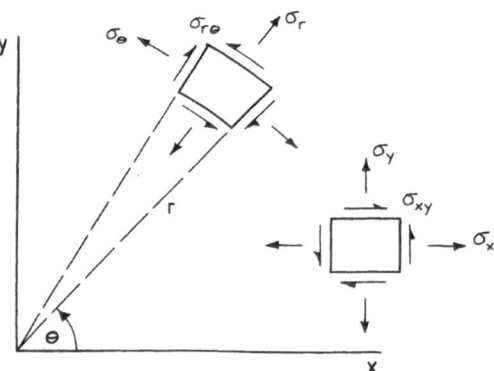

Fig. (2) - Coordinate systems and components of stress

The analysis is made in terms of an Airy stress function, denoted by ϕ and defined by

$$\sigma_x = \partial^2\phi/\partial y^2 \tag{1a}$$

$$\sigma_y = \partial^2\phi/\partial x^2 \tag{1b}$$

$$\sigma_{xy} = -\partial^2\phi/\partial x \partial y \tag{1c}$$

This definition ensures that the equations of equilibrium are satisfied automatically. In terms of polar co-ordinates, the stresses may be derived as follows:

$$\sigma_r = \sigma_x\cos^2\theta + \sigma_y\sin^2\theta + 2\sigma_{xy}\cos\theta\sin\theta \tag{2a}$$

$$\sigma_\theta = \sigma_x\sin^2\theta + \sigma_y\cos^2\theta - 2\sigma_{xy}\cos\theta\sin\theta \tag{2b}$$

$$\sigma_{r\theta} = (\sigma_y-\sigma_x)\cos\theta\sin\theta + \sigma_{xy}(\cos^2\theta-\sin^2\theta) \tag{2c}$$

For linearly elastic orthotropic material, the equation of compatibility may be written [1]:

$$\partial^4\phi/\partial x^4 + (2k/\varepsilon^2)\partial^4\phi/\partial x^2\partial y^2 + (1/\varepsilon^4)\partial^4\phi/\partial y^4 = 0 \tag{3}$$

where k and ε are elastic constants. Examples of elastic constants for various orthotropic materials are given in a previous paper by Leicester [1]. For example, the constants have the typical values $k=\varepsilon=2$ for wood with the X-axis lying along the direction of the wood grain.

It is convenient to introduce a new pair of elastic constants α_I and α_{II} defined as follows:

$$\alpha_I^2 = (1/\varepsilon^2)[k + (k^2-1)^{0.5}] \tag{4a}$$

$$\alpha_{II}^2 = (1/\varepsilon^2)[k - (k^2-1)^{0.5}] \tag{4b}$$

These constants are used to define two further co-ordinate systems which will be indicated by the subscripts I and II. The Cartesian co-ordinates of these systems are defined by

$$x = x_I = x_{II} \tag{5a}$$

$$y = \alpha_I y_I = \alpha_{II} y_{II} \tag{5b}$$

The corresponding polar co-ordinates will be denoted by r_I, θ_I and r_{II}, θ_{II}. Equations (3), (4) and (5) then lead to

$$[\partial^2/\partial x_I^2 + \partial^2/\partial y_I^2][\partial^2/\partial x_{II}^2 + \partial^2/\partial y_{II}^2]\phi = 0 \tag{6}$$

An obvious solution to equation (6) is

$$\phi = \phi_I + \phi_{II} \tag{7}$$

where ϕ_I and ϕ_{II} are harmonic functions of the co-ordinates in the I and II systems respectively.

B. Harmonic Functions for Eigenfields

A suitable pair of harmonic functions for eigenfields is the following:

$$\phi_I = A_1 r_1{}^\lambda \cos(\lambda\theta_I) + A_2 r_1{}^\lambda \sin(\lambda\theta_I) \tag{8a}$$

$$\phi_{II} = A_3 r_{II}{}^\lambda \cos(\lambda\theta_{II}) + A_4 r_{II}{}^\lambda \sin(\lambda\theta_{II}) \tag{8b}$$

where A_1 to A_4 and λ are arbitrary constants. It is outside the scope of this paper to discuss whether other types of stress fields can exist in the vicinity of a notch root.

Substitution of equations (8) into (1) leads to the following:

$$\sigma_x = \lambda(\lambda-1)r_I{}^{\lambda-2}(1/\alpha_I^2)[-A_1\cos(\lambda\theta_I-2\theta_I) - A_2\sin(\lambda\theta_I-2\theta_I)]$$

$$+ \lambda(\lambda-1)r_{II}{}^{\lambda-2}(1/\alpha_{II}^2)[-A_3\cos(\lambda\theta_{II}-2\theta_{II}) - A_4\sin(\lambda\theta_{II}-2\theta_{II})]$$

$$\sigma_y = \lambda(\lambda-1)r_I{}^{\lambda-2}[A_1\cos(\lambda\theta_I-2\theta_I) + A_2\sin(\lambda\theta_I-2\theta_I)] \tag{9}$$

$$+ \lambda(\lambda-1)r_{II}{}^{\lambda-2}[A_3\cos(\lambda\theta_{II}-2\theta_{II}) + A_4\sin(\lambda\theta_{II}-2\theta_{II})]$$

$$\sigma_{xy} = \lambda(\lambda-1)r_I{}^{\lambda-2}(1/\alpha_I)[A_1\sin(\lambda\theta_I-2\theta_I) - A_2\cos(\lambda\theta_I-2\theta_I)]$$

$$+ \lambda(\lambda-1)r_{II}{}^{\lambda-2}(1/\alpha_{II})[A_3\sin(\lambda\theta_{II}-2\theta_{II}) - A_4\cos(\lambda\theta_{II}-2\theta_{II})]$$

C. Eigenfield Equations

Eigenfield equations are obtained by the substitution of equations (9) into (2) to obtain stresses in terms of polar co-ordinates and then substituting these into the boundary conditions $\sigma_\theta = \sigma_{r\theta} = 0$ at the notch edges along $\theta = \psi_A$ and $\theta = \psi_B$ as shown in Figure 3. These four conditions lead to a matrix equation

$$\begin{bmatrix} a_{11} & a_{12} & a_{13} & a_{14} \\ a_{21} & a_{22} & a_{23} & a_{24} \\ a_{31} & a_{32} & a_{33} & a_{34} \\ a_{41} & a_{42} & a_{43} & a_{44} \end{bmatrix} \begin{bmatrix} A_1 \\ A_2 \\ A_3 \\ A_4 \end{bmatrix} = 0 \tag{10}$$

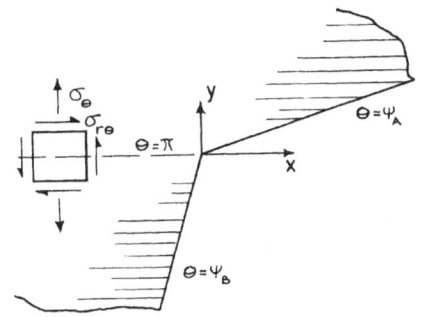

(a) Proposed definition

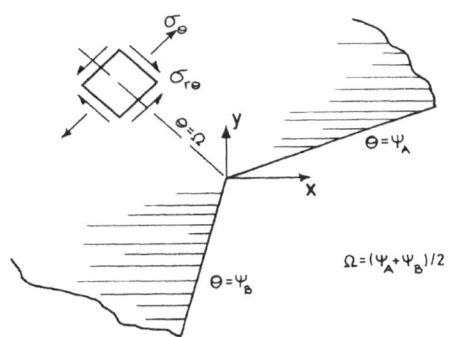

(b) Alternative definition

Fig. (3) - Notation for definition of stress intensity factors

where the terms a_{ij} are functions of α_I, α_{II}, ψ_A, ψ_B and λ. Equation (10) is satisfied only for discrete values of λ, and hence may be considered to be a non-linear eigenvalue equation. The required eigenvalues of λ are those that lean to a zero determinant, i.e., values that lead to:

$$|a_{ij}| = 0 \tag{11}$$

Since the terms a_{ij} are non-linear functions of λ, the solution of equation (11) is obtained most simply by trial and error.

From equation (9), it is apparent that if λ is less than 2.0, then a stress singularity will exist at the origin. Furthermore, it can be shown that if λ is less than 1.0, then a singularity will also occur with respect to the displacements. Thus, if the displacements are to remain finite, the eigenvalues of a singular eigenfield are limited to the range $1<\lambda<2$. Within this range, it is found that there is always at least one eigenvalue, and that if the notch angle is sufficiently small, there will be two eigenvalues [1].

The two eigenvalues will be denoted λ_A and λ_B where $\lambda_A<\lambda_B$. The eigenfields associated with these eigenvalues will be denoted the "primary" and "secondary" eigenfields respectively. It will be apparent from equation (9) that, except for the special case of $\lambda_A=\lambda_B$, which is the condition that occurs for a sharp crack, the stresses associated with the primary eigenfield will always dominate in the immediate vicinity of a notch root.

The substitution of an eigenvalue of λ into equation (10) leads to the determination of the constants A_1 to A_4 to within one arbitrary constant.

D. Unit Eigenfields

Because five arbitrary constants must be determined in order to describe the unit eigenfield completely according to equation (9), an additional criterion to the four boundary conditions which describe stress free notch edges is necessary. The magnitude of the stress intensity factor provides this required criterion.

For the type of notch shown in Figure 3a, the stress intensity factors for the primary and secondary stress fields, denoted by K_A and K_B respectively, will be defined as follows [4]:

$$\sigma_\theta |_{\theta=\pi} = K_A/(2\pi r)^{2-\lambda_A} \tag{12a}$$

$$\sigma_{r\theta} |_{\theta=\pi} = K_B/(2\pi r)^{2-\lambda_B} \tag{12b}$$

These definitions of K_A and K_B have been chosen because the failure of many orthotropic materials occurs by splitting along the X-axis. However, a more rational definition, illustrated in Figure 3b, may possibly be the following:

$$\sigma_\theta |_{\theta=\Omega} = K_A/(2\pi r)^{2-\lambda_A} \tag{13a}$$

$$\sigma_{r\theta} |_{\theta=\Omega} = K_B/(2\pi r)^{2-\lambda_B} \tag{13b}$$

where

$$\Omega = (\psi_A+\psi_B)/2 \tag{14}$$

in which $\theta = \psi_A$ and $\theta = \psi_B$ are the lines along which the notch edges lie. With this definition, the stress intensity factors K_A and K_B would be the values associated with symmetrical and antisymmetrical eigenfields in the special case of isotropic materials. Moreover, they would coincide with the conventional definitions of stress intensity factors in the case of sharp cracks [3]. Many other definitions of stress intensity factors are also possible, and so it is necessary to state the definition of the stress intensity factor in any discussion on non-zero angle notches.

Unit eigenfields will be defined as stress fields for which $K_A = 1$ or $K_B = 1$.

E. Examples of Unit Eigenfields

The example given will be that of a wood element containing a right angle notch with edges lying along $\theta = 0°$ and $\theta = 270°$. For this case, the parameters of the unit primary eigenfield defined according to equation (12a) are the following:

$$\lambda = \lambda_A = 1.5498$$
$$A_1 = 0.1833 \qquad A_2 = -0.7095$$
$$A_3 = -0.1833 \qquad A_4 = 0.1901$$

(15)

and for the unit secondary eigenfield, the parameters are

$$\lambda = \lambda_B = 1.8972$$
$$A_1 = 0.5419 \qquad A_2 = 5.0888$$
$$A_3 = -0.5419 \qquad A_4 = -1.3635$$

(16)

Substitution of these parameters into equations (9) leads to the complete stress distribution of the unit eigenfields. As an example, the stress distribution of the unit primary eigenfield is illustrated in Figure 4. This figure shows the stresses acting on sections located 10 mm from the notch root.

If the stress intensity factors were to be defined according to equations (13), then the factors A_1 to A_4 would be reduced to 0.384 and 0.180 of the values cited in equations (15) and (16) respectively.

NUMERICAL ANALYSIS

The method of numerical analysis of a notched element will be illustrated by application to the example shown in Figure 5a. This example comprises a glued lap-joint made of wood and subjected to a nominal axial tensile stress of 10 MPa. It is assumed that the stiffness of the glue has a negligible effect on the stress distribution, and that failure is related to fracture at the root of the notch denoted "A" in the figure. The possibility of failure is assessed from the magnitude of the stress intensity factors at this notch root. To do this, the joint is represented by the finite element mesh shown in Figure 5b and detail of the distortion mesh at the notch root is shown in Figure 5c.

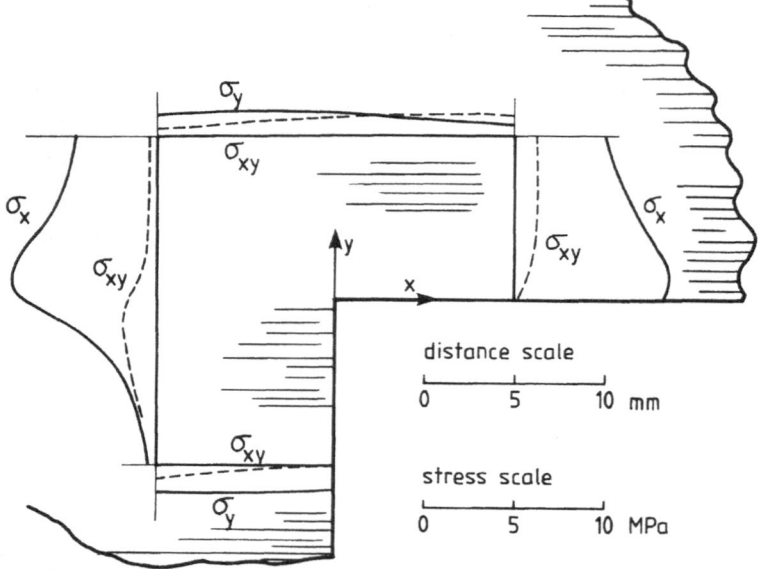

Fig. (4) - Boundary stresses for an element of the primary stress field

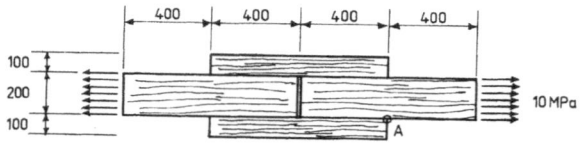

(a) Dimensions of wood lap-joint (mm)

(b) Finite element model of the joint

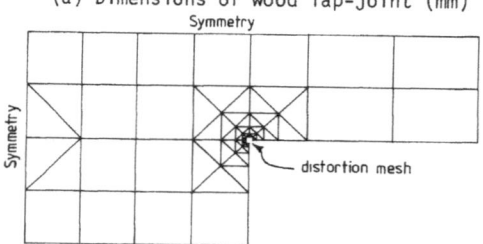

(c) Detail of distortion mesh

Fig. (5) - Details of worked example

The distortion mesh is calibrated by applying boundary stresses due to unit eigenfields similar to those shown in Figure 4. The associated displacements relative to the notch root are then computed in the X and Y directions for the eight locations indicated in Figure 5c. These relative displacements will be denoted by a_i and b_i for the primary and secondary stress fields respectively.

The distortion mesh is then inserted into the model for the lap-joint shown in Figure 5b, and the relative displacements that occur when the tensile stress of 10 MPa is applied are computed. These displacements will be denoted by u_i. Neglecting the effects of numerical approximation, these displacements may be written

$$u_i = a_i K_A + b_i K_B + c_i \omega \tag{17}$$

where i = 1,2,...16, ω is the average rotation of the elements around the notch root, and c_i is the rotation constant for displacement u_i. The relationship in equation (17) may be used to define an error or residual function Φ as follows:

$$\Phi = \sum_{i=1}^{16} (u_i - a_i K_A - b_i K_B - c_i \omega)^2 \tag{18}$$

Hence, a good estimate of K_A, K_B and ω will be given through minimization of Φ by:

$$\partial \Phi / \partial K_A = \partial \Phi / \partial K_B = \partial \Phi / \partial \omega = 0 \tag{19}$$

For the example of the lap-joint shown in Figure 5a, the values of stress intensity factor obtained are:

$$K_A = 12.0 \text{ Nmm}^{-1.55} \tag{20a}$$

$$K_B = 0.21 \text{ Nmm}^{-1.90} \tag{20b}$$

The units of stress intensity factor are $\text{Nmm}^{2-\lambda}$ and may be deduced from equation (12) or (13). The unusual units are related to the size effect associated with the elastic equations of fracture mechanics [2]. The use of elastic fracture mechanics, as illustrated here, to measure the strength of wood lap-joints has been undertaken by Walsh et al [6].

CONCLUDING COMMENTS

The procedure described provides a simple method for incorporating the eigenfields around a notch root into a conventional finite element technique to evaluate the stress intensity factors of notches having non-zero notch angles. Useful eigenfunctions for this have been given in an earlier paper by Leicester [1]. Stress intensity factors for several practical types of structural configurations have been computed by Walsh [4].

Finally, it should be mentioned that a useful refinement to the procedure described herein is to make use of the finite element grading technique described by Walsh [5]. When used in the vicinity of the notch root, this technique can lead to very accurate solutions with only modest computing effort.

REFERENCES

[1] Leicester, R. H., "Some aspects of stress fields at sharp notches in ortho-tropic materials, I. Plane stress", CSIRO Australia Division of Forest Products Technological Paper No. 57, 1971.

[2] Leicester, R. H., "Effect of size on the strength of structures", CSIRO Australia Division of Building Research Technological Paper No. 71, 1973.

[3] Paris, P. C. and Sih, G. C., "Stress analysis of cracks", ASTM Special Technical Publication No. 381, Symposium on Fracture Toughness Testing and its Applications, Chicago, pp. 30-81, 1964.

[4] Walsh, P. F., "Linear fracture mechanics solutions for zero and right angle notches", CSIRO Australia Division of Building Research Technical Paper, Second Series, No. 2, 1974.

[5] Walsh, P. F., "Intensive finite element grading for stress concentrations", Journal of Engineering Fracture Mechanics, Vol. 10, pp. 211-213, 1978.

[6] Walsh, P. F., Ryan, A. and Leicester, R. H., "The strength of glued lap joints in timber", Forest Products Journal, Vol. 23, No. 5, pp. 30-33, May 1973.

[7] Williams, M. L., "Stress singularities resulting from various boundary conditions in angular corners of plates in extension", Journal of Applied Mechanics, Vol. 19, pp. 526-528, 1952.

PREDICTION OF FAILURE SITES AHEAD OF MOVING ENERGY SOURCE

G. C. Sih

Institute of Fracture and Solid Mechanics, Lehigh University
Bethlehem, Pennsylvania 18015 USA

and

C. I. Chang

Marine Technology Division, Naval Research Laboratory
Washington, D.C. 20375 USA

INTRODUCTION

High concentrations of energy can affect the local material properties and cause irreversible material damage. The degree of damage depends on the intensity and rate at which energy is being transferred to the system and the ability of the material to dissipate the excessive energy in the form of irrecoverable deformation and/or fracture. Predictions of the mechanical behavior of materials damaged by localized energy sources are difficult for two reasons. Firstly, no reliable methods are available for determining the change in material properties in relation to certain threshold energy level. Secondly, damage threshold of materials, say for crystalline solids, in terms of yielding and fracture must also be accounted for. The former and latter are distinguished by the size scale at which damage is being done to the material. For instance, yielding and fracture are associated with material damage at the microscopic and macroscopic scale level [1] while a change in material properties* corresponds to disturbances at the atomic level.

From the engineering application point of view, it is essential to develop the means of assessing the ability of structural components to withstand high concentrations of energy. A simple model of a moving heat source is considered in this work for discussion. To begin with, it is important to know the combination of heat source intensity and velocity and material properties that correspond to material damage threshold. This is accomplished by solving for the thermal stresses ahead of the moving heat source. Possible sites of crack initiation

---

*This effect will not be considered in this communication.

172

and yielding are examined by application of the strain energy density theory[*] [2,3] which has been used successfully for predicting failure due to yielding and/or fracture. The theory assumes that the crack can initiate in regions of high energy density, dW/dV. The location corresponds to dW/dV attaining a relative minimum such that the dilatational energy component dominates as compared to the distortional energy component. The results for the aluminum material are displayed graphically and suggest possible failures that could be examined experimentally.

MOVING HEAT SOURCE: TEMPERATURE DISTRIBUTION

Consider a heat source of intensity Q(X,Y,t) moving with a constant velocity v in the X-direction of a two-dimensional domain (X,Y) that is infinite in extent, Figure 1. The temperature gradients and stresses at large distances away

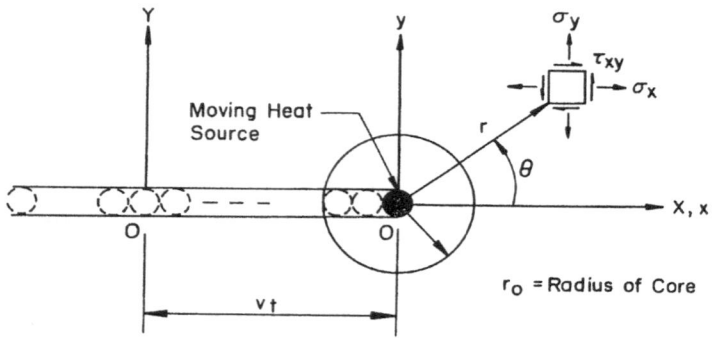

Fig. (1) - Thermal stresses on element ahead of moving heat source

from the source are assumed to vanish[**]. The temperature distribution T(X,Y,t) within the XY-plane may be determined from the equation [4]:

$$\frac{\partial^2 T}{\partial X^2} + \frac{\partial^2 T}{\partial Y^2} = \frac{1}{\kappa} \left(\frac{\partial T}{\partial t} + Q \right) \tag{1}$$

in which $\kappa = k/\rho C$ stands for the thermal diffusivity with k being the coefficient of thermal conductivity, C the specific heat and ρ the material density. Because of the steady state nature of the problem, it is simpler to solve equation (1) in

---

[*] This theory can predict failure initiating from local disturbances due to geometric discontinuities and/or energy concentrations.

[**] Since the problem is linear, the stress solution resulting from externally applied mechanical loads may be added into that due to thermal disturbances by superposition.

a moving coordinate system (x,y) defined by

$$x = X + vt, \quad y = Y \tag{2}$$

With a heat source of constant intensity Q_0 located at the origin of the xy-co-ordinates, equation (1) becomes [5]

$$\frac{\partial^2 T}{\partial x^2} + \frac{\partial^2 T}{\partial y^2} + 2c \frac{\partial T}{\partial x} = -\frac{1}{\kappa} Q_0 \delta(x)\delta(y) \tag{3}$$

where $2c = v/\kappa$ and δ is the Dirac Delta function. A solution of equation (3) is of the form [4]

$$T(x,y) = \frac{Q_0}{2\pi\kappa} \exp(-cx)K_0(cr), \quad r^2 = x^2+y^2 \tag{4}$$

The modified Bessel function of the second kind of order zero is denoted by $K_0(cr)$ with argument cr.

Once the temperature distribution is known, the corresponding stresses follow from the equations of thermoelasticity.

THERMOELASTIC STRESS FORMULATION AND SOLUTION

The thermoelastic problem may be formulated in terms of a potential $\phi(x,y)$ such that the rectangular displacement components $u(x,y)$ and $v(x,y)$ may be expressed as [6]

$$u(x,y) = \frac{\partial\phi}{\partial x}, \quad v(x,y) = \frac{\partial\phi}{\partial y} \tag{5}$$

Upon satisfaction of the equation of motion, an equation of the form

$$\frac{\partial^2\phi}{\partial x^2} + \frac{\partial^2\phi}{\partial y^2} = \frac{1+\nu}{1-\nu} \alpha T \tag{6}$$

is obtained for the case of plane strain. The Poisson's ratio is ν and the co-efficient of linear thermal expansion is α. Inserting equations (5) into the stress and displacement relations, it is found that

$$\sigma_x(x,y) = -2\mu \frac{\partial^2\phi}{\partial y^2}$$

$$\sigma_y(x,y) = -2\mu \frac{\partial^2\phi}{\partial x^2} \tag{7}$$

$$\tau_{xy}(x,y) = 2\mu \frac{\partial^2\phi}{\partial x \partial y}$$

with μ being the shear modulus of elasticity. The stress component normal to the xy-plane is

$$\sigma_z(x,y) = -2\mu \left(\frac{\partial^2\phi}{\partial x^2} + \frac{\partial^2\phi}{\partial y^2}\right) = -2\mu\alpha \left(\frac{1+\nu}{1-\nu}\right)T \tag{8}$$

Following a procedure similar to that described in [5,7], the displacement potential $\phi(x,y)$ is obtained:

$$\phi(x,y) = \frac{\alpha}{2c} \left(\frac{1+\nu}{1-\nu}\right) \left[\int T(x,y)dx + \phi_0(x,y)\right] \tag{9}$$

in which $\phi_0(x,y)$ is given by

$$\phi_0(x,y) = \frac{Q_0}{2\pi\kappa} \log r \tag{10}$$

It follows from equations (7) and (9) that[*]

$$\sigma_x(x,y) = -N\{[K_0(cr) - \frac{x}{r} K_1(cr)]\exp(-cx) + \frac{1}{c} \frac{x}{r^2}\}$$

$$\sigma_y(x,y) = -N\{[K_0(cr) + \frac{x}{r} K_1(cr)]\exp(-cx) - \frac{1}{c} \frac{x}{r^2}\} \tag{11}$$

$$\tau_{xy}(x,y) = N \frac{y}{r} [K_1(cr)\exp(-cx) - \frac{1}{cr}]$$

and equation (8) gives

$$\sigma_z(x,y) = -2NK_0(cr)\exp(-cx) \tag{12}$$

The parameter N is defined as

$$N = \frac{\mu\alpha}{2\pi\kappa} \left(\frac{1+\nu}{1-\nu}\right)Q_0 \tag{13}$$

Note that the thermal stresses decay rapidly along the path travelled by the heat source. In other words, the heat disturbance confines locally to a core region near the source as shown in Figure 1. The normal stresses contain a logarithmic singularity as source tip is approached. This character is embedded in the modified Bessel function $K_0(cr)$.

---

[*]The solution to this problem was first obtained in [8] and quoted incompletely in [7] that assumed $\phi_0(x,y)$ in equation (10) to be zero. As a result, the term $1/r^2$ and $1/r$ were missing in the normal and shear stresses given by [7], respectively.

In the limit as the velocity of the moving heat source approaches zero or $c \to 0$, the logarithmic stress singularity in terms of the radial distance is retained:

$$\sigma_X \text{ or } \sigma_Y = N \log R + \text{---} \tag{14}$$

except that $R^2 = X^2 + Y^2$ is now referred to the stationary coordinates (X,Y).

The stresses σ_x, σ_y and τ_{xy} in equations (11) are plotted as a function of the moving polar coordinate θ in Figure 1 for the 2024-T3 aluminum alloy with the following material properties:

$$k = 1.903 \times 10^2 \text{ w/m°K}, \qquad C = 9.623 \times 10^2 \text{ J/Kg°K}$$

$$\rho = 2.770 \times 10^3 \text{ kg/m}^3, \qquad \alpha = 13 \times 10^{-6} \text{ m/m/°F} \tag{15}$$

$$\nu = 0.33 \qquad \mu = 2.748 \times 10^4 \text{ MPa}$$

Three radial distances $r = 10^{-1}$, 10^{-2} and 10^{-3} cm are selected such that they are all larger than the radius of the core region r_o outside of which the microstructure of the material is assumed to be negligible. The numerical results are displayed in Figures 2 to 4. Near the source at a distance $r = 10^{-3}$ cm, Figures 2(a) to (d) show that both σ_x and σ_y are compressive as θ is varied from 0° to 180° in the counterclockwise direction as indicated in Figure 1 and v takes the values of 1.0 cm/sec and 10.0 cm/sec. The material element directly ahead of the source at $\theta = 0°$ is in compression with $\sigma_x = -8.37701$ N and $\sigma_y = -6.37220$ N for v = 1.0 cm/sec, and $\sigma_x = -6.05772$ N and $\sigma_y = -4.02592$ N where N is given by equation (13). The degree of compression tends to decrease with increasing speed of the heat source. At a larger distance $r = 10^{-2}$ cm, the situation remained qualitatively unchanged except that the degree of compression is further decreased. This is illustrated in Figures 3(a) to 3(d) for v = 1.0 cm/sec and 10.0 cm/sec. As r is further increased to 10^{-1} cm, the σ_y stress component finally becomes tensile for $\theta = 0°$ as shown in Figure 4(d) while the other stresses still remain as compressive, Figures 4(a) to 4(c). The magnitude of the stresses, however, has decreased rapidly as a function of r. This effect is shown in Figures 5(a) and (b) where the normalized stresses σ_x/N and σ_y/N are plotted as a function of the dimensionless distance parameter cx. Note that σ_y becomes tensile for approximately cx > 0.2.

From the variations of the stresses with r and θ as displayed in Figures 2 to 5, it is not obvious how the material ahead of the heat source would fail in yielding or fracture. The elements surrounding the moving heat source are mostly in a state of compression. More insight on material damage can be gained by turning to a criterion based on the fluctuation of the energy density within the material.

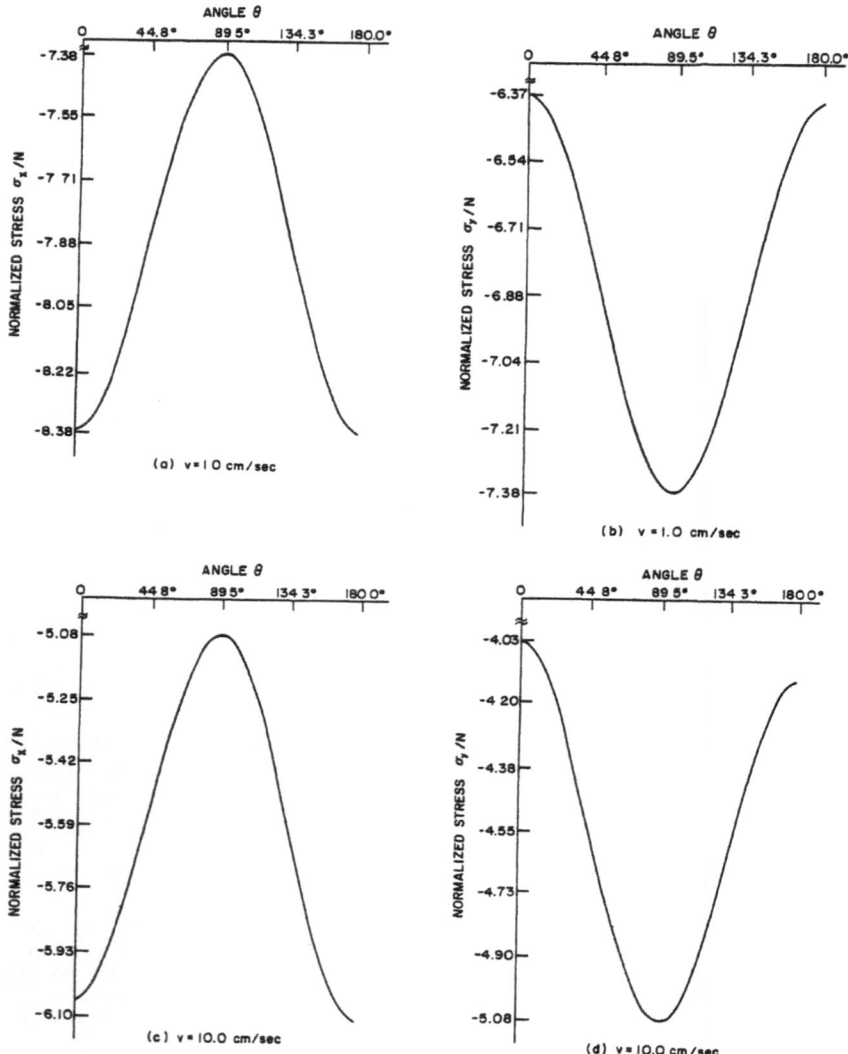

Fig. (2) - Angular distribution of normal stress components
σ_x and σ_y for r = 10^{-3} cm

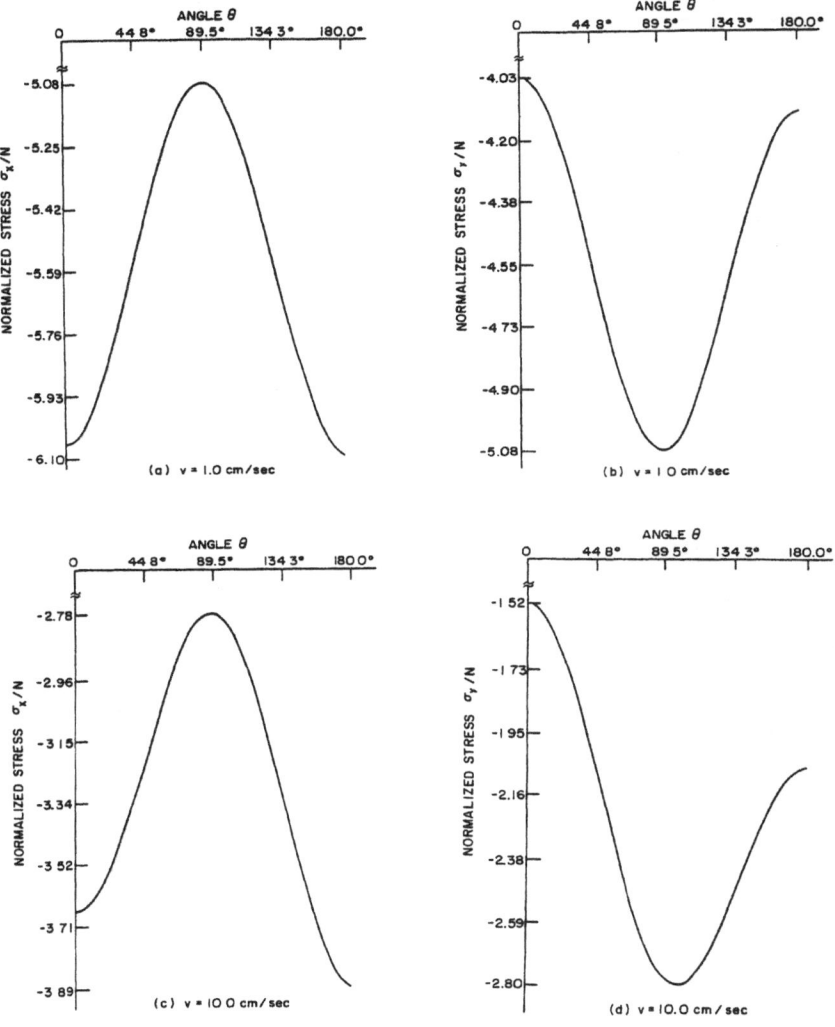

Fig. (3) - Angular distribution of normal stress components
σ_x and σ_y for $r = 10^{-2}$ cm

178

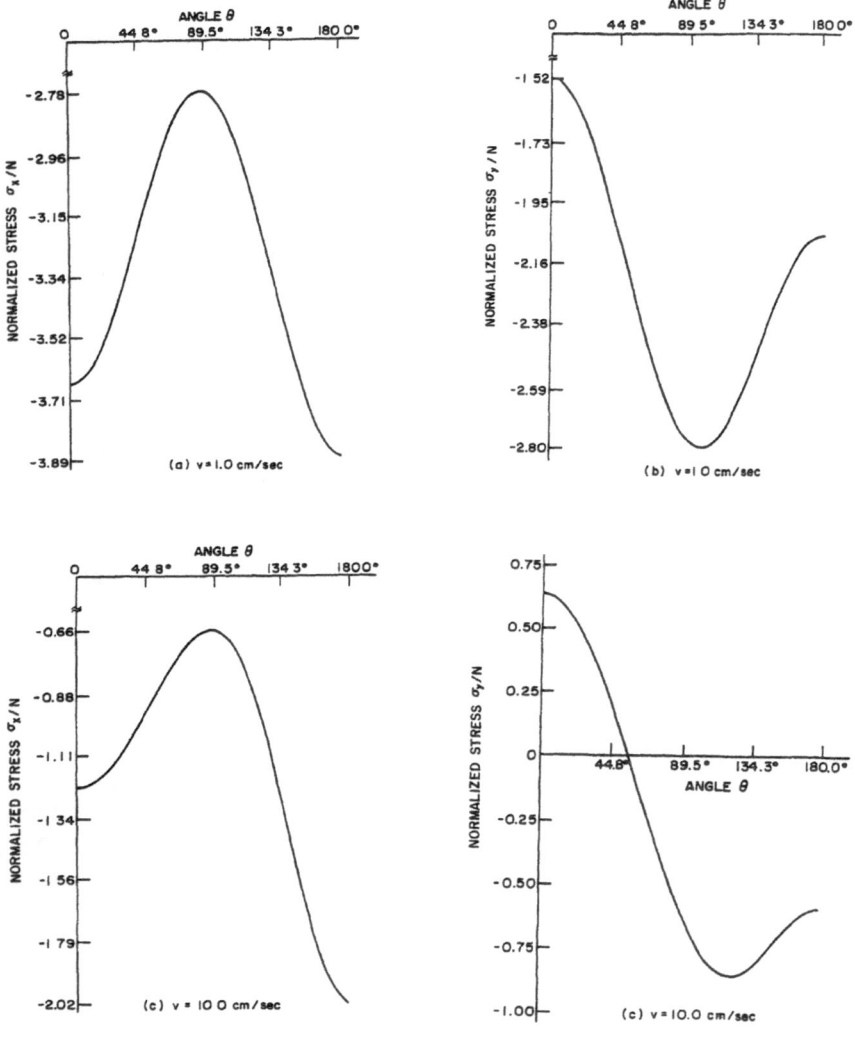

Fig. (4) - Angular distribution of normal stress components
σ_x and σ_y for $r = 10^{-1}$ cm

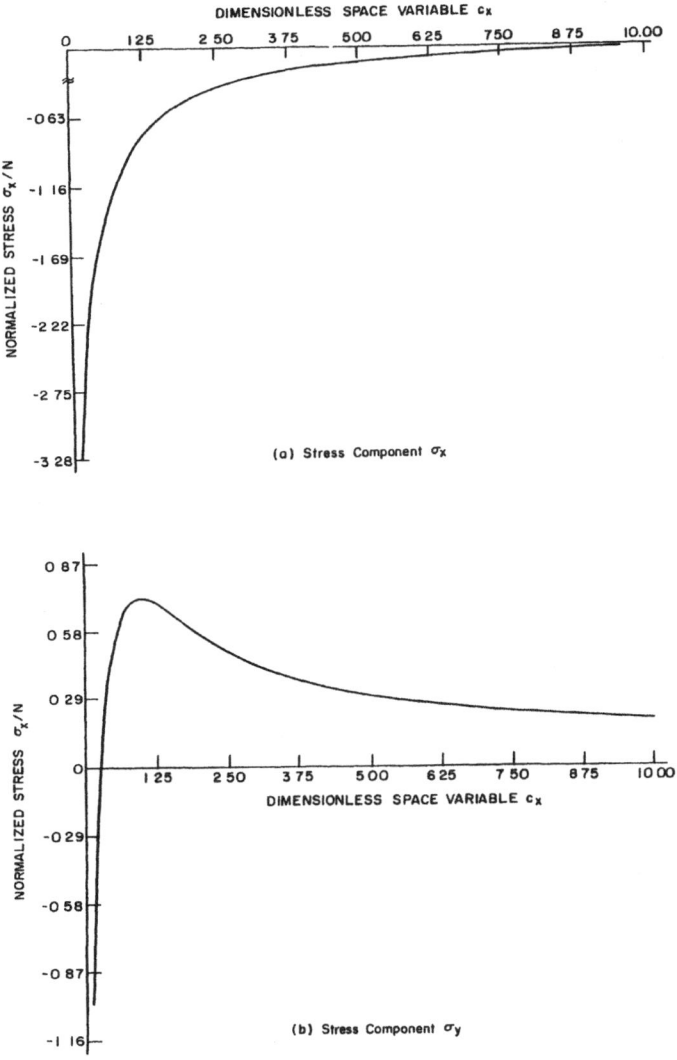

Fig. (5) - Variation of thermal stresses at θ = 0° with dimensionless
distance for 2024-T3 aluminum

180

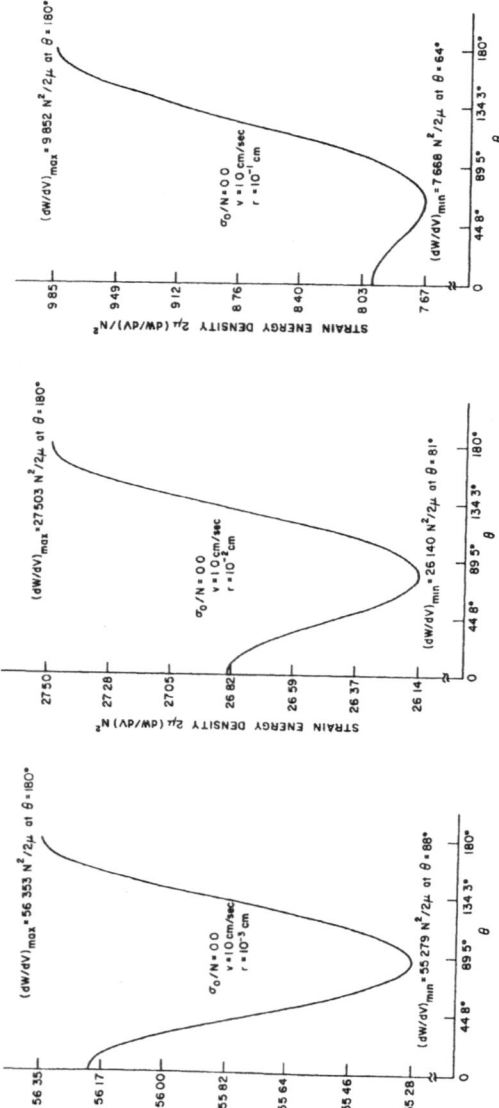

Fig. (6) - Normalized strain energy density function versus the angle θ for $r = 10^{-3}$, 10^{-2} and 10^{-1} cm and $v = 1.0$ cm/sec

181

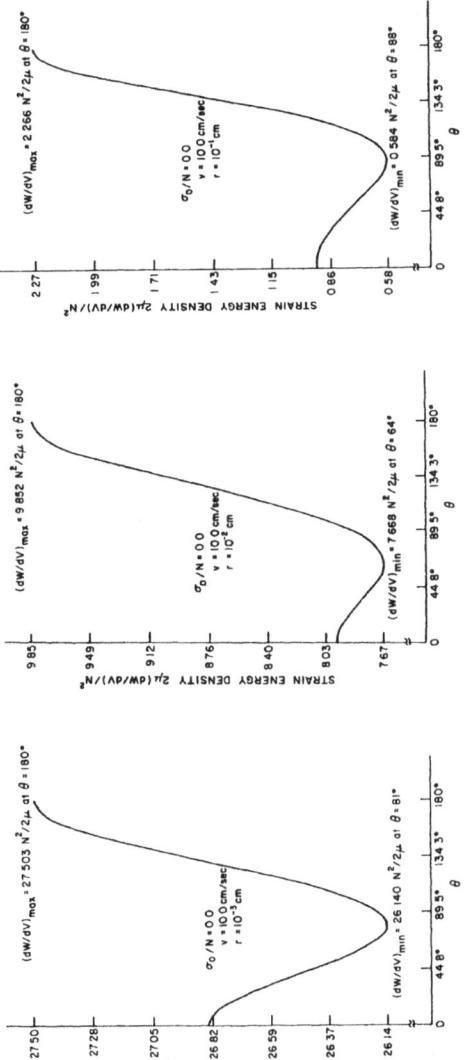

Fig. (7) - Normalized strain energy density function versus the angle θ for $r = 10^{-3}$, 10^{-2}, 10^{-1} cm and $v = 10.0$ cm/sec

FAILURE PREDICTION: STRAIN ENERGY DENSITY CRITERION

Assuming that the material ahead of the moving heat source will be damaged by a combined process of yielding and/or fracture, the strain energy density theory will be applied to determine the sites of possible failure. In this theory, attention is focused on the fluctuation of strain energy density dW/dV throughout the system. The relative maxima and minima of dW/dV can be associated with the locations of yielding and fracture in metals [2,3], respectively.

Failure Criterion. More specifically, the strain energy density theory relies on the following assumptions:

(1) Yielding and fracture are assumed to coincide with locations of $(dW/dV)_{max}$ and $(dW/dV)_{min}$, respectively.

(2) The respective critical values of $(dW/dV)_{max}$ and $(dW/dV)_{min}$ are assumed to denote the commencement of yielding and fracture.

In general, dW/dV may be computed from the stress and strain components σ_{ij} and ε_{ij} as

$$\frac{dW}{dV} = \int_0^{\varepsilon_{ij}} \sigma_{ij} d\varepsilon_{ij} \tag{16}$$

For a linear elastic material in a state where $\sigma_z = \nu(\sigma_x + \sigma_y)$, equation (16) reduces to

$$\frac{dW}{dV} = \frac{1}{4\mu} [(1-\nu)(\sigma_x+\sigma_y)^2 - 2(\sigma_x\sigma_y - \tau_{xy}^2)] + \mu(1+\nu)(\alpha T)^2 \tag{17}$$

Substituting equations (11) into (17) yields

$$\frac{2\mu}{N^2} \frac{dW}{dV} = \frac{1}{(cr)^2} \{\frac{3-5\nu}{1+\nu} [\exp(-cx)K_0(cr)]^2 + [\exp(-cx)K_1(cr) - \frac{1}{cr}]^2\} \tag{18}$$

which varies as a function of r and θ since $x = r\cos\theta$ and the material properties.

Stationary Values. Plotted in Figures 6 and 7 are the normalized strain energy density function $2\mu(dW/dV)/N^2$ as a function of θ. The curves oscillate and show that $(dW/dV)_{max}$ occurred at $\theta = 180°$ and $(dW/dV)_{min}$* occurred off to the axis

---

*It should be mentioned that for $v = 1.0$ cm/sec and r of the order of 10^{-8} cm, $(dW/dV)_{min}$ occurs at $\theta = 0°$, i.e., directly ahead of the heat source. This distance, however, is extremely small and is outside the region of validity of the continuum mechanics approach.

of the traveling heat source. The results are summarized in Table 1 for v = 1.0 and 10.0 cm/sec. It is seen that the magnitude of the energy density decays rap-

TABLE 1 - MAXIMUM AND MINIMUM VALUES OF $2\mu(dW/dV)/N^2$
FOR 2024-T3 ALUMINUM

	r (cm)		
	10^{-3}	10^{-2}	10^{-1}
I. v = 1.0 cm/sec			
$2\mu(dW/dV)_{max}/N^2$	56.353 (θ = 180°)	27.503 (θ = 180°)	9.852 (θ = 180°)
$2\mu(dW/dV)_{min}/N^2$	55.279 (θ = 88°)	26.140 (θ = 81°)	7.668 (θ = 64°)
II. v = 10.0 cm/sec			
$2\mu(dW/dV)_{max}/N^2$	27.503 (θ = 180°)	9.852 (θ = 180°)	2.266 (θ = 180°)
$2\mu(dW/dV)_{min}/N^2$	26.140 (θ = 81°)	7.668 (θ = 64°)	0.584 (θ = 88°)

idly with the distance r as illustrated in Figure 8 in which $2\mu(dW/dV)/N^2$ is plotted against the variable cx. Less energy is transferred to the solid as the heat source velocity is increased from 1.0 cm/sec to 10.0 cm/sec. It is to be expected. The magnitude of dW/dV is proportional to the square of the heat source intensity Q_o and may be computed. For v = 1.0 cm/sec and r = 10^{-3} cm, Table 1 gives

$$\left(\frac{dW}{dV}\right)_{max} = 7.044 \ \mu Q_o^2 \ \left[\frac{\alpha}{\pi\kappa}\left(\frac{1+\nu}{1-\nu}\right)\right]^2 \qquad (19)$$

The values of the physical parameters for the 2024-T3 aluminum are given in equation (15) with $\kappa = k/\rho C = 7.139 \times 10^{-5}$ m²/sec. According to the strain energy density criterion, yielding will first occur at $(dW/dV)_{max}$ directly behind the heat source at θ = 180° and fracture in elements off to the x-axis where the minima of dW/dV or $(dW/dV)_{min}$ are located. For appropriate values of Q_o and v, both $(dW/dV)_{max}$ and $(dW/dV)_{min}$ can exceed their respective threshold values to cause material damage in a strip zone of height 2r along the path of the heat source by a combined process local of yielding and fracture.

 Damaged Zone. A more detailed account of the damage zone caused by the moving heat source can be visualized by considering the specific case of v = 1.0 cm/sec and r = 10^{-2} cm. Figure 9 displays the stress states on material elements corresponding to $(dW/dV)_{max}$ at θ = 180° and $(dW/dV)_{min}$ at θ = ±88° as indicated

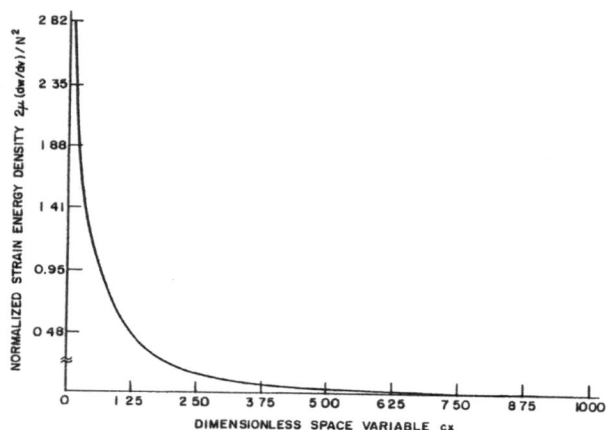

Fig. (8) - Normalized strain energy density versus cx for 2024-T3 aluminum

in Table 1. The stresses on the element ahead of the heat source at $\theta = 0°$ is also shown to illustrate that the elements at $\theta = 0°$ and $180°$ are highly distorted due to the differences in σ_x and σ_y. The stress states at $\theta = \pm88°$ are nearly hydrostatic as $\sigma_x \simeq \sigma_y$ and τ_{xy} is one order of magnitude smaller in magnitude.

If the threshold values of dW/dV in yielding and fracture are exceeded, say at $r = 10^{-2}$ cm, then the material trailing the heat source will be damaged locally by intense yielding in the center strip and fracture outside of this strip. The profile of $(dW/dV)_{min}$ near the heat source is shown in Figure 10 and is assumed to correspond with the frontal shape of the damaged zone. Starting at $r = 10^{-4}$ cm, the profile tends to lean ahead of the source and gradually sweeps back as r is increased to $10°$ cm. For a given velocity and heat source intensity, the magnitude of dW/dV is seen from Figure 8 to increase rapidly as the heat source is approached. The degree of damage becomes less severe at higher heat source velocities since less time will be available to transfer energy to do damage to the material. This can be seen from the results on dW/dV in Table 1.

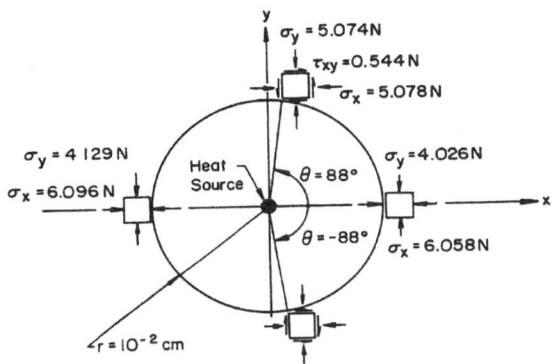

Fig. (9) - Stress state around heat source for v = 1.0 cm/sec and r = 10^{-2} cm

CONCLUDING REMARKS

Material damage due to local energy concentration has been analyzed by applying the strain energy density criterion to the thermoelastic stress solution of a moving heat source. The material around the source is subjected to high energy density concentration and can be damaged locally by a combination of yielding and fracture if the threshold values of $(dW/dV)_{max}$ and $(dW/dV)_{min}$ for the material are exceeded. The locations of $(dW/dV)_{max}$ and $(dW/dV)_{min}$ suggest that intense yielding could occur at the center strip of the damaged zone while fracture could prevail to both sides depending on the speed and intensity of the heat source.

If mechanical loads were also applied in a direction normal to the traveling heat source, a small crack could develop near the source and change the logarithmic stress singularity into an inverse square root stress singularity in terms of the radial distance r. This could conceivably affect the structural integrity of the system and cause unstable crack propagation, a situation which will be investigated in a forthcoming communication.

ACKNOWLEDGEMENT

This research was supported by the Naval Research Laboratory under Contract N00014-82-K-2006 with the Institute of Fracture and Solid Mechanics at Lehigh University, Bethlehem, Pennsylvania.

186

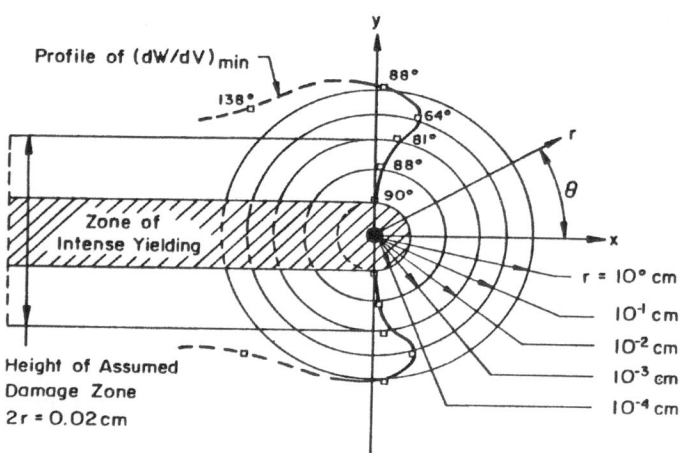

Fig. (10) - Damage zone behind the traveling heat source traveling
at v = 1.0 cm/sec

REFERENCES

[1] Sih, G. C., "Experimental fracture mechanics: strain energy density cri-
terion", Mechanics of Fracture, Vol. 7: Experimental Evaluation of Stress
Concentration and Intensity Factors, edited by G. C. Sih, Martinus Nijhoff
Publishers, The Netherlands, pp. XVII-LVI, 1981.

[2] Sih, G. C., "A special theory of crack propagation", Mechanics of Fracture,
Vol. 1: Methods of Analysis and Solutions of Crack Problems, edited by
G. C. Sih, Noordhoff International Publishers, The Netherlands, pp. XXI-
XLV, 1973.

[3] Sih, G. C., "A three-dimensional strain energy density factor theory of
crack propagation", Mechanics of Fracture, Vol. 2: Three-Dimensional Crack
Problems, by M. K. Kassir and G. C. Sih, Noordhoff International Publishers,
The Netherlands, pp. 15-53, 1975.

[4] Carslaw, H. S. and Jaeger, J. C., Conduction of Heat in Solids, Oxford University Press, 1959.

[5] Nowacki, W., Thermoelasticity, Pergamon Press - PWN, 1962.

[6] Goodier, J. N., "On the integration of the thermo-elastic equations", Phil. Mag., Vol. 23, pp. 1017-1032, 1937.

[7] Parkus, H., Instationäre Wärmespannungen, pp. 76-87, 1959.

[8] Melan, E., Wärmespannungen in einer Scheibe infolge einer wanderenden Wärmequelle, Ing.-Arch. 20, 1, pp. 46-48, 1952.

TENSION AND BENDING OF A SEMI-ELLIPTICAL SURFACE CRACK

M. Isida and H. Noguchi

Faculty of Engineering, Kyushu University
Hakozaki, Fukuoka 812, Japan

THEORETICAL ANALYSIS

Consider the tension and the bending of a finite thickness plate with a semi-elliptical surface crack. Denote the coordinate axes and the geometrical parameters as shown in Figure 1. Preliminary calculations for a flawed semi-infinite solid have shown that the stress intensity factor at the deepest

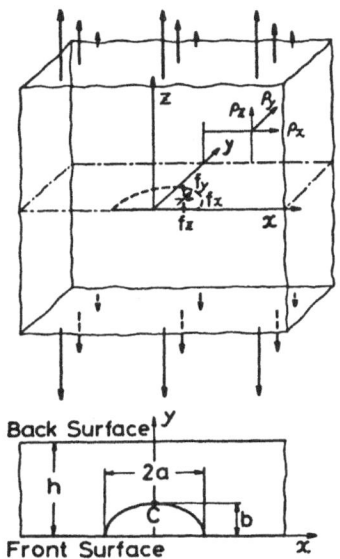

Fig. (1)

point on the crack front, which is the main concern of this paper, is little affected by whatsoever is happening in the boundary layer along the front surface. On this basis, the boundary layer effect [1] is ignored throughout the analysis.

The analysis is done by generalizing the body force method [2], in which the boundary conditions of the front surface are completely satisfied, and those of the crack and the back surface are treated by superposing the stress states of a semi-infinite solid subjected to concentrated forces derived by Mindlin [3].

The crack region is divided into elements as shown in Figure 2. Force doublets and forces in x-y plane are then distributed along those elements with the following densities:

$$f_z^{ij} = W_{ij} \, f_{zo} \; , \; f_x^{ij} = W_{ij} \, f_{xo} \; , \; f_y^{ij} = W_{ij} \, f_{yo}$$

$$f_{zo} = \frac{4(1-\nu)^2 \sigma b}{(1-2\nu)E(k)} \sqrt{1 - \left(\frac{\xi}{a}\right)^2 - \left(\frac{\eta}{b}\right)^2} , \, k = \sqrt{1 - \frac{b^2}{a^2}} \tag{1}$$

$$f_{xo} = \frac{\nu}{1-\nu} \frac{\partial f_{zo}}{\partial \xi} \; , \; f_{yo} = - \frac{\nu}{1-\nu} \frac{\partial f_{zo}}{\partial \eta}$$

where f_{zo}, f_{xo} and f_{yo} are alternative expressions for the body force densities derived by Nisitani and Murakami [2], and W_{ij} are the unknown weights to be determined from the boundary conditions. As for the back surface, take a sufficiently large portion and divide it into rectangular elements as shown by Figure 3, and body forces are distributed along those elements with densities

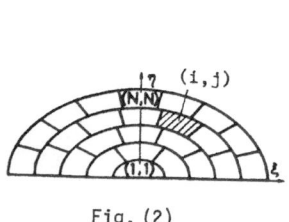

Fig. (2)

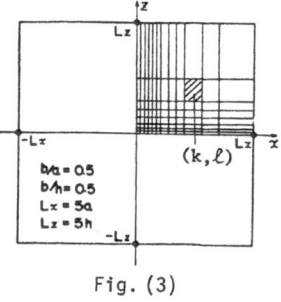

Fig. (3)

$\rho_x^{k\ell}$, $\rho_y^{k\ell}$ and $\rho_z^{k\ell}$ in x, y and z-directions respectively. The boundary conditions along the crack surface are approximated by the conditions for the stress σ_z at the central points of all the crack elements. Those along the back surface are expressed in terms of the resultant forces for each of the back surface elements, in order to improve the accuracy of the numerical results.

The unknown weights W_{ij} and densities $\rho_x^{k\ell}$, $\rho_y^{k\ell}$ and $\rho_z^{k\ell}$ are determined from the above boundary conditions, and the stress intensity factor at the deepest point C on the crack front is calculated from the formula

$$K_I = \frac{W_{NN}}{E(k)} \sigma\sqrt{\pi b} \tag{2}$$

NUMERICAL RESULTS

In demonstrating the numerical results of the stress intensity factor, the following notations will be used, where σ_B is the bending fiber stress:

$$M_T = \frac{K_{I,T}}{\sigma_T\sqrt{\pi b}/E(k)} \quad , \quad F_T = \frac{K_{I,T}}{\sigma_T\sqrt{\pi b}} \qquad \text{(tension)}$$

$$\tag{3}$$

$$M_B = \frac{K_{I,B}}{\sigma_B\sqrt{\pi b}/E(k)} \quad , \quad F_B = \frac{K_{I,B}}{\sigma_B\sqrt{\pi b}} \qquad \text{(bending)}$$

Table 1 gives M_T and F_T for the limiting case of semi-infinite solids, showing the effect of the Poisson's ratio. Results of tension and bending of finite thickness plates are summarized in Table 2 and Figures 4 and 5.

TABLE 1 - M_T AND F_T OF SURFACE FLAWED SEMI-INFINITE SOLIDS

	b/a \ ν	0.2	0.3	0.4
M_T	0	1.122	1.122	1.122
	0.125	1.099	1.106	1.116
	0.25	1.080	1.096	1.117
	0.5	1.047	1.070	1.103
	1.0	1.016	1.036	1.064
F_T	0	1.122	1.122	1.122
	0.125	1.074	1.081	1.091
	0.25	1.007	1.022	1.042
	0.5	0.865	0.883	0.911
	1.0	0.647	0.660	0.677

TABLE 2 - STRESS INTENSITY MAGNIFICATION FACTORS ($\nu = 0.3$)

	b/a \ b/h	0	0.1	0.2	0.3	0.4	0.5	0.6
M_T	0	1.122	1.187	1.361	1.643	2.071	2.728	3.787
	0.125	1.106	1.121	1.188	1.305	1.456	1.631	
	0.25	1.096	1.102	1.138	1.204	1.294	1.395	1.494
	0.5	1.070	1.073	1.088	1.117	1.156	1.201	1.240
	0.75	1.049	1.052	1.060	1.076	1.096	1.119	1.138
	1.0	1.036	1.037	1.041	1.051	1.063	1.076	1.087
M_B	0	1.122	1.042	1.044	1.108	1.242	1.479	1.901
	0.125	1.106	0.984	0.906	0.855	0.814	0.768	
	0.25	1.096	0.966	0.861	0.775	0.697	0.614	0.516
	0.5	1.070	0.935	0.811	0.696	0.584	0.469	0.346
	0.75	1.049	0.911	0.777	0.648	0.521	0.392	0.258
	1.0	1.036	0.891	0.751	0.613	0.476	0.337	0.195

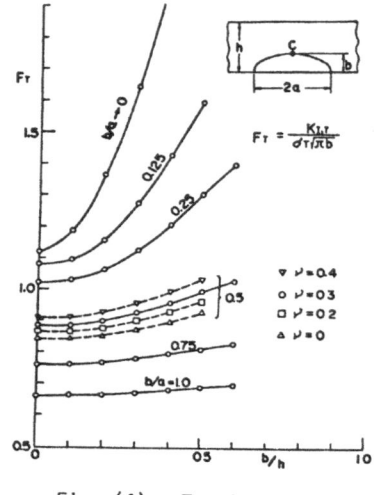

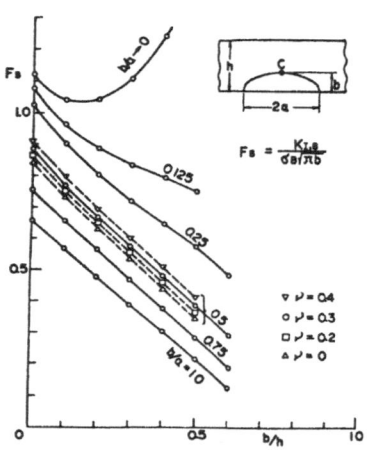

Fig. (4) - Tension   Fig. (5) - Bending

REFERENCES

[1] Hartranft, R. J. and Sih, G. C., Mechanics of Fracture I, edited by Sih, G. C., 224, Noordhoff, 1973.

[2] Nisitani, H. and Murakami, Y., Int. J. Frac., 10-3, 353, 1974.

[3] Mindlin, R. D., Physics, 7, 195, 1936.

STRESS INTENSITY FACTORS FOR EXTERNALLY AND INTERNALLY CRACKED PRESSURIZED
THICK CYLINDERS WITH RESIDUAL AND THERMAL STRESSES

C. P. Andrasic and A. P. Parker

Royal Military College of Science, Shrivenham
Swindon, Wiltshire, SN6 8LA, England

ABSTRACT

A Modified Mapping Collocation (MMC) program is used to calculate stress intensity factors (K) for cracked tubes of various wall ratios, R_2/R_1, where R_1 and R_2 are inner and outer radii respectively. In particular, wall ratios of 1.25, 1.5, 1.75, 2.00, 2.5 and 3.00 are considered. For the case of from one to fifty internal radial cracks, the two loading systems are:

a. Pressure acting in the bore and on the crack surfaces.

b. 100% overstrain. (Overstrain is the proportion of the cylinder wall thickness that is subjected to plastic strain during the autofrettage process).

Similarly, for the case of from one to forty external cracks, the loading systems are:

a. Pressure acting on the bore.

b. 100% overstrain.

Errors are estimated at less than 1% overall, and less than ½% at short crack lengths.

K for a multiply-cracked, partially autofrettaged, internally pressurized thick cylinder may be obtained by appropriate superposition of the solution for the pressurized, non-autofrettaged cylinder and the solution for the unpressurized, fully autofrettaged cylinder. Since the loading due to full autofrettage is essentially equivalent to that produced by steady-state thermal loading, the former solution may be modified exactly to obtain thermal solutions.

INTRODUCTION

Fatigue crack growth arising from the cyclic pressurization of thick cylinders tends to produce a regular array of from one to fifty equal length, internal

radial cracks [1], Figure 1. In order to predict the fatigue growth rate of
such cracks, and hence the safe-life, a knowledge of the opening mode-crack tip

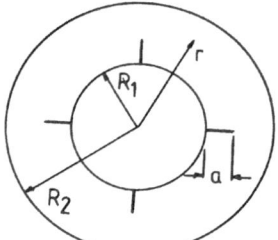

Fig. (1) - Internally cracked thick cylinder

stress intensity factor, K_I, is essential. K_I may be used to predict crack growth
rate via the well-known Paris equation [2] or an alternative crack growth rate
formulation which is applicable to values of K_I approaching the plane strain fracture toughness, K_{IC} [1]. Most of the fatigue lifetime of such cylinders is expended at short crack lengths.

It is common practice to produce a more advantageous stress distribution near
the bore by inducing yielding of the thick cylinder by suitable pressurization,
or swaging prior to use. The residual stress field arising from this so-called
"autofrettage" process is well-known [3], as is the stress field after further removal of cylinder material [4]. However, the optimum autofrettage condition may
not be full autofrettage (100% overstrain*) since fatigue cracks may develop at
the outside radius as a result of the residual tensile stress, Figure 2.

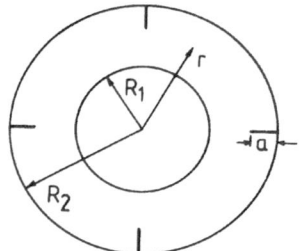

Fig. (2) - Externally cracked thick cylinder

The stress intensity factor in a cracked, fully or partially autofrettaged,
internally pressurized thick cylinder may be obtained by appropriate superposi-

---

*Overstrain is the proportion of the cylinder wall thickness which is subjected
to plastic strain during the autofrettage process.

tion of the solution for the pressurized, non-autofrettaged cylinder, K_I^P, and the solution for the unpressurized, fully autofrettaged cylinder, K_I^A, [5]. In general, it may be anticipated that the latter contribution will be negative (i.e., will act so as to close the cracks) for internal cracks, and positive for external cracks.

Several solutions to K_I^P are available, [1,6-8]. However, each of these solutions has some drawback. The solution due to Goldthorpe [1] is based on an accurate infinite sheet solution for 1 or 40 internal cracks [9], but is only valid at short crack lengths. Pu and Hussain's solution [6] is based on a Finite Element analysis using cubic, isoparametric elements for from 1 to 40 internal cracks. The errors in such a Finite Element solution may be of order 7%. Results are not presented for cracks shorter than 1/10 of the bore radius. Tracy's solution [8] employs a Modified Mapping-Collocation (MMC) technique, with an estimated accuracy of 1%, but is only available for from 1 to 4 cracks.

Parker and Farrow [9] have presented an approximate solution for K_I^A based on the use of Load Relief Factors. Accuracy in this case is estimated at 8%. Since the stress distribution arising from 100% overstrain is equivalent to that arising from steady-state thermal loading of the tube [10], K values for one of these loadings may be obtained from the other by a simple scaling operation.

Thus there is a requirement for an accurate, comprehensive set of stress intensity factor results for thick cylinders of various radii ratios with either internal or external, single or multiple, radial cracks. These results should include both internal pressure and autofrettage loading, and should be of high accuracy at short crack lengths. The results presented herein are intended to fill this requirement.

MODIFIED MAPPING-COLLOCATION (MMC)

The MMC method [11] can be used to compute stress intensity factors and displacements for different geometries and loadings. Complex variable methods, due to Muskhelishvili [12] are utilized. Stresses and displacements within a body are given in terms of the complex stress functions $\phi(z)$ and $\psi(z)$ by:

$$\sigma_x + \sigma_y = 4 \; \text{Re}\{\phi'(z)\} \tag{1a}$$

$$\sigma_y - \sigma_x + 2i\tau_{xy} = 2[\bar{z}\phi''(z) + \psi'(z)] \tag{1b}$$

$$2G(u+iv) = \kappa\phi(z) - z\overline{\phi'(z)} - \overline{\psi(z)} \tag{1c}$$

where the complex variable $z = x+iy$, x and y are the physical coordinates, prime denote differentiation and bars represent the complex conjugate. σ and τ represent direct and shear stress respectively, while u and v are the x and y direction displacements. Also:

$$G = \frac{E}{2(1+\nu)} \quad \text{(E is the Elastic modulus and } \nu \text{ is Poisson's ratio)}$$

$$\kappa = 3-4\nu \quad \text{(plane strain)}, \quad \kappa = \frac{3-\nu}{1+\nu} \text{ (plane stress)}$$

while the resultant force over an arc s is:

$$f_1 + if_2 = i \int_s (X_n + iY_n)ds = \phi(z) + z\overline{\phi'(z)} + \overline{\psi(z)} \qquad (2)$$

where $X_n ds$ and $Y_n ds$ are the horizontal and vertical components of force acting on ds.

The solution of the cracked thick cylinders was carried out on similar lines to that of Tracy [8]. A sketch of a multiply cracked thick cylinder is shown in Figure 3(a). Figure 3(b) shows the geometry that is used when three or more cracks

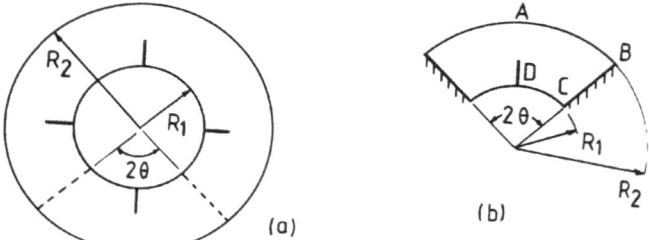

(a)　　　　　(b)

Fig. (3) - Model of segment of cracked thick cylinder

are modelled. The number of cracks, n, will depend on the value of θ, i.e., for n cracks $\theta = \pi/n$, $(50 \geq n \geq 3)$. To model Figure 3(a) using Figure 3(b), it is necessary to impose symmetry conditions along edge BC. Thus along BC

$u \cos\theta - v \sin\theta = 0$

$$f_1 \cos\theta - f_2 \sin\theta = 0 \qquad (3)$$

The first stage of solution involves the conformal mapping

$$z = \omega_1(\gamma) = R_1 e^{(-i\gamma + i\alpha)} \qquad (4)$$

This maps straight lines parallel to the real axis in the γ-plane to curved lines in the physical (z) plane, in particular the real axis in the γ plane is mapped to an arc of radius R_1, centered at the origin in the z-plane, Figure 4. For the cracked region α is usually set to $\pi/2$. Substitute the mapping (4) into (1) and (2) to obtain:

$$f_1 + if_2 = \phi(\gamma) + \frac{\omega_1(\gamma)}{\overline{\omega_1'(\gamma)}} \overline{\phi'(\gamma)} + \overline{\psi(\gamma)} \qquad (5)$$

$$2G(u+iv) = \kappa\phi(\gamma) - \frac{\omega_1(\gamma)}{\overline{\omega_1'(\gamma)}} \overline{\phi'(\gamma)} - \overline{\psi(\gamma)} \qquad (6)$$

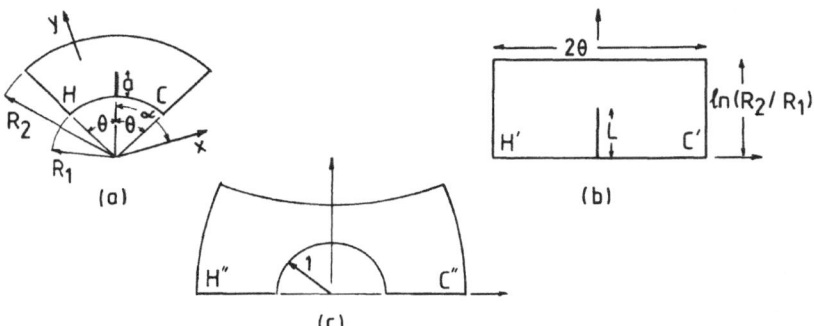

Fig. (4) - Physical and mapped planes. (a) Physical plane;
(b) γ plane; (c) ζ plane

where, to avoid new notation, we define:

$$\phi(\gamma) = \phi(\omega_1(\gamma)) \text{ and } \psi(\gamma) = \psi(\omega_1(\gamma))$$

The analytic continuation arguments of Muskhelishvili are used to ensure traction-free conditions along H'C' and hence HC in the physical plane, Figure 4. Let S^+ and S^- denote the region in the γ-plane above and below the real axis respectively, and define:

$$\phi(\gamma) = -\bar{\psi}(\gamma) - \frac{\omega_1(\gamma)}{\bar{\omega}_1'(\gamma)} \bar{\phi}'(\gamma), \quad \gamma \epsilon S^- \tag{7}$$

where the bar notation is defined by

$$\bar{f}(\gamma) = \overline{f(\bar{\gamma})} \tag{8}$$

Then, $\psi(\gamma)$ can be written as

$$\psi(\gamma) = -\bar{\phi}(\gamma) - \frac{\bar{\omega}_1(\gamma)}{\omega_1'(\gamma)} \phi'(\gamma), \quad \gamma \epsilon S^+ \tag{9}$$

and equations (5) and (6) become

$$f_1 + if_2 = \phi(\gamma) - \phi(\bar{\gamma}) + \frac{[\omega_1(\gamma)-\omega_1(\bar{\gamma})]}{\overline{\omega_1'(\gamma)}} \phi'(\gamma) \tag{10}$$

$$2G(u+iv) = \kappa\phi(\gamma) + \phi(\bar{\gamma}) - \frac{[\omega_1(\gamma)-\omega_1(\bar{\gamma})]}{\overline{\omega_1'(\gamma)}} \overline{\phi'(\gamma)} \tag{11}$$

When γ is real, it is seen that equation (10) gives zero resultant force along H'C' and HC as desired. Note that the analytic continuation argument reduces the problem to finding only one unknown function, $\phi(\gamma)$.

A further mapping is introduced which maps the unit semi-circle plus its exterior in the ζ-plane to the crack plus its exterior in the z-plane, see Figure 4 (b,c).

$$\gamma = \omega_2(\zeta) = \frac{L}{2} (\zeta - \zeta^{-1})$$ (12)

where $L = \ln(1 + a/R_1)$ is the length of the crack in the γ-plane. Substituting (12) into (10) and (11):

$$f_1 + if_2 = \phi(\zeta) - \phi(\bar{\zeta}) + \frac{[\omega_1(\gamma) - \omega_1(\bar{\gamma})]}{\overline{\omega_1'(\gamma)} \; \omega_2'(\zeta)} \; \overline{\phi'(\zeta)}$$ (13)

$$2G(u+iv) = \kappa\phi(\zeta) + \overline{\phi(\zeta)} - \frac{[\omega_1(\gamma) - \omega_1(\bar{\gamma})]}{\overline{\omega_1'(\gamma)} \; \omega_2'(\zeta)} \; \overline{\phi'(\zeta)}$$ (14)

where $\phi(\zeta) = \phi(\omega(\zeta))$, etc.

For certain geometries, it is necessary to introduce partitioning [8] to obtain the desired accuracy. The partitioning plan used for large numbers of cracks (>20) of relatively short crack lengths ($a/(R_2 - R_1) \leq 0.1$) is shown in Figure 5 [13].

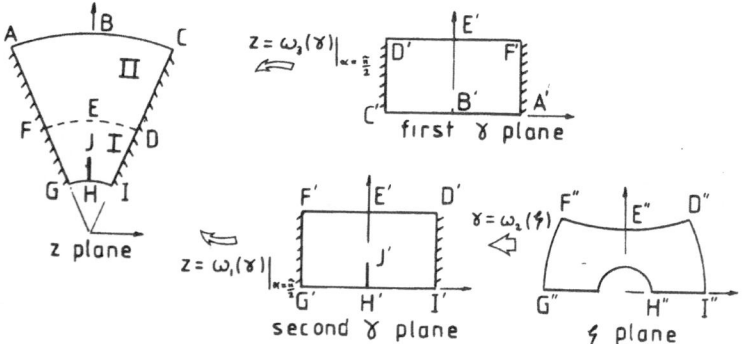

Fig. (5) - Partitioning and mapping scheme for multiply-cracked thick cylinder

For the upper half of the section (region II) in Figure 5, the mapping

$$z = \omega_3(\gamma) = R_2 e^{(i\gamma + i\alpha)}$$ (15)

is used with $\alpha = \pi/2$. When the cracked ring contains one or two cracks, the partitioning plans are as shown in Figures 6 and 7. For the two crack case, only two

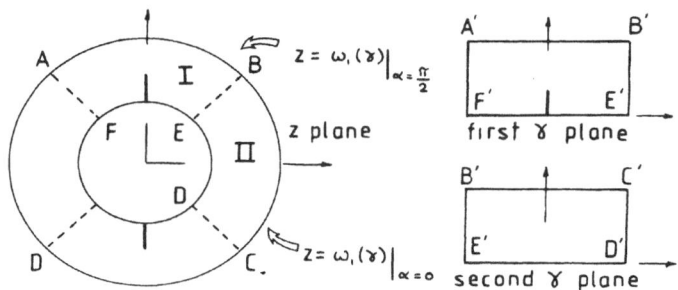

Fig. (6) - Partitioning and mapping scheme for two cracks in a thick cylinder

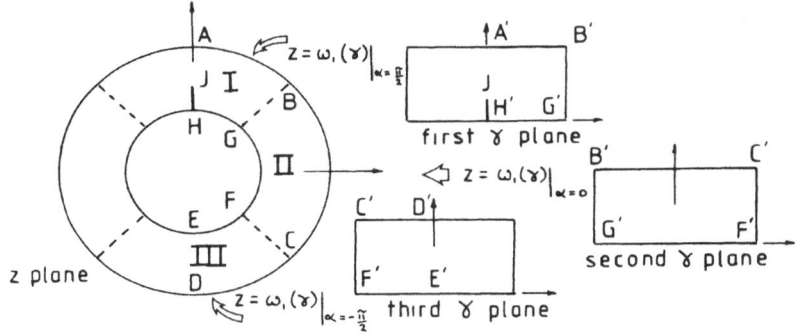

Fig. (7) - Partitioning and mapping scheme for a single crack in a thick cylinder

regions need be considered because of symmetry about both real and imaginary axes. For the single crack, three regions are necessary because the imaginary axis is the only axis of symmetry. In general, each region has its own complex stress and mapping function. For region I (the cracked region), the stress function is $\phi(\zeta)$ while for regions II and III, the stress function is $\phi(\gamma)$. When partitioning is used, it is necessary to "switch" along common boundaries, by imposing equilibrium and compatibility of displacements, i.e.,:

$$(f_1+if_2)_{II} = (f_1+if_2)_I$$

$$(u+iv)_{II} = (u+iv)_I$$

(16)

For symmetry about the imaginary axis, the stress function for the region containing the crack takes the following form [8].

$$\phi(\zeta) = \sum_{n=-\infty}^{\infty} \frac{(a_{2n}\zeta^{2n}+ia_{2n+1}\zeta^{2n+1})}{\zeta+i} \tag{17}$$

where a_{2n} and a_{2n+1} are real coefficients. Also, for regions II and/or III, it is assumed that:

$$\phi(\gamma) = \sum_{n=0}^{\infty} c_n \gamma^n \tag{18}$$

where c_n are unknown complex coefficients. When the uncracked region has symmetry about an axis which is at an angle α to the positive x-axis, (18) can be modified to:

$$\phi(\gamma) = \sum_{n=0}^{\infty} d_n [(-1)^n]^{\frac{1}{2}} e^{-i\alpha_\gamma n} \tag{19}$$

where d_n are unknown real coefficients.

In the MMC method, the infinite series representations of $\phi(\zeta)$ and (if appropriate), $\phi(\gamma)$ are truncated to a finite number of terms. Force conditions are imposed at selected boundary points (along JH, AB, BC and CD in the single crack case, Figure 7), via equations (10) and (13), which gives conditions on the unknown coefficients for each stress function. Thus, each boundary point produces two rows in a real main matrix A, and two corresponding elements in the boundary conditions vector b, where:

$$A\underset{\sim}{x} = \underset{\sim}{b} \tag{20}$$

and x is the vector of unknown coefficients. When necessary, symmetry conditions, given by equation (3) are imposed, as is the case when the thick cylinder has 3 or more cracks, see Figures 2 and 3. Again, these conditions each give rise to two rows in the matrix equation (20).

When partitioning is used, the common boundary points (along BG and CF in the single crack case) are used to obtain conditions relating coefficients of the stress function for each region, via equation (16), in conjunction with (10), (11), (13) and (14). Each common boundary point gives four rows in A and four corresponding zeroes in b. In general, A is a matrix of ℓ rows and m columns, where ℓ and m depend upon the number of boundary points and unknown coefficients respectively. It was found that convergence is generally better when $2m < \ell < 2.5m$, this conforms with other workers [14]. A least-square error minimization procedure was used to solve the overdetermined set of linear equations.

For the case of the single crack, there is an indeterminate contribution to the resultant force (f_1+if_2) in summing up from the inner to the outer radius, e.g., along edge ED in Figure 7. When the crack face is loaded, the resultant force applied to the outer edge cannot be put to zero as in the case of two or more cracks. As there is only one line of symmetry for the single crack problem, f_1 is zero but f_2 is not (=c, say, a real constant). It is necessary to treat this constant c as an unknown and incorporate it into the system of equations to be solved.

Knowing the coefficients for the stress function in the crack region, the stress intensity factor, K, is given by [15].

$$K = K_I - iK_{II} = 2(2\pi)^{\frac{1}{2}} \underset{z \to z_c}{\text{Limit}} (z-z_c)^{\frac{1}{2}} \cdot \phi'(z) \tag{21}$$

where K_I and K_{II} are opening and sliding mode stress intensity factors respectively, and z_c is the location of the crack tip. Thus, substituting from (17) into (21):

$$K_I = iK_{II} = 2(2\pi)^{\frac{1}{2}} \underset{\substack{\zeta \to i \\ \gamma \to iL}}{\text{Limit}} (-i(\omega_1(\gamma) - \omega_1(iL))^{\frac{1}{2}} \frac{\phi(\zeta)}{\omega_1'(\gamma) \cdot \omega_2'(\zeta)} \tag{22}$$

giving:

$$K_I - iK_{II} = 2 \frac{\pi}{LR_1 e^L}^{\frac{1}{2}} \cdot \phi'(i) \tag{23}$$

where points $\zeta = i$ and $\gamma = iL$ correspond to the location of the crack tip in ζ and γ planes respectively.

The method employed for externally cracked cylinders is almost identical to that for internal cracking. The major difference is in the choice of complex mappings, the mapping (15) is normally used instead of (4) for externally cracked configurations. Mapping (15) maps a rectangle in the γ-plane into a ring segment in the z-plane, Figure 8.

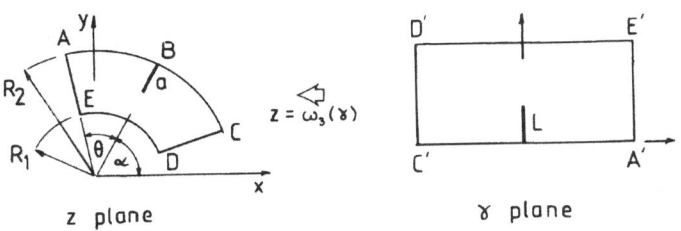

$$z = \omega_3(\gamma)$$

z plane γ plane

Fig. (8) - General mapping scheme for an externally cracked thick cylinder

STRESS INTENSITY FACTOR SOLUTIONS FROM MMC

In all cases considered, there is symmetry of loading and geometry about the crack line, the only non-zero stress intensity factor being K_I.

A. Internal Pressure

Consider a tube, internal radius R_1 and external radius R_2 which is subjected to an internal pressure p. The distribution of hoop (σ_θ) stress in this case is given by Lame's equation as:

$$\sigma_\theta = \frac{R_1^2 p}{R_2^2 - R_1^2} \left[1 + \frac{R_2^2}{r^2}\right] \tag{24}$$

where r is the radius at which the stress is defined.

In order to determine K_I, the appropriate crack line loading for an external crack in such a thick cylinder consists of a loading on the crack surface equal and opposite to that predicted by equation (24), [5]. For the case of an internal crack, it is necessary to add a constant crack line loading p to account for the pressure which has infiltrated the crack from the bore.

The opening mode stress intensity factor K_I^p for internally cracked, pressurized cylinders with R_2/R_1 ratios (R_2 = outer radius, R_1 = inner radius) of between 1.25 and 3 are presented in Figures 9 to 14 for from 2 to 50 cracks. Note that in this case, K_I is non-dimensionalized using:

$$K_o = p(\pi a)^{\frac{1}{2}} \frac{2R_2^2}{(R_2^2 - R_1^2)} \tag{25}$$

where p is the internal pressure, and a is the crack length. The limiting value for all these results at short crack lengths is known to be 1.12. The form of convergence at short crack lengths is shown in Figure 15, and indicates an important aspect of the MMC technique adopted, namely its accuracy at short crack lengths.

The equivalent results for the case of a single internal crack are presented in Figure 16.

In the case of external cracking, with two or three radial cracks, opening mode stress intensity factors for internal pressure are presented in Figures 17 to 22. Here, K_I is non-dimensionalized using:

$$K_o = p(\pi a)^{\frac{1}{2}} \frac{2R_1^2}{(R_2^2 - R_1^2)} \tag{26}$$

Equivalent results for the case of a single external crack are given in Figure 23.

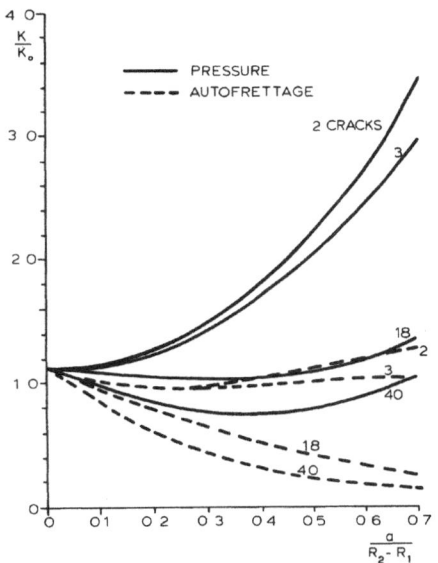

Fig. (9) - Stress intensity factors for an internally cracked thick cylinder, $R_2/R_1 = 1.25$

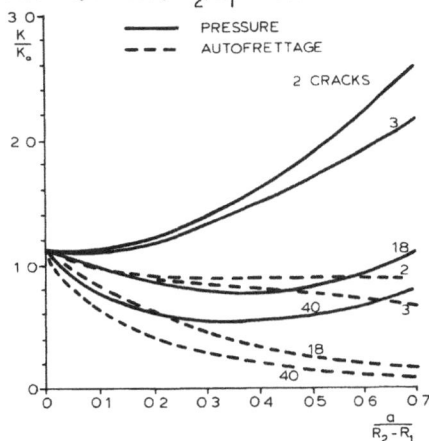

Fig. (10) - Stress intensity factors for an internally cracked thick cylinder, $R_2/R_1 = 1.5$

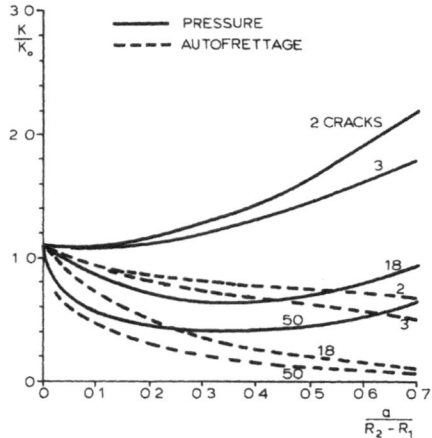

Fig. (11) - Stress intensity factors for an internally cracked thick cylinder, $R_2/R_1 = 1.75$

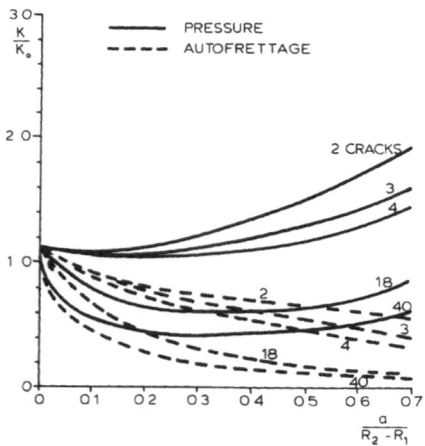

Fig. (12) - Stress intensity factors for an internally cracked thick cylinder, $R_2/R_1 = 2.0$

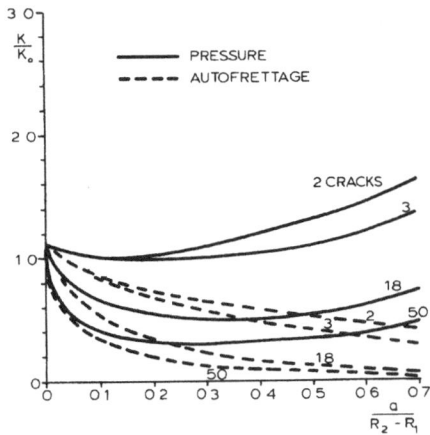

Fig. (13) - Stress intensity factors for an internally cracked
thick cylinder, R_2/R_1 = 2.5

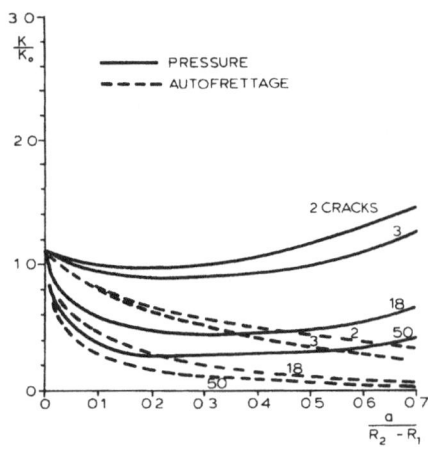

Fig. (14) - Stress intensity factors for an internally cracked
thick cylinder, R_2/R_1 = 3.0

206

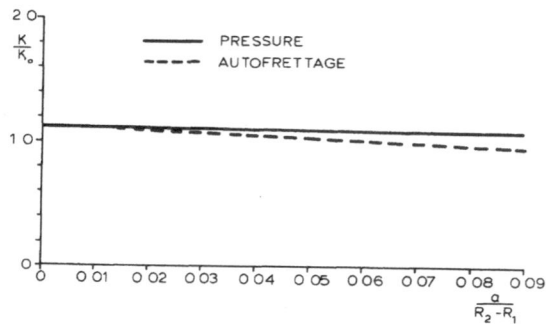

Fig. (15) - Stress intensity factors for an internally cracked
thick cylinder, R_2/R_1 = 1.25, 18 cracks - short
crack length convergence

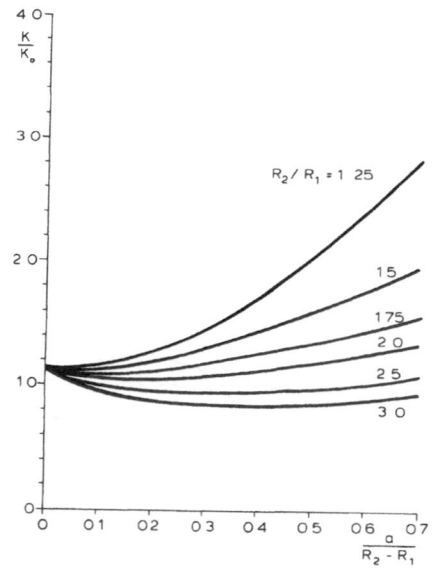

Fig. (16) - Stress intensity factors for pressurized thick cylinders
with single internal cracks

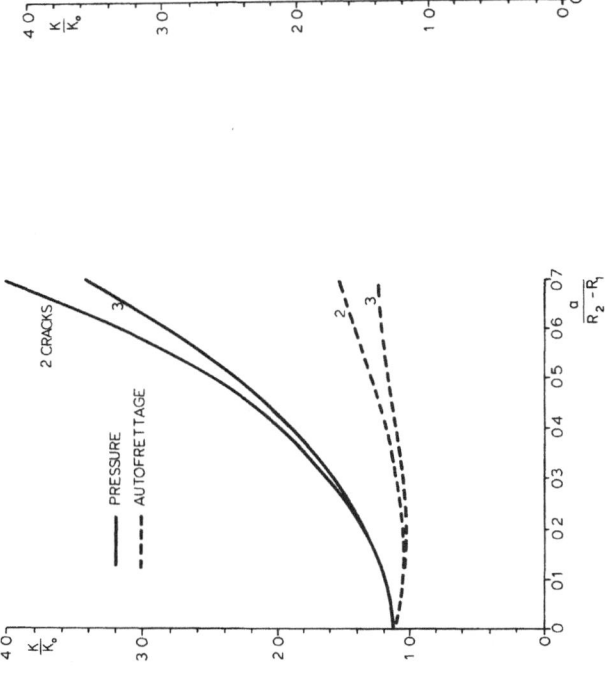

Fig. (17) - Stress intensity factors for an externally cracked thick cylinder, $R_2/R_1 = 1.25$

Fig. (18) - Stress intensity factors for an externally cracked thick cylinder, $R_2/R_1 = 1.5$

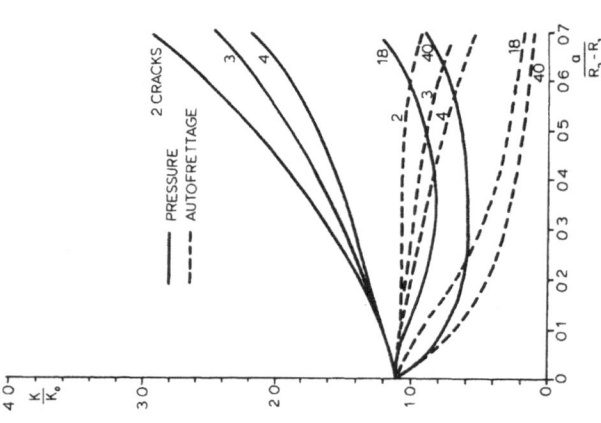

Fig. (20) – Stress intensity factors for an externally cracked thick cylinder, $R_2/R_1 = 2.0$

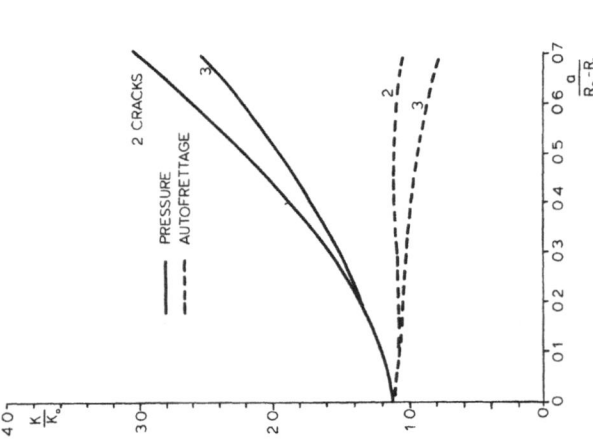

Fig. (19) – Stress intensity factors for an externally cracked thick cylinder, $R_2/R_1 = 1.75$

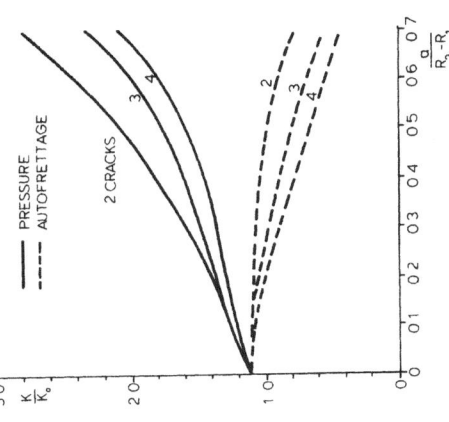

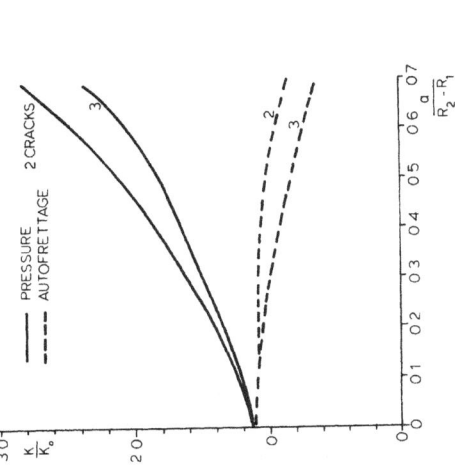

Fig. (21) - Stress intensity factors for an externally cracked thick cylinder, R_2/R_1 = 2.5

Fig. (22) - Stress intensity factors for an externally cracked thick cylinder, R_2/R_1 = 3.0

210

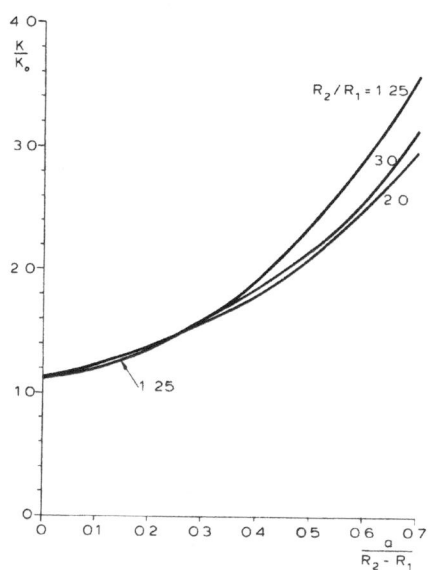

Fig. (23) - Stress intensity factors for pressurized thick cylinders
with single external cracks

The results presented for up to four internal or external cracks may be
compared with those due to Tracy [8]. Agreement is generally within 1%.

B. Autofrettage Stresses (100% Overstrain)

For the case of a fully autofrettaged tube, internal radius R_1, external
radius R_2 the distribution of hoop (σ_θ) stress is given by [5]:

$$\sigma_\theta = -Y\ln(R_2/R_1)(1 + \frac{R_1^2}{R_2^2-R_1^2} [1 + \frac{R_2^2}{r^2}]) + Y(1+ n(r/R_1)) \tag{27}$$

where Y is the yield strength of the material in uniaxial tension.

In the determination of the K_I contribution arising from such a residual
stress distribution, it is necessary to load the crack surface with tractions
equal and opposite to those given in (27). The K_I^A solutions for the case of an
internally cracked fully autofrettaged tube are given in Figures 9 to 14 and 24.
In this case, K_I is non-dimensionalized using:

$$K_o = Y(\pi a)^{\frac{1}{2}} \left[1 - \left(\frac{2R_2^2 \ln(R_2/R_1)}{R_2^2 - R_1^2} \right) \right] \tag{28}$$

The limiting value at short crack lengths is again 1.12, and at long crack lengths is zero for two or more cracks. The form of convergence at short crack lengths is shown in Figure 15. (<u>Note</u>: While stress intensity factors arising from the auto-frettage distribution <u>have</u> been calculated for $R_2/R_1 \geq 2.22$, these are cases in which reversed yielding will occur during unloading following autofrettage. These results should not be used to calculate K_I in this case, although they are still applicable to the "equivalent" case of steady-state thermal loading [10], and for appropriate superpositions to model partial autofrettage [5]).

The results for the case of full autofrettage may be compared with an earlier, approximate solution [9]. Agreement is generally within the estimated 8% accuracy of [9].

The K_I^A solutions for the externally cracked fully autofrettaged tube are given in Figures 17 to 22 and 25. To obtain the limiting value of 1.12 at short crack lengths, K_I is non-dimensionalized using:

$$K_o = Y(\pi a)^{\frac{1}{2}} \left[1 - \left(\frac{2R_1^2 \ln(R_2/R_1)}{R_2^2 - R_1^2} \right) \right] \tag{29}$$

Once again, the limiting value at long crack lengths tends to zero.

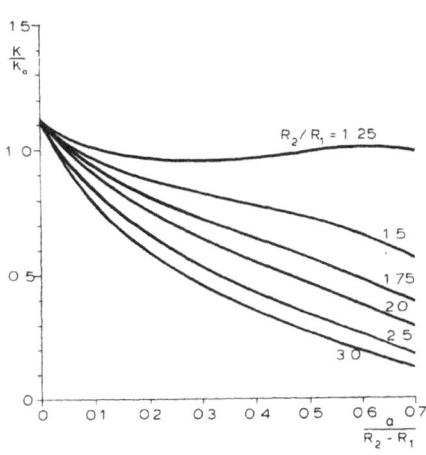

Fig. (24) - Stress intensity factors for fully autofrettaged thick cylinders with single internal cracks

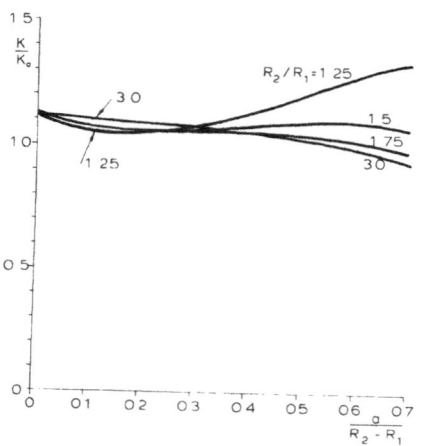

Fig. (25) - Stress intensity factors for fully autofrettaged thick cylinders with single external cracks

INTERNAL PRESSURE AND PARTIAL AUTOFRETTAGE

A. Internal Crack(s)

The stress intensity factor for an internally cracked, pressurized, partially autofrettaged tube K_I^* is given by [5]:

$$K_I^* = [1 + \frac{Y}{p} \ln(R_2/R_A) - \frac{Y}{2pR_2^2} (R_2^2 - R_A^2)]K_I^P + K_I^A, \quad 0 < a < R_A - R_1 \tag{30}$$

where a is the crack length and R_A is the radius within which plastic flow occurs during the autofrettage process. (For example, $100\ R_A/(R_2 - R_1) \equiv$ percentage overstrain).

B. External Crack(s)

The stress intensity factor for an externally cracked pressurized, partially autofrettaged tube, K_I^{**}, is obtained by a simple scaling of the results for internal pressure, namely [5]:

$$K_I^{**} = [1 + \frac{Y}{p} (\frac{R_A^2 - R_1^2}{2R_1^2}) - \frac{Y}{p} \ln(R_A/R_1)]K_I^P, \quad 0 < a < R_2 - R_A \tag{31}$$

Thus, stress intensity factors for cracked, pressurized, partially autofrettaged thick cylinders may be obtained directly from the results presented herein, via equation (30) or (31).

SUMMARY

The Modified Mapping Collocation (MMC) method has been used in the derivation of K_I values for a range of internally pressurized or fully autofrettaged thick cylinders with single or multiple cracks. Cases of internal cracking and of external cracking have been considered. Where comparison is possible, the results are in good agreement with those of other workers.

A particular feature of the MMC method employed is high accuracy at very short crack lengths, the regime wherein most fatigue lifetime is expended.

A straightforward exact superposition may be employed to calculate K_I values for cracked, pressurized, partially autofrettaged thick cylinders with any proportion of overstrain from 0 to 100%.

The stress intensity factor presented in this paper may be employed in the calculation of residual strength, or of fatigue crack growth rates and lifetimes. Numerical values for the results are given in [16,17].

ACKNOWLEDGEMENTS

This work was supported by a UK Ministry of Defence research contract. One of us (CPA) acknowledges support as a Research Scientist for the period of the work.

REFERENCES

[1] Goldthorpe, B. D., "Fatigue and fracture of thick walled cylinders and gun barrels", Case Studies in Fracture Mechanics, AMMRC MS77-5, US Army Materials and Mechanics Research Center, 1977.

[2] Paris, P. C. and Erdogan, F., "A critical analysis of crack propagation laws", Trans. ASME, J. Bas. Engng., 85, pp. 528-534, 1963.

[3] Godfrey, D. E. R., "Theoretical Elasticity and Plasticity for Engineers", Thames and Hudson, 1959.

[4] Kendall, D. P., "The effect of material removal on the strength of autofrettaged cylinders", AD-701-049, Watervliet Arsenal, New York, January 1970.

[5] Parker, A. P., "Stress intensity and fatigue crack growth in multiply-cracked, pressurized, partially autofrettaged thick cylinders", Fatigue of Engineering Materials and Structures, 4, 4, pp. 321-330, 1981.

[6] Pu, S. L. and Hussain, M. A., "Stress intensity factors for a circular ring with uniform array of radial cracks using cubic isoparametric singular elements", Transactions of 24th Conference of Army Mathematics, ARO Report 79-1, 1979.

[7] Baratta, F. I., "Stress intensity factors for internal multiple cracks in thick-walled cylinders stressed by internal pressure using load relief factors", Engng. Frac. Mech., 10, pp. 691-697, 1978.

214

[8] Tracy, P. G., "Elastic analysis of radial cracks emanating from the outer and inner surfaces of a circular ring", Engng. Frac. Mech., 11, pp. 291-300, 1979.

[9] Parker, A. P. and Farrow, J. R., "Stress intensity factors for multiple radial cracks emanating from the bore of an autofrettaged or thermally stressed, thick cylinder", Engng. Frac. Mech., 14, pp. 237-241, 1981.

[10] Parker, A. P. and Farrow, J. R., "On the equivalence of axi-symmetric bending, thermal and autofrettage residual stress fields", J. Strain Analysis, 15, 1, pp. 51-52, 1980.

[11] Bowie, O. L. and Neal, D. M., "A modified mapping collocation technique for accurate calculation of stress intensity factors", Int. J. Frac. Mech., 6, pp. 199-206, 1970.

[12] Muskhelishvili, N. I., "Some Basic Problems of the Mathematical Theory of Elasticity", Noordhoff, 1973.

[13] Andrasic, C. P., "Numerical methods applied to the solution of problems in fracture mechanics", CNAA Ph.D. Thesis, Royal Military College of Science, Shrivenham, 1981.

[14] Eason, E. D., "A review of least squares methods for solving partial differential equations", Int. J. Num. Meth. Engng., 10, pp. 1021-1046, 1976.

[15] Sih, G. C., Paris, P. C. and Erdogan, F., "Crack tip stress intensity factors for plane extension and plate bending problems", J. Appl. Mech., 29, pp. 306-312, 1962.

[16] Parker, A. P. and Andrasic, C. P., "Stress intensity factors for multiply cracked thick cylinders and cracked ring segments", Technical Note MAT/28, Royal Military College of Science, Shrivenham, 1981.

[17] Parker, A. P. and Andrasic, C. P., "Stress intensity factors for thick cylinders with single radial cracks", Technical Note MAT/33, Royal Military College of Science, Shrivenham, 1982.

INFLUENCE OF CRACK CLOSURE ON THE STRESS INTENSITY FACTOR FOR CYLINDRICAL
SHELLS SUBJECTED TO BENDING

K. N. Ramachandran Nambissan and R. S. Alwar

Indian Institute of Technology
Madras 600036, India

ABSTRACT

Cylindrical shell with axial through crack and subjected to moment load is
analyzed taking into account the closure of the crack faces on the compression
side. A three-dimensional finite element method employing degenerate quarter-
point brick elements to model the crack tip region is used for the analysis. The
crack faces are modelled such that they come in contact over an area on the compression side and interact with each other. The influence of the crack closure on
the variation of stress intensity factor along the crack front is obtained.

INTRODUCTION

The problem of cylindrical shells with axial and circumferential through-the-
thickness cracks was solved by Folias [1], Erdogan et al [2,3] and Duncan-Fama and
Sanders [4] using shallow thin shell theory and imposing Kirchhoff's boundary conditions on the crack faces. These boundary conditions are, however, inadequate
to determine the intensity of the stress field accurately in the neighbourhood of
the crack tip. As in the case of plates, the approximate satisfaction of the
boundary conditions along the crack faces gives rise to a difference in the angular distribution of the extensional and bending stress fields near the crack tip
and hence they cannot be conveniently combined except along the line of crack extension.

Sih and Hagendorf [5] developed a tenth order system of shell equations for
spherical shells which accounted for the effects of transverse shear. Their solution satisfied all the five physically natural boundary conditions on a free edge
of the shell and acquired a three-dimensional character in the crack front stresses
resembling those in [9]. Erdogan and Dalale [6] in 1978, presented solutions for
a circumferentially cracked cylindrical shell, taking the transverse shear effects
into account. Numerical solutions for both axial and circumferential cracks in
cylindrical shells were given by Barsoum et al [7] using thick shell elements developed by them. Ayres [8] gave a three-dimensional finite element analysis for
an axially cracked cylindrical shell.

None of the above solutions take into consideration the crack closure which
may arise on the compression side for bending loads. In the case of internal pres-

sure, the hoop stresses in the shell will be predominant and an axial crack may not close for this case. But for moment load case which will arise in evaluating cracks in cylinders with discontinuities due to external loads, attachments and cutouts and also for curved panels subjected to bending, the crack closure takes place and hence will influence the response of the structure.

The object of the present investigation is to find out the region of crack closure and to obtain the influence of this closure on the stress intensity factors for axially cracked cylindrical shells subjected to moment loads. The crack faces are allowed to close on the compression side of the shell and to interact with each other over an area. The effect of crack closure should be considered to get the true nature of the stress field near the crack front.

EXTRACTION OF STRESS INTENSITY FACTORS

The three-dimensional stress and displacement fields in the close neighbourhood of the crack front have the following form. (See Figure 1 for crack-tip coordinates).

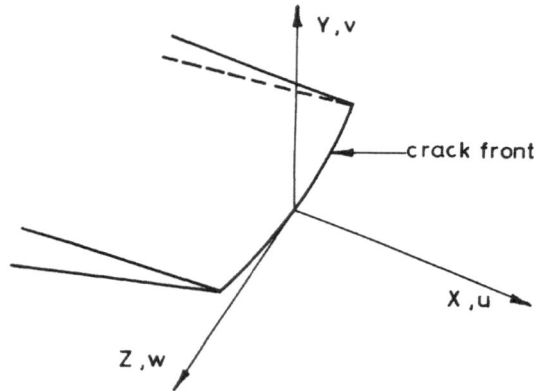

Fig. (1) - Coordinate system near the crack front

$$\sigma_x = \frac{K_1(z)}{(2r)^{1/2}} \cos \frac{\theta}{2} [1 - \sin \frac{\theta}{2} \sin \frac{3\theta}{2}]$$

$$- \frac{K_2(z)}{(2r)^{1/2}} \sin \frac{\theta}{2} [2 + \cos \frac{\theta}{2} \cos \frac{3\theta}{2}] + 0(1) \qquad (1a)$$

$$\sigma_y = \frac{K_1(z)}{(2r)^{1/2}} \cos \frac{\theta}{2} [1 + \sin \frac{\theta}{2} \sin \frac{3\theta}{2}]$$

$$+ \frac{K_2(z)}{(2r)^{1/2}} \sin \frac{\theta}{2} \cos \frac{\theta}{2} \cos \frac{3\theta}{2} + 0(1) \tag{1b}$$

$$\sigma_z = \nu(\sigma_x + \sigma_y) \tag{1c}$$

$$\tau_{xy} = \frac{K_1(z)}{(2r)^{1/2}} \sin \frac{\theta}{2} \cos \frac{\theta}{2} \cos \frac{3\theta}{2}$$

$$+ \frac{K_2(z)}{(2r)^{1/2}} \cos \frac{\theta}{2} [1 - \sin \frac{\theta}{2} \sin \frac{3\theta}{2}] + 0(1) \tag{1d}$$

$$\tau_{xz} = - \frac{K_3(z)}{(2r)^{1/2}} \sin \frac{\theta}{2} + 0(1) \tag{1e}$$

$$\tau_{yz} = - \frac{K_3(z)}{(2r)^{1/2}} \cos \frac{\theta}{2} + 0(1) \tag{1f}$$

$$4Gu = K_1(z)(2r)^{1/2} \cos \frac{\theta}{2} [\kappa - 1 + 2 \sin^2 \frac{\theta}{2}]$$

$$+ K_2(z)(2r)^{1/2} \sin \frac{\theta}{2} [\kappa + 1 + 2 \cos^2 \frac{\theta}{2}] + 0(r) \tag{2a}$$

$$4Gv = K_1(z)(2r)^{1/2} \sin \frac{\theta}{2} [\kappa + 1 - 2 \cos^2 \frac{\theta}{2}]$$

$$- K_2(z)(2r)^{1/2} \cos \frac{\theta}{2} [\kappa - 1 - 2 \sin^2 \frac{\theta}{2}] + 0(r) \tag{2b}$$

$$Gw = K_3(z)(2r)^{1/2} \sin \frac{\theta}{2} + 0(r) \tag{2c}$$

:re $\kappa = 3 - 4\nu$ for plane strain and

$\kappa = \frac{3 - \nu}{1 + \nu}$ for generalized plane stress

ν = Poisson's ratio

An exact three-dimensional solution [10] incorporates $\kappa = 3-4\nu$ and hence this value has been used in the present analysis.

The finite element values of displacements or stresses are matched with the elastic singular solutions in the vicinity of the crack tip, equations (1) and (2), yielding the stress intensity factors as a function of distance from the crack tip. Extrapolation to the crack tip using the exact stress or displacement expressions, yields accurate values for the stress intensity factors. Stresses are calculated at two consecutive Gauss points very close to the crack tip and lying on a radial line from the crack front, using a 10x10x10 Gauss point rule. Substituting the computed values of σ_y in equation (1b) and extrapolating using square root singular behaviour of the stresses, the value of SIF at r=0 is obtained. Alternatively, the crack opening displacements (COD) at the quarter-node and end-node of the crack tip element may be used to calculate the stress intensity factors $K_1^{(1)}$ and $K_1^{(2)}$ respectively using equation (2b) and the extrapolated value of the stress intensity factor K_1 at r=0 is given by $K_1 = 2K_1^{(1)} - K_1^{(2)}$ [11]. The evaluation of K_1 in the manner incorporates the $\sqrt{r}$ displacement variation near the crack tip and eliminates the rigid body motion and linear terms from the displacements. The K_1 variation along the crack front is obtained by calculating the continuous displacement field on the traction-free surface of the crack tip element from the element nodal displacement values and the shape functions of the element. Substituting $\eta = -1$ in the shape functions of the element and substituting for ξ from the quarter point singularity relationship viz. $r = \ell(\xi-1)^2/4$, the displacement field on the free surface of the crack tip element for any r and z can be written as follows, see Figure 2.

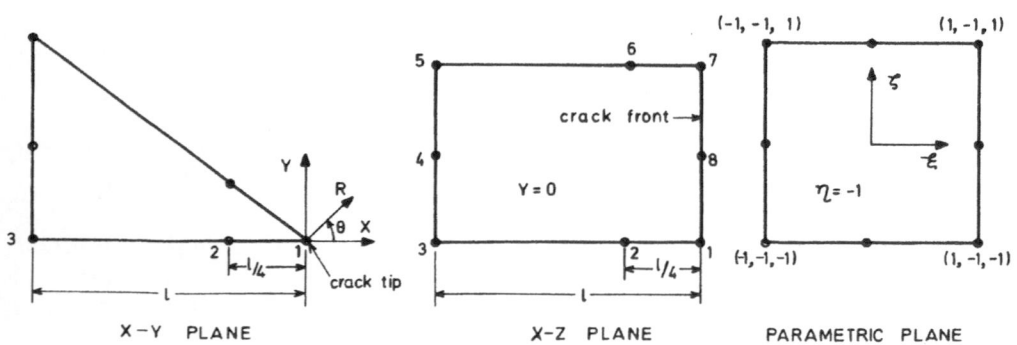

X−Y PLANE   X-Z PLANE   PARAMETRIC PLANE

Fig. (2) - Crack tip element node numbers used in equation (3)

$$v(r,z) = \{(2v_2 - v_3 + v_4 - v_5 + 2v_6) - (\frac{z}{h} - \frac{1}{2}) \times$$

$$\times (4v_2 - v_3 + v_5 - 4v_6) + \frac{1}{2}(\frac{2z}{h} - 1)^2(v_3 - 2v_4 + v_5)\} \sqrt{r/\ell}$$

$$+ \{2(\frac{z}{h} - 1)(2v_2 - v_3) - \frac{2z}{h}(2v_6 - v_5)\} \frac{r}{\ell} \tag{3}$$

The crack front displacements v_1, v_7 and v_8 are zero. It can be seen that equation (3) reduces to the nodal displacements on substitution of the corresponding coordinates. u displacements can also be obtained in a similar way.

NO-CLOSURE ANALYSIS

To check the accuracy of the analysis, no-closure solutions have been obtained for axially cracked cylindrical shells subjected to internal pressure and bending moment and compared with the available results. Using the method in [2], non-dimensional values of extensional and bending stress intensity factors denoted by K_e and K_b respectively have been calculated for various shell parameters λ and R_0/h ($\lambda = [12(1-v^2)]^{1/4}$ $a/(R_0h)^{1/2}$). The non-dimensionalizing factors are $\sigma_0\sqrt{a}$, representing the SIF for an infinite plate with a crack length "2a" and subjected to the same remote stress σ_0 as the normal hoop stress in the cylinder for the case of internal pressure and $6M_0\sqrt{a}/h^2$ for the case of applied bending moment M_0 on the remote boundary of the shell.

Figure 3 shows the comparison of the present three-dimensional results for the case of internal pressure loading with the numerical results obtained by Erd-

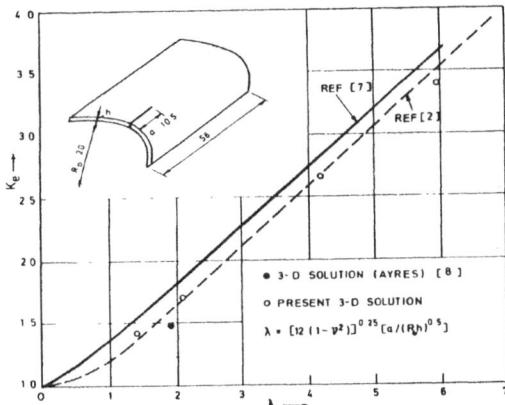

Fig. (3) - Non-dimensional extensional SIF (internal pressure)

ogan et al [2] using classical thin shell theory. Barsoum et al [7] using shear deformation theory and by Ayres [8] using three-dimensional finite element analysis. The shell geometry analyzed is also shown in Figure 3. Only single layer of elements has been taken across the thickness of the shell. The shell thickness is varied to get various values for the shell parameter λ. It is shown in [7] that K_e, the extensional stress intensity factor, is independent of R_o/h ratio. The significant difference between the results using three-dimensional analysis [8] and the shear deformation theory [7] was attributed by Barsoum et al [7] to the end effect of the shell. The three-dimensional analysis by Ayres used considerably shorter length for the shell (= 28) which was less than the decay length L_c (= $2.5\sqrt{R_o h}$). Barsoum et al used a much larger length for the cylinder which was several times the decay length. The present three-dimensional analysis with a length equal to 56 gives results which are close to the results by Barsoum et al. The existing difference between the two results may be due to the end effects since Barsoum et al chose larger lengths in order to simulate the conditions of an infinite cylinder.

Table 1 compares the results of the present analysis with those by Barsoum for the case of moment loading on the boundary of the shell. It may be seen that there

TABLE 1 - NON-DIMENSIONAL K_b AND K_e VALUES (MOMENT LOAD)

λ	$\dfrac{R_o}{h}$	K_b Barsoum et al	K_b present	K_b Barsoum et al	K_e present
0.135	20	0.94	0.936	Extremely small	Extremely small
0.192	10	0.96	0.958	-do-	-do-
0.235	6.6	0.975	0.979	-do-	-do-
1.0	10	0.77	0.761	0.04	0.0281

is good agreement between the two results. For the shell parameters considered, the K_e values for the bending load are extremely small compared to K_b values. It is also verified as a check that the analysis gives results corresponding to the plate solution by making the radius of the shell extremely larger compared to other geometrical dimensions of the shell.

The above analysis deals with the no-closure solution where crack faces are considered as stress free. A three-dimensional analysis for the shell geometry shown in Figure 4 taking into account the closure of crack faces during bending is given below.

DETERMINATION OF CLOSURE AREA

In reality, during bending, the crack faces on the compression side will come into contact over an area and this area of crack closure depends upon the shell dimensions and the crack length. A three-dimensional finite element analysis is well suited to model the crack faces such that a desired area of crack closure may be imposed while specifying the boundary conditions on the crack faces. From the symmetry of the problem, the "v" displacements of the crack faces in the region where the crack faces close, will be zero. The closure area is obtained by the following procedure.

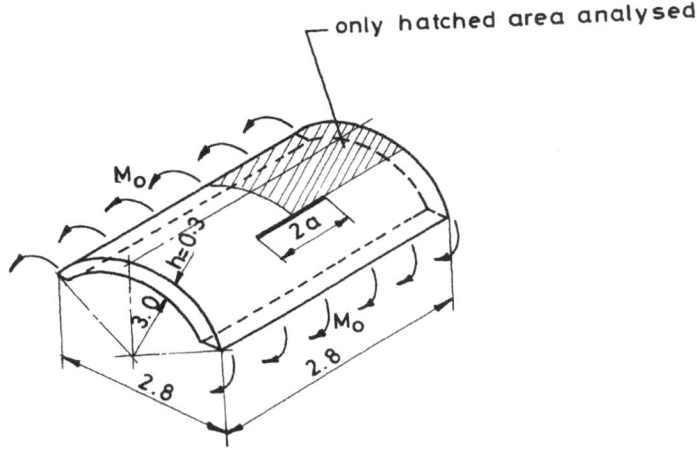

Fig. (4) - Geometry of the shell

The shell geometry is discretized by a number of layers across the shell thickness. The closure height h_1 is represented by the layer (A), Figure 5a.

All the nodes of layer (A) lying on the crack face are constrained to arrest the "v" displacements (COD) and the displacements and stresses are determined for this condition. The crack opening displacements are determined not only at nodal points, but also at extremely close intervals across the plate thickness and at different distances from the crack front using the shape functions appropriate for the element. Figure 5 shows schematically how these COD values may be interpreted to arrive at as correct a closure height as possible in the neighbourhood of the crack front using the numerical method employed. If the layer height h_1 is less than the actual closure height, an interpenetration of the surfaces in layer (B) near the interface of the layers (A) and (B) is observed, Figure 5c. For a given value of the applied bending moment, closure height h_1 is gradually varied and the finite element analysis performed, so that a displacement configuration like what is shown in Figure 5d is obtained. The COD distribution for this case changes gradually from positive value in layer (B) to zero just at the interface and remains zero throughout in layer (A). The closure height h_1 which is evidently a maximum at the crack front is thus fixed based on the COD distribution in the neighbourhood of the crack front. This method is an iterative procedure. After fixing h_1 in the neighbourhood of the crack front as explained above, the arrested nodes in layer (A) lying on the crack face are released in turn starting from the centre of the crack length, to arrive at the minimum number of nodes that need to be arrested so that there is no interpenetration of the crack faces anywhere along the crack length. If more number of nodes than necessary are released, the crack faces are found to interpenetrate indicated by negative COD values. It is also found that the release of the nodes in the region away from the crack front has no effect on the already

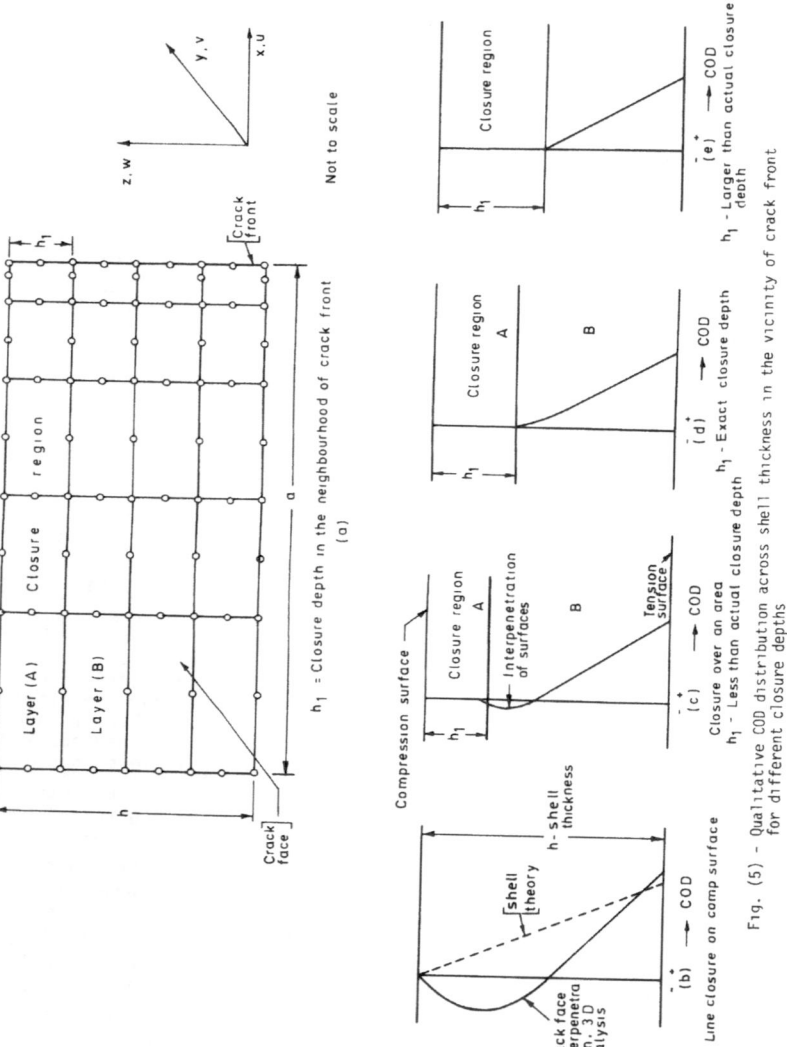

h_1 = Closure depth in the neighbourhood of crack front

(a)

Not to scale

(b)

Line closure on comp surface

(c)

Closure over an area
h_1 - Less than actual closure depth

(d)

h_1 - Exact closure depth

(e)

h_1 - Larger than actual closure depth

Fig. (5) - Qualitative COD distribution across shell thickness in the vicinity of crack front for different closure depths

determined closure zone in the neighbourhood of the crack front. Thus, the minimum value for the layer height h_1 and the least number of nodes to be arrested on the crack face so that there is no crack face interpenetration is taken to represent the region of crack closure. The procedure is repeated by increasing the number of layers across the shell thickness until convergence is achieved. Care is also taken to ensure that a greater area of closure is not imposed which will alter the problem as one of a shell with part-through crack, Figure 5e.

NUMERICAL ANALYSIS

The geometry of the shell and the mesh pattern adopted for the analysis are shown in Figures 4 and 6. The crack tip element size r_m is taken to be 0.01 for the semi-crack length "a" = 0.1. The ability of the isoparametric element for representing curved boundaries has been utilized to model the shell curvature. Care has been taken to properly locate the mid-nodes of regular elements and quarter-point nodes of the singular elements lying on the curved surface. Making use of the symmetry only 1/4 of the shell is considered for the analysis. Figure 7 shows the singular elements used in the analysis [12].

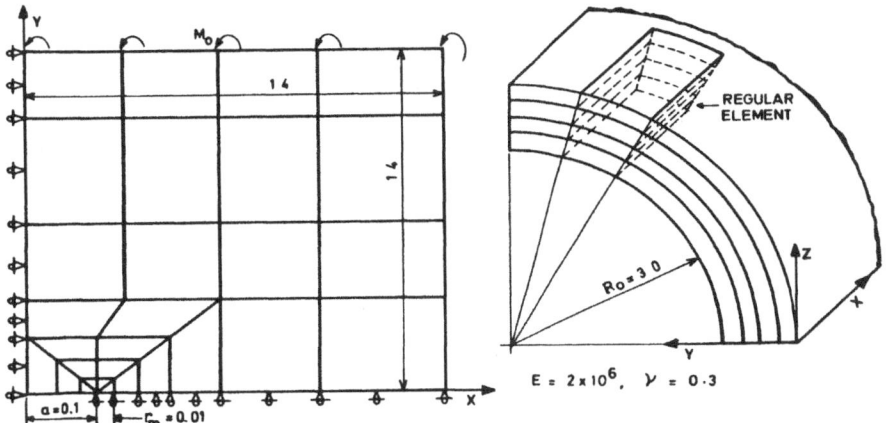

Fig. (6) - Mesh pattern (not to scale)

CONVERGENCE

Figure 8 shows the convergence of the SIF variation across the shell thickness for two, three and four layers of elements. The coordinate directions near the crack front are as shown in Figure 1 with the x-axis along the line of crack extension (parallel to the axis of the cylinder), y-axis along the tangent to the cross section of the cylinder and the z-axis along the radial direction of the shell. It can be seen from the figure that there is excellent agreement for three and four layer solutions. The variation of SIF along the crack front for three and four layer solutions is almost indistinguishable throughout the shell thickness except at the tension surface where the difference is about 1%. The figure also

224

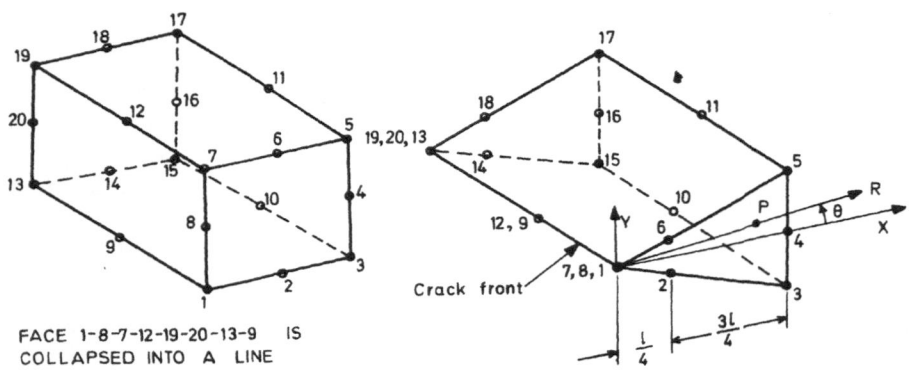

FACE 1-8-7-12-19-20-13-9 IS
COLLAPSED INTO A LINE

Crack front

Nodes 2, 6, 14 and 18 are shifted to quarter points

Fig. (7) - Three-dimensional degenerate quarter-point prismatic element

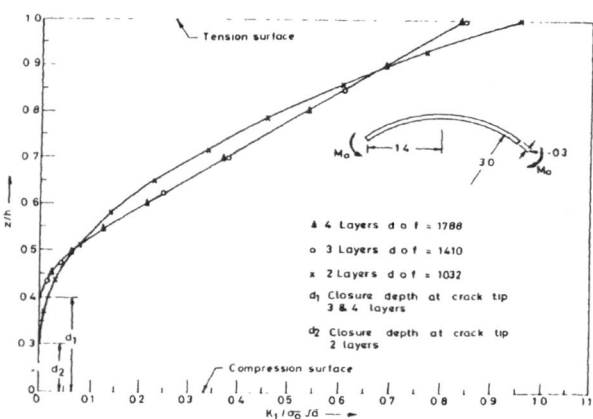

Fig. (8) - Convergence of SIF and closure depth for shell subjected to bending

shows good convergence regarding closure height in the close vicinity of the crack front. Hence, three layer solutions are taken to be sufficiently accurate.

RESULTS AND DISCUSSION

Figure 9 compares the non-dimensional crack opening displacement $v \cdot 2D(1+\nu)/a^2M_o$ at the tension surface of the shell for three-dimensional closure and no-closure

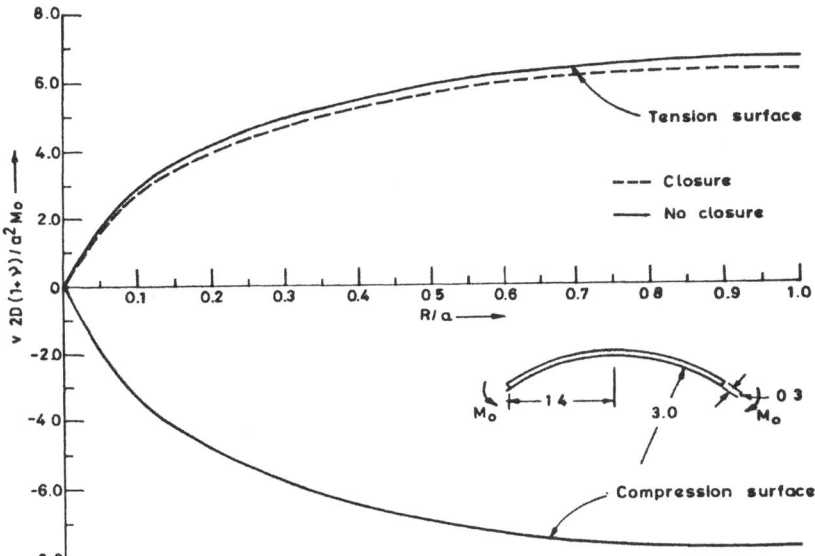

Fig. (9) - Variation of COD along the crack length

case. R represents the distance along the crack from the crack tip and D is the flexural rigidity. It is seen from the figure that the influence of closure is to reduce the COD on the tension surface by about 8% compared to no-closure case. This is due to the additional constraint imposed on the closure region which makes the shell stiffer. There is a steep increase in the crack opening displacement in the neighbourhood of the crack tip and for R/a greater than 0.2, the increase is gradual. It can be seen that the crack opening displacements for the no-closure case are not symmetric about the mid-plane of the shell. An increase in the magnitude of about 15% is observed on the compression surface compared to the tension surface. This is due to the curvature of the shell and the resulting shifting of the neutral plane from the mid-plane of the shell.

Figures 10, 11 and 12 show the variation of the crack opening displacements across the shell thickness for different R/a values for closure and no-closure cases. It is seen from the no-closure solution that there is warping of the crack faces and the variation is not exactly linear. The closure case shows a linear variation of the COD across the shell thickness for z/h - 1.0 to 0.6 and assumes

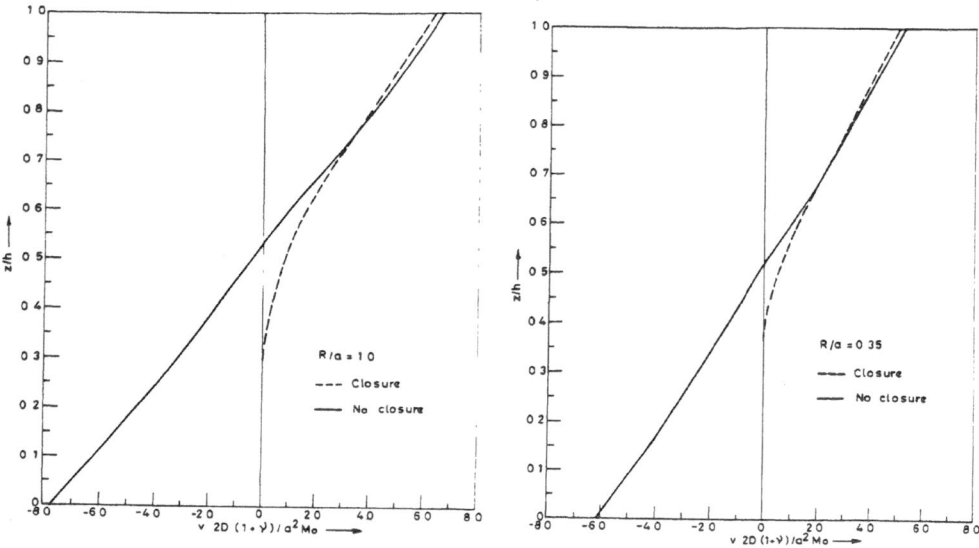

Fig. (10) - Variation of COD across
shell thickness

Fig. (11) - Variation of COD across
shell thickness

the shape of a smooth curve and becomes zero tangentially at the closure region.
It may be observed that the skew-symmetry of the COD distribution about the mid-
plane, present in the case of plates, no longer exists in the case of shells.

Figure 13 shows the closure area for the shell geometry considered in the
analysis. The shape of the closure area compares well with those obtained for
plates in [13]. The closure height is a maximum near the crack front and a minimum at the centre of the crack length. Closure region is observed to be constant
from R/a = 0 to R/a = 0.7 and at R/a = 1, viz. at the centre of the crack, the
value decreases by about 12%.

Figure 14 represents the variation of the non-dimensional mid-plane transverse
deflection $w \cdot 2D(1+\nu)/a^2M_0$ along the crack edge. The closure effect is to reduce
the transverse displacements along the crack edge by about 40%. The no-closure
solution gives a negative value for the transverse deflection up to R/a = 0.07.
This is not physically reasonable and such type of results is possibly obtained
due to the fact that closure which is present in reality, is not taken into consideration. This is justified with the fact that this negative w value is absent
when closure effect is taken into consideration.

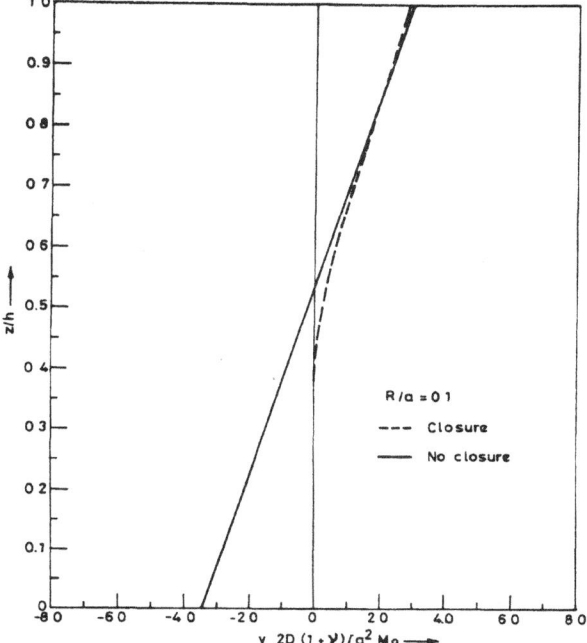

Fig. (12) - Variation of COD across shell thickness

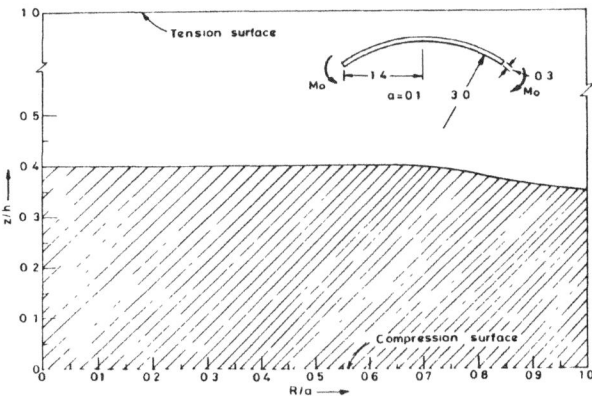

Fig. (13) - Closure area for shell subjected to bending

228

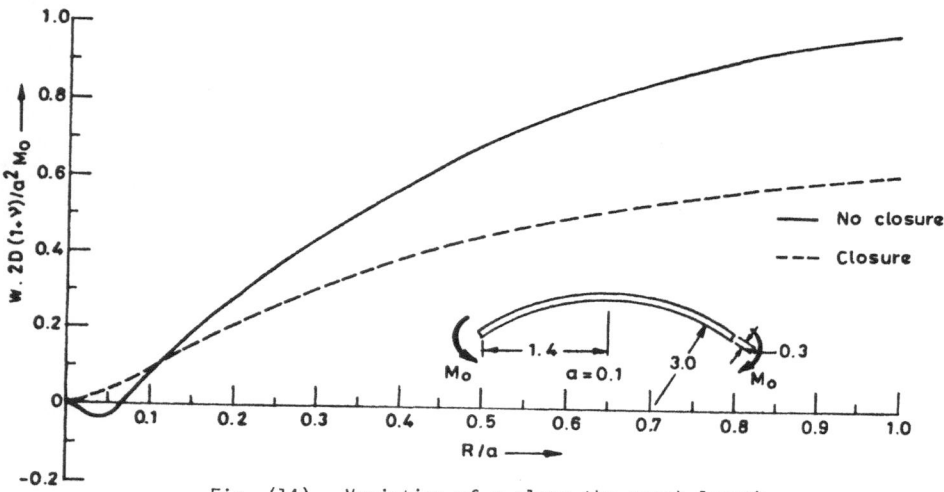

Fig. (14) - Variation of w along the crack length

Stresses are calculated at 10x10x10 Gauss points of the crack front elements. Figure 15 shows the value of σ_y at the nearest Gauss point from the crack front

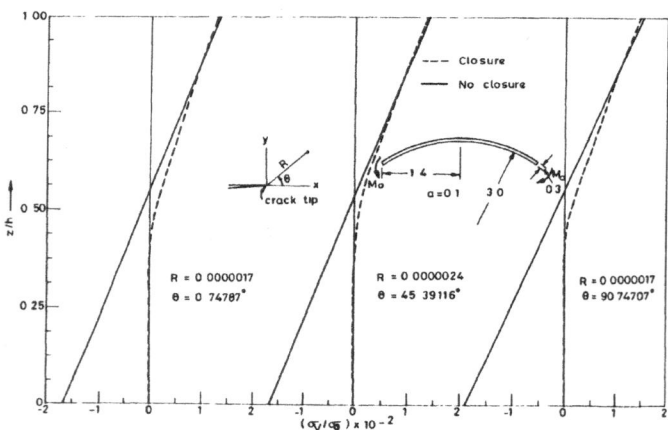

Fig. (15) - Variation of σ_y across shell thickness for closure and no-closure cases

for θ = 0.74787°, 45.39116° and 90.74707°. The figure also gives the values for the no-closure solution. σ_y variation is found to be almost linear along the shell thickness for the no-closure solution. For the closure solution, the stresses vary linearly from z/h = 0.6 to z/h = 1.0 and near the closure region, the variation becomes non-linear. The values of σ_y in the closure region in the neighbourhood of the crack front are found to take negative values of very small magnitude. It is observed that σ_x = σ_y along the line of crack extension in the tension zone and σ_z = $\nu(\sigma_x+\sigma_y)$ at all points in the neighbourhood of the crack front. The stress-based values of stress intensity factors have also been found out and the agreement with the COD-based values is very good.

Figure 16 shows the SIF variation across the shell thickness for closure and no-closure cases. The closure effect is to reduce the SIF by about 8% on the ten-

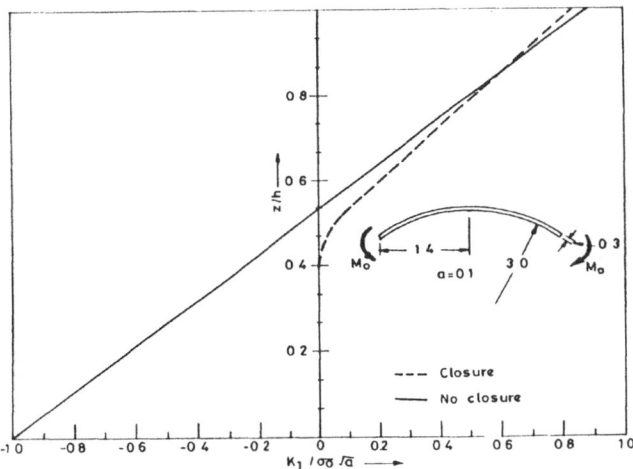

Fig. (16) - Variation of SIF across shell thickness

sion surface. The variation is almost linear for the no-closure case for the geometry of the shell analyzed. The negative value of the SIF shown in the figure on the compression zone of the shell for no-closure case does not have any physical significance. For the closure case, the SIF is non-linear in the vicinity of the closure region and becomes zero tangentially at the junction of the closure and non-closure regions and remains zero throughout the closure region of the crack front.

CONCLUSION

The inclusion of crack closure reduces the SIF which is about 8% for the shell geometry studied.

The variation of SIF along the crack front for the closure solution deviates from the no-closure solution near the closure region.

The crack opening displacements and transverse deflections along the crack length decrease with the inclusion of closure effect. The closure effect is more for the transverse deflection compared to crack opening displacements.

The closure height is a maximum near the crack front and a minimum at the centre of the crack length.

REFERENCES

[1] Folias, E. S., "A circumferential crack in a pressurized cylindrical shell", Int. J. Fract. Mech., 3, pp. 1-12, 1967.

[2] Erdogan, F. and Kibler, J. J., "Cylindrical and spherical shells with cracks", Int. J. Fract. Mech., 5, pp. 229-237, 1969.

[3] Erdogan, F. and Ratwani, M., "A circumferential crack in a cylindrical shell under torsion", Int. J. Fract. Mech., 8, pp. 87-95, 1972.

[4] Duncan-Fama, M. E. and Sanders, J. L., "A circumferential crack in a cylindrical shell under tension", Int. J. Fract. Mech., 8, pp. 15-20, 1972.

[5] Sih, G. C. and Hagendorf, H. C., "A new theory of spherical shells with cracks", Thin Shell Structures: Theory, Experiment and Design, edited by Y. C. Fung and E. E. Sechler, Prentice Hall, New York, pp. 519-545, 1974.

[6] Erdogan, F. and Dalale, F., "Transverse shear effects in a circumferentially cracked cylindrical shell", 8th U.S. National Congress of Appl. Mech., UCLA, California, 1978.

[7] Barsoum, R. S., Loomis, R. W. and Stewart, B. D., "Analysis of through cracks in cylindrical shells by the quarter-point elements", Int. J. Num. Meth. Engng., 15, pp. 259-280, 1979.

[8] Ayres, D. J., "Determination of the largest stable suddenly appearing axial and circumferential through-cracks in ductile pressure pipe", Int. Cong. on Struct. Mech. in Reactor Technology, San Francisco, 1977.

[9] Sih, G. C., "A review of the three-dimensional stress problem for a cracked plate", Int. J. Fract. Mech., 7, pp. 39-61, 1971.

[10] Sih, G. C., Williams, M. L. and Swedlow, J. L., "Three-dimensional stress distribution near a sharp crack in a plate of finite thickness", Air Force Materials Lab., Wright-Patterson Air Force Base, AFML-TR-66-242, 1966.

[11] Barsoum, R. S., "Author's reply to the discussion by Tracey", Int. J. Num. Meth. Engng., 11, pp. 401-402, 1977.

[12] Barsoum, R. S., "Triangular quarter-point elements as elastic and perfectly plastic crack tip elements", Int. J. Num. Meth. Engng., 11, pp. 85-98, 1977.

[13] Smith, D. G. and Smith, C. W., " A photoelastic evaluation of the influence of closure and other effects upon the local bending stresses in cracked plates", Int. J. Fract. Mech., 6, pp. 305-317, 1970.

ANALYSIS OF A SINGULAR SOLUTION IN THREE DIMENSIONAL SURFACE CRACK PROBLEM

Y. Fujitani

Hiroshima University
Shitami Saijo-cho Higashihiroshima 724, Japan

ABSTRACT

This study deals with the three dimensional stress singularity of the surface crack especially its peculiar behavior at the free end of the crack front line in an elastic body. The surface crack problem is reduced to an eigen problem in the deformation theory of elasticity. First, this study is another challenge to this surface crack proposed by Benthem and intends to make formulations analytically by using the so-called spherical coordinates Boussinesq's function. Second, this paper intends to make formulations and analyze numerically by Rayleigh-Ritz method which is widely used in the field of structural mechanics.

INTRODUCTION

As regards a theoretical study of three dimensional crack problem, there is Sih and Hartranft's pioneering work [1], which made clear, on the cylindrical coordinates, the form of solution in the vicinity of the crack with infinite depth. Also, there are Benthem's analytical study [2] and Bazant and Estenssoro's finite element solution [3] concerning the singular solution of surface crack problem. Their study deals with the stress singularity of the surface crack. This paper shows the two solutions of this surface crack problem obtained by another analytical method and another numerical method. The outline of which will be described in the following sections.

SURFACE CRACK PROBLEM

Surface crack problem is the problem of finding, in the half-infinite three dimensional elastic body defined in the domain of the following equation:

$$0 \leq r < \infty$$
$$0 \leq \theta \leq \frac{\pi}{2} \tag{1}$$
$$\pi \leq \phi \leq \pi$$

the form of solution which meet the following boundary conditions:

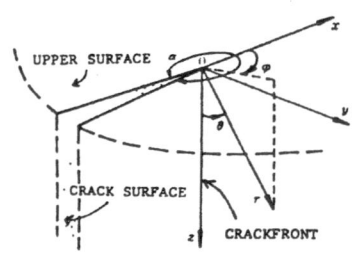

Fig. (1) - Surface crack problem

$$\tau_{\phi r} = \tau_{\theta\phi} = \sigma_\phi = 0 \quad \text{at crack surfaces} \ (\phi = \pm \tfrac{\alpha}{2}) \tag{2}$$

$$\tau_{r\theta} = \sigma_\theta = \tau_{\theta\phi} = 0 \quad \text{at upper surface} \ (\theta = \tfrac{\pi}{2}) \tag{3}$$

and examining the stress singularity in the vicinity of the crack.

ANALYTICAL SOLUTION

A. Expression of Displacements and Stresses by Spherical Coordinates Boussinesq's Function

Based on the three dimensional theory of elasticity, the displacements and stresses of the Boussinesq's function are expressed in the following equations (as for details, refer to item [4]). Hereafter, formulations are made of the symmetric mode.

$$2Gu_r(r,\theta,\phi) = r^{\lambda-1} [\lambda P_\lambda^\mu(p)a_\mu + 2\mu P_\lambda^\mu(p)b_\mu + (\lambda-4+4\nu)pP_{\lambda-1}^\mu(p)c_\mu]\cos\mu\phi \tag{4}$$

$$\sigma_r(r,\theta,\phi) = r^{\lambda-2} [\lambda(\lambda-1)P_\lambda^\mu(p)a_\mu + 2\mu(\lambda-1)P_\lambda^\mu(p)b_\mu + \{(\lambda^2-5\lambda+4-2\nu)pP_{\lambda-1}^\mu(p)$$

$$+ 2\nu(\lambda-\mu)P_\lambda^\mu(p)\}c_\mu]\cos\mu\phi \tag{5}$$

where, $p = \cos\theta$, $P_\lambda^\mu(p)$ is the first Legendre function and the term (a_μ, b_μ, c_μ) shows unknown coefficients. The other displacement components u_θ, u_ϕ and stress components σ_θ, σ_ϕ, $\tau_{\theta\phi}$, $\tau_{\phi r}$, $\tau_{r\theta}$ can be expressed in the same way. All the displacements have the term $r^{\lambda-1}$ and all the stresses have the term $r^{\lambda-2}$.

B. Derivation of Displacements and Stresses in the Vicinity of Crack Front

$P_\lambda^\mu(p)$ and others can be expanded to the power series of θ as follows:

$$P_\lambda^\mu(p) = \frac{2^\mu}{\Gamma(1-\mu)} \, \theta^{-\mu} + \frac{2^\mu}{\Gamma(1-\mu)} \{- \frac{\mu}{12} - \frac{\lambda(\lambda+1)}{4(1-\mu)}\}\theta^{-\mu+2} + O(\theta^{-\mu+4}) + \ldots$$

$$p = 1 - \frac{1}{2}\theta^2 + \ldots, \quad \bar{p} = \sin\theta = \theta - \frac{1}{6}\theta^3 + \ldots \tag{6}$$

When these are substituted into the equations (4) and (5), displacements and stresses can be expanded asymptotically in the vicinity of the crack front in the following way:

$$2Gu_r(r,\theta,\phi) = r^{\lambda-1} \, [\theta^{-\mu}f(\mu) + \theta^{-\mu+2}\bar{f}_r(\mu) + O(\theta^{-\mu+4}) + \ldots]\cos\mu\phi \tag{7}$$

$$2Gu_\theta(r,\theta,\phi) = r^{\lambda-1} \, [-\theta^{-\mu-1}\bar{g}(\mu) + \theta^{-\mu+1}\bar{f}_\theta(\mu) + O(\theta^{-\mu+3}) + \ldots]\cos\mu\phi$$

$$\sigma_r(r,\theta,\phi) = r^{\lambda-2} \, [\theta^{-\mu}f_r(\mu) + O(\theta^{-\mu+2}) + \ldots]\cos\mu\phi$$

$$\sigma_\theta(r,\theta,\phi) = r^{\lambda-2} \, [\theta^{-\mu-2}g(\mu) + \theta^{-\mu}f_\theta(\mu) + O(\theta^{-\mu+2}) + \ldots]\cos\mu\phi \tag{8}$$

$$\tau_{\phi r}(r,\theta,\phi) = r^{\lambda-2} \, [\theta^{-\mu-1}h(\mu) + \theta^{-\mu+1}f_{\phi r}(\mu) + O(\theta^{-\mu+3}) + \ldots]\sin\mu\phi$$

where

$$g(\mu) = \kappa_\mu \mu(\mu+1)(a_\mu - 2b_\mu + c_\mu)$$

$$h(\mu) = \kappa_\mu \mu(1-\lambda)a_\mu + (\lambda-\mu-2)b_\mu - (\lambda-3+2\nu)c_\mu$$

$$\bar{g}(\mu) = \frac{1}{\mu+1} g(\mu) \tag{9}$$

$$f(\mu) = -\frac{1}{\mu}[\frac{1}{\mu+1}(\lambda-2)g(\mu) + 2h(\mu)] \quad \kappa_\mu = \frac{2^\mu}{\Gamma(1-\mu)}$$

$\Gamma(x)$ is Gamma function. Now, in the areas except the surface $(r\neq0)$ there must exist, in the primary solution of displacement and stress in the vicinity of the crack front, the solution of the two dimensional crack (Mode I, the plane strain field), that is, the solution of the stress singularity $\theta^{-1/2}$. By this, the lower limit value which the parameter μ must have, is fixed. When it is assumed $\mu = -3/2$, the maximum stress singularity is $\theta^{-1/2}$ according to the equation (8), but, when the crack surface boundary condition (2) is applied, we have $g(-3/2) = 0$ and the term $\theta^{-1/2}$ disappears, and net stress singularity is $\theta^{1/2}$, therefore, $\mu = -3/2$ could not be adopted. Therefore, $\mu = -1/2$ is adopted. At this time, the strong singularity of $\theta^{-5/2}$ and $\theta^{-3/2}$ appears, but by having their coefficient $=0$, and also by the boundary condition of the equation (2), we have the following equations. Here, the maximum stress singularity is $\theta^{-1/2}$.

$$g(1/2) = 0, \quad f_{\theta\phi}(1/2) + g(-3/2) = 0$$

$$\tag{10}$$

$$h(1/2) = 0, \quad f_{\phi r}(1/2) + h(-3/2) = 0$$

This equation provides the relationship among $a_{1/2}$, $b_{1/2}$, $c_{1/2}$, $a_{-3/2}$, $b_{-3/2}$ and $c_{-3/2}$.

$$a_{1/2} = \frac{-2\lambda+7-8\nu}{2\lambda+1}\,c_{1/2}, \quad c_{1/2} = \frac{2}{2\lambda-1}\,(a_{-3/2}-2b_{-3/2}+c_{-3/2})$$

$$\tag{11}$$

$$b_{1/2} = \frac{4-4\nu}{2\lambda-1}\,c_{1/2}, \quad c_{-3/2} = -\frac{2\lambda-3}{2(\lambda-2)}\,a_{-3/2} + \frac{(2\lambda-3)(7-8\nu)}{8(\lambda-2)(1-\nu)}\,b_{-3/2}$$

When equations (7) and (8) are simplified by using this relationship, the displacement and stress equations can be expressed in the following equations. Their equations give the first primary solution of the displacement and stress in the vicinity of the crack front of the infinite depth without considering upper surface.

$$2Gu_r(r,\theta,\phi) = r^{\lambda-1}\theta^{3/2}\left[(\lambda-4+4\nu)\cos\frac{\phi}{2} - \left(\frac{7-8\nu}{3}\lambda - 4+4\nu\right)\cos\frac{3\phi}{2}\right] \times$$

$$\times \sqrt{\frac{2}{\pi}}\,(\lambda - \frac{1}{2})c_{1/2} \tag{12}$$

$$\sigma_r(r,\theta,\phi) = r^{\lambda-2}\theta^{-1/2}\left[-2\nu\cos\frac{\phi}{2}\right]\sqrt{\frac{2}{\pi}}\,(\lambda - \frac{1}{2})c_{1/2} \tag{13}$$

When in the coordinates transformation, the equations(12) and (13) to the cylindrical coordinates and we assume:

$$r = \sqrt{\rho^2+z^2} \div z, \quad \theta = \arctan\frac{\rho}{z} \div \frac{\rho}{z} \tag{14}$$

it coincides with the solution obtained by Sih and Hartranft [1].

The second, the third and so on primary solution can be obtained by using the above method.

C. Analysis of Singularity Factor

Singularity factor λ in the crack front on the stress free surface is calculated by using the several stress solutions expanded asymptotically in the vicinity of the crack front that is satisfied the condition of stress free crack surface. This eigen equation is constructed by Fourier integral method by using the stress free conditions in upper surface as follows:

$$\int_{-\pi}^{\pi} \tau_{r\theta}(r, \frac{\pi}{2}, \phi)\cos(k-1)\phi d\phi = 0$$

$$\int_{-\pi}^{\pi} \sigma_{\theta}(r, \frac{\pi}{2}, \phi)\cos(k-1)\phi d\phi = 0 \tag{15}$$

$$\int_{-\pi}^{\pi} \tau_{\theta\phi}(r, \frac{\pi}{2}, \phi)\sin k\phi d\phi = 0 \qquad (k = 1,2,3,...)$$

Figure 2 shows the convergency of the singularity factor λ which are calculated as eigen value of equations (15), where m, $\bar{m}$ is the number of unknown coefficients, primary solutions, respectively. From Figure 2, it is observed that singularity factor λ converge to the Benthem's solution [2] in each Poisson's ratio ν.

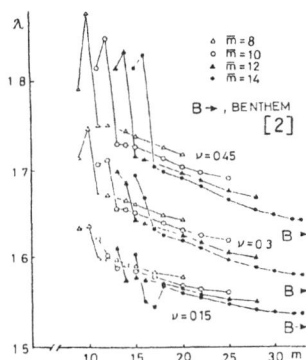

Fig. (2) - Convergency of the singularity factor λ

NUMERICAL SOLUTION

Based on the energy theorem of elasticity, the eigen solution λ can be obtained by solving the following equation [3,5]:

$$\int_{0}^{\infty} r^{2\lambda}dr \iint [\{(\lambda+1)\bar{p}\sigma_r^* - \bar{p}(\sigma_\theta^* + \sigma_\phi^*)\}\delta u_r^* - \bar{p}\tau_{r\theta}^* \frac{\partial}{\partial\theta}\delta u_r^* - \tau_{r\phi}^* \frac{\partial}{\partial\phi}\delta u_r^*$$

$$+ \{(\lambda+2)\bar{p}\tau_{r\theta}^* - p\sigma_\phi^*\}\delta u_\theta^* - \bar{p}\sigma_\theta^* \frac{\partial}{\partial\theta}\delta u_\theta^* - \tau_{\theta\phi}^* \frac{\partial}{\partial\phi}\delta u_\theta^*$$

$$+ \{(\lambda+2)\bar{p}\tau_{r\phi}^* + p\tau_{\theta\phi}^*\}\delta u_\phi^* - \bar{p}\tau_{\theta\phi}^* \frac{\partial}{\partial\theta}\delta u_\phi^* - \sigma_\phi^* \frac{\partial}{\partial\phi}\delta u_\phi^*]d\theta d\phi = 0 \tag{16}$$

where

$$u_r(r,\theta,\phi) = r^\lambda u_r^*(\theta,\phi) \tag{17}$$

$$\sigma_r(r,\theta,\phi) = r^{\lambda-1}\sigma_r^*(\theta,\phi)$$

As the domain of θ and φ in the surface crack problem is given with a simple rectangular form (see equation (1) and Figure 3), I analyze by using Rayleigh-Ritz method instead of finite element method developed by Bazant and Estenssoro [3].

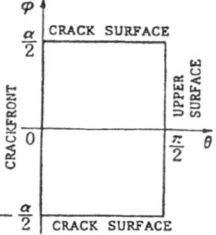

Fig. (3) - Domain of θ-φ

Displacement fields can be assumed as following equation in case of Mode I and Mode II respectively:

$$u_r^*(\theta,\phi) = a_{00} + \sum_{m=1}\sum_{n=0} a_{mn}\theta^m\phi^{2n}$$

$$u_\theta^*(\theta,\phi) = \sum_{m=1}\sum_{n=0} b_{mn}\theta^m\phi^{2n} \left.\begin{array}{c}\\\\\\\end{array}\right\} \text{Mode I}$$

$$u_\phi^*(\theta,\phi) = \sum_{m=1}\sum_{n=0} c_{mn}\theta^m\phi^{2n+1}$$

$$u_r^*(\theta,\phi) = \sum_{m=1}\sum_{n=0} a_{mn}\theta^m\phi^{2n+1}$$

$$u_\theta^*(\theta,\phi) = \sum_{m=1}\sum_{n=0} b_{mn}\theta^m\phi^{2n+1} \left.\begin{array}{c}\\\\\\\end{array}\right\} \text{Mode II}$$

$$u_\phi^*(\theta,\phi) = \sum_{m=1}\sum_{n=0} c_{mn}\theta^m\phi^{2n}$$

(18)

By considering the physical meaning of the crack front, these displacement fields have been so assumed that following conditions are satisfied:

$$u_r^*(0,\phi) = c, \quad u_\theta^*(0,\phi) = 0, \quad u_\phi^*(0,\phi) = 0 \tag{19}$$

Figure 4 shows the convergence of singularity factor λ in case of Mode I and Mode II. The present solution in case of Mode I is very close to the solution obtained by Benthem [2], Bazant and Estenssoro [3]. Figure 5 shows the relation between the singularity factor λ and the crack opening angle α. From Figure 5, the singularity factor λ in case of Mode II is stronger than that of two-dimensional case, so that its singularity in $335° < \alpha \leq 360°$ is stronger than that of the well-known $r^{-1/2}$ singularity. Figure 6 shows the relation between the singularity factor λ and Poisson's ratio ν.

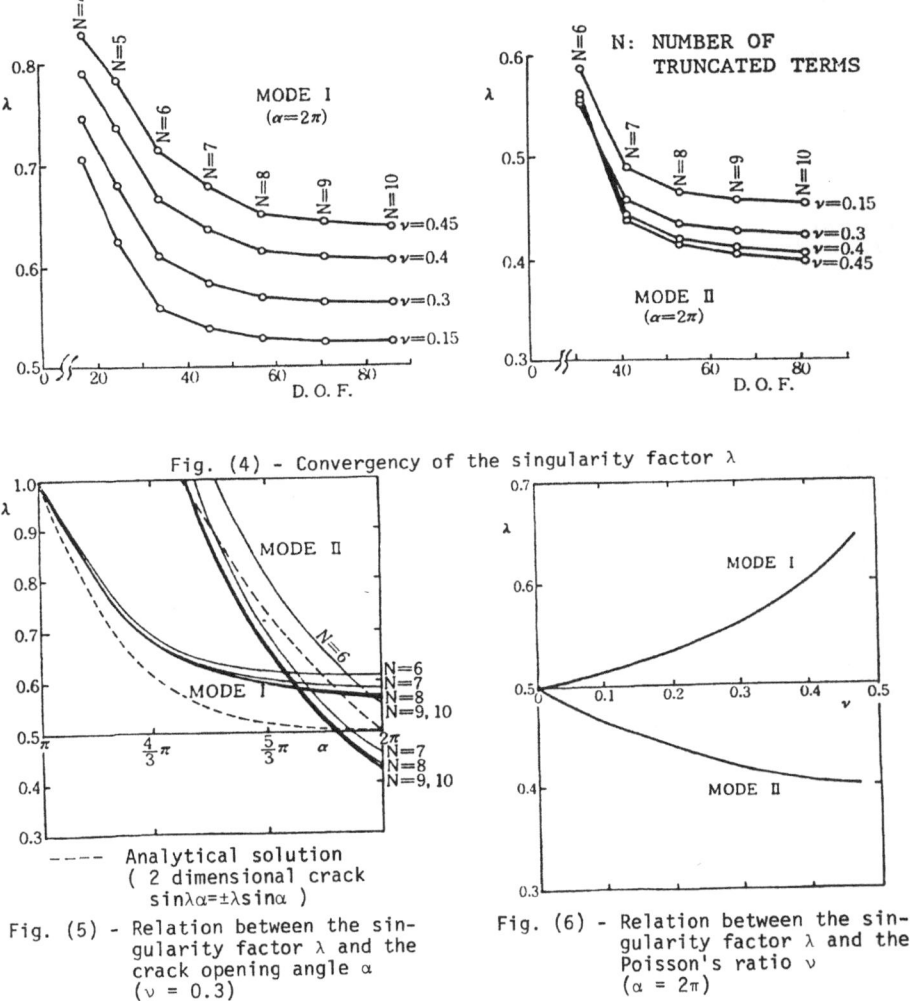

Fig. (4) - Convergency of the singularity factor λ

---- Analytical solution
(2 dimensional crack
sinλα=±λsinα)

Fig. (5) - Relation between the singularity factor λ and the
crack opening angle α
(ν = 0.3)

Fig. (6) - Relation between the singularity factor λ and the
Poisson's ratio ν
(α = 2π)

CONCLUSION

This analytical method has enabled me to obtain, in explicit form, the several
primary solutions of stress in the vicinity of the crack front and has enabled me

to make clear the "structure" in the analytical aspect of the surface crack problem. And this numerical method by Rayleigh-Ritz enabled me to obtain the solutions for Mode I and Mode II of the surface crack problem with good convergency by using less degree of freedoms.

ACKNOWLEDGEMENT

I want to thank Professor Tadahiko Kawai, The University of Tokyo, for several suggestions in early stages of this work.

REFERENCES

[1] Hartranft, R. J. and Sih, G. C., "The use of eigen function expansions in the general solution of three-dimensional crack problem", J. Math. Mech. 19, p. 123, 1969.

[2] Benthem, J. P., "States of stress at the vertex of a quarter-infinite crack in a half-space", J. Solid and Structure, 13, p. 479, 1977.

[3] Bazant, Z. P. and Estenssoro, L., "General numerical method for three-dimensional singularity in cracked or notched elastic solids", Fracture, 3, ICF 4, p. 371, 1977.

[4] Miyamoto, H., "Theory of 3D Elasticity", Shoukabo, 1967 (in Japanese).

[5] Fujitani, Y., "Finite element analysis of singular solutions in cracked problem", Numerical Methods in Fracture Mechanics, Luxmoore and Owen, eds., 1978.

[6] Fujitani, Y., "Analysis of the stress singular solution in the three dimensional surface crack problem by Rayleigh-Ritz method", Bulletin of the Faculty of Eng., Hiroshima University, 28, No. 2, 1980 (in Japanese).

SECTION III

FRACTURE AND MATERIAL TESTING

BASIC TENSILE PROPERTIES OF A LOW-ALLOY STEEL BS4360-50D

C. Q. Zheng and J. C. Radon

Imperial College of Science and Technology
London SW7 2BX, England

ABSTRACT

Tensile tests were carried out on a low-alloy steel BS4360-50D used for the construction of North Sea oil platforms. In particular, static yield stress, σ_y, the strain hardening exponent n, fracture strain, ε_f and the relationship between a/R and ε, were investigated using cylindrical specimens tested at low loading rates (0.0025 in/min). Additional cyclic tests were performed in tension between zero and a range of loads. It was observed that the value of the cyclic strain hardening exponent n, decreased from 0.22 to 0.156, while the fracture strain remained reasonably constant at 1.26.

Both static and cyclic parameters were used for the fatigue crack growth at very low values of stress intensity by means of a formula recently developed:

$$\frac{da}{dN} = \frac{2^{(1+n)}(1-2\upsilon)^2(\Delta K_{eff}^2 - \Delta K_{c,eff}^2)}{4(1+n)\pi\sigma_{yc}^{(1-n)}E^{(1+n)}\varepsilon_f^{(1+n)}}$$

where K are stress intensity factors and υ is the Poisson's ratio. It was shown that for all values of da/dN (from 10^{-8} to 10^{-5} mm/cycle), the cyclic parameters provided a better crack growth prediction than the static factors.

INTRODUCTION

Basic material parameters, such as yield stress, fracture strain and strain hardening, are essential in the analysis of subcritical crack growth. Unfortunately, not all of these are freely available and extensive experimentation may be necessary. This report describes some monotonic testing methods which may be used for their evaluation.

In a previous study of fatigue crack growth [1] of a low-alloy steel BS4360-50D, it was proposed that the propagation process at very low ΔK values might be predicted using the expression:

$$\frac{da}{dN} = \frac{2^{(1+n)}(1-2\nu)^2(\Delta K^2_{eff} - \Delta K^2_{c,eff})}{4(1+n)\pi\sigma_{yc}^{(1-n)}E^{(1+n)}\varepsilon_f^{(1+n)}} \tag{1}$$

where n is a cyclic strain hardening exponent, σ_{yc} is a cyclic yield stress at the appropriate strain rate, ε_f is a true fracture strain, and stress intensity factors have their usual meaning. Apart from limited static data, not all the mentioned parameters are known. In practical engineering design, cyclic yield data are usually replaced by their static values, while the strain rates are frequently only approximated. Other expressions of similar form to equation (1) have been derived for the same purpose, but again cannot be evaluated simply. Similarly, dynamic values of flow stress are necessary for the computation of dynamic toughness, while the use of the usual static value derived in standard laboratory tests is reserved for quality assurance, material selection, etc. Although the description of the stress and strain fields in which cracks may initiate is important, there appears to be only limited information available regarding the monotonic or cyclic behaviour of engineering materials. There are two possible methods of attack: an approximate and an exact evaluation of the required properties. The approximations can be arrived at by using chemically and physically similar materials. For example, this method is suitable for most structural steels, where E, ν and γ do not vary significantly. An exact evaluation of these parameters is, of course, preferable.

The tensile test is one of the most frequently used mechanical tests, but its complete theoretical analysis is not yet available. One of the first studies of stress in the neck of a tensile specimen was that by Bridgman [2], who investigated the effect of stress non-uniformities. Approximate solutions for the stress and strain distribution across the neck section were obtained and these were found to be adequate for most engineering applications, provided that the strain rate was not very high. Here also Bridgman's approximate solution is used, mainly for its simplicity. Nevertheless, two types of progressive deformation occurring during the tensile test require further study, namely, the formation of voids and the strain hardening process; the appropriate tests are nearly completed and will be reported shortly.

Metallurgical observations confirmed that the fracture in a ductile specimen initiated in the form of voids in the necked region and the strain hardening showed the extent of the plastic zone. Both processes will be discussed later.

The usual properties recorded during the laboratory tensile test are the yield strength, ultimate tensile strength, elongation and area reduction; appropriate standards are available [3]. Theoretical analyses of tensile tests are also available; the use of elasto-plastic strain hardening material was reported in [4].

The present study was motivated by a shortage of data on the low-alloy steel BS4360-50D, used in the construction of the North Sea oil platforms. Preliminary static fracture tests have recently been completed [5]. Also, the effects of R-ratio, plate thickness, frequency and environment on fatigue crack propagation have been analyzed. However, for the application of the new predictive model for fatigue crack growth, equation (1), as discussed earlier [1], detailed information of the mechanical behaviour of this steel at appropriate strain rates is necessary.

ANALYSIS

Phenomenological models of homogeneous deformation have been proposed by many workers. These relationships are usually expressed in the form of true stress, $\bar{\sigma}$, and natural plastic strain, $\bar{\varepsilon}$, sometimes including strain rate effects, $\dot{\varepsilon}$. The stress $\bar{\sigma}$ is frequently dependent on the previous strain history and is defined as:

$$\bar{\sigma} = f(\bar{\varepsilon}, \dot{\varepsilon}) \tag{2}$$

In order to simplify the stress-strain relationship of ductile materials, it may be assumed that the true stress, $\bar{\sigma}$, is a simple parabolic function of the natural strain, $\bar{\varepsilon}$, in the form:

$$\bar{\sigma} = C\bar{\varepsilon}^n \tag{3}$$

where C and n are constants, and n<1. In this expression, the equivalent plastic strain is replaced by total strain. This is acceptable due to large plastic deformations in the tensile test. Many other expressions are available. A more complicated relationship, yet essentially identical in form with equation (3), is equation (4):

$$\bar{\sigma} = \frac{\bar{\sigma}_y}{B^n} (\varepsilon + B)^n \tag{4}$$

where $\bar{\sigma}_y$ is the yield stress, and B is the material parameter. This relationship may often be found advantageous in the formulation of material behaviour where two or more yielding and fracture processes operate simultaneously, as in steels. Note that for some steels, the product of n and σ_y may be a constant:

$$n\sigma_y = \text{const.} \tag{5}$$

the value of this constant increasing with the carbon content. The data for mild steels and low-alloy steels supporting relation (5) were reported by Rosenfield and Hahn [6], but the general validity of equation (5) remains questionable.

It will be realized that the formation of a neck will complicate the stress state in a cylindrical specimen. The necking process begins with tensile instability.

At the maximum load:

$$\frac{d\bar{\sigma}}{d\bar{\varepsilon}} = \bar{\sigma} \tag{6}$$

$$Cn\bar{\varepsilon}^{n-1} = C\bar{\varepsilon}^n \tag{7}$$

therefore:

$$n = \bar{\varepsilon} \tag{8}$$

As the necking begins, the uniform stretching of the specimen will cease. The respective value of $\bar{\varepsilon}$ may be accurately determined from $\bar{\sigma}$ versus $\bar{\varepsilon}$ diagram, provided that the single power function mentioned above is used.

It has been observed that the onset of necking, or of ductile fracture, may occur just before reaching the maximum tensile load. Immediately after the onset of necking, the longitudinal stresses close to the surface cease to be parallel with the axis of the specimen and this deviation results in a radial pull directed outwards in the minimum cross-section and a non-uniform stress distribution in the axial direction. Thus, a triaxial stress field will develop from the originally uniaxial field of the cylindrical specimen. Radial stresses decrease from the center of the specimen to zero at the neck circumference.

As suggested by Bridgman, the true tensile stress at the surface of the neck, $\bar{\sigma}_a$, where a is the minimum radius, is lower than the average true stress, σ_m, and its corrected value is:

$$\sigma_a = \frac{\sigma_m}{(1 + 2\frac{R}{a})\log(1 + \frac{1}{2}\frac{a}{R})} \tag{9}$$

where R is the radius of the neck. It was observed that during the necking process, the ratio a/R increased with the true strain due to the R decreasing more rapidly than a. At the center of the minimum cross-section, the axial stress can be expressed as a function of the average stress σ_m and, on attaining the maximum, σ_{max}, a fracture will occur at the center of the neck, as observed already by Ludwik [7]. Again, more advanced expressions than equation (9) are now available, for example, [8].

Parker et al [9] investigated the fracture of a steel bar in tension at a range of temperatures, and in particular, the transition from shear to cleavage. Above the transition temperature, the shear fractures dominated across the entire cross-section of the specimen, including the center portion of the cup and cone which appears macroscopically to be flat.

TENSILE TESTS

The material used was a low-alloy steel, BSI, KS4360-50D, 1979, supplied in the form of 50 mm thick plate and normalized. The chemical composition is shown in Table 1. The cylindrical specimens, having the gauge daimeter d_o equal to

TABLE 1 - CHEMICAL COMPOSITION (WEIGHT) OF 50D

Element	C	Si	Mn	Ni	Cr	Mo	P	S	Cu	Nb	Al
%	0.180	0.36	1.40	0.095	0.11	0.020	0.018	0.003	0.16	0.039	0.035

11.28 mm, were manufactured with their axes parallel to the rolling direction of the plate, in accordance with BSI, BS18, Part 2, 1971. Eighteen tensile tests were performed using a Tinius Olsen machine, 120,000 lbs capacity, at a loading rate of 0.0025 in/min and the same testing conditions, at 20°C and 50% RH.

The following results were obtained, see Table 2:

TABLE 2

Yield stress, σ_y	383 MPa
Ultimate tensile strength, σ_u	543 MPa
Young's modulus, E	213 GPa
Elongation, A	33%
Reduction in area, ψ	65.6%

Gauge length = 5.65 $\sqrt{S_o}$, where S_o is the cross-sectional area of the specimen

The elongation was measured on the gauge length L_o = 56 mm. The distribution of the elongation is shown in Table 3. The specimen gauge length was divided into eleven sections and their relative elongations recorded using a workshop microscope (x40). The maximum relative elongation of the unbroken specimen was about 90%, increasing to approximately 120% when the specimen suddenly and completely separated into two parts. The plot of the relative strain energy which is similar to that of the elongation is also included in Table 3.

The longitudinal distribution of the strain was then calculated and the respective values for a broken specimen no. 2 as well as for the unbroken one, no. 3, are presented in Figure 1. It was noted that the natural strain reached the value of 0.85 in the necked section and this value increased to 1.06 at fracture. However, outside the necked section, the strain decreased rapidly to approximately 0.15 - 0.24; note that this value corresponds with the strain hardening exponent n.

The total elongation of the gauge length L may be described as the function (here referred to as a Barba's law):

$$e = bL + c\sqrt{A} \tag{10}$$

where A is cross-sectional area, and b and c are constants. For the investigated steel 50D (0.18% C), b = 0.23 and c = 0.5, and these may be compared with those known for a medium-carbon steel (0.3% C), 0.16 and 0.454, respectively. The relevant comparative values for a low-carbon steel (0.08% - 0.20% C) were somewhat higher, namely, 0.279 and 0.73.

STRAIN HARDENING EXPONENT, n

1. As already mentioned, the true stress in a ductile polycrystalline material may be expressed as function of true strain using, for example, equation (3). In general, the true stress is found to increase with strain, and this process is referred to as strain hardening. At the critical point, the rate of the cross-section change is so high that the load begins to drop despite the hardening. The values of strain $\bar{\varepsilon}$ were experimentally determined and the mean value n, equation (8), is:

TABLE 3 – THE DISTRIBUTION OF RELATIVE ELONGATION IN SECTIONS OF SPECIMEN NO. 3

Marked Line Position	1	2	3	4	5	6	7	8	9	10	11	12
Distance, ℓ (mm)	0	5.09	10.18	15.27	20.36	25.45	30.54	35.63	40.72	45.81	50.90	56.00
Section Number	1	2	3	4	5	6	7	8	9	10	11	
Relative Elongation in the Section (%)	15.3	19.5	19.9	24.6	38.2	88.0	39.0	22.0	19.5	19.1	15.3	
Relative Strain Energy (%)	4.9	6.1	6.2	7.7	12.0	27.8	12.3	6.9	6.1	5.2	4.8	

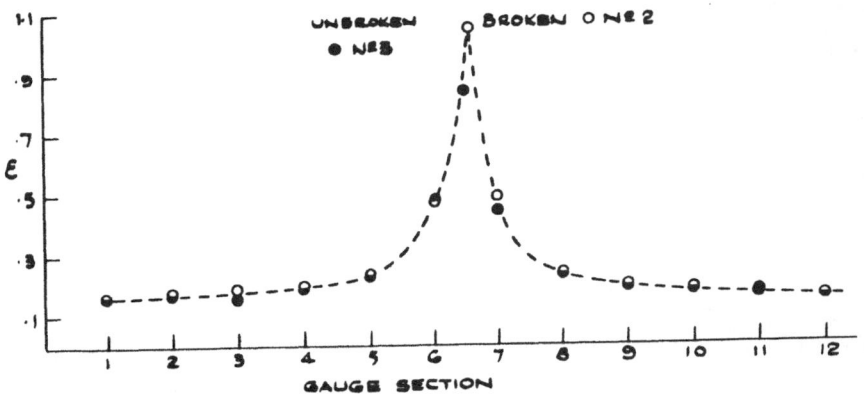

Fig. (1) - Strain distribution along the gauge length

$$n = \bar{\epsilon} = 0.174 \pm 13\% \tag{11}$$

However, the neck formation may influence the precise determination of the true strain.

2. Another experimental method involving a log σ versus log ϵ plot is shown in Figure 2. It is usual to present these results on the logarithmic scales.

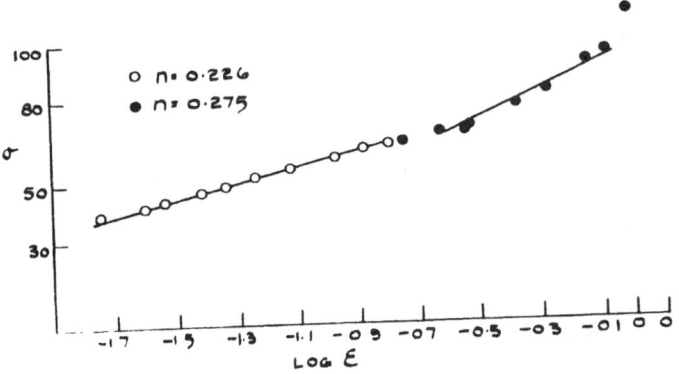

Fig. (2) - Steel 50D. σ versus log ϵ

However, in Figure 2, the stress scale is linear for clearer representations. Using the least square method, the results in (1) correlate well with the slope measured in Figure 2, i.e., before necking:

$$n_B = 0.22 \qquad\qquad (12)$$

and after necking:

$$n_A = 0.27$$

These results may be compared with the values for steel En 32B quoted in [10]: Close to the crack tip the true strain was found to be 0.43 - 0.75, but decreased rapidly within a short distance to 0.34 - 0.42. The higher value of n, namely n_A = 0.27, quoted here, should be used in the elasto-plastic analysis. On reaching the maximum tensile load, the true stress versus natural strain graph of steel 50D became a straight line, as shown in Figure 3. This part of the tensile curve represents the main portion of the strain hardening range.

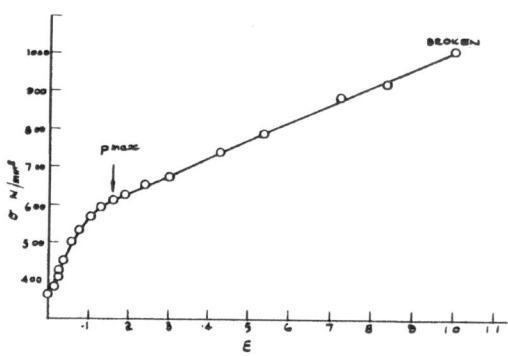

Fig. (3) - Steel 50D. True stress-strain curve

3. Using 10 cyclic excursions at the same loading rate as in the static tests, additional short cyclic tests were performed in pure tension, Figure 4. Ten specimens were cycled between zero and a range of plastic strains, up to ε_p = 0.20.

Figure 4 shows an example of a cyclic test at a plastic strain range ε_p equal to 0.05. Subsequently, the specimen was loaded to fracture. It was noted

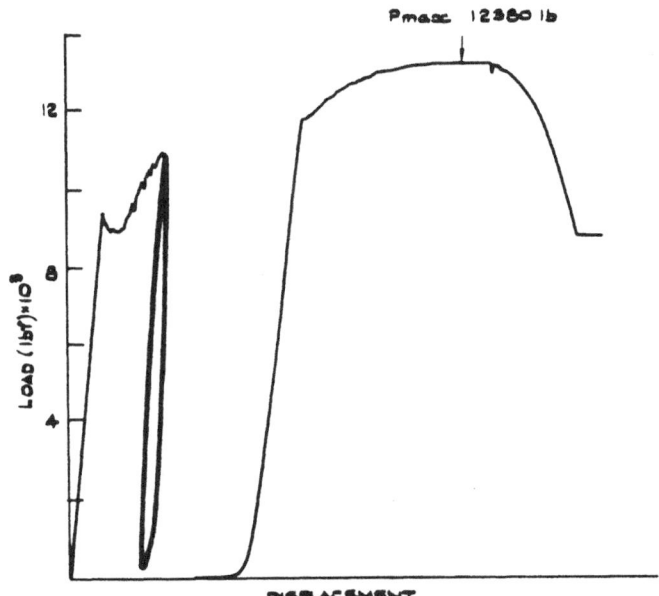

Fig. (4) - Steel 50D. Cyclic test of ten strain cycles between zero
and $\varepsilon_p = 0.05$ followed by static fracture in tension

that the instability strain ε_i was invariably lower than the true fracture strain.
The difference amounted to approximately 10% for all the specimens tested. Thus,
for example, the mean instability strain for the second group of tests amounted to
1.17, while the true fracture strain was 1.26; this difference requires further
analysis.

With increasing cyclic strain range ε_p, the strain hardening exponent n
substantially decreased from 0.22 to 0.06, Figure 5. These values of n were then
used for the construction of crack growth rate curves as described later, Figure 6.

4. Relationship between a/R and ε: Bridgman conducted tensile tests under a
range of hydrostatic pressure, and proposed a linear relationship of the form:

$$a/R = k(\varepsilon - \varepsilon_u) \qquad (13)$$

where ε_u is the strain at maximum load, and k is a constant. Comparing these results with the tests performed at atmospheric pressure, it was concluded that the
ratio a/R was pressure independent. However, further analysis showed a definite
influence of the hydrostatic pressure. Apart from the simplicity of equation (13),
it should be noted that the value of k may vary according to the investigated ma-

252

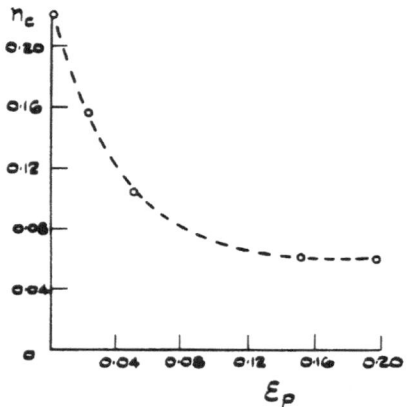

Fig. (5) - Steel 50D. n versus ε_p

terial. Similarly, it has been suggested that the strain at maximum load may be sometimes replaced by an even higher value.

The tests reported here suggest that the simple linear relationship as shown in Figure 7 is not adequate. The complete curve consists of three parts; the central portion, being linear, may be described for the strain values between 0.4 and 0.9 as:

$$a/R = 0.88 \ (\varepsilon - 0.34) \tag{14}$$

The regions I and III are highly non-linear. It was shown that at the low values of strain, the ratio a/R increased slowly. However, at high tensile stresses, the strains also reached high values (approximately 1) and the change of strain was localized on the plane of the crack. Here, voids formed rapidly, Figure 8, and the ratio a/R increased, as shown in Table 4.

5. Physical processes of deformation during loading of a cylindrical specimen in a tensile test can be described in the following terms:

    a. At very low stresses, below general yield, a limited number of dislocations moves within the grains. Here, the stress does not depend on grain size. With increasing stress, some dislocations will reach the grain boundaries.

    b. On reaching the yield stress, which is grain size-dependent, the dislocation sources at the grain boundaries are dominant. Their number increases with increasing stress.

    c. With a further increase of the strain, two processes have to be considered. When the cross-slip is easy, the slip length decreases rapidly with the strain. As the cross-slip becomes more difficult,

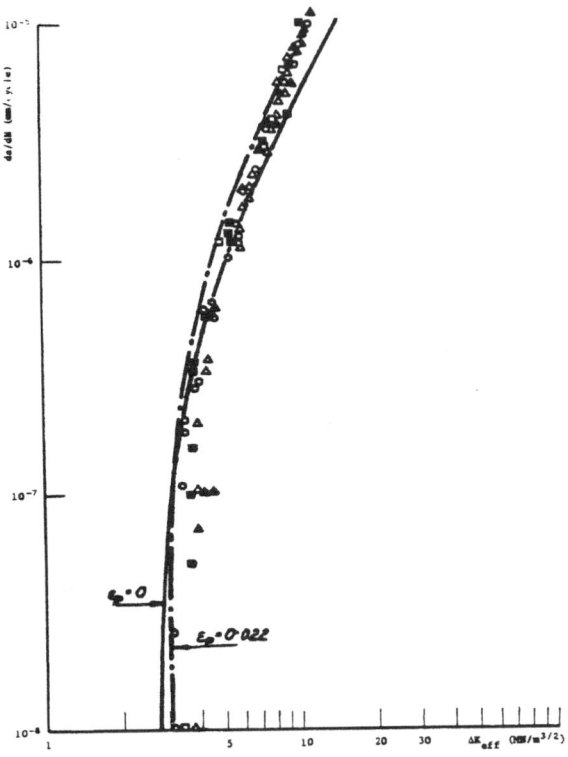

Fig. (6) - BS4360-50D: da/dN versus ΔK: B = 24 mm, 30 Hz, air

the slip length decreases with increasing strain very slowly and the influence of grain size becomes relevant. The influence of the grain size on the hardening exponent n in the typical intermediate strain range (up to 10%) appears to be strong for many metals. However, the significance of this inverse square root relationship has not yet been elucidated.

EXPERIMENTAL

In order to investigate cyclic crack growth, standard fatigue tests were performed at 30 Hz frequency in air and at a range of R values [5]. The experimental results are shown in Figure 6 for low growth rates between 10^{-5} and 10^{-8} mm/cycle. These results are compared with the predicted values using equation (1) developed

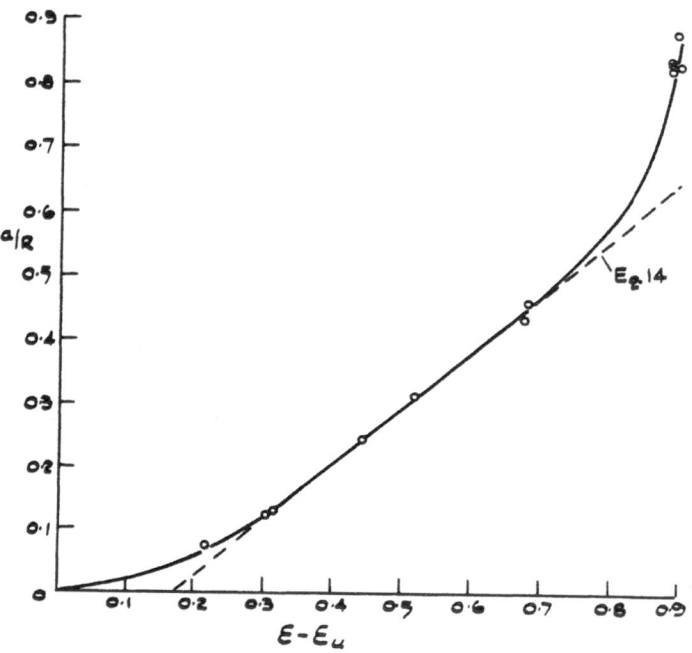

Fig. (7) - Steel 50D. a/R versus strain ($\varepsilon - \varepsilon_u$)

recently [1]. It is common practice to use statically derived parameters for the
calculation of fatigue crack growth curves and the static prediction curve $\varepsilon_p = 0$
(Figure 6) represents the crack growth rate using the static parameters n = 0.22
and ε_f = 1.26. It appears that at higher growth rates the static curve substantially underestimates the da/dN values, while below 10^{-7} mm/cycle, the growth rates are
overestimated.

The second curve was constructed from the cyclically stabilized parameters n
and ε_f (at constant strain). In this method, specimens were subjected to ten cycles
in tension between zero and a load at which ε_p = 0.022 at the same strain rate as
that used in the previous static tests. The strain hardening exponent, calculated
after the cyclic test, was 0.156 and the strain ε_f was 1.26. It appears that the
curve derived from a cyclically stabilized material is much nearer to the experimental fatigue results than that from static tests. Therefore, it is suggested
that the cyclically stabilized parameters should be used when predicting crack propagation curves in preference to statically derived values.

Fig. (8) - Steel 50D. Voids in the neck of a tensile specimen

TABLE 4

Specimen Number	Fracture (f) or Not (nf)	$\epsilon - \epsilon_u$	a/R
101	nf	0.217	0.070
102	nf	0.310	0.123
103	nf	0.316	0.127
104	nf	0.445	0.242
105	nf	0.519	0.312
106	nf	0.677	0.433
107	nf	0.682	0.459
108	f	0.885	0.818
109	f	0.886	0.826
110	f	0.895	0.875
111	f	0.899	0.824

CONCLUSIONS

Fatigue tests have been performed on a low-alloy steel BS4360-50D used for the construction of North Sea oil platforms. Crack propagation rates at very low value of stress intensity were investigated and a propagation law of the form:

$$\frac{da}{dN} = \frac{2^{(1+n)}(1-2\nu)^2(\Delta K^2_{eff}-\Delta K^2_{c,eff})}{4(1+n)\pi\sigma_{yc}^{(1-n)}E^{(1+n)}\varepsilon_f^{(1+n)}}$$

was developed. Experimental results confirmed that the cyclic strain hardening exponent n is necessary for the prediction of the fatigue crack growth rate.

Methods of deriving the static and cyclic strain hardening exponents and the relationships between the strain and the neck geometry of a tensile specimen were also described. It was concluded that a further study of the relevant parameters is needed for a better fatigue and fracture analysis.

REFERENCES

[1] Radon, J. C., "A model for fatigue crack growth in a threshold region", Int. J. Fatigue, July 1982.

[2] Bridgman, P. W., "Studies in Large Plastic Flow and Fracture", Harvard University Press, 1964.

[3] ASTM-E8-77a.

[4] Tvergaard, V., Needleman, A. and Lo, K. K., "Flow localisation in the plane strain tensile test", J. Mech. Phys. Solids 29, pp. 115-142, 1981.

[5] Musuva, J. D. and Radon, J. C., "Fatigue crack growth at low stress intensities", in Proc. Fatigue 81, F. Sherratt, ed., SEE, pp. 106-111, 1981.

[6] Rosenfield, A. R. and Hahn, G. T., "Numerical descriptions of the ambient low temperature and high strain rate flow and fracture behaviour of plain carbon steel", Trans. ASM, 59, pp. 962-977, 1966.

[7] Ludwik, P., "The importance of spatial tension for the testing of metals", Zeitschrift VDI, 71, 1927.

[8] Lee, D. and Zaverl, F., "The influence of material parameters on non-uniform plastic flow in simple tension", Acta Met., 28, pp. 1415-1426, 1980.

[9] Parker, E. R., David, H. E. and Flanigan, A. E., "A study of the tension test", Proc. ASTM, 46, p. 1159, 1946.

[10] Willoughby, A. A., Pratt, P. L. and Turner, C. E., "The meaning of elastic-plastic fracture criteria during slow crack growth", Int. J. Fracture, 17(5), pp. 449-466, 1981.

FRACTURE TOUGHNESS ACCEPTANCE TESTING OF ALUMINUM ALLOYS FOR
AIRCRAFT APPLICATIONS

K. R. Brown

Kaiser Aluminum & Chemical Corporation, Center for Technology
P. O. Box 877, Pleasanton, California 94566

ABSTRACT

High-strength aluminum alloys that are used for critical aircraft components are screened by fracture toughness testing in addition to other mechanical, stress corrosion and fatigue testing that may be specified by the aircraft
manufacturer.

Sections that are sufficiently thick can be tested by a standard plane
strain fracture test (e.g., ASTM E399), but most alloys and sections cannot
be tested in strict compliance with the test method. Most aircraft company
and government specifications allow minor deviations in particular test
validity requirements, when experience has shown that the test result will be
conservative; however, there is a lack of standardization, and specifications
differ depending on the experiences of the writers and the state-of-the-art at
the time of writing. Further, because of the expense of fracture toughness
testing, simpler screening tests are often applied before K_{Ic} testing. The
type and acceptability of screening tests varies from specification to
specification, and from supplier to supplier.

There is even less coherency in plane stress testing, due largely to the
immature state of the science of elastic-plastic fracture. Each specification
uses a plane stress test criterion selected by experience. Successful attempts
have been made to standardize most test methods, but the interpretation of the
test record can vary from one specification to another.

This paper summarizes the fracture toughness testing requirements of a
number of relevant U. S. and European aerospace specifications, and describes
the test methods, geometries and interpretations used. The testing difficulties are described, and the need for continuing research into plane stress
test methods is pointed out.

INTRODUCTION

A large proportion of high-strength aluminum alloy products that are used for aircraft structures are procured to specified fracture toughness minima in addition to the usual mechanical and other property limits. The products are usually heat-treated plate, extrusions, forgings, or sheet.

Although there are limited ranges of alloys and tempers supplied to the aerospace industry, it is common that each customer specifies the product differently. Different specifications usually require a different set of testing procedures, and require different toughness minima. This lack of standardization results from parallel engineering development within different aircraft companies or government agencies, so that each has had a different history of experience with testing methods, and each has accumulated data bases for different tests. It also reflects in part, differences in the preferences and judgments of individual materials and testing engineers.

Very successful attempts have been made by the Aluminum Association [1] and the American Society for Testing Materials (ASTM) [2] to guide the industry, but the field of fracture mechanics has evolved so rapidly that in some areas their efforts now lag practice.

Most specifications base their fracture toughness test procedures on ASTM standard test methods or practices, or their European equivalents. These standards have been established in response to the needs of the aerospace, nuclear or other industries, and with the cooperation of these industries. Thus standards relevant to aluminum alloys are based heavily on data supplied by aluminum users and producers, and represent an industry concensus.

The divergences in specifications are sometimes in areas where standards organizations have not yet established a concensus or where the result of a standardized test may be subject to a variety of interpretations. Some of these areas are on the frontiers of elasto-plastic fracture mechanics technology and are the subject of current research and standards development. Here the aircraft engineers, supported by large individual pools of empirical data, are ahead of the scientists and standards writers.

Another area of divergence results from the use of inexpensive screening tests to reduce the amount of testing by a costly standard method. To obtain sufficient statistical confidence, large data bases are required to establish the correlation between the cheap and the expensive test. The limits imposed in the user's specification may vary from supplier to supplier, and depend on the size, and hence statistical accuracy, of each supplier's data base.

TEST METHODS

The aerospace industry is concerned with the prevention of both plane strain fractures, that involve little plastic deformation near the crack tip, and plane stress fractures that may be associated with extensive plasticity. Plane strain fracture testing is generally required on the materials for the thicker sections of an aircraft, such as rolled plate wing boxes and bulkheads, forged undercarriage legs and heavy longeron extrusions in which plane strain conditions can occur at a crack tip. Plane stress testing is usually specified on thinner sections such as rolled sheet or plate wing skins or flanges and webs on some extrusions.

It is convenient for discussion to divide the individual test methods into those relating to plane strain, mixed mode and plane stress. The division is, of course, arbitrary, as there is a continuum in the fracture behavior in this range, and a considerable overlap in the range of suitability of some test methods.

A. Plane Strain Tests

Plane strain fractures are those involving little or no macroscopic yielding around the crack, and are characterized by a flat fracture face normal to the applied stress. They are promoted by relative high strength, relatively low toughness and are more readily established in thick sections. In most heat-treated aluminum alloys used in aircraft, plane strain tests are normally relevant only to section thicknesses greater than about 20 mm (1 inch); however, in some -T6 temper alloys, this may be as low as about 5 mm.

All major aircraft specifications which require plane strain testing invoke the ASTM E399 test method [3], or closely related foreign standards. This test method is also used for metals other than aluminum and measures the plane strain fracture toughness, K_{Ic}. Within the aluminum industry, it is usually determined using the compact tension test sample shown in Figure 1. A fatigue crack is made in the sample, and a tensile load applied. The record of load and crack mouth opening displacement (CMOD) is analyzed to obtain the crack tip stress intensity after a 2% or less increase in crack length. This critical stress intensity is termed K_{Ic}.

Fig. (1) - Compact tension fracture toughness specimen (ASTM E399 test method)

The K_{Ic} test result can be of use in studies of the fracture mechanics of aircraft structures, if plane strain conditions prevail in the structure. However, it is common for an aircraft manufacturer to accept a material with a plane strain test, yet to machine the material to a thickness in which a fracture would be by plane stress. For example, heavy section plate (> 10 cm) may often be machined into waffle structures, in which the thickness of no part exceeds 1 cm. Wing extrusions and skins are frequently machined to taper from the fuselage to the wing tip, so that the potential fracture mode could vary continuously along the wing.

The E399 test method incorporates a number of after-the-fact checks on the test data, to determine if the measured fracture toughness is valid. If it is not, the test result is termed K_Q. These include checks to ensure that the sample size was sufficiently large that the test fracture occurred under essentially plane strain conditions. Many test results fail the validity checks, however, invalidity rarely implies an inferior product or an improper test procedure. Most aluminum alloy products are not ordered, or cannot be made, in a sufficiently thick section to allow plane strain crack growth in a test, and hence it is often not possible to measure a valid K_{Ic}. Further, when dealing with a material of unknown K_{Ic}, it is not possible to predict the sample size necessary to obtain a valid test.

Test results that are marginally invalid are often meaningful, and they may extend the useful range of applicability of the E399 test result. Some validity checks in E399 are appropriate for some metals, but are unnecessarily restrictive for aluminum alloys. Further, where experience has shown that a particular test invalidity causes the measured toughness (K_Q) to be lower than K_{Ic}, then the material can be considered acceptable if the conservative invalid result exceeds the specified minimum toughness.

For these reasons the ASTM B7 committee has developed the document ASTM B645 [4] especially for aluminum alloys. This document specifies which invalidities are acceptable, and which invalid E399 test results can be considered to be meaningful. The terms "valid" and "meaningful" have strict legal definitions in the ASTM documents.

A growing number of specifications permit the use of B645 to extend the range of applicability of the E399 test method into the mixed mode regime. Some specifications which do not, simply have not been revised since B645 was written; others continue to accept specific invalidities without reference to B645.

Plane Strain Screening Tests

As the test specimen used in the ASTM E399 method is expensive and time-consuming to machine and to fatigue pre-crack, the aluminum producers and users have cooperated to reduce costs and speed the shipment of material by using a relatively inexpensive notched round tensile test to screen material before E399 testing. Material which fails is retested to determine K_{Ic} or K_Q.

The test sample (Figure 2) and method are described in ASTM E602 [5]; however, this standard does not cover the interpretation of the test result. The industry has generally adopted the interpretation described in ASTM B646 [2] which refers to the ratio of the notched tensile strength (NTS) to the tensile yield strength (TYS) of a conventional smooth tensile specimen. This is termed the Notched Yield Ratio (NYR) or NTS/TYS ratio which has been shown to correlate with K_{Ic} or K_Q.

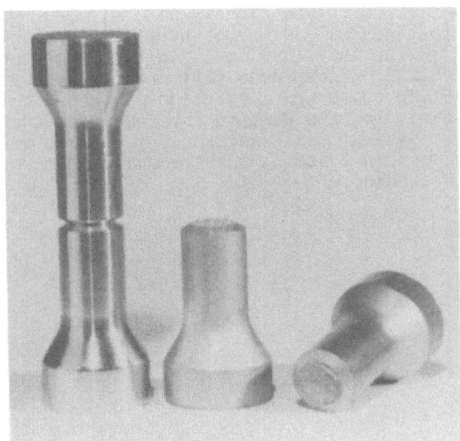

Fig. (2) - Round notched tensile specimen
(ASTM E602 test method)

When the E602 sample is used as a screening test, the minimum acceptable NYR values are selected by the customer to correspond to a particular statistical confidence that the specified minimum K_{Ic} will be met [1]. The probability of not passing the limit thus depends on the number of K_{Ic}-NYR correlations in the supplier's data base for that particular product. When a lower NYR is obtained, a K_{Ic} or K_Q retest must be made by ASTM E399. The frequency of retesting also depends on the size of the data base. When only a small data base has been established, a high re-test rate may make the use of the screening test uneconomical or too slow. Conversely, the larger the data base, the greater the savings achieved by the use of the screening test, but this must be balanced against the cost of establishing the data base and the producer's expectation of future orders for the product. In many cases different producers must pass different NYR minima for the same product as a direct result of differences in the size of their data bases.

In the United States the obvious need for standardization has prompted the Metals Properties Council, the Aluminum Association, and committees for the Military Handbook [6] and Damage Tolerant Design Handbook [7] to cooperate and gather industry-wide data bases for NYR/K_{Ic} correlations. The effect of this action will shortly be seen in revised procurement specifications.

The chevron-notched short rod and short bar samples are used in
another recent plane strain fracture test that is currently under development
and which will probably become significant in the future. This sample
requires no fatigue pre-cracking and the stress intensity, K_{SR} or K_{SB}, is
measured at the front of a crack that has grown from the tip of a chevron in
a heavily side-grooved rod or bar (Figure 3). Tests on aluminum alloys have
shown that these parameters correlate closely with K_{Ic} [8], and that useful
plane strain fracture toughnesses can be measured with a sample that is
smaller than that used in the ASTM E399 test. Although there is a Society
of Automotive Engineers (SAE) recommended test practice [9] for these sample
geometries, and the testing costs are a fraction of those of the ASTM E399
test, it is still too early for the test to have been included in any aircraft
company specifications. Several major aerospace companies are interested in
the method for use as a screening test, but are awaiting more data, and the
results of current ASTM studies.

Fig. (3) - Chevron notched short bar toughness specimens
B = 25 mm (1") left and 12.5 mm right. A
broken half of a compact tension sample is
shown for comparison (center).

B. Mixed Mode Tests

A number of specifications have extended into the mixed mode regime,
the usefulness of plane strain fracture mechanics for material testing by
using "R" curve techniques. An "R" curve for a material describes the
increase in fracture toughness, K_R, as a crack grows, and is usually presented as a plot of K_R versus crack extension, Δa. Crack growth in
relatively high toughness materials is associated with increasing plasticity,

decreasing crack sharpness and the loss of plane strain conditions at the crack tip. This in turn increases the measured toughness. "R" curves are used for a wide range of mixed mode and plane stress fractures and their use for relatively thin sections will be described under plane stress testing.

Several standard methods for "R" curve determination are described in ASTM E561 [10]; however, the interpretation of the "R" curve is not, and the interpretation may vary from one specification to another, and from one test method to another.

In relatively heavy sections, "R" curves may be determined using a compact tension specimen and techniques very similar to those used for plane strain E399 tests (Figure 1). However, the E561 specimen has a relatively shorter initial crack length, minor geometry differences, and a different set of validity criteria. It is the intent of some specifications that when an E399 test is found to be invalid, the E399 test record is analyzed by the "R" curve methods of E561. However, the strict compliance with the rules of E561 precludes this. The approach is clearly desirable for ease of testing, and for lower testing costs; however, its application would require data to ensure that the departures from E561 geometry produce conservative results.

The "R" curve itself is the fracture property of the material in the thickness tested, but it is inconvenient to use the curve in its entirety in a specification. It is standard practice to measure the toughness, K_c, after a particular increment of crack growth, or in some plane stress tests, at maximum load. The crack-tip stress intensity after a particular crack growth increment, Δa, is determined by constructing secants on the load/CMOD record (Figure 4). These slopes are, by convention, either 5% or 25% lower than the

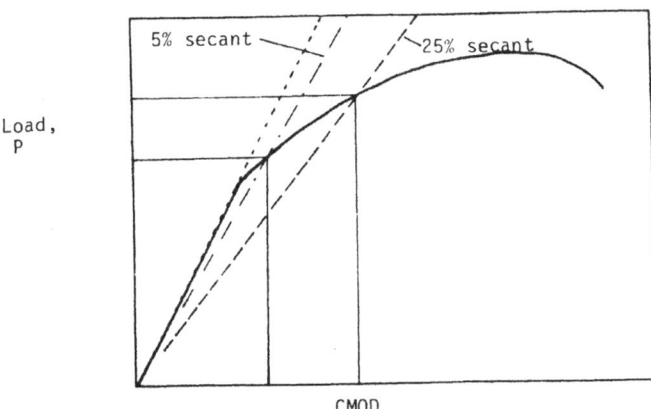

Load,
P

5% secant

25% secant

CMOD

Fig. (4) - A typical load versus crack mouth opening curve for a compact tension specimen used for generating an "R" curve. The 5% and 25% secant offsets are indicated (ASTM E561 test method)

elastic portion of the record. The 5% slope is the same as that used in the E399 plane strain test which corresponds to a 2% increase in crack length, whereas the 25% secant approximates a 30% crack growth. The particular value of K_c is often identified by using the subscripts $K_{5\%}$ or $K_{25\%}$.

It is a frequent source of confusion to the novice that different toughness values will be quoted for the same material, and even for the same test, depending on the convention adopted by the specification. In the one material the 5% secant toughness determined from the "R" curve, $K_{5\%}$, may be higher or lower than K_{Ic}, and will usually be much lower than the 25% secant value, $K_{25\%}$. Relative differences for one lot of a high strength alloy tested with typical sample geometries are shown in Table 1.

TABLE 1 - TYPICAL FRACTURE TOUGHNESSES MEASURED ON THE SAME
LOT OF A HIGH-TOUGHNESS, 7475-T7351 ALUMINUM
ALLOY PLATE BY DIFFERENT TEST METHODS*

Specified Test Method	Typical L-T** Toughness MPa m$^{1/2}$ (ksi√in.)
ASTM E399 - K_{Ic} - [3]	55 (50)
ASTM E561 - "R" curve - [10] (a) 5% secant	45-65 (40-60)
ASTM E561 - "R" curve - [10] (b) 25% secant	75-110 (70-100)
(c) Center cracked panel 150-mm (6") wide	130 (120)
(d) Center cracked panel 400-mm (16") wide	200 (180)

*Test sample thicknesses are those typically specified. This data is approximate, and is to contrast different test results. It should not be used for design purposes.
**The test sample orientation code is described in ASTM E399 [3]. The first letter represents the direction of applied tensile stress, the second is the direction of crack growth. L = longitudinal, T = transverse, S = thickness direction.

Mixed Mode Screening Tests

There are no extensive data bases for screening tests in this fracture regime, and relatively little use is made of screening tests.

In thicker sections some use is made of round notched tensile testing, and it is likely that in the future NYR values will be correlated with a selected "R" curve parameter, probably $K_{25\%}$, and the notched yield ratio will be used to reduce the amount of "R" curve testing.

Some specifications permit machining of some intermediate thickness materials to thicknesses suitable for plane stress testing, but this has little impact on testing costs.

C. Plane Stress Tests

Plane stress fractures involve extensive plasticity which does not permit sufficient restraint to develop high lateral stresses in addition to the applied stress. In common usage, this includes flat fractures that are accompanied by significant plasticity, and shear-mode fractures. The fracture in many "plane-stress" tests may be of mixed mode, typically a flat fracture with marginal shear lips, and the relative proportion of each cracking mode may change as the crack propagates. Commonly the fracture appearance changes from a flat fracture, with increasing shear lips, to a fully shear failure mode. The occurrence of these failure modes depends on material properties and sample thickness.

Common plane-stress tests are based on ASTM E561 [8] "R" curve techniques as described earlier for mixed mode tests, but the center-cracked panel sample that is tested in tension (Figure 5) is more commonly used. A 400-mm (15-inch) wide by 1000-mm (40-inch) long panel is the industry standard plane stress test for high strength aluminum alloy sheet, and plate less than about 25 mm (1 inch) in thickness.

Fig. (5) - A 405-mm-wide (16") center notched panel test
sample used for plane stress fracture toughness
testing (ASTM E561 test method)

The load versus crack opening test record may be interpreted by 5% or 25% secants (Figure 4) as for the compact tension geometry sample described earlier, but in the center-cracked panel test, by convention, the K_C value is usually calculated at the maximum load during the test. This toughness measure also differs from K_{IC}, $K_{5\%}$ and $K_{25\%}$ (Table 1).

To simplify testing, most specifications allow several departures from the ASTM E561 test method.

(1) They permit the use of sawn slots instead of a fatigue crack.

(2) They permit the net section stress in the panel at maximum load to exceed the yield stress.

(3) They permit machining of the sample thickness to 6.3 mm (0.25 inch) to allow the use of testing machines of normal capacity.

These departures greatly reduce the difficulty of testing for material acceptance, but severely limit the usefulness to the aircraft designer of the K_C data, and reduce the center-cracked panel test to the status of an empirical screening test.

Plane Stress Screening Tests

The center-cracked panel described in E561 is an expensive test sample to prepare, and the removal of one L-T specimen from a production plate produces a piece of scrap about 1-meter long by the plate width. Consequently, less expensive screening tests that use smaller specimens have been developed.

The most common screening test is the flat, edge-notched tensile specimen described in ASTM E338 [11], and shown in Figure 6. This is used in a similar manner to the notched round specimen [5] described earlier, and also necessitates the establishment of a data base for the empirical correlation of edge-notched tensile data and center-cracked panel data.

Some specifications permit the use of a smaller 150-mm (6-inch) wide and 400-mm (16-inch) long center-cracked panel as a screening test for the similar 400-mm-wide test. In these cases a conservative result is assured by maintaining the specified lower limit of toughness K_C at the same level as for the larger sample, even though the test usually measures a lower K_C in the smaller panel (Table 1).

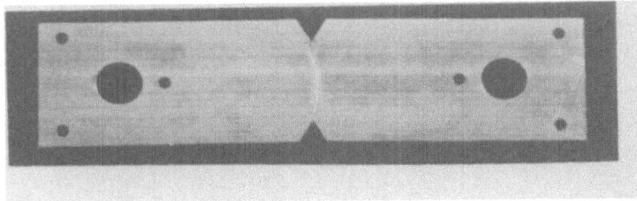

Fig. (6) - A double-edged notched tension sample used
in the ASTM E338 test method

SUMMARY OF CURRENT SPECIFICATIONS

The general approaches taken, and the different fracture toughness tests
favored in different specifications issued by ten major aircraft manufacturers
are summarized in Table 2. These include six United States corporations, two
German manufacturers and one each from the UK and France, and are those with
which the author has had experience. This list is not exhaustive, but it is
believed sufficient to summarize the fracture toughness requirements imposed
on most aluminum alloys used for aircraft construction.

Because new and old specifications coexist, and because they are con-
tinually being revised in the light of new technology, the summary will not
be correct for every toughness specification within the one aerospace company;
the most recent trends are emphasized where possible.

The similarities shown between some different specifications are not always
coincidental; it is common for some aerospace customers to use another
company's specifications, particularly when they are cooperating on a joint
project. It is also normal to refer to government or society specifications.

Where a customer is attempting to set up a data base for a particular
test method, it is common to require that the test be performed, but without
setting a minimum test result. In those cases it is to be expected that in
a few years a lower limit will appear in revised specifications, and this test
will be permitted in lieu of a more expensive or less satisfactory test.

TABLE 2 - GENERAL REVIEW OF AIRCRAFT SPECIFICATION REQUIREMENTS

Aerospace Company	Plane Strain				Compact Tension		Mixed Mode and Plane Stress				
	ASTM E399+	ASTM B645	Other Invalidity Acceptance	Notched Round Tensile ASTM E602	Accepted	$K_{25\%}$	Center Notched Panel			Notched Flat Tensile ASTM E338	Others
							400 mm (16")	150 mm (6")	Max Load K_c		
1	*	*		*	*	*	*		*		
2	*	*		*			*	A	*	A	
3	*	*		*				A	*	A	
4	*		*	*	*	*	*	A	*	*	
5	*			*	D		*	D	*	*	B
6	*		*	*						*	
7	*				D		D				
8++	*		*		*	C				*	
9	*		*	*							
10	*			*	D		D			D	

Notes: An asterisk indicates that the test is required
or is acceptable on appropriate products.

A   qualified suppliers only
B   fatigue pre-cracked charpy
C   extrapolates K_R data to estimate K_{Ic}
D   no reference found in limited available specifications
+   or BS5447 or other equivalent
++ fracture specifications currently under review

CONCLUSIONS

The field of fracture testing for the quality assurance of aluminum alloys for the aerospace industry is complicated, and is constantly changing. There is a clear, but recognized need for standardization in test methods and data interpretation, and for cooperation in data collection throughout the industry.

Toughness data generated for material acceptance is essentially empirical and very little is of direct use to the aircraft designer or NDT engineer who uses fracture mechanics. For example, materials are rarely tested in the same thickness in which they are used in the aircraft; materials tested in a plane strain test may be machined extensively to a section in which failure would be in plane stress. In plane stress tests, the toughness measured for quality assurance relates to a particular crack growth increment that may be irrelevant to the material in most aircraft structures.

There is a relatively satisfactory and understood test method for plane strain fracture, but it is expensive. Screening tests are being used increasingly for material acceptance testing, with economic benefit, and this trend is expected to continue as correlation data is accumulated.

There is no really satisfactory measure of plane stress fracture toughness at present, with the result that different aluminum alloy customers specify their requirements differently. There is active research in the field of elastic-plastic fracture mechanics, but it is unlikely that a satisfactory test and interpretation will be developed and accepted by the industry in the near future.

Fracture toughness acceptance testing, however, is essential for aircraft safety as it can reject materials of low fracture resistance that could otherwise find their way into critical structures.

REFERENCES

[1] "The Aluminum Association Position on Fracture Toughness Requirements and Quality Control Testing", Bulletin T-5, Aluminum Association, New York, September 1974.

[2] "Standard Practice for Fracture Toughness Testing of Aluminum Alloys", ASTM B646-78 in 1981 Annual Book of ASTM Standards, Part 7, ASTM, Philadelphia, 1981.

[3] "Standard Test Method for Plane-Strain Fracture Toughness of Metallic Materials", ASTM E399-81 in 1981 Annual Book of ASTM Standards, Part 10, ASTM, Philadelphia, 1981.

[4] "Standard Practice for Plane Strain Fracture Toughness Testing of Aluminum Alloys", ASTM B645-78 in 1981 Annual Book of ASTM Standards, Part 7, ASTM, Philadelphia, 1981.

[5] "Sharp-Notch Tension Testing with Cylindrical Specimens", ASTM E602-81 in 1981 Annual Book of ASTM Standards, Parts 7 & 10, ASTM, Philadelphia, 1981.

[6] "Military Standardization Handbook, Metallic Materials and Elements for Aerospace Vehicle Structures", MIL-HDBK-5B. Dept of Defense, Washington, D.C., current update.

[7] "Damage Tolerant Design Handbook", MCIC-HB-01, Metals and Ceramics Information Center, Battelle, Columbus, Ohio, January 1975 and current updates.

[8] Brown, K. R., "An Evaluation of the Chevron-Notched Short-Bar Specimen for Fracture Toughness Testing in Aluminum Alloys", Kaiser Aluminum & Chemical Corp., Report, October 1, 1981.

[9] "Determination of Short Bar Fracture Toughness of Metallic Materials", Aerospace Recommended Practice, ARP 1704, Society of Automotive Engineers, Inc., Warrendale, PA, July 1, 1981.

[10] "Recommended Practice for "R" Curve Determination", ASTM E561-81 in 1981 Annual Book of ASTM Standards, Part 10, ASTM, Philadelphia, 1981.

[11] "Standard Method of Sharp-Notch Tension Testing of High-Strength Sheet Materials", ASTM E338-81 in 1981 Annual Book of ASTM Standards, Part 10, ASTM, Philadelphia, 1981.

RELATION BETWEEN CRACKS AND ELECTRON BEAM WELDING CONDITION OF SUS304L STAINLESS SINTERED STEEL

Y. Suezawa

College of Science and Technology, Nihon University
Tokyo, 101 Japan

ABSTRACT

Experimental data on the effects of melting Aluminum 5183 filler wire and SUS308L stainless filler wire into 0.2% carbon sintered steel and SUS304L stainless sintered steel in connection with weldments by electron beam are presented. The heat input per unit length is maintained constant at 3kJ/cm.

The result of the experiment showed that porosity is reduced because of deoxidizing of Aluminum or Chromium at the welded joint. In case of SUS304L sintered steel added with SUS308L and A5183 filler wire, A-porosity was reduced remarkably over all the range of welding speed from 150 to 600 mm/min. In case of SUS304L sintered stainless steel added with SUS308L filler wire, a necklace crack was observed at the welding speed at 150 mm/min. However, at the welding speed of 75 mm/min, the necklace crack disappeared and only R-porosity remained. No effect was observed when using A5183 filler wire.

In addition, it was found that low welding speed was essential for obtaining porosity-free weld even if filler metal was used. X-ray micro analyser was used to detect Chromium and Nickel traces which were contained in SUS308L filler wire in the fused area of sintered steel. It was found that Chromium content enhances diffusion into base metal better than Nickel content.

INTRODUCTION

This investigation is concerned with the experimental study of melting Aluminum A5183 filler wire and SUS308L stainless filler wire into 0.2% carbon sintered steel and SUS304L stainless sintered steel weldments by electron beam.

One of the main problems which had to be solved in welding these sintered materials was to reduce porosity in the fused zone, and to investigate the parameters that control porosity. Special attention had to be given to the welding speed, heat input, focus position, etc. [1-4]. After the first test trial, the range of welding conditions for preventing porosities were found to be severely limited. In order to broaden this limitation, it became apparent that filler metals such as SUS308L and A5183 during welding had to be added to obtain porosity-free weld. It was also found that low welding speed was essential for obtaining porosity-free weld.

The cracks and porosities of these specimens were measured from X-ray photographs as shown in Figure 1. The welding conditions are summarized in Table 2. a_b^* is defined as $a_b^* = D_0/D_F^*$, D_0 is a distance from the lens to the work surface and D_F^* is a distance from the lens to the focused point [6].

In this experiment, D_F^* was kept constant at 200 mm and D_0 was changed each time according to a_b^*.

TABLE 2 - WELDING CONDITIONS FOR BEAD ON
PLATE WELDING WITH FILLER METAL

t_b (mm/min)	P_b (kW)	Q (kJ/cm)	a_b^* $(D_F^* = 220$ mm)
600	3	3	1 12~0 82
450	2 25	3	1 12~0 82
300	1 5	3	1 12
220	1 1	3	1 11
150	0 75	3	1 12
110	0 55	3	1 12

Notes t_b Welding speed
P_b Beam power
Q Weld heat input
a_b^* Modified beam active parameter
D_F^* Focal distance determined by observation with naked eye

SINTERED STEEL

Figure 2 shows the relation between the porosity and a_b^* when welding speed changed with and without filler wire. In every case when $a_b^* = 1$, the porosity shows maximum percentage. The application of filler wires tends to reduce porosity while even better reduction can be achieved by lowering the welding speed.

Generally speaking, the observed porosity in sintered steel on X-ray can be divided into two categories: R-porosity and A-porosity. R-porosity applies to porosity observed at the spiking or root section and its diameter is in general less than 0.5 mm. A-porosity applies to a larger diameter porosity that occurs generally at the central part of the penetration.

Figure 3 shows an X-ray picture of sintered steel with filler wire SUS308L in case of $a_b^* = 0.82$ and $a_b^* = 1.12$ maintaining welding speed $V_b = 450$ mm/min. It is seen that the lower values of A-porosity and R-porosity are observed respectively for $a_b^* = 1.12$ than $a_b^* = 0.82$. Adopted, is $a_b^* = 1.12$ as the focus distance, i.e. the beam focused at 30 mm above the surface of the work. Figure 4 shows an X-ray picture of the sintered steel with SUS308L stainless steel filler wire at the welding speeds of 300 mm/min, 450 mm/min and 600 mm/min respectively maintaining heat input at 3 kJ/cm.

EXPERIMENTS

The commercial steel powder (c = 0.2% etc., 200 mesh) was compressed into a 120 x 10 x 20 mm rectangular piece by a 50 ton hydraulic press and the compressed powder pieces were sintered in 35 kw resistance furnace operated in hydrogen gas atmosphere (dew point -42° C) at 1150° C.

In the case of stainless, the commercial SUS304L stainless powder (Ni 12%, Cr 19% etc., 200 mesh) was compressed to the above mentioned size by a 100 ton capacity hydraulic press and the compressed powder pieces were sintered in 35 kw resistance furnace operated in resolved ammonia (NH_3) gas at 1170° C.

The density value ρ of these manufactured steel and stainless sintered steel were measured as

$$\rho = \frac{\text{weight of specimen}}{\text{cubic volume}} \tag{1}$$

Table 1 shows the density values of these sintered materials and chemical compositions of various materials used in this experiment.

TABLE 1 - CHEMICAL COMPOSITIONS OF VARIOUS MATERIALS USED

		Density (g/cm)	Dia (mm)	chemical compositions(wt %)												
				C	Si	Mn	P	S	Ni	Cr	Cu	Fe	Mg	Zn	Ti	Al
BASE METAL	Sintered steel	6.8		0.2			0.008	0.001								
	Sintered SUS304L steel	6.6		0.019	0.41	1.05	0.027	0.009	10.15	19.21						
FILLER METAL	SUS308L	–	2.6	0.018	0.44	1.77	0.020	0.010	10.80	19.63						
	A4043WY		1.6		0.14	0.69						0.11	0.010	0.214	0.020	0.01	RF

As illustrated in Figure 1, the 2.6φ SUS308L filler wire was fixed tightly on the surface of the sintered steel and its roughness value was 25-s. Maintaining the heat input constant at 3 kJ/cm and the accelerating potential at 150 kV, electron beam current, welding speed and focus position were adjusted so that the penetrated depth was always kept about 10 mm. These operations were carried out in a vacuum chamber at 5 x 10^{-4} Torr.

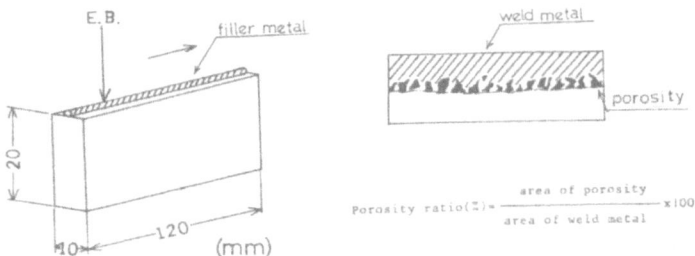

Fig. (1) - Method for electron beam welding with filler metal and definition of porosity ratio

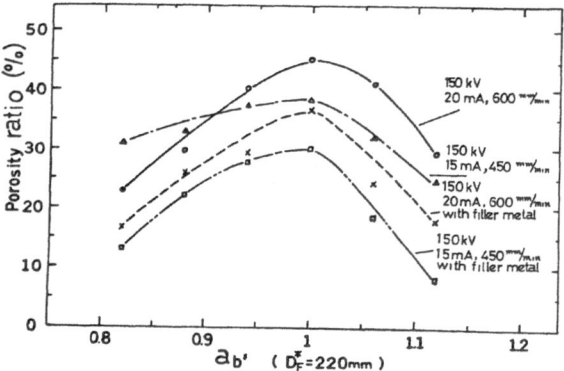

Fig. (2) - Relationship between porosity and a_b^* value

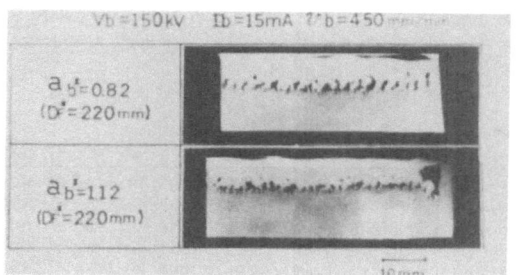

Fig. (3) - Side X-ray picture of sintered steel with
SUS308L filler metal in case of $a_b^* = 0.82$
and $a_b^* = 1.12$

In the case of V_b = 300 mm/min, most of A-porosity disappears even if there remains some R-porosity while they appeared for V_b = 450 mm/min and 600 mm/min.

In order to reduce A and R porosities even if the welding speed is 450 mm/min and 600 mm/min respectively, the addition of Aluminum filler wire were tried. In this case, 1.6 mm diameter Aluminum filler wire was pressed into a 0.5 mm thickness plate and inserted between the base metal and SUS308L filler wire. Figure 5 shows the results, i.e. X-ray picture of sintered steel with filler wire SUS308L on left half side and on right half side the pressed Aluminum filler wire A5183 which was inserted between the base sintered metal and SUS308L filler wire. The result shows a remarkable reduction of R and A porosities when Aluminum filler was added.

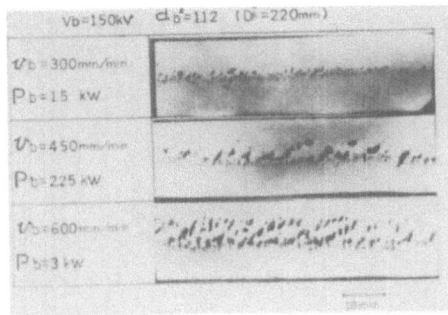

Fig. (4) - X-ray picture of sintered steel with SUS308L
stainless steel filler metal at various welding speeds maintaining heat input constant

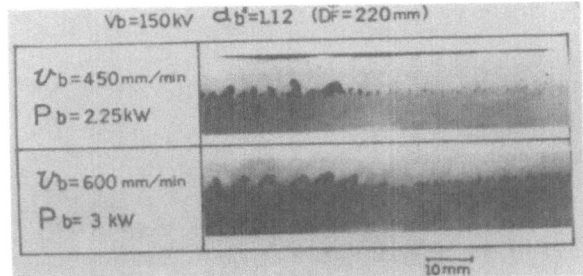

Fig. (5) - X-ray picture of sintered steel with SUS308L
stainless steel filler metal on left half
side and SUS308L plus A5183 on right half side

Figure 6 illustrates the relation between porosity and welding speed by
keeping the heat input constant at 3 kJ/cm for the sintered steel with and
without filler wire.

SINTERED STAINLESS STEEL

Figure 7 shows an X-ray picture of SUS304L sintered stainless steel
the welding speeds of 75 mm/min, 150 mm/min and 300 mm/min maintaining heat
input constant at 3 kJ/cm. From Figure 7, it is seen that the results
obtained for the stainless sintered steel are completely different from those
for the sintered steel, i.e. A-porosity varied discontinuously in accordance
with the welding speed.

At the welding speed of 300 mm/min, large A-porosity and R-porosity coexist, but, at the speed of 150 mm/min, a necklace crack appears at the

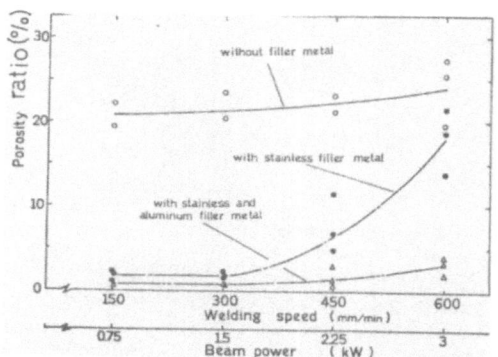

Fig. (6) - Relation between porosity and welding speed
maintaining heat input constant in case of
sintered steel with and without filler metal

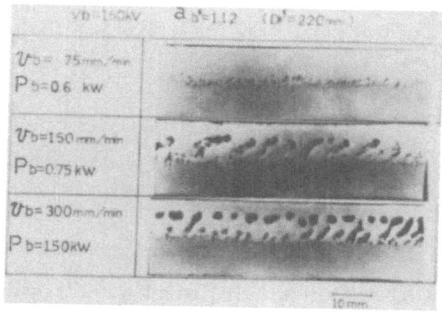

Fig. (7) - X-ray picture of SUS304L sintered stainless
steel at various welding speeds maintaining
heat input constant

section of A-porosity and at the speed of 75 mm/min A-porosity almost
disappears, only R-porosity remaining. Figure 8 shows an enlarged section
of A-porosity and microphotographs A-A cross section and B-B cross section.
When one A-porosity is closely inspected, the upper part of the porosity
is seen to consist of a blow hole, and from the B-B cross section the lower
part of the porosity assumed a horseshoe form. From the A-A cross section
the lower part of the porosity shows a so-called necklace crack (a kind of a
shut).

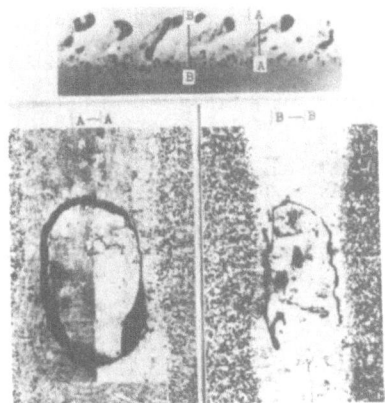

Fig. (8) - X-ray picture for necklace crack of SUS304L
stainless sintered steel with filler metal and
and microphotographs of its cross section

The reason why the necklace cracks can be observed only in the stainless
steel, titanium alloy, nickel alloy and carbon steel is not yet clarified.
Figure 9 shows the relation between the porosity and the welding speed

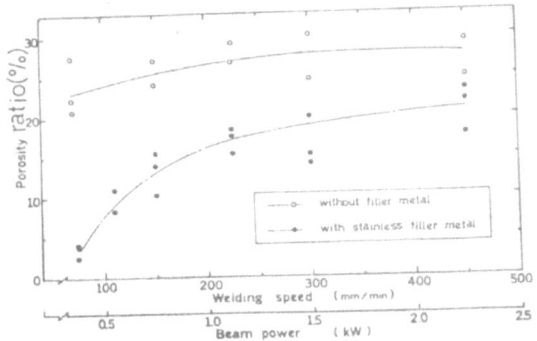

Fig. (9) - Relation between porosity and welding speed
maintaining heat input constant in case of
SUS304L sintered stainless steel with and
without SUS308L stainless steel filler wire

maintaining heat input at 3 kJ/cm in case of SUS304L sintered stainless steel
with and without SUS308L stainless filler wire. Generally speaking, for
stainless sintered steel, porosity reduction can be achieved by lowering the
welding speed below that used for sintered steel even if SUS308L filler wire
was used. The slower the welding speed, the less porosity was observed.
This is one of the major conclusions of the experiment.

INVESTIGATION BY X-RAY MICROANALYSES

In order to investigate how the filler material diffuses into the weld
metal, EPMA was used to trace Cr, Ni, and Al components in the welded zone.
The beam diameter of EPMA's electron beam is about 0.1 mm. Figure 10 shows

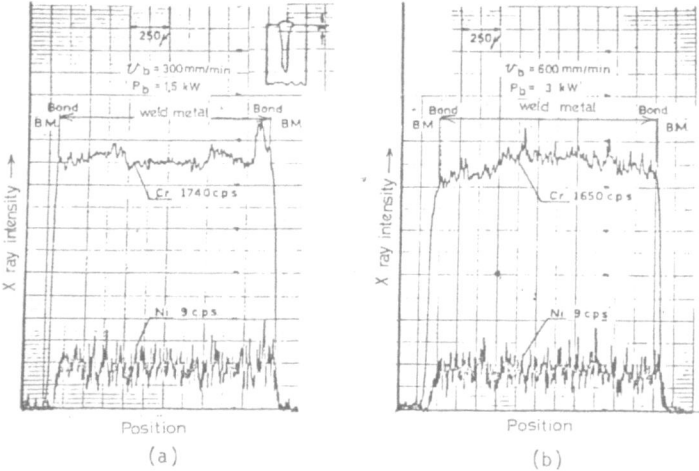

Fig. (10) - Line scanning analysis for Cr and Ni elements in upper
           part of fused sintered steel weld with SUS308L stain-
           less steel filler metal by EPMA.

the results for sintered steel added with SUS308L filler wire the welding
speed being 600 mm/min and 300 mm/min. EPMA's electron beam was scanned at
the position of 1 mm below the surface and 3 mm above the root bottom of the
weld metal.

Cr component of upper part of weld metal (1 mm below the surface) is about
1740 cycles per second and 1650 cps depending upon the welding speed. This
means Cr component is almost the same even if the passing time of welding
electron beam, i.e. the welding speed is different.

Ni component is only 9 cps in both cases. There are no Ni and Cr components at the base metal and bond section. This means Ni and Cr component did
not penetrate into the base metal area.

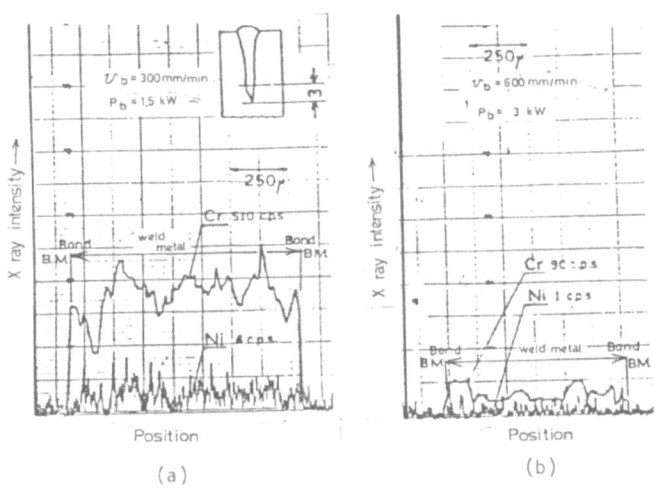

Fig. (11) - Line scanning analysis for Cr and Ni elements in lower
part of fused sintered steel with SUS308L stainless
steel filler metal by EPMA

Figure 11 shows the results for a line scanned 3 mm above roots bottom.
In this case, there is a difference between 600 mm/min and 300 mm/min of
welding speed when Cr is traced. The Cr component is 90 cps in the former
case and 510 cps in the latter case. This shows that Cr content decreased as
V_b value increased.

CONCLUSIONS

1. In all cases, when $a_b^* = 1$, porosity showed maximum and the porosity
decreased as a_b^* value approached to 1.12 or 0.82.

2. When heat input was maintained constant 3 kJ/cm, porosity of sintered
steel with SUS308L filler wire decreased remarkably for V_b = 600 mm/min and
reduced to V_b = 300 mm/min even if no A5183 was added. When A5183 filler wire
was added, porosity decreased remarkably all over the range from V_b = 600 mm/
min to 150 mm/min.

3. In the case of SUS304L sintered stainless steel, porosity did not
decrease in the range of V_b = 600 mm/min to 150 mm/min even if SUS308L filler
wire was added, but when the welding speed was reduced to 75 mm/min, porosity
decreased regardless of the existence of A5183 filler wire.

4. In the case of SUS304L stainless sintered steel, a necklace crack was
observed at V_b = 150 mm/min when SUS308L filler wire was added. However, no

necklace crack was observed at V_b = 300 mm/min and V_b = 75 mm/min.

5. According to the EPMA analysis, Ni component did not penetrate into the fused zone. On the other hand Cr component did penetrate into the fused area of the sintered steel added with SUS308L filler wire. The content of Cr in the upper part of material was larger than that in the lower part. However, Cr content decreased in accordance with a higher welding speed.

REFERENCES

[1] Suezawa, Y., et al., "A study on electron beam welding of sintered iron", J. of JWS., Vol. 47, No. 7, p. 22, 1978.

[2] Suezawa, Y., Kuroda, H., "An investigation of E.B. welding of SUS304L stainless sintered steel:, 2nd Int. Colloquium E.B. welding and melting, Avignon, p. 215, 1978.

[3] Suezawa, Y., Kuroda, H., "Relation between the E.B. welding conditions and its joint structure of the SUS304L stainless sintered steel", Proceeding of the 21st Japan Congress on Material Research, The Society of Materials Science, Japan, Kyoto, Japan, p. 135, 1978.

[4] Suezawa, Y., Kuroda, H., "A metallurgical study of E.B. welding of sintered iron", The International Conference of Fracture Mechanics in Engineering Application, Bangalore, India, 1979.

[5] Matsuda, Y., et al., "Powder Metallurgy", NIGANKOGYO SHINBUN, p. 195.

[6] Arata, Y., "Terms and definitions proposed from Japan, IIW, 1972.

FRACTURE TOUGHNESS OF A533 B STEEL: ELECTRICAL POTENTIAL METHOD

H. Matsushita

Mitsui Engineering and Shipbuilding Company, Ltd.
Ichihara, Chiba, 290 Japan

and

T. Miyoshi

University of Tokyo
Tokyo 113, Japan

ABSTRACT

The electrical potential method for detecting the initiation of slow crack growth is studied in relation to fracture toughness testing of three-point bend and compact tension specimens made of A533 B steel. Experimentally obtained values of δ_i and J_i are also discussed. The main experimental results can be summarized as:

(1) the inflection point in the potential difference change and clip gauge displacement indicate the initiation of slow crack growth;

(2) the J_i/σ_y values correlated well with the δ_i values; and

(3) J_i and δ_i values obtained by the electrical potential method were in the same region as those by the R-curve method.

INTRODUCTION

During fracture toughness testing of small specimens, slow crack growth prior to unstable fracture is often observed. In such cases, the crack opening displacement, δ_i, and the J-integral, J_i, at initiation has been proposed for assessing the fracture toughness of the material. Both δ_i and J_i are approximately constant when J_i satisfies the condition $B = 25\ J_i/\sigma_y$, where B is thickness and σ_y is the yield strength of the material. J_i is considered to be equivalent to J_{IC} under the condition of plane strain.

Methods such as the electrical potential method [1-4], AE method [5], R-curve method [6,7] and stretched zone method [18] are available to detect the initiation of slow crack growth. Among these methods, the electrical potential method seems the most promising for industrial use in that it can easily be applied to detect crack growth and avoids complicated measurement equipments.

This paper discusses the fracture toughness values δ_i and J_i obtained from the above mentioned method for the A533 B steel.

EXPERIMENTAL PROCEDURES

Material and Specimen. The surface side of the nuclear vessel steel A533 Gr. B cl.1 with the thickness B = 165 mm was used for making the specimens. The chemical composition is shown in Table 1. Figure 1 shows the relationship between tem-

TABLE 1 - CHEMICAL COMPOSITION

Steel	Thickness mm	σ_y kg/mm^2	Chemical Composition (wt%)						
			C	Si	Mn	P	S	Ni	Mo
A533 B	165	49	0.18	0.18	1.45	0.012	0.006	0.58	0.52

perature and mechanical properties in the rolling direction. All specimens were cut with the longitudinal direction parallel to the rolling direction and machined in accordance with BS.DD19 [1] and/or ASTM standards [25]. Shapes and dimensions are shown in Figure 2. Notches were machined with the long axis perpendicular to

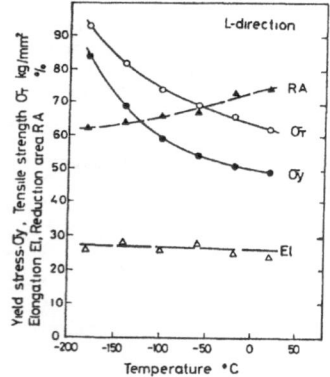

Fig. (1) - Relationship between mechanical properties and temperature

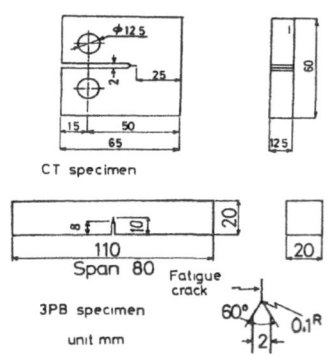

Fig. (2) - Shapes and dimensions of test specimen

the rolling direction. The notch tip was located at either 6/100 B or 24/100 B from the surface for three-point bending specimens and at 24/100 B for CT specimens. A fatigue crack of about 2 mm was induced at the notch tip under tensile loading in accordance with BS.DD19 so as to set the ratio of total notch length to the specimen width at 0.5.

Experimental Apparatus. An Instron 25 ton testing machine was used for the three-point bending and tensile loading. Crosshead speed was 2 mm/min. The schematic diagram for loading and measuring the three-point bending test is shown in Figure 3. The same method was used for the CT test. Applied load was measured by a load cell and opening displacement by a clip gauge which was hooked on knife edges mounted on the specimen. Changes in electrical potential difference due to crack growth were detected through probes which were fixed on the specimen so that the imaginary line between them crossed the notch obliquely. The details are shown in Figure 4.

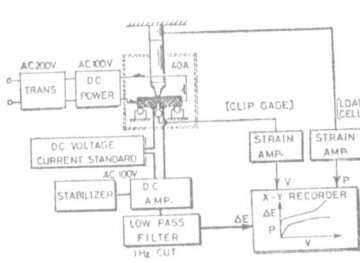

Fig. (3) - Schematic diagram for loading and measurement of three-point bending test

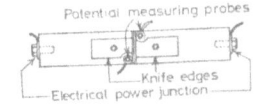

(a) Three-point Bending Specimen

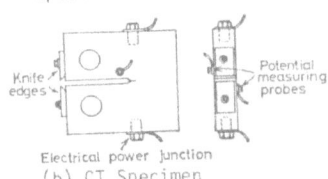

(b) CT Specimen

Fig. (4) - Details of detective tips for measurement of electrical potential

The basic concept of the electrical potential method is to detect minute voltage changes caused by the resistance change due to micro crack creation under a constant supply of DC current. As BS.DD19 recommends a current density of more than 10 A/cm^2 [1,2], a current of 40 A was used in this study. The initial electrical potential difference was cancelled out by loading the same voltage reversely on the electric circuit so that only the change of electrical potential difference due to crack growth was measured through a DC amplifier. Since specimens were required to be isolated electrically, the loading system and lead wires were insulated with Teflon. The area shown by the dotted line in Figure 3.

In low temperature tests, the temperature was detected by copper-constantan thermocouple and controlled to within ±0.5°C by spraying liquid nitrogen into a test chamber. The test was conducted after the specimen was kept at the temperature for 30 minutes.

EXPERIMENTAL RESULTS AND DISCUSSION

Detection of the Initiation of Slow Crack Growth. BS.DD19 recommends considering the inflection point in the relationship between electrical potential difference ΔE and knife edge opening displacement V as indicating the initiation of slow crack growth [1-2].

To verify this recommendation, several specimens were unloaded before and after the inflection point, cut at the middle of the thickness and examined optically

for slow crack growth. This verification was conducted for both three-point bending and CT specimens. Examples of three-point bending specimens are shown in Figure 5. The generation of ductile cracks are observed slightly after the inflection point and obvious growth is observed thereafter. The reason the change of

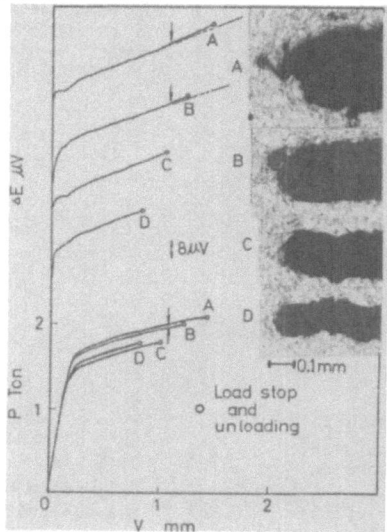

Fig. (5) - Relationship between load, electrical potential difference
and knife edge opening displacement (at R.T. three-point
bending specimen)

ΔE at the inflection point in Figure 5 is not clear is that since the electrical
potential method is based on the measurement of resistance change caused by the
change in the current path, it is harder to detect the microscopic resistance
change caused by the generation of a ductile crack than to detect a macroscopic
change caused by crack growth. The detection sensitivity could be a reason as
well.

*Relationship between Load, Electrical Potential Difference Change and Knife
Opening Displacement at Low Temperature.* The relationship between electrical
potential difference change and knife edge opening displacement for three-point
bending specimens between room temperature (R.T.) and -60°C is shown in Figure 6.
Figure 6 shows that the relationship at low temperatures is similar to that at
R.T. in that they have inflection points. The influence of temperature on the
relationship before the inflection points is small. The knife edge opening displacements at the inflection points are slightly different at different test temperatures. The relationship between load and knife edge opening displacement and
the inflection points for three-point bending and CT tests are shown in Figure 7.

Fracture Toughness Values. Measured knife edge opening displacement V was
transformed into COD, δ, by Wells' equation, equation (1) [1]. The constant γ

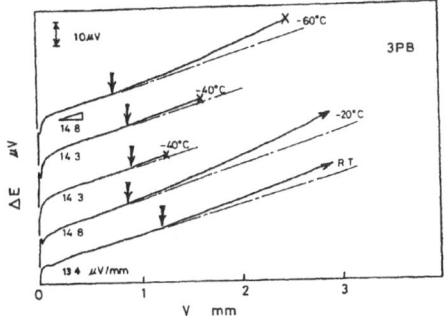

Fig. (6) - Relationship between electrical potential difference ΔE and knife edge opening displacement V (three-point bending specimen)

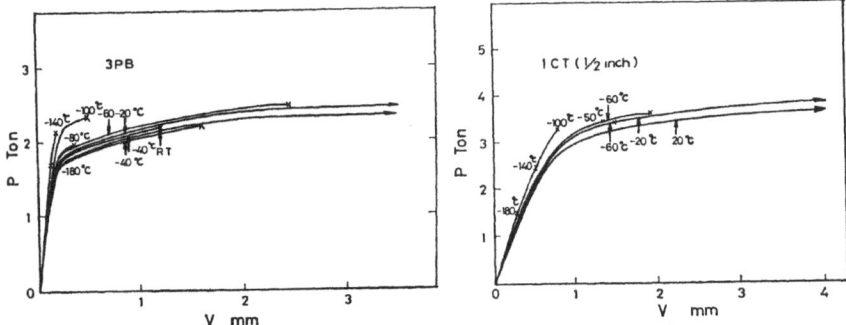

Fig. (7) - Relationship between load P and knife edge opening displacement V

in equation (1) is 1.54 [1] and 2.34 [8] for three-point bending and CT specimens respectively with a/w = 0.5, where σ_y is yield strength, E is Young's modulus, ν is Poisson's ratio. W, B and a are width, thickness notch length respectively and Z is either the distance of the clip gage location from the specimen surface or from the center of a pin hole in the specimen, see Figure 9.

$$\delta = \frac{0.45(W-a)}{0.45W+0.55+Z} [V - \frac{V}{2}], \quad V \geq V$$

$$\delta = \frac{0.45(W-a)}{0.45W+0.55+Z} [\frac{V^2}{2V}], \quad V \leq V$$

(1)

in which

$$V = \frac{2\gamma\sigma_y W(1-\nu^2)}{E}$$

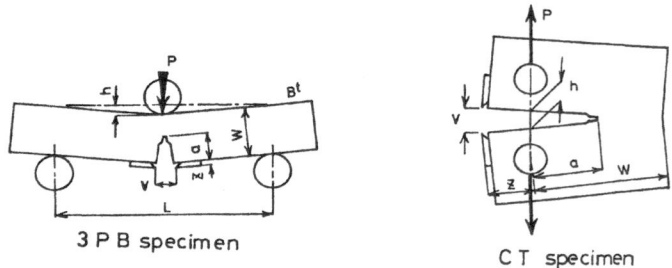

3 P B specimen   C T specimen

Fig. (8) - Schematic view of deformed specimen

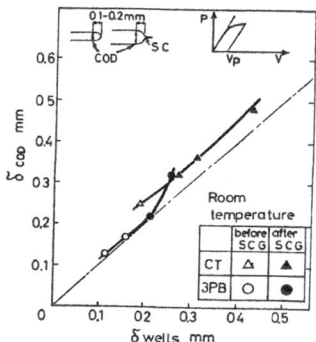

Fig. (9) - Relationship between measured COD and δ calculated
from Wells' equation

To calculate the J-integral, Rice's equation, equation (2), [9], and Merkle's equation, equation (3), [10,11], in which the tensile component was used for the three-point bending and CT specimens. A_h is potential energy, A_v is area beneath the P-V curve and P is load.

$$J = \frac{2A_h}{B(W-a)} = \frac{\zeta}{B(W-a)} [\eta_A A_v + \eta_c (PV - A_v)]$$

(2)

The other quantities in equation (2) are defined as

$$A_h = \zeta A_v = \zeta \int_0^v P dV, \quad \zeta = \frac{h}{V}, \quad \eta_A = \frac{2(1+\alpha)}{1+\alpha^2}$$

(3)

$$\eta_c = \frac{2\alpha(1-2\alpha-\alpha^2)}{(1+\alpha^2)^2}, \quad \alpha = \sqrt{\beta^2+2\beta+2} - (\beta+1), \quad \beta = \frac{2a}{W-a}$$

The relationship between h and V is given by equation (4) and equation (5) for three-point bending specimens and CT specimens, respectively. L is the span between loads in three-point bending tests and h is the displacement or distance shown in Figure 8.

$$\frac{h}{V} = \frac{L}{4(a+Z)} [1 - \frac{\delta}{V}], \text{ (Three-Point Bending)} \qquad (4)$$

$$\frac{h}{V} = 1 - \frac{Z}{a+Z} [1 - \frac{\delta}{V}], \text{ (Compact Tension)} \qquad (5)$$

Figure 9 shows the relationship between the measured crack tip COD after unloading and calculated from equation (1) using the plastic component V_p of V. Calculated and measured values are in good agreement. When fracture occurs before slow crack growth, COD is defined as δ_c and J as J_c. This is to be distinguished from δ_i and J_i. J_c and J_i values showed no significant difference for the three-point bending and CT specimens as shown in Figure 10. The three-point bending and CT specimens did show a difference in their δ_c and δ_i values as shown in Figure 11. The reason for the good agreement between J_c and J_i is that the tensile com-

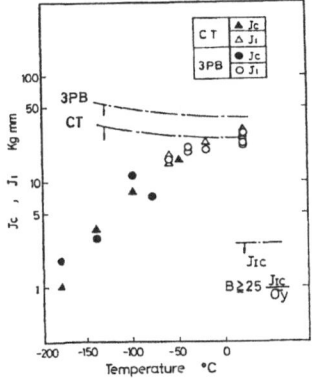

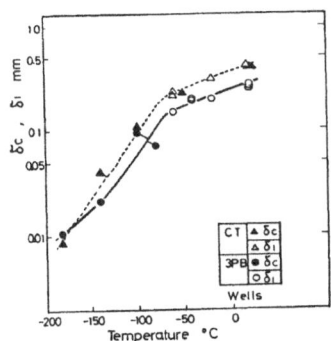

Fig. (10) - Relationship between J_c, J_i and test temperature

Fig. (11) - Relationship between δ_i, δ_c and test temperature

ponent was taken into consideration in the J calculation for the CT specimens while the slight disagreement in δ_c and δ_i is attributed to the difference in the resistance to deformation at the crack tip between the theee-point bending and CT specimens. This requires an additional tensile component of δ in the CT specimens. The relationship between J_i/σ_y and δ_i is shown in Figure 12 and expressed by equation (6), where C is a constant which was a different value according to the specimen type.

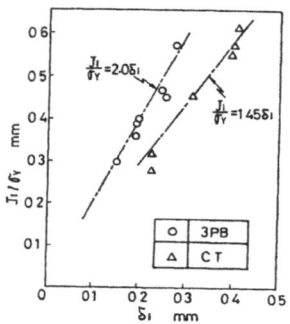

Fig. (12) - Relationship between J_i/σ_y and δ_i

$$\frac{J_i}{\sigma_y} = C\delta_i \qquad (6)$$

The upper limit of the J_i value which satisfies the thickness condition of J_{IC}, equation (7), [14], is shown in Figure 10.

$$B \geq 25 \frac{J_{IC}}{\sigma_y} \qquad (7)$$

J_i values for CT specimens at room temperature (RT) do not satisfy the above condition. Only at low temperatures was equation (7) satisfied. For the three-point bending specimens, J_i values satisfied equation (7) at any temperature. Figure 13 compares the R-curve method and the electrical potential method for J_i and δ_i at RT using the three-point bending specimens. The former is based on Griffis' method [7] and the latter is based on the method in BSI (1979), [12,13]. Almost the same values for J_i and δ_i are obtained by both methods. Hijikata et al reported that the electrical potential method, R-curve method and stretched zone method gave the same value for J_{IC} for 3.5 Ni-Mo-V steel [26]. Figures 14 and 15 compare the values of K_{IC} or $K_{IC}(J)$ given by equation (8) for the data in [14-24] and the present data for A533 B steel.

$$K_{IC}(J) = \sqrt{\frac{J_{IC}E}{1-\nu^2}} \qquad (8)$$

The present data show the upper values at various temperatures. Perhaps the wide variation in value is due to differences in the quality of the materials used, differences in charge, differences in thickness of the plates or differences in the toughness distribution towards plate thickness. These possibilities are examined

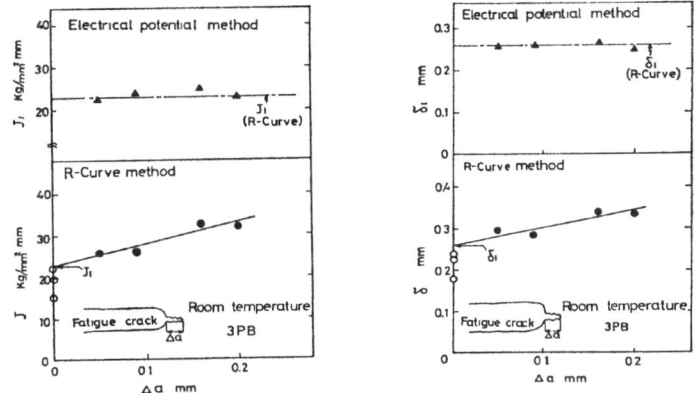

Fig. (13) - Comparisons of R-curve method and electrical potential method for J_i values and δ_i values

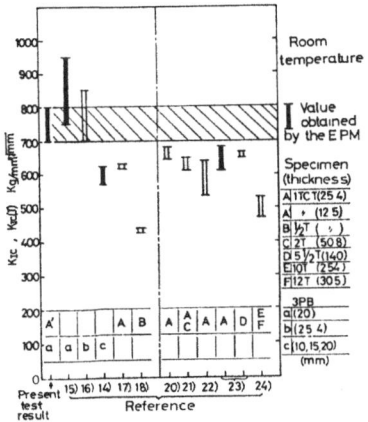

Fig. (14) - Comparison of data between present study and reference (RT)

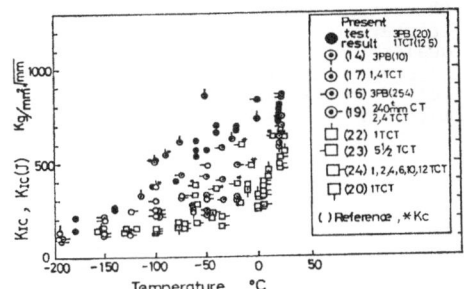

Fig. (15) - Comparison of data between present study and references (-196°C - -20°C)

in Figures 14 and 16. The K_{IC} and $K_{IC}(J)$ values measured by the electrical potential method are different for specimens of the same shape in various references. Also, Figure 17 [19] suggests that the Charpy absorbed energy, which is a parameter related to toughness, and $K_{IC}(J)$ differ according to sampling location. The material on the surface was tougher than that of the center. The second explanation

290

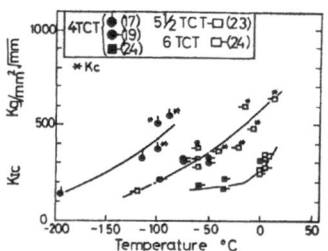

Fig. (16) - Scatter of K_{IC} among materials prepared

is that Rice's equation, equation (2), was used in the references to calculate
$K_{IC}(J)$ for CT specimens giving values about 7% lower than Merkle's equation, equation (3), [11,26]. As mentioned before, Merkle's equation takes the tensile component into account. A third possibility is that the sensitivity in detecting slow
crack growth may be different due to two definitions of slow crack growth. One
is the moment when a microcrack initiates at the very tip of a fatigue crack. The
other is the moment when the microcrack spreads to some distance along the tip of
the fatigue crack. The electrical potential method detects the latter moment due
to characteristics mentioned in 3.1. Of the three possible explanations, the
quality of materials used is considered to have the most influence on the data in
Figures 16 and 17. The conclusion is that the high fracture toughness data obtained in this paper was due to specimens taken from the plate surface where the
material is expected to be tougher.

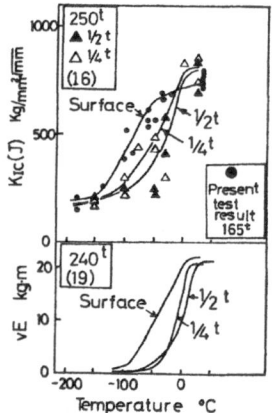

Fig. (17) - Shift of toughness in thickness direction in thick plates

CONCLUSIONS

Fracture toughness values δ_i and J_i of A533 B steel were experimentally obtained with the electrical potential method. Slow crack growth initiated at the inflection point in the relation between the electrical potential change ΔE and the knife edge opening displacement V. The δ_c and δ_i values of three-point bending specimens and CT specimens showed some differences, while the J_c and J_i values showed no significant differences. There was a correlation between J_i/σ_y and δ_i. The J_i and δ_i values obtained with the electrical potential method were almost the same as those obtained with the R-curve method.

REFERENCES

[1] British Standards Institution DD19, 1972.

[2] Lowes, J. M. and Fearnehough, G. D., Eng. Fract. Mech., 3, p. 103, 1971.

[3] Miyoshi, T. and Miyamoto, H., J. Facul. Eng., University Tokyo (B), 33-2, p. 185, 1975.

[4] Tanaka, S., Akita, S. and Takamatsu, T., J. JSFM 12-4, p. 143, (1978-1), in Japanese.

[5] Tetelman, A. S., Proc. 1st U.S. Japan Joint Symp. on Acoustic Emission, p. 1, 1972.

[6] Begley, J. A. and Landes, J. D., ASTM STP 560, p. 170, 1974.

[7] Griffis, C. A., Trans. ASME, Ser. J, 94-4, p. 278, 1975-11).

[8] Pense, A. W. and Stout, R. D., WRC Bull. 205, p. 278, 1975.

[9] Rice, J. R., Paris, P. C. and Merkle, J. G., ASTM STP 536, p. 231, 1973-8.

[10] Merkle, J. G. and Corten, H. T., Trans. ASME, Ser. J, 96-4, p. 286, 1974-11.

[11] Shiratori, M. and Miyoshi, T., Trans. JSME, Ser. A, 45-389, p. 50, 1979, in Japanese.

[12] British Standards Institution BS5762, 1979.

[13] Smith, R. F. and Knott, J. F., Inst. Mech. Eng., London, 1971.

[14] Kodaira, T. and Nakajima, N., Proc. 3rd Int. Symp. Japan Welding Soc., p. 185, 1978-9.

[15] See Reference [3].

[16] Nakano, Y., Sano, K., Tanaka, M. and Ohashi, N., Kawasaki Steel Giho, 12-4, p. 593, 1980, in Japanese.

[17] Japan Welding Eng. Society, JI Committee Report, p. 53, 1979, in Japanese.

[18] Kobayashi, H., Nakamura, H. and Nakazawa, H., Proc. 3rd Int. Symp. Japan Welding Soc., p. 191, 1978-9.

[19] Susukida, H., Sato, M., Takano, G., Uebayashi, T. and Yoshida, K., Mitsubishi Juko Giho, 13-1, p. 21, 1976, in Japanese.

[20] Hahn, G. T., et al, BMI, 1973, 1975.

[21] Begley, J. A. and Landes, J. D., ASTM STP 514, p. 1, 1972.

[22] Buchalet, C. and Mager, T. R., ASTM STP 536, p. 281, 1973-8.

[23] Sumpter, J. D. G., Metal Science, p. 354, 1976-10.

[24] Shabbits, W. O. and Wessel, E. T., Trans. ASME, Ser. E, 93-2, p. 231, 1971-6.

[25] ASTM E399-70T.

[26] Miyamoto, H., Hijikata, A., Yoshioka, S., Kumasawa, M. and Tohda, H., Trans. JSME, Ser. A, 45-396, p. 869, 1979, in Japanese.

SECTION IV

ANALYTICAL AND EXPERIMENTAL MODELLING

MODELLING OF CRACK GROWTH RESISTANCE CURVES

R. M. L. Foote and G. P. Steven

The University of Sydney
Sydney, Australia

ABSTRACT

It has been postulated that cracking in fibre reinforced cement composites occurs when the effective stress intensity factor at the tip of a crack reaches a critical value K_i corresponding to the initiation of fracture in the matrix. This stress intensity factor can be separated into two components $K = K_R + K_r = K_i$ where K_R is the stress intensity factor at the tip of the cracked matrix if there were no fibres bridging the crack, but the load to generate K_i is still applied, and K_r is the stress intensity factor due to the bridging fibres closing the crack faces.

Two techniques are presented to model the fibre pull-out behaviour. In the first, the fibres are modelled as beams bridging the crack. Those beams have negative stiffness which becomes zero if the fibre is pulled out. The combination of negative stiffness and force gives the appropriate representation for the fibres. In the second modelling technique, an influence coefficient method is used whereby the effect on K of unit point loads at nodes along the crack is individually determined. These are assembled into a matrix which can be operated upon by the appropriate influence vector, and K_R is obtained by iteration for any given crack length.

INTRODUCTION

In previous work [1], it has been shown that the K_R-curve is useful to describe the fracture behaviour of fibre-reinforced cements and that this curve appears independent of both size and starter notch depth. It was also shown that the K_R-curve increased gradually from a matrix initiation value, K_i, to a plateau value, K_m, when the fibres at the starter notch began to pull clear of the crack faces. The crack extension to reach this plateau K_m value was about 70 mm to 80 mm.

Previous work by Foote et al [2] has presented experimental work on crack growth resistance curves for cement reinforced with asbestos and cellulose fibres in compact tension, double cantilever beam and notch bend configurations. This work also presented theoretical results for an infinite sheet with a crack and fibres across part of its length. It is postulated that the K_R-curve is geometrically invariant to the test specimen, although there is certain experimental evidence against this [3-5]; and the purpose of this present work is to detail methods of determining the validity of this hypothesis and, at the same time, provide an analytical check against the experimental work of [2].

For random fibre reinforced cement composites, it has been shown [6-8] that linear elastic fracture mechanics is approximately valid. This enables standard linear finite element models to be used. Much has been presented in the literature concerning the use of the finite element method, both for linear and nonlinear elastic fracture mechanics. Special features have generally been used to model the crack tip in order to represent the stress singularity there; such things as special elements incorporating the Westergaard elasticity solution or quarter point quadratic isoparametric elements which emulate the $r^{\frac{1}{2}}$ displacement behaviour at the crack tip. In this present work, it has been determined that, using simple elements of adequate resolution and extrapolation techniques, satisfactory results can be obtained when compared with Rook and Cartwright [9], see Figure 1.

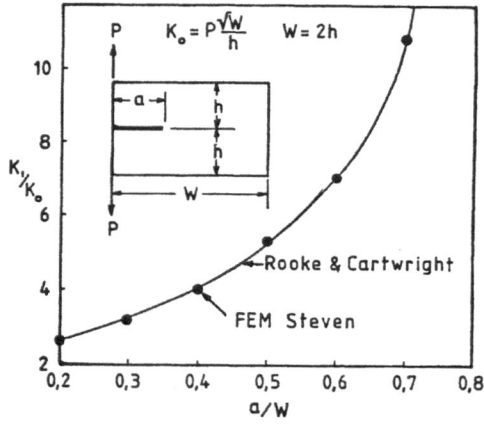

Fig. (1) - Comparison of stress intensity factors for CT

FEM EVALUATION OF K

The determination of the stress intensity factor at the crack tip in this work was achieved by the use of a standard finite element software package without modification, but ensuring that there was adequate mesh resolution at the tip. The plane stress element used is a four node quadrilateral made up from four constant strain triangles. Centroidal stresses are the only ones used and a regular mesh is always present at the crack tip. The problems solved always have an axis of symmetry colinear with the crack. Consider the group of elements ahead of the crack tip as shown in Figure 2.

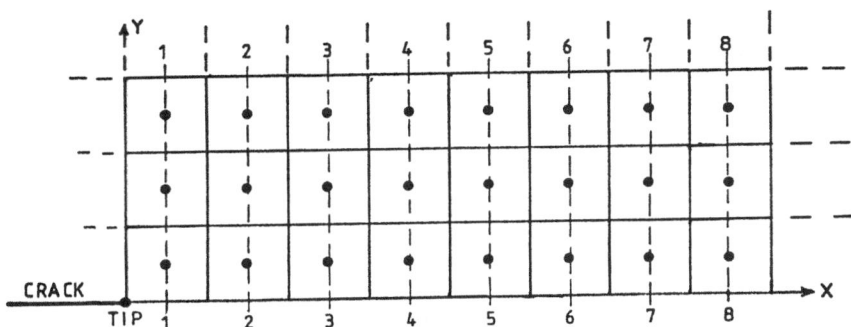

Fig. (2) - Arrangement of elements at crack tip

At the sections 11, 22,...88, the three centroidal $\sigma_{\bar{y}\bar{y}}$ stresses are extrapolated to find the $\sigma_{\bar{y}\bar{y}}$ value at $\bar{y} = 0$. Intensity factors are calculated using $\sqrt{2\pi\bar{x}}\ \sigma_{\bar{y}\bar{y}}$, and extrapolated to $\bar{x} = 0$, which is taken as the stress intensity factor K_1. Each author used a slightly different extrapolation formula to find $\sigma_{\bar{y}\bar{y}}$ at $\bar{y} = 0$, which are:

$$\sigma_{\bar{y}\bar{y}} = a + b\bar{y}^2 + c\bar{y}^3 \qquad \text{(GPS)}$$

$$\sigma_{\bar{y}\bar{y}} = a' + b'\bar{y}^2 + c'\bar{y}^4 \qquad \text{(RMLF)}$$

Having obtained stress intensities at stations 1, 2 and 3 in Figure 2, a quadratic curve was fitted through these values by GPS. This procedure is validated by comparison with results presented by Rooke and Cartwright [9], see Figure 1. The other method used by RMLF was to linearly extrapolate the stress intensities of stations 3 to 8. This method gives a small error when compared to the solution of Srawley and Gross [10] but compares favourably with intensities calculated from the displacement, V_y, of the crack face nodes and linear extrapolation, see Figure 3.

NEGATIVE STIFFNESS MODEL

In this model, the fibres are modelled as beams having negative stiffness which becomes zero if the fibre is pulled out. At the same time, a positive outward normal force is applied to the crack face at the point of beam attachment. Let the equivalent beam elements have zero flexural stiffness and a negative extensional stiffness (k_f) together with an extensional limit which coincides with the mean . fibre pull-out length (δ_0). For extensions greater than δ_0, the stiffness of the

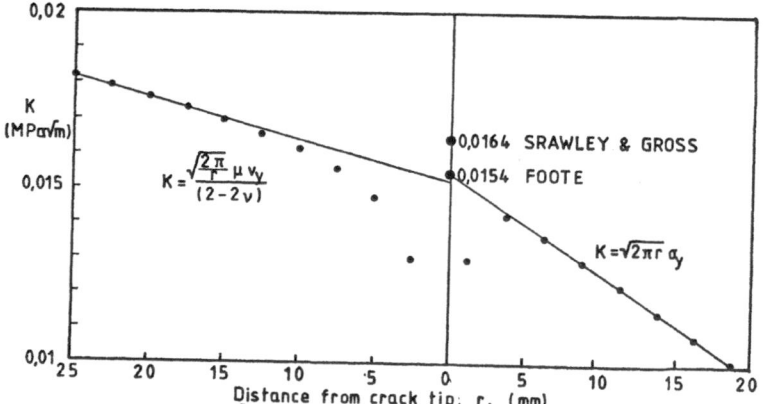

Fig. (3) - Extrapolation of K values calculated from $\sigma_{\bar{y}\bar{y}}$ and $V_{\bar{y}}$

fibres is zero. Thus, the relationship between axial force (P_f) and extension (δ) for the equivalent beam element is,

$$P_f = k_f(\delta_0-\delta), \quad \delta < \delta_0$$

$$= 0 \qquad \delta \geq \delta_0 \tag{1}$$

This relationship is shown in Figure 4a and denoted by fibre. Such a relationship can only be regarded as approximate since not all fibres may be of the same type nor will they all pull out at the same extension, i.e., at $\delta = \delta_0$. To test the validity of using such a beam in a finite element model, the following simple example is used as a demonstration. A cantilever beam with stiffness due to tip transverse load of k_b is resisted by a spring whose characteristic is that of equation (1), see Figure 4a and 4b. The potential Π for an arbitrary displacement δ in the direction of the load P is given by,

$$\Pi = -P\delta + k_b\delta^2/2 + k_f\delta_0\delta - k_f\delta^2/2$$

and the principle of Minimum Potential gives for an arbitrary variation $d\delta$ away from a compatible state of deformation,

$$\frac{d\Pi}{d\delta} = k_b\delta + k_f\delta_0 - k_f\delta - P = 0$$

Thus, for equilibrium of the system,

$$\delta = (P-k_f\delta_0)/(k_b-k_f) \text{ for } \delta < \delta_0 \tag{2}$$

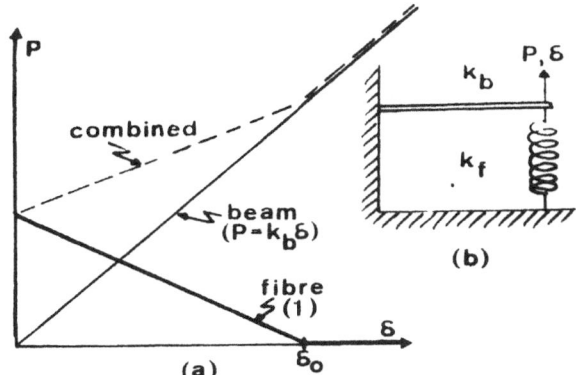

Fig. (4) - Extensional characteristic for equivalent beam and behaviour
of beam spring model

When $\delta \geq \delta_0$, $\delta = P/k_b$. For a stable equilibrium path, the second variation in the potential must be greater than zero, i.e.,

$$\frac{d^2\Pi}{d\delta^2} = k_b - k_f > 0 \tag{3}$$

Thus, for stability, k_b must be greater than k_f and in equation (2) for a positive displacement, P must be greater than $k_f\delta_0$. The quantity $k_f\delta_0$ is that force required to get the fibre (to which the spring in this demonstration problem is equivalent) to start moving, somewhat like pulling the cork out of a wine bottle. The load displacement curve for the combined structure is shown in Figure 4a.

In the finite element model for two standard fracture testing specimens, namely compact tension and double cantilever bending, a standard mesh size, based on 5 mm square elements is used. A plane stress system is analysed with material thickness being constant at 1 mm. On the basis of this and the properties of the fibres, the stiffness characteristic of the equivalent beams, which will be pitched every 5 mm along the crack interface, can be established as follows. The tests mentioned in Foote et al [2] used a mixture of asbestos and cellulose fibres and from this source, the following data has been taken.

Asbestos

(length/diameter)	= 80
mean length (l_a)	= 2 mm
fibre density	= 163*/mm^2
limiting shear stress in cement matrix	= 0.8 MPa
nominal stress	= 163*(π*.025*0.8)

$$*[\frac{l_a}{4} - 2\delta]$$

= 5.12-20.5δ N/mm^2

Cellulose

(length/diameter)	= 135
mean length (l_c)	= 3.5 mm
fibre density	= 133*/mm^2
limiting shear stress in cement matrix	= 0.35 MPa
nominal stress	= 133*(π*.0259*.35)

$$*[\frac{l_c}{4} - 2\delta]$$

= 3.31-7.56δ N/mm^2

Since one beam element has to represent 5 mm^2, it will be regarded as having five times the sum of the nominal stresses for both fibre types, i.e.,

$$P_f = 42.16 - 139.9\delta \qquad (4)$$

This expression can only be regarded as approximate as it pays no regard to the fact that the shorter asbestos fibres will pull out before the cellulose ones, and also does not take the fibre orientations into account nor any anisotrope due to rolling. However, it will be used in the first instance to demonstrate the efficiency of the modelling process.

Two geometries of test specimens have been examined, the compact tension (CT) and the double cantilever beam (DCB), the geometries of which are shown in Figure 5.

The finite element models for these consist in both cases, of a regular mesh of square elements of 5 mm side for one symmetrical half with a series of beam elements connected rigidly at one end and to appropriate nodes on the crack face between AF and A, see Figure 5. The stress intensity factor for crack initiation has been determined to be 1.8 MPa$\sqrt{m}$ [2] and the object of the computer experiments is, for any given crack length A, determine the value of AF and P which will give this value of K_i while having fibres at AF just about to pull out. This may take several iterations, but since the problem is still a linear one for any given model and load, interpolation techniques can be used.

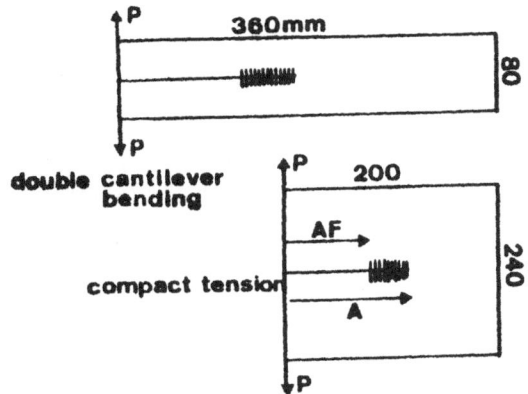

Fig. (5) - Geometry of experimental and computer model specimens

Having determined the value of P and AF for a given crack length, the analysis is rerun with the same P but no fibres, giving the value of K_R at the crack tip. Figure 6 shows the deformed outline of a typical mesh for the CT specimen.

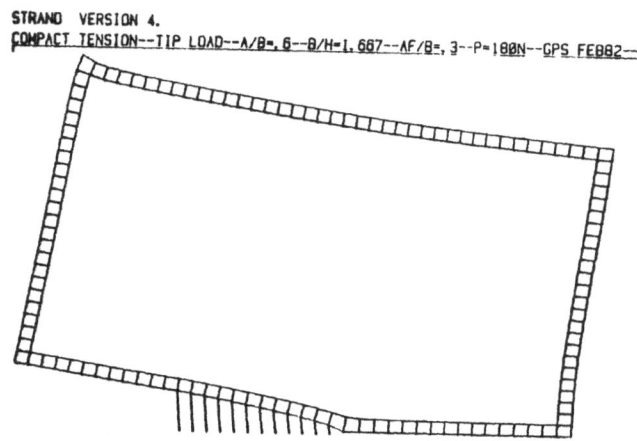

Fig. (6) - Deformed mesh outline for CT specimen

For the negative stiffness model, the extrapolation technique of GPS was used.

INFLUENCE COEFFICIENT MODEL

The compact tension specimen was modelled using this method, with a regular array of 2.5 mm x 5 mm quadrilateral elements. A unit load P was applied at the corner shown in Figure 5b and a unit load perpendicular to the crack at each node on the fibre bridging zone. The effect of each unit load was solved separately and the $\sigma_{\overline{yy}}$ stresses from the 24 elements near the crack tip, see Figure 2, and the vertical displacements of the nodes on the crack face were assembled respectively into a row of a matrix A. The first row was multiplied by a factor F to give a K_i of 1.8. The displacements at each node for this case were used to calculate the closing force at each node from the equation given by [2]:

$$P_f = N\tau\pi d[\tfrac{1}{4} - 2V_y]$$ (5)

Each subsequent row of matrix A was then multiplied by the relevant P_f, and each column added to enable the stress intensity factor at the crack tip to be evaluated for the superimposed loads. Because the fibres help close the crack, the K value obtained would be less than $K_i = 1.8$, so the first row was multiplied by a factor K_i/K and the process repeated until the K value for the model was equal to K_i. Row 1 was then used to find the value of K_R, using the extrapolation method of RMLF.

The influence coefficient method was selected because it could provide an easy method of studying different load distributions on the crack face once matrix A was filled. The computer time involved in solving for each separate load case was not significantly increased because the stiffness matrix of the structure has to be assembled only once for each crack length. A different load vector was required for each solution, but the solution time is small compared to the assembly time involved.

RESULTS

Figures 7 and 8 show values of K_R for different crack extensions, Δa, as well as experimental results from [2]. In Figure 7, the transverse direction compact tension case, the agreement with the experimental results of the negative stiffness model is sufficiently close to suggest that the value of τ used in previous work is low. Also, the agreement of the influence coefficient model (using the orientation factor of 0.81 for the transverse direction) with the theoretical K_R curve for the infinite plane is sufficiently close to validate this model.

In the longitudinal direction, the experimental values for CT and DCB specimens are shown in Figure 8, together with the infinite plane theory from [2] and the results of the two FEM models. It can be seen that the agreement between the infinite plane theory and the influence coefficient method is good, but that τ is still undervalued. The agreement of the negative stiffness method would be better if the effect of fibre orientation had been included. However, a comparison be-

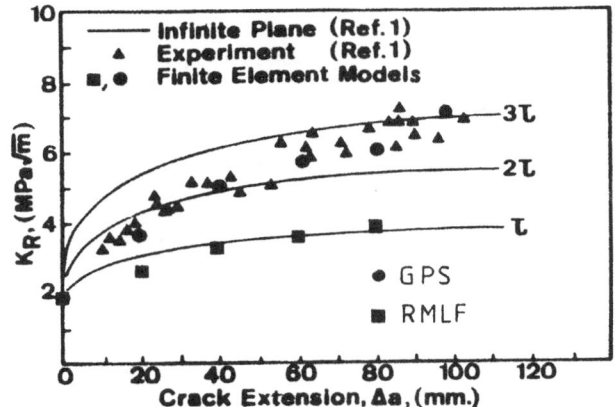

Fig. (7) - K_R-curve for transverse Ct specimen showing experimental and modelled values

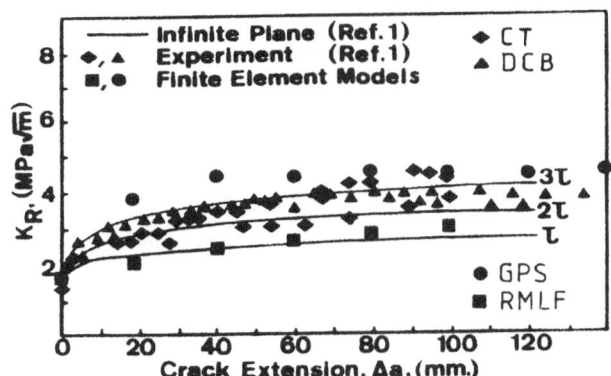

Fig. (8) - K_R-curve for longitudinal CT and DCB specimens showing experimental and modelled values

tween the curve for CT and DCB specimens is possible with this method and it can be seen that the shapes of the K_R-curves for these two specimen geometries are different. This suggests that the K_R-curve is dependent on geometry, but it may be possible to include some conditions on the range of the K_R-curve to enable it to be useful as a material property. The curves for the CT geometry in both directions are very close to the infinite plane solution, which suggests that at least some regions of the CT K_R-curve are independent of geometry.

CONCLUDING REMARKS

It has been shown that the finite element models suggested in this work, incorporating the negative stiffness beam elements or using the influence coefficients and calculating stress intensity factors by the polynomial extrapolation techniques are capable of emulating experimental results for the K_R-curves in fibre reinforced cement where LEFM prevails. The value of τ obtained from [2] is shown to be undervalued, and a better value could be obtained by fitting a FEM derived K_R-curve to the experimental results. The finite element modelling presented in this paper was conducted using the STIN/STRAND/STOUT structural finite element package developed at the Department of Aeronautical Engineering, The University of Sydney by Grant Steven and Doug Auld.

REFERENCES

[1] Mai, Y. W., Foote, R. M. L. and Cotterell, B., "Size effects and scaling laws of fracture in asbestos cement", Int. J. Cement Composites, 2, p. 23, 1980.

[2] Foote, R. M. L., Cotterell, B. and Mai, Y. W., "Crack growth resistance curve for a cement composite", Advances in Matrix Composites, Mat. Res. Soc. Ann. Meeting, p. 135, 1980.

[3] Adams, N. J. I. and Munro, H. G., "The relationship of the resistance curve and G_c to specimen configuration", Aeronaut. Quart., 28, p. 28, 1977.

[4] Bradshaw, F. J. and Wheeler, C., "The crack resistance of some aluminium alloys and the prediction of thin section failure", Royal Aircraft Establishment Technical Report Tr73191, 1974.

[5] Wilson, W. K., "Geometry and loading effects on elastic stress at crack lips", Westinghouse Research Report, 67-Id7-BTLPV-RI, 1967.

[6] Mindess, S. and Nadeau, J. S., "Effect of notch width on K_{1c} for mortar and concrete", Cement and Concrete Res., 6, p. 529, 1976.

[7] Harris, B., Varlow, J. and Ellis, C. D., "The fracture behaviour of fibre reinforced concrete", Cement and Concrete Res., 2, p. 447, 1972.

[8] Radjy, F. and Hansen, T. C., "Fracture of hardened cement paste and concrete", Cement and Concrete Res., 3, p. 343, 1973.

[9] Rooke, D. P. and Cartwright, D. J., "Compendium of stress intensity factors", H.M.S.O., 1974.

[10] Srawley, J. E. and Gross, B., "Stress intensity for bend and compact specimens", Eng. Fract. Mech., 4, p. 587, 1972.

FINITE ELEMENT ANALYSIS OF NON-SELF-SIMILAR CRACK GROWTH PROCESS

J. D. Lee[*]

Department of Mechanical and Aerospace Engineering
West Virginia University, Morgantown, West Virginia 25606

S. Du and H. Liebowitz

School of Engineering and Applied Science,
The George Washington University, Washington, D.C. 20052

ABSTRACT

A finite element analysis of the process of non-self-similar crack growth has been made for a cracked specimen subjected to monotonically increasing biaxial loading until the point of fast fracture is reached. The orientation and the amount of crack growth are determined by the equivalent maximum opening stress criterion and the linear relation between plastic energy and crack size, respectively. The analysis is based on an incremental theory of plasticity with kinematic hardening. Numerical results of several cases are presented.

INTRODUCTION

It is observed that cracked specimens made of ductile materials subjected to monotonically and slowly increased loading show a considerable amount of crack tip plasticity and significant amount of stable crack growth prior to the onset of fast fracture. In the past few years, several papers have been published on the finite element analysis of self-similar crack growth by the authors of this paper [1-5]. A computer program was developed at The George Washington University to analyze two dimensional stable crack growth in a self-similar manner based on the incremental theory of plasticity with isotropic hardening rule. During the process of two dimensional self-similar crack growth, the crack size becomes an unknown variable. Strictly speaking, one more unknown variable corresponds to one more governing equation for the system. In order to establish the governing equation for the crack size, various kinds of relation between the crack size and other fracture parameter have been proposed and tested. Namely, Newman [6], de Koning [7], Shih [8], Sih [9], Lee and Liebowitz [1], and Kanninen et al. [10]

[*]Present address: Research Division, The General Tire & Rubber Company, Akron, Ohio 44305.

proposed to take the crack tip strain, crack tip opening angle (CTOA), J-integral, strain energy density, plastic energy, and the combination of J and CTOA, respectively as the parameter to govern the amount of crack growth.

However, in reality, not all the problems are symmetric. Besides, it was found that the symmetry can be easily upset by eccentricity which is due to either material imperfections or misalignment of loading and that results in non-self-similar crack growth [11]. Therefore, it motivates the research work presented in this paper.

In the two dimensional non-self-similar crack growth problem, the amount and the orientation of crack growth are two unknown variables. It is proposed to determine the orientation of crack growth by using the maximum opening stress criterion. In other words, the direction of incremental crack growth is assumed to be perpendicular to the direction of nodal force at the current crack tip. To do so, a subroutine called REMESH, which modifies the finite element mesh after each incremental crack growth, is incorporated into the computer program. The amount of crack growth is determined by the experimental load-crack size curve or by the linear plastic energy-crack size relation.

Also, the fundamental theory, on which the analysis is based, has been upgraded to the incremental theory of plasticity with kinematic hardening rule. The numerical results of a few typical cases will be presented to show the paths of crack growth up to the onset of fast fracture.

CONSTITUTIVE EQUATIONS

The constitutive equations for an elastic-plastic theory may be briefly introduced as follows. To begin with, let elastic, plastic, and total strains be denoted as e'_{ij}, e''_{ij}, and e_{ij}, respectively. Then the strain-displacement relation can be written as:

$$e_{ij} = e'_{ij} + e''_{ij} = (u_{i,j} + u_{j,i})/2. \tag{1}$$

The stress-elastic strain relation for homogeneous and isotropic solids may be expressed by

$$\sigma_{ij} = \lambda e'_{kk} \delta_{ij} + 2\mu e'_{ij} , \tag{2}$$

where λ and μ are Lame constants. Let the yield surface be defined as

$$f(\underset{\sim}{\sigma}, e'') = \chi , \tag{3}$$

where f is the yield function and χ is called the hardening parameter. The loading rate, ξ, is defined as

$$\xi = \frac{\partial f}{\partial \sigma_{ij}} \dot{\sigma}_{ij} . \tag{4}$$

There are three distinctive cases which can be characterized as: (1) loading (f = χ and ξ > 0), (2) neutral loading (f = χ and ξ = 0), and (3) unloading (f < χ or ξ < 0). The constitutive relatives for e''_{ij} and χ in these three cases are postulated as:

$$\dot{e}''_{ij} = \dot{\chi} = 0 \quad \text{(neutral loading, unloading)} \ , \tag{5}$$

$$\dot{e}''_{ij} = \beta_{ij} \ \xi \ , \ \dot{\chi} = \gamma \ \xi \ \text{(loading)} \ . \tag{6}$$

Due to the requirement of continuity in the loading case, namely

$$\dot{f} = \frac{\partial f}{\partial \sigma_{ij}} \ \dot{\sigma}_{ij} + \frac{\partial f}{\partial e''_{ij}} \ \dot{e}''_{ij} = \dot{\chi} \ , \tag{7}$$

the following relation has to be satisfied

$$1 + \frac{\partial f}{\partial e''_{ij}} \ \beta_{ij} = \gamma \ . \tag{8}$$

A special elastic-plastic theory with kinematic hardening rule could be developed by making the following assumptions:

$$f = \frac{1}{2} \ (s_{ij} - c \ e''_{ij})(s_{ij} - c \ e''_{ij}) \ , \tag{9}$$

$$\chi = \frac{1}{3} \ \sigma_Y^2 \ , \ \dot{\chi} = 0 \ , \tag{10}$$

$$\beta_{ij} = \alpha \ \frac{\partial f}{\partial \sigma_{ij}} \ , \tag{11}$$

where σ_Y is the yield stress and s_{ij} is the stress deviator, i.e.,

$$s_{ij} = \sigma_{ij} - \frac{1}{3} \ \sigma_{kk} \ \delta_{ij} \ . \tag{12}$$

It can be readily shown that, in the loading case,

$$\dot{e}''_{ij} = 1.5 \ (s_{ij} - c \ e''_{ij})(s_{k\ell} - c \ e''_{k\ell}) \ \dot{s}_{k\ell}/c \ \sigma_Y^2 \ . \tag{13}$$

It is noticed that (1) the Prager-Ziegler type of assumption, usually called "normality", has been adopted in equation (11), (2) the size of the yield surface in stress space is kept constant, (3) the initial yield surface is the same as the one proposed by von Mises, and (4) the plastic volume change is zero, i.e.,

$$\dot{e}''_{kk} = e''_{kk} = 0 \ . \tag{14}$$

In uniaxial simple tension test, the stress-strain relation according to the constitutive equations developed above will be bilinear, i.e.,

$$\frac{d\sigma}{de} = E \ , \ \text{if} \ \sigma \le \sigma_Y \ , \tag{15}$$

$$\frac{d\sigma}{de} = \frac{E}{1 + \frac{2}{3} E/c} \equiv S \text{ , if } \sigma > \sigma_Y \quad . \tag{16}$$

The true-stress-true strain curve for aluminum alloy 2024-T3 obtained experimentally is shown in Figure 1. Let it be idealized by the bilinear relation from which the Young's modulus, E, the yield stress, σ_Y , and the slope in the plastic range, S, can be obtained. Then the material constant, c, which appears in equations (9) and (13), can be obtained as

$$c = bE \text{ ,} \tag{17}$$

where

$$b = \frac{2}{3} \frac{S/E}{(1-S/E)} \quad . \tag{18}$$

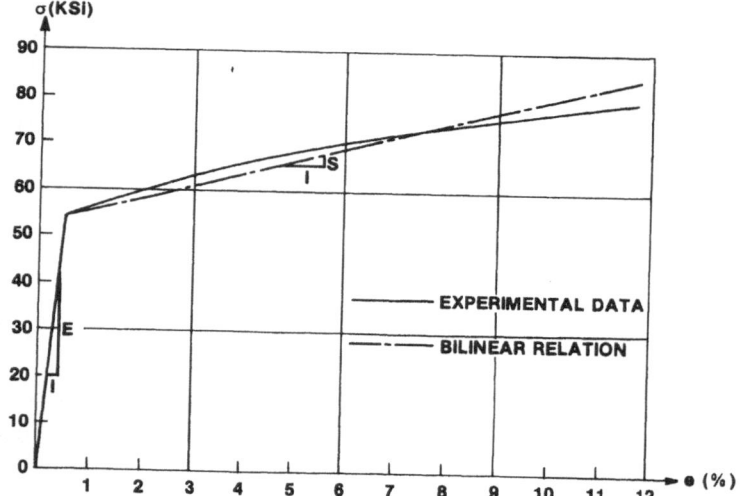

Fig. (1) - True stress-true strain curve for 2024-T3 aluminum alloy

PLANE STRESS

In this work, let the attention be focused on the cases of generalized plane stress, in which

$$\sigma_{13} = \sigma_{23} = \sigma_{33} = 0 \quad , \tag{19}$$

$$e'_{13} = e'_{23} = e''_{13} = e''_{23} = 0 \quad , \tag{20}$$

$$e'_{33} = - \frac{\lambda}{\lambda + 2\mu} (e'_{11} + e'_{22}) \quad , \tag{21}$$

$$e''_{33} = - (e''_{11} + e''_{22}) \quad , \tag{22}$$

$$s_{13} = s_{23} = 0 \quad , \tag{23}$$

$$s_{12} = \sigma_{12} \quad , \tag{24}$$

$$s_{33} = - (\sigma_{11} + \sigma_{22})/3 \quad , \tag{25}$$

$$s_{11} = (2\sigma_{11} - \sigma_{22})/3 \quad , \tag{26}$$

$$s_{22} = (2\sigma_{22} - \sigma_{11})/3 \quad . \tag{27}$$

The corresponding yield function and the loading rate are reduced to

$$f = H_1^2 + H_2^2 + H_1 H_2 + H_3^2/4 \quad , \tag{28}$$

$$\xi = H_1 \dot{\sigma}_{11} + H_2 \dot{\sigma}_{22} + H_3 \dot{\sigma}_{12} \quad , \tag{29}$$

where

$$\begin{aligned} H_1 &= s_{11} - bEe''_{11} \quad , \\ H_2 &= s_{22} - bEe''_{22} \quad , \\ H_3 &= 2(\sigma_{12} - bEe''_{12}) \quad . \end{aligned} \tag{30}$$

Then the stress-elastic strain relation in its incremental form may be written as

$$\begin{vmatrix} \delta\sigma_{11} \\ \delta\sigma_{22} \\ \delta\sigma_{12} \end{vmatrix} = \frac{E}{1 - \nu^2} \begin{vmatrix} 1 & \nu & 0 \\ \nu & 1 & 0 \\ 0 & 0 & (1 - \nu)/2 \end{vmatrix} \begin{vmatrix} \delta e'_{11} \\ \delta e'_{22} \\ \delta\gamma'_{12} \end{vmatrix} \quad , \tag{31}$$

where ν is the Poisson's ratio and $\gamma'_{12} = 2e'_{12}$. In the case of neutral loading or unloading, the incremental plastic strains are zero. In the case of loading, i.e., $f = \sigma_Y^2/3$ and $\xi > 0$, the stress-plastic strain relation in its incremental form may be written as

$$
\begin{vmatrix} \delta e''_{11} \\[2mm] \delta e''_{22} \\[2mm] \delta \gamma''_{12} \end{vmatrix} = \frac{1.5}{bE\,\sigma_Y^2} \begin{vmatrix} H_1^2 & H_1\,H_2 & H_1\,H_3 \\[2mm] H_1\,H_2 & H_2^2 & H_2\,H_3 \\[2mm] H_1\,H_3 & H_2\,H_3 & H_3^2 \end{vmatrix} \begin{vmatrix} \delta\sigma_{11} \\[2mm] \delta\sigma_{22} \\[2mm] \delta\sigma_{12} \end{vmatrix} , \tag{32}
$$

where $\gamma''_{12} = 2e''_{12}$. The rate of plastic energy density is defined as

$$
\dot{p} = \sigma_{ij}\,\dot{e}''_{ij} \tag{33}
$$

and, in the loading case, the incremental plastic energy density may be expressed as

$$
\delta p = \frac{1.5}{bE\,\sigma_Y^2}\,(\sigma_{11}\,H_1 + \sigma_{22}\,H_2 + \sigma_{12}\,H_3)(H_1\,\delta\sigma_{11} + H_2\,\delta\sigma_{22} + H_3\,\delta\sigma_{12}). \tag{34}
$$

The general stress-strain relation in its incremental form may be written as

$$
[\delta\sigma] = [h]^{-1}\,[\delta e] , \tag{35}
$$

where

$$
[\delta\sigma] = [\delta\sigma_{11},\ \delta\sigma_{22},\ \delta\sigma_{12}]^T ,
$$

$$
[\delta e] = [\delta e_{11},\ \delta e_{22},\ 2\delta e_{12}]^T , \tag{36}
$$

and $[h]$ is a 3 x 3 symmetric matrix with

$$
h_{11} = (1 + gH_1^2)/E ,
$$

$$
h_{12} = (-\nu + gH_1\,H_2)/E ,
$$

$$
h_{13} = gH_1\,H_3/E ,
$$

$$h_{22} = (1 + gH_2^2)/E \quad ,$$

$$h_{23} = gH_2 H_3/E \quad ,$$

$$h_{33} = [2(1 + \nu) + gH_3^2]/E \quad , \tag{37}$$

and $g = 1.5/b \ \sigma_Y^2$ only if $f = \sigma_Y^2/3$ and $\xi > 0$ (otherwise, $g = 0$).

THE CRACKED SPECIMEN

Let a rectangular plate of length 2L, width 2W, and thickness B, with a centered and inclined line crack be subjected to symmetric loading conditions. Figure 2 indicates the geometric configuration and the loading condition of the cracked specimen in its initial state. The initial crack size is $2a_o$ and its orientation is denoted by angle ϕ. The purpose of this work is to analyze the stable crack growth process from its initial state (a_o, σ_o) to its critical state (a_c, σ_c) beyond which the fast fracture occurs. It is noticed that there is a periodicity of π about z-axis and, because of that, only half of the plate, $R = [x, y| \ 0 \leq x \leq W, \ - L \leq y \leq L]$, needs to be analyzed. Region R is is further separated into two parts, U and D, which are shown in Figure 3. Region U is bounded by S_1, S_2, S_3, S_4, S_5, and S_6. Correspondingly, S_1', S_2', S_3', S_4', S_5', and S_6' are the boundaries of region D. Suppose the stresses, displacements, and the applied stress of the i-th state are known, i.e., $s_{k\ell} = \sigma_{k\ell}^{(i)}$, $u_k = u_k^{(i)}$, $\sigma = \sigma^{(i)}$, and the crack tip of the i-th state is located at point $p_i(x_i, y_i)$. Then the incremental boundary conditions between the i-th state and the j-th state (j = i+1), of which the crack tip is located at point $p_j(x_j, y_j)$, may be specified as (Figure 3):

(1) on S_1 and S_1'

$$\delta\sigma_{11} = k(\sigma^{(j)} - \sigma^{(i)}) = k \ \delta\sigma, \ \delta\sigma_{12} = 0 \ ; \tag{38}$$

(2) on S_2 and S_2'

$$\delta\sigma_{22} = \sigma^{(j)} - \sigma^{(i)} = \delta\sigma, \ \delta\sigma_{12} = 0 \ ; \tag{39}$$

(3) on S_3 and S_3'

$$\delta\sigma_{11}(I) = \delta\sigma_{11}(J), \ \delta\sigma_{12}(I) = \delta\sigma_{12}(J) \ ,$$

$$\delta u_1(I) = - \delta u_1(J), \ \delta u_2(I) = - \delta u_2(J) \ , \tag{40}$$

312

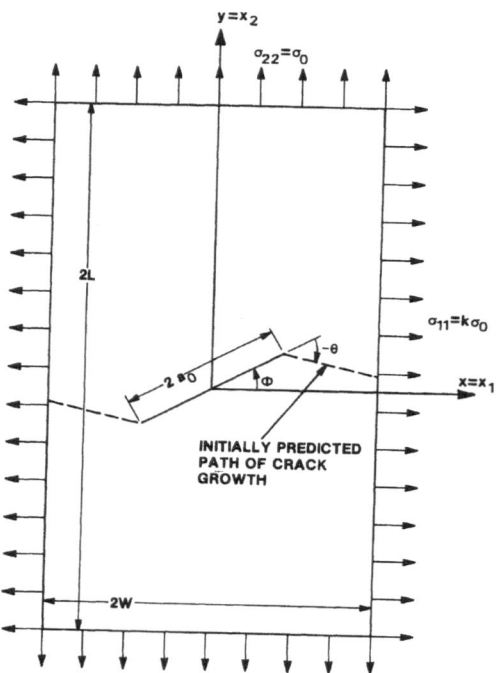

Fig. (2) - The geometric configuration and loading condition
of the cracked specimen initial state

where point I and point J are a pair of conjugate points on S_3 and S_3',
respectively;

(4)   on S_4 and S_4'

$$\delta\sigma_{11}\, n_1 + \delta\sigma_{12}\, n_2 = 0, \quad \delta\sigma_{22}\, n_2 + \delta\sigma_{12}\, n_1 = 0 \quad , \tag{41}$$

where (n_1, n_2) specifies the normal of the cracked surfaces, S_4 and S_4';

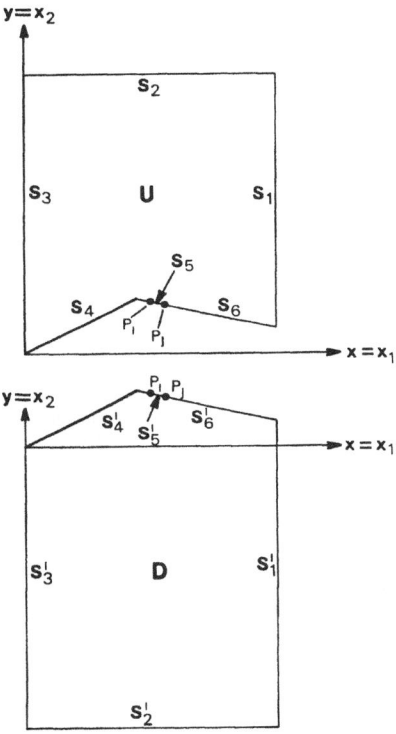

Fig. (3) - Region of solution, R = U+D, and its boundaries

(5) on S_5 and S_5'

$$\delta\sigma_{11} \, n_1 + \delta\sigma_{12} \, n_2 = - (\sigma_{11}^{(i)} n_1 + \sigma_{12}^{(i)} n_2) \quad ,$$

$$\delta\sigma_{22} \, n_2 + \delta\sigma_{12} \, n_1 = - (\sigma_{22}^{(i)} n_2 + \sigma_{12}^{(i)} n_1) \quad ,$$

(42)

where S_5 (S_5') is a line connecting p_i and p_j and $\overline{p_i p_j}$ is perpendicular to the resultant nodal force at p_i and (n_1, n_2) is the normal to S_5 (S_5') ;

(6) on S_6 and S_6'

$$[\delta\sigma_{11}(I) - \delta\sigma_{11}(J)] n_1 + [\delta\sigma_{12}(I) - \delta\sigma_{12}(J)] n_2 = 0 \quad,$$

$$[\delta\sigma_{22}(I) - \delta\sigma_{22}(J)] n_2 + [\delta\sigma_{12}(I) - \delta\sigma_{12}(J)] n_1 = 0 \quad,$$

$$\delta u_1(I) = \delta u_1(J) \quad,$$
(43)

$$\delta u_2(I) = \delta u_2(J) \quad,$$

where point I and point J are a pair of conjugate points on S_6 and S_6', respectively and (n_1, n_2) is the normal to $S_6(S_6')$.

The boundary condition specified on S_5 and S_5' simply means that, when the crack tip advances from p_i to p_j, the surface traction of the i-th state distributed along S_5 and S_5' should be released in order to make $\overline{p_i p_j}$ as part of the cracked surface for the j-th state. In terms of finite-element terminology, this means when the crack tip advances to its adjacent nodal point the nodal force at the previous crack tip should be released. After the incremental boundary value problem (from i-th state to j-th state) being solved, the stress field, displacement field, and the applied stress of the j-th state may be obtained as

$$\sigma_{k\ell}{}^{(j)} = \sigma_{k\ell}{}^{(i)} + \delta\sigma_{k\ell} \quad,$$

$$u_k{}^{(j)} = u_k{}^{(i)} + \delta u_k \quad,$$
(44)

$$\sigma^{(j)} = \sigma^{(i)} + \delta\sigma \quad.$$

Also, the crack size and the plastic energy (per unit thickness) of the j-th state can be calculated as

$$a^{(j)} = a^{(i)} + \sqrt{(x_j - x_i)^2 + (y_j - y_i)^2} \quad,$$
(45)

$$p^{(j)} = p^{(i)} + \int_R \delta p \, dA \quad.$$
(46)

FINITE ELEMENT ANALYSIS

In this work, the two dimensional finite element mesh consists of a family of triangular elements. There are 682 (=N_e) elements and 396 (=N_p) nodal points for region R with same numbers of elements and nodal points for regions U and D. For any nodal point on S_3, S_4, S_5, or S_6, there is a conjugate nodal point on S_3', S_4', S_5', or S_6', respectively.

The stiffness matrix of the triangular element linking the incremental nodal forces (per unit thickness) and the incremental nodal displacements may be obtained as [12,13]:

$$[K]_e = [B]^T [h]^{-1} [B] A \quad , \tag{47}$$

where A is the area of the element, [B] is the matrix which links the incremental strains and the incremental nodal displacements. Finally, the governing equation for the incremental boundary value problem of region R (=U + D) can be written as

$$\sum_{\beta=1}^{2N_p} K_{\alpha\beta} \, \delta u_\beta = \delta F_\alpha \quad , \qquad \alpha = 1, 2, 3, \ldots 2N_p \tag{48}$$

where the $2N_p$ x $2N_p$ matrix [K] is the sum of N_e local stiffness matrices $[K]_e$, e = 1, 2,...N_e .

However, it is noticed that, although the applied stress is increasing monotonically, during the process of crack growth some elements behind the advancing crack tip experience unloading or neutral loading. The matrix [h], which appears in equations (35,47), depends on the current stresses, plastic strains, and the loading rate. Suppose, for each element, at the i-th state the stresses, plastic strains are denoted by σ_{ij}, e_{ij}'', respectively and the incremental stresses between the i-th state and the j-th state are estimated to be $\delta\sigma_{ij}$, then the matrix [h] can be calculated from one of the following three cases:

Case 1: $f(\sigma_{ij}, e_{ij}'') = \frac{1}{3} \sigma_Y^2$, $f(\hat{\sigma}_{ij}, e_{ij}'') > \frac{1}{3} \sigma_Y^2$

$$h_{ij} = h_{ij}(\underset{\sim}{\sigma}, \underset{\sim}{e}'') \quad , \tag{49}$$

which means [h] is calculated according to equation (37) based on the current stresses with g = $1.5/b \ \sigma_Y^2$.

Case 2: $f(\sigma_{ij}, e''_{ij}) \leq \frac{1}{3}\sigma_Y^2$, $f(\hat{\sigma}_{ij}, e''_{ij}) \leq \frac{1}{3}\sigma_Y^2$

$$h_{ij} = h_{ij}(0,0) \quad , \tag{50}$$

which means [h] is calculated according to equation (37) with g = 0.

Case 3: $f(\sigma_{ij}, e''_{ij}) < \frac{1}{3}\sigma_Y^2$, $f(\hat{\sigma}_{ij}, e''_{ij}) > \frac{1}{3}\sigma_Y^2$

$$h_{ij} = h_{ij}(0,0) \, r + h_{ij}(\overset{*}{\underset{\sim}{\sigma}}, e'')(1-r) \quad , \tag{51}$$

where

$$r = [\sqrt{\tfrac{1}{3}}\,\sigma_Y - \sqrt{f(\underset{\sim}{\sigma}, e'')}]/[\sqrt{f(\hat{\underset{\sim}{\sigma}}, e'')} - \sqrt{f(\underset{\sim}{\sigma}, e'')}] \quad ,$$

$$\hat{\sigma}_{ij} = \sigma_{ij} + \delta\sigma_{ij} \quad , \tag{52}$$

$$\overset{*}{\sigma}_{ij} = \sigma_{ij} + r\,\delta\sigma_{ij} \quad .$$

After solving equation (48), one may obtain the calculated incremental stresses $\delta\sigma_{ij}$. The iteration process will be continued until, for each element, the estimated and the calculated incremental stresses are approximately the same.

NUMERICAL RESULTS, DISCUSSIONS

For general two dimensional fracture problems, during the process of crack growth the location of the crack tip should be regarded as unknown. Suppose all the informations about the i-th state are known, in order to solve the incremental boundary value problem between the i-th state and the j-th state (j = i+1), it is necessary to specify the incremental applied stress $\delta\sigma$ and the location of the crack tip at the j-th state, $p_j(x_j, y_j)$.

In this work, the orientation of crack growth is assumed to be perpendicular to the resultant nodal force at point $p_i(x_i, y_i)$. Therefore, for a given crack size increment, if the incremental applied stress, $\delta\sigma$, is known, then it is straightforward to solve the incremental boundary value problem (Figure 3).

Lee and Liebowitz [1] utilize the experimental curve, which relates the applied stress σ and the crack size a as the input data to analyze the process of self-similar crack growth. However, for the non-self-similar crack growth problem, the experimental σ-a curve is difficult to obtain, to say the least. Anyway, such experimental data at this moment is not available. But, from the numerical solutions obtained by analyzing the center-cracked specimens made of 2024-T3 aluminum alloy with flat crack (ϕ=0), it was found that, during the entire process of stable crack growth, the plastic energy is linearly related to crack size with the correlation coefficient being as high as 0.99. Moreover, Liebowitz et al. [2,3] further developed the computer program such that the linear relation between plastic energy and crack size can be taken as the governing equation for crack growth. Based on that idea, the analysis of non-self-similar crack growth problem may be proceeded as follows. First, let the governing equation for the amount of crack growth be represented by

$$P = P_0 + m(a-a_0) \quad , \tag{53}$$

where P_0 is the value of plastic energy at the onset of stable crack growth and m is the slope of the P-a line. For a given initial configuration of the cracked specimen (Figure 2), the initial angle of crack extension, θ, may be predicted as [14]:

$$\theta = - \cos^{-1} \{[3K_2^2 + K_1 \sqrt{K_1^2 + 8K_2^2}]/[K_1^2 + 9K_2^2]\} \quad , \tag{54}$$

where the stress intensity factors, K_1 and K_2, are approximated to be

$$K_1 = \sigma\sqrt{a_0} \, (\cos^2 \phi + k \sin^2 \phi) \quad , \tag{55}$$

$$K_2 = \sigma\sqrt{a_0} \, (1-k) \sin \phi \cos \phi \quad ,$$

and k is called the biaxial load factor. Then the initial line separating the regions, U and D, is determined. Increase the applied stress from $\sigma = 0$ to $\sigma = \sigma_0$ at which the plastic energy reaches P_0. Then the modified orientation of crack growth is obtained as

$$\theta_m = \tan^{-1}(F_x/F_y) \quad , \tag{56}$$

where θ_m is the angle measured clockwise from the x-axis, F_x and F_y are the two components of nodal force at the crack tip. After that, the subroutine REMESH is called to modify the finite element mesh accordingly. For the subsequent steps, any crack size increment $\delta a = a^{(j)} - a^{(i)}$ requires the increment of plastic energy to be

$$\delta P = m \, \delta a \quad . \tag{57}$$

In order words, keeping the applied stress unchanged, i.e., $\delta\sigma = 0$, the increment of plastic energy due to the increment of crack size alone, δP_a, can be obtained; if δP_a is less than δP, then keep the crack size at $a^{(j)}$ and gradually increase the applied stress until the increment of plastic energy due to the increment of the applied stress alone, δP_σ, satisfies the following relation

$$\delta P_\sigma = m \, \delta a - \delta \, P_a .$$ (58)

If, at certain stage, $\delta \, P_a \geq \delta \, P$, which means the increment of plastic energy due to the increment of crack size alone is large enough for crack growth, then the onset of fast fracture has been reached and hence, the maximum applied stress σ_c which the specimen can take, is obtained.

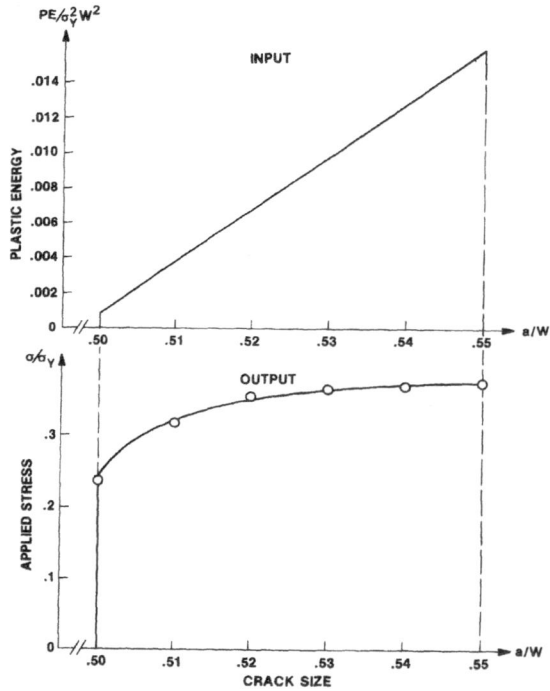

Fig. (4) - Relation among plastic energy, applied stress, and crack size; $\phi = 0°$, k=0

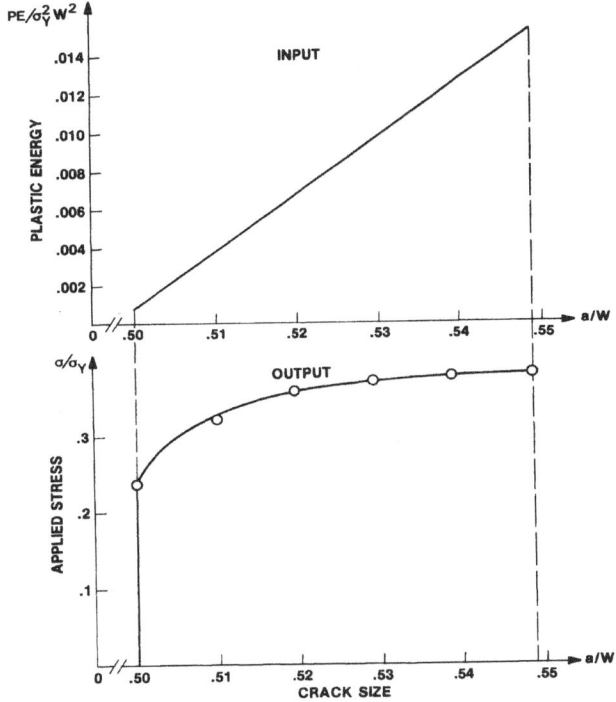

Fig. (5) - Relation among plastic energy, applied stress, and crack size;
$\phi = 15°$, k=0

For illustrative purpose, the numerical results of four cases are reported as follows:

Case 1: $\phi = 0°$, k = 0 .

This means the crack is parallel to the x-axis and the specimen is subjected
to uniaxial loading. In this case, it is found that the onsets of stable
crack growth and fast fracture are represented by

$$\sigma_o = 0.237 \; \sigma_Y \; , \quad a_o = 0.5 \; W \; ,$$

$$\sigma_c = 0.372 \; \sigma_Y \; , \quad a_c = 0.55 \; W \; . \tag{59}$$

The relations among plastic energy, applied stress, and crack size are shown in Figure 4. Because of symmetry, the predicted and the actual paths of crack growth are identical, in other words, the path of crack growth is along x-axis.

Case 2: $\phi = 15°$, $k = 0$.

In this case, the orientation of the initial crack is inclined by 15 degrees with respect to x-axis and the specimen is subjected to uniaxial loading. The onsets of stable crack growth and fast fracture are

$$\sigma_o = 0.236 \; \sigma_Y \; , \quad a_o = 0.5 \; W \; ,$$

$$\sigma_c = 0.378 \; \sigma_Y \; , \quad a_c = 0.549 \; W \; . \tag{60}$$

The relations among plastic energy, applied stress, and crack size are shown in Figure 5. The predicted and the actual paths of crack growth are shown in Figure 6.

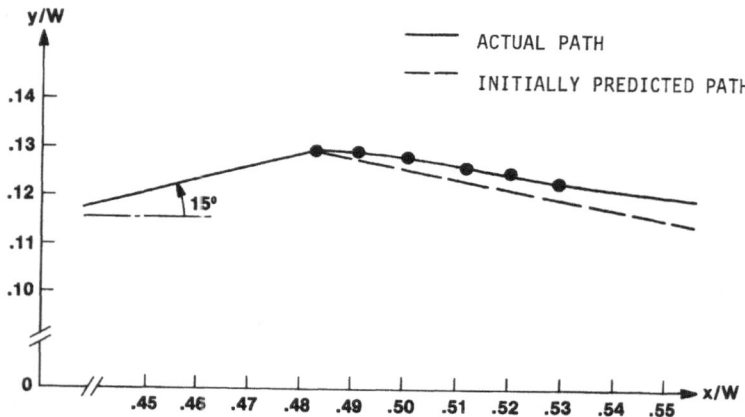

Fig. (6) - The initially predicted and the actual path of crack growth; $\phi = 15°$, k=0

Case 3: $\phi = 30°$, $k = 0$

In this case, it is found that

$$\sigma_0 = 0.247 \ \sigma_Y \ , \qquad a_0 = 0.5 \ W,$$

$$\sigma_c = 0.396 \ \sigma_Y \ , \qquad a_c = 0.544 \ W, \tag{61}$$

and the results are shown in Figures 7 and 8.

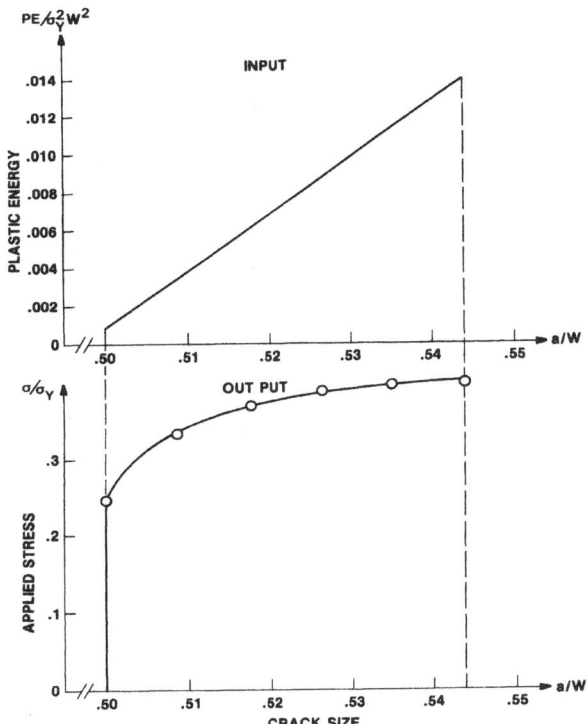

Fig. (7) - Relation among plastic energy, applied stress, and crack size; $\phi = 30°$, $k=0$

322

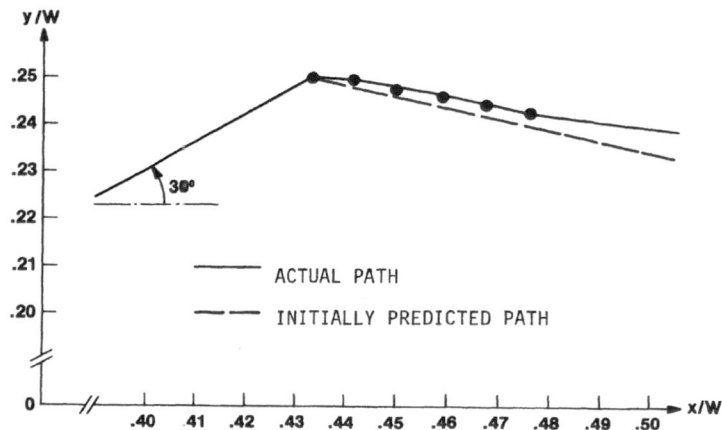

Fig. (8) - The initially predicted and the actual path of crack growth; $\phi = 30°$, $k=0$

Case 4: $\phi = 15°$, $k = 1$

This means the initial crack is inclined by 15 degrees and the specimen is subjected to equal biaxial loading. The onsets of stable crack growth and fast fracture are

$$\sigma_o = 0.233 \, \sigma_Y \quad , \quad a_o = 0.5 \, W \quad ,$$

$$\sigma_c = 0.385 \, \sigma_Y \quad , \quad a_c = 0.551 \, W. \tag{62}$$

The relations among plastic energy, applied stress, and crack size are shown in Figure 9. The predicted and the actual paths of crack growth are shown in Figure 10. The differences between Case 2 and Case 4 are due to the effect of biaxial loading. All these four cases have the following input parameters in common:

$$a_o/W = 0.5 \quad , \quad L/W = 2.25 \quad ,$$

$$S/E = 0.21 \quad , \quad \nu = 0.33 \tag{63}$$

$$P_o E/(\sigma_Y W)^2 = 0.0003 \quad , \quad mE/\sigma_Y^2 W = 0.3 \quad .$$

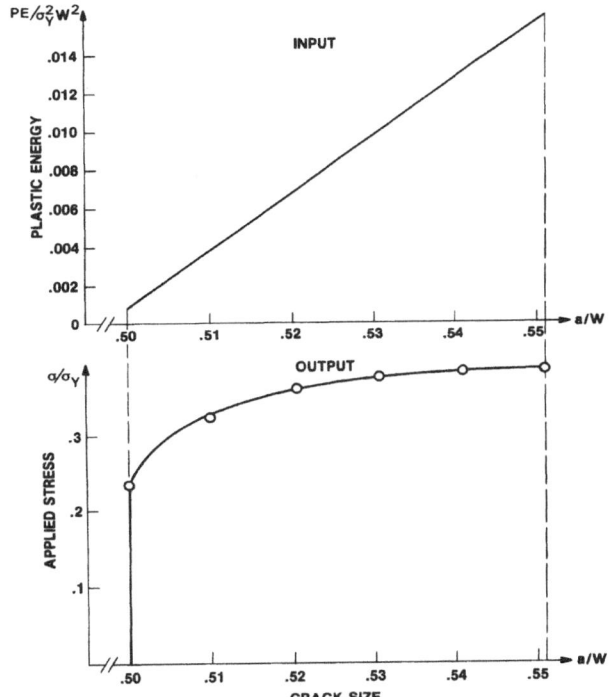

Fig. (9) - Relation among plastic energy, applied stress, and crack size; $\phi = 15°$, k=1

From the numerical results of these cases, it is seen that the path of crack growth predicted by equations (54,55), which is equivalent to the corresponding elastic solution for specimen having a very small crack, is relatively close to that based on the finite element results obtained during the process of crack growth.

The authors would like to take this opportunity to emphasize that it is more realistic and straightforward to take experimental load-crack size curve as input data to analyze the process of stable crack growth; on the other hand, the computer program developed at this stage can take any specified plastic energy - crack size relation (not necessarily linear) as the governing equation for the analysis of stable crack growth.

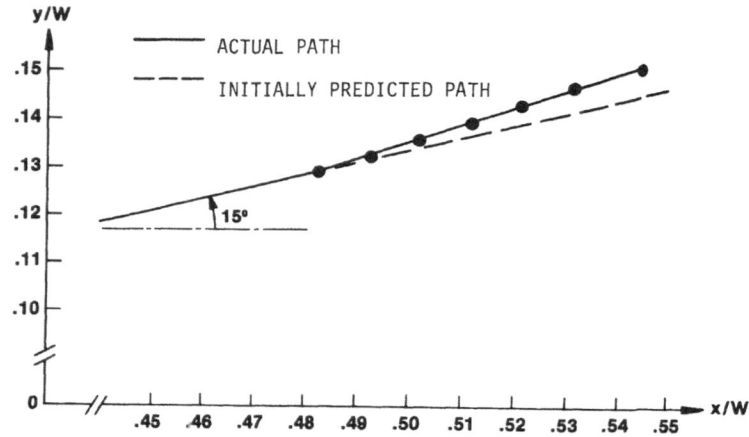

Fig. (10) - The initially predicted and the actual path of crack growth; θ = 15°, k=1

Acknowledgement -- The authors wish to acknowledge the financial support for this work from the Office of Naval Research (#N00014-75-C-0946).

REFERENCES

[1] Lee, J. D. and Liebowitz, H., Computers & Structures, 8, p. 403, 1978.

[2] Liebowitz, H., Lee, J. D. and Subramonian, N., Trans. 5th Int. Conf. on Structural Mechanics in Reactor Technology, G6/2, Berlin, 1979.

[3] Liebowitz, H., Lee, J. D. and Subramonian, N., Nonlinear and Dynamic Fracture Mechanics (Edited by Perrone and Atluri), ASME, New York, 1979.

[4] Du, S. and Lee, J. D., Engineering Fracture Mechanics, 16, p. 229, 1982.

[5] Du, S. and Lee, J. D., "Variations of Various Fracture Parameters During the Process of Subcritical Crack Growth," accepted for publication on Engineering Fracture Mechanics.

[6] Newman, Jr., J. S., ASTM STP 637, American Society for Testing and Materials, Philadelphia, 1976.

[7] de Koning, A. U., Report NLR MP 75035 U, National Aerospace Laboratory (NLR), The Netherlands, 1975.

[8] Shih, C. F., _Trans. 5th Int. Conf. on Structural Mechanics in Reactor Technology_, G6/5, Berlin, 1979.

[9] Sih, G. C., _Fracture Mechanics_ (Edited by Perrone, Liebowitz, Mulville, and Pilkey), University Press of Virginia, Charlottesville, 1978.

[10] Kanninen, M. F. et al., Final Report to EPRI on RP 601-1, Battelle's Columbus Laboratories, 1980.

[11] Liebowitz, H., Lee, J. D. and Subramonian, N., _Proceedings of Int. Conf. on Analytical and Experimental Fracture Mechanics_, Rome, Italy, 1980.

[12] Zienkiewicz, O. S., _The Finite Element Method in Engineering Science_, McGraw-Hill, London, 1971.

[13] Segerlind, L. J., _Applied Finite Element Analysis_, Wiley, New York, 1976.

[14] Shen, W. and Lee, J. D., "The Nonlinear Energy Method for Mixed Mode Fracture," accepted for publication in Engineering Fracture Mechanics.

ENERGY-RELEASE RATES IN DYNAMIC FRACTURE: PATH-INVARIANT INTEGRALS, AND SOME
COMPUTATIONAL STUDIES

S. N. Atluri and T. Nishioka

Georgia Institute of Technology
Atlanta, Georgia 30332

ABSTRACT

In this paper, the subject of path-invariant contour integrals that quantify
the rate of energy release at a crack-tip, propagating under mixed-mode, unsteady
conditions, with a non-constant velocity, is critically examined. Such an integral, over a space-fixed contour is newly introduced, and its utility in computing
mixed-mode stress-intensity factors for propagating cracks, using finite element
methods, is demonstrated.

INTRODUCTION

It is well-known that the so-called J-integral introduced by Eshelby [1] and
Rice [2] is: (i) the first component along the crack-axis of a vector integral,
(ii) is applicable only to elasto-statics and (iii) in elasto-statics, it has the
meaning of rate of energy release per unit quasistatic crack extension.

Recently, one of the authors [3] has derived some very general conservation
laws for both finite-elastic solids, as well as for those described by rate-sensi-
tive or rate-insensitive incremental constitutive laws, wherein body forces, inertia, and arbitrary crack-face conditions were accounted for. On the basis of
these conservation laws, [3] also investigated path-invariant integrals in the
case of dynamic crack propagation in elastic as well as inelastic solids.

The contour-integral in [3], even though path-independent, was shown not to
be equivalent, in general, to the rate of energy release for a propagating crack
under non-steady-state conditions, and non-constant velocity of propagation. The
integration path for the contour-integral defined in [3] is fixed in space. On
the other hand, Kishimoto et al [4] introduced an integral, $\hat{J}_k$ which was argued

to have the meaning of an energy release rate. This is critically examined and
negated in the present paper.

In this paper, we also examine the result in [5] for a contour-integral involving a path which is rigid in shape and translates along with the crack-tip at
the same velocity.

Finally, we introduce a new path-invariant vector integral, involving a space-fixed contour, which has the meaning of energy-release under general mixed-mode conditions for a crack propagating with a non-constant velocity under unsteady conditions. This integral is shown to reduce to that in [6] which is valid for steady-state crack propagation at constant velocity. Further, even though the subject of path-independent integrals is not discussed in [7], it is shown that the energy-release-rate expression given therein [7], is in agreement with the present result.

We conclude the paper with an illustration of the use of the new integral in numerical evaluation of time-dependent stress-intensity factors for propagating cracks.

PATH-INDEPENDENT INTEGRALS FOR DYNAMIC FRACTURE

We consider, without loss of generality, only two-dimensional dynamic fracture problems, even though the direct-tensor formalism employed in [3] makes extensions to three-dimensional situations rather transparent. We consider a fixed cartesian coordinate system x_i ($i = 1,2,3$) such that x_1 is along the crack surface, x_2 normal to the crack face, and x_3 along the crack front. We also restrict ourselves here to linear elasto dynamics even though extensions to finite deformation cases of nonlinear materials is discussed in sufficient detail in [3]. We consider the general, unsteady, propagation of the crack at a non-constant velocity, $\underline{c}(t)$. At any instant t, the field equations and boundary conditions are:

$$\sigma_{ij,j} + \rho f_i = \rho \ddot{u}_i; \quad \sigma_{ij} = \sigma_{ji} \tag{1}$$

$$\varepsilon_{ij} = \frac{1}{2} [u_{i,j} + u_{j,i}] \tag{2}$$

$$\sigma_{ij} = \partial W / \partial \varepsilon_{ij} \tag{3}$$

$$\sigma_{ij} n_j = \bar{t}_i \text{ at } S_t \tag{4}$$

$$u_i = \bar{u}_i \text{ at } S_u \tag{5}$$

where: σ_{ij} is the stress tensor; ε_{ij} the strain tensor; $(\ )_{,j}$ denotes partial differentiation with respect to x_j; u_i is the displacement; $\dot{u}_i$ is the absolute velocity; and $\ddot{u}_i$ is the absolute acceleration; W is the strain energy density per unit volume, which for the linear elastic homogeneous material is a single-valued function of ε_{ij}; $\bar{t}_i$ for prescribed tractions at S_t, and $\bar{u}_i$ are prescribed displacements at S_u.

Now, we consider a closed volume V which is free from singularities or other defects. Provided the field equations (1-5) above are valid, the following conservation law holds for this volume V [3]:

$$0 = \int_V [\frac{\partial W}{\partial x_k} - \frac{\partial}{\partial x_i} (\sigma_{ij} u_{j,k}) - \rho_0 (f_j - \ddot{u}_j) u_{j,k}] dV$$

$$+ \int_{S_t} [n_i \sigma_{ij} - \overline{t}_j] u_{j,k} + \int_{S_u} n_i \sigma_{ij} (u_{j,k} - \overline{u}_{j,k}) ds \qquad (6)$$

Equation (6) can be verified from equations (1-5) and the additional identities:

$$\frac{\partial W}{\partial x_k} = \sigma_{ij} \epsilon_{ij,k} = \sigma_{ij} \frac{1}{2} [u_{i,jk} + u_{j,ik}] = \sigma_{ij} u_{j,ik} \qquad (7)$$

Consider the case when V includes the crack-tip, as shown for instance in Figure 1. Near the crack-tip, i.e., the point of singularity, W is singular and is of

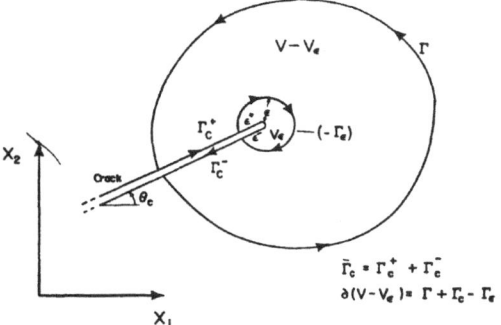

Fig. (1)

order r^{-1} where r is the radial distance from the crack-tip. Thus, $\partial W / \partial x_k$ is order r^{-2}. Likewise, near the crack-tip, σ_{ij} is $O(r^{-1/2})$ and $u_{j,k}$ is $O(r^{-1/2})$, while $\ddot{u}_j$ is of $O(r^{-3/2})$ [8]. Thus, the integrands in the volume integrals in equation (6) are non-integrable. Thus, the divergence theorem cannot be applied to the entire volume V, but only to the domain $V-V_\epsilon$ in the limit as $\epsilon \to 0$ as indicated in Figure 1.

Before applying the divergence theorem to the conservation-law of the type of equation (6) for a volume $V-V_\epsilon$, we note the identity:

$$\int_{V-V_\epsilon} \rho \ddot{u}_i u_{i,k} dV = \int_{V-V_\epsilon} \{\frac{d}{dt} (\rho \dot{u}_i u_{i,k}) dV - \frac{\partial}{\partial x_k} (\frac{1}{2} \rho \dot{u}_i \dot{u}_i)\} dV \qquad (8)$$

Using (8) in (6) and applying the divergence theorem to the domain $(V-V_\epsilon)$, we obtain, by definition, the kth component of a path-independent vector, denoted here as J_k, as:

$$J_k \overset{def}{\equiv} \underset{\varepsilon \to 0}{Lt} \int_{\Gamma_\varepsilon} [n_k(W-T) - n_j\sigma_{ji}u_{i,k}]ds$$

$$= \underset{\varepsilon \to 0}{Lt} \int_{\Gamma+\Gamma_c} [n_k(W-T) - n_j\sigma_{ji}u_{i,k}]ds$$

$$+ \underset{\varepsilon \to 0}{Lt} \int_{V-V_\varepsilon} \frac{d}{dt}(\rho\dot{u}_i u_{i,k})dV - \int_{\Gamma_c} [\bar{t}_i u_{i,k}]ds$$

$$- \int_{\Gamma_c} [n_j\sigma_{ji}\bar{u}_{i,k}]ds \tag{9}$$

where T denotes the kinetic energy density ($\equiv 1/2\rho\dot{u}_i\dot{u}_i$), n_j are direction cosines of the unit outward normal, and the definitions of paths Γ_ε, Γ, and Γ_c and the volumes V and V_ε are shown in Figure 1.

It should be noted that the contour Γ of the far-field integral in equation (9) is <u>fixed</u> in space, and the crack-tip moves into this fixed contour.

Using equation (8) in (9) we may obtain a slightly different path-independent integral, denoted here by $\hat{J}_k$, as

$$\hat{J}_k \overset{def}{\equiv} \underset{\varepsilon \to 0}{Lt} \int_{\Gamma_\varepsilon} [Wn_k - n_j\sigma_{ji}u_{i,k}]ds$$

$$= \underset{\varepsilon \to 0}{Lt} \int_{\Gamma+\Gamma_c} (Wn_k - n_j\sigma_{ji}u_{i,k})ds$$

$$+ \underset{\varepsilon \to 0}{Lt} \int_{V-V_\varepsilon} \rho\ddot{u}_i u_{i,k}dV - \int_{\Gamma_c} [\bar{t}_i u_{i,k} + n_j\sigma_{ji}\bar{u}_{i,k}]ds \tag{10}$$

Equation (10) was given in a slightly less general form in [4], even though the definition of $\hat{J}_k$ as the limit of integral over Γ does not appear to have been stressed in [4]. Once again Γ in equation (10) is a space-fixed contour.

On the other hand, considering a far-field contour Γ to be a rigid path surrounding the crack-tip and in translation at the same velocity, c_1 (along the x_1 axis) as the crack-tip, a path-independent integral, denoted here by J_1^*, was derived in [5]:

$$J_1^* = \int_\Gamma [Wn_1 - n_k\sigma_{kj}u_{j,1} - \frac{1}{2}\rho\dot{u}_k\dot{u}_i n_1 - \rho\dot{u}_j u_{j,1}c_1 n_1]ds$$

$$+ \frac{D}{Dt} \int_V \rho\dot{u}_j u_{j,1}dV \tag{11}$$

where V is the area enclosed by Γ. Since $\dot{u}_j u_{j,1}$ is $0(r^{-1})$ near the crack-tip, one may show [9] that the time derivative of the integral over the moving volume V becomes:

$$\frac{D}{Dt} \int_V \rho \dot{u}_j u_{j,1} dV = \underset{\varepsilon \to 0}{Lt} \int_{V-V_\varepsilon} \frac{d}{dt} (\rho \dot{u}_j u_{j,1}) dV + \int_\Gamma \rho \dot{u}_j u_{j,1} c_1 n_1 ds$$

$$- \underset{\varepsilon \to 0}{Lt} \int_{\Gamma_\varepsilon} \rho \dot{u}_j u_{j,1} c_1 n_1 ds \tag{12}$$

The last integral on the right-hand-side of equation (12) can be evaluated from the asymptotic (singular) field near the propagating crack-tip:

$$\dot{u}_i \simeq - c_1 u_{i,1}; \; - \rho \dot{u}_j u_{j,1} c_1 n_1 \simeq n_1 \rho \dot{u}_j \dot{u}_j = 2Tn_1 \tag{13}$$

Using (12) and (13) in (11) and comparing with (9) [upon noting that (11) is written for the case when crack faces are free from any prescribed conditions] we obtain:

$$J_1^* = \int_\Gamma [Wn_1 - \sigma_{jk} n_k u_{j,1} - Tn_1] ds + \underset{\varepsilon \to 0}{Lt} \int_{V-V_\varepsilon} \frac{d}{dt} (\rho \dot{u}_j u_{j,1}) dV$$

$$+ \underset{\varepsilon \to 0}{Lt} \int_{\Gamma_\varepsilon} 2Tn_1 ds$$

$$\equiv J_1 + \underset{\varepsilon \to 0}{Lt} \int_{\Gamma_\varepsilon} 2Tn_1 ds \tag{14}$$

We emphasize again that in the direct evaluation of J_1^* from equation (11) for a propagating crack, one must choose a rigid Γ that _moves_ along with the crack-tip.

It has been shown [3] from first principles, that the rate of energy-release to a propagating crack-tip can be expressed by:

$$G = (c_k/c)G_k$$

$$G_k = \underset{\varepsilon \to 0}{Lt} \int_{\Gamma_\varepsilon} [(W+T)n_k - t_i u_{i,k}] ds \tag{15}$$

where c_k denotes the component of crack velocity in the x_k direction, and $T = 1/2(\rho \dot{u}_i \dot{u}_i)$ where $\dot{u}_i$ is the absolute velocity of a material particle.

As this point it is worth comparing the current energy-release rate expression, equation (15), with those derived in [6] and [7] for a mode I problem under steady state, constant velocity propagation (with velocity c_1). For a mode I crack-propagation at constant velocity c_1 along x_1 axis, one may introduce a moving coordinate system, $X_1 = x_1 - c_1 t$, $X_2 = x_2$ etc. Thus the displacements u_i can be expressed as $u_i(X_1, X_2)$. The absolute velocity and acceleration can be written, in general, respectively, as:

$$\dot{u}_i = \frac{\partial u_i}{\partial t} + \frac{\partial u_i}{\partial X_1} \frac{\partial X_1}{\partial t} = \frac{\partial u_i}{\partial t} - c_1 \frac{\partial u_i}{\partial X_1} \tag{16}$$

and

$$\ddot{u}_i = \frac{\partial^2 u_i}{\partial t^2} + c_1^2 \frac{\partial^2 u_i}{\partial X_1^2} - 2c_1 \frac{\partial^2 u_i}{\partial X_1 \partial t} \tag{17}$$

Under steady-state conditions at constant crack-velocity c_1, approximated everywhere in the domain as:

$$\dot{u}_i \approx - c_1 \frac{\partial u_i}{\partial X_1}; \; \ddot{u}_i \approx c_1^2 \frac{\partial^2 u_i}{\partial X_1^2} \tag{18a,b}$$

Further, the asymptotic (singular) velocity near the crack-tip can be approximated even in the non-steady case as:

$$\dot{u}_i \approx - c_1 \frac{\partial u_i}{\partial X_1} \tag{19}$$

The energy release-rate expression given by Sih [6] for mode I crack-propagation is [using the present notation]:

$$G_1 = \underset{\varepsilon \to 0}{Lt} \int_{\Gamma_\varepsilon} (Wn_1 + \frac{1}{2} \rho c_1^2 \frac{\partial u_i}{\partial X_1} \frac{\partial u_i}{\partial X_1} n_1 - n_j \sigma_{ji} \dot{u}_{i,1}) ds \tag{20a}$$

$$\equiv \int_{\Gamma} (Wn_1 + \frac{1}{2} \rho c_1^2 \frac{\partial u_i}{\partial X_1} \frac{\partial u_i}{\partial X_1} n_1 - n_j \sigma_{ji} \dot{u}_{i,1}) ds \tag{20b}$$

Equation (20a) can be seen to be a special case of the present equation (15) for steady-state crack-propagation at constant velocity c_1. Further, the path-independency, viz, the equality of integrals over Γ_ε and Γ as in (20a,b) is predicated on the satisfaction of the equation of motion

$$\sigma_{ij,j} = \rho c_1^2 \frac{\partial^2 u_i}{\partial X_1^2} \text{ in } V \tag{21}$$

as shown in [6]. Equation (21) is clearly valid for steady-state conditions under constant velocity propagation.

On the other hand, the energy-release rate expression given in [7] is:

$$G = \underset{\varepsilon \to 0}{Lt} \int_{\Gamma_\varepsilon} (\sigma_{ij} n_j \dot{u}_i + \frac{1}{2} \sigma_{ij} u_{i,j} c_1 n_1 + \frac{1}{2} \rho \dot{u}_i \dot{u}_i c_1 n_1) ds \tag{22}$$

Upon substituting the asymptotic values,

$$\dot{u}_i = - c_1 u_{i,1}$$

it is seen that (22) reduces to:

$$G = c_1 \underset{\varepsilon \to 0}{Lt} \int_{\Gamma_\varepsilon} [(W+T)n_1 - n_j \sigma_{ji} u_{i,1}] ds \tag{23}$$

which agrees with the corresponding mode I result in equation (15). However, the subject of path-independent integrals is not discussed in [7].

Now, comparing equations (9,10,14 and 15) one finds that:

$$J_k = G_k - \underset{\varepsilon \to 0}{Lt} \, 2 \int_{\Gamma_\varepsilon} (\frac{1}{2} \rho \dot{u}_i \dot{u}_i) n_k ds \tag{24}$$

$$\hat{J}_k = G_k - \underset{\varepsilon \to 0}{Lt} \int_{\Gamma_\varepsilon} (\frac{1}{2} \rho \dot{u}_i \dot{u}_i) n_k ds \tag{25}$$

$$J_1^* = G_1 \tag{26}$$

For stationary cracks in dynamic elastic fields the kinetic energy density, $T \equiv \frac{1}{2} \rho \dot{u}_i \dot{u}_i$ is non-singular, since the absolute velocity $\dot{u}_1$ is still of the order of $O(r^{1/2})$. Thus the last terms in equations (24,25) vanish as Γ_ε shrinks to the crack-tip. Thus for <u>stationary</u> cracks in dynamic elastic fields, J_k, $\hat{J}_k$, as well as J_1^* have the meaning appropriate energy release rates.

On the other hand, for dynamically <u>propagating</u> crack the kinetic energy density become singular since $\dot{u}_1$ is of the order of $(r^{-1/2})$ near the crack-tip. Thus the last integrals in equations (24,25) lead to finite values for <u>propagating</u> cracks. Thus $\hat{J}_k$ does not have the meaning of an energy-release rate despite claims, to the contrary, made in [4]. On the other hand J_k, while also not having the meaning of an energy-release rate, does have the meaning of the rate of change of the Lagrangian of the dynamic system due to unit crack growth [3].

An examination of equations (9,10 and 24) reveals the possibility of deriving a path-independent vector integral, using space-fixed contours, which has the meaning of an energy release rate in the general case of non-constant velocity, non-steady crack propagation. To this end, we first observe that:

$$\int_{V-V_\epsilon} \frac{\partial}{\partial x_k} (\frac{1}{2} \rho \dot{u}_i \dot{u}_i) \equiv \int_{V-V_\epsilon} \rho \dot{u}_i \dot{u}_{i,k} dV$$

$$= \int_{\Gamma+\Gamma_c} (\frac{1}{2} \rho \dot{u}_i \dot{u}_i) n_k ds - \int_{\Gamma_\epsilon} (\frac{1}{2} \rho \dot{u}_i \dot{u}_i) n_k ds \qquad (27)$$

Likewise,

$$\int_{V_2-V_1} \frac{\partial}{\partial x_k} (\frac{1}{2} \rho \dot{u}_i \dot{u}_i) dV = \int_{\Gamma_2+\Gamma_{C2}} (\frac{1}{2} \rho \dot{u}_i \dot{u}_i) n_k ds - \int_{\Gamma_1+\Gamma_{C1}} (\frac{1}{2} \rho \dot{u}_i \dot{u}_i) n_k ds \qquad (28)$$

Considering, without loss of generality, the entire crack-faces to be traction-free, one may, from an observation of equations (10) and (27), define a path-independent integral J_k' such that [8]

$$J_k' \equiv G_k = \mathop{Lt}_{\epsilon \to 0} \int_{\Gamma_\epsilon} [(W+T)n_k - n_j \sigma_{ji} u_{i,k}] ds$$

$$\equiv \mathop{Lim}_{\epsilon \to 0} \{ \int_{\Gamma+\Gamma_c} [(W+T)n_k - n_j \sigma_{ji} u_{i,k}] ds + \int_{V-V_\epsilon} [\rho \ddot{u}_i u_{i,k} - \rho \dot{u}_i \dot{u}_{i,k}] dV \} \qquad (29)$$

Path-independency of the extreme right hand side of (29) is clear when one verifies from the elastodynamic equilibrium, equation (1) and equation (28) that

$$\int_{\Gamma_2+\Gamma_{C2}} [(W+T)n_k - n_j \sigma_{ji} u_{i,k}] ds + \int_{V_2-V_1} [\rho \ddot{u}_i u_{i,k} - \rho \dot{u}_i \dot{u}_{i,k}] dV$$

$$- \int_{\Gamma_1+\Gamma_{C1}} [(W+T)n_k - n_j \sigma_{ji} u_{i,k}] ds \equiv 0 \qquad (30)$$

It should be remarked that J_k' evaluated from the far-field contour as in equation (29) is valid for non-steady propagation at non-constant velocity. If one invokes the steady-state assumption and constant-velocity propagation (with velocity C_1), one may set

$$T = \frac{1}{2} \rho \dot{u}_i \dot{u}_i \approx \frac{1}{2} \rho c_1^2 \frac{\partial u_i}{\partial X_1} \frac{\partial u_i}{\partial X_1}$$

$$\rho \ddot{u}_i u_{i,k} \approx \rho c_1^2 \frac{\partial^2 u_i}{\partial X_1^2} \frac{\partial u_i}{\partial X_k} \qquad \rho \dot{u}_i \dot{u}_{i,k} \approx \rho c_1^2 \frac{\partial u_i}{\partial X_1} \frac{\partial^2 u_i}{\partial X_1 \partial X_k} \qquad (31)$$

Thus, for steady-state crack propagation at constant velocity c_1 along x_1 axis, equations (29) can be simplified for mode I, as:

$$(J_1')_{\text{steady-state}} = (G_1)_{\text{steady-state}}$$

$$= \int_{\Gamma} [(W + \frac{1}{2} \rho c_1^2 \frac{\partial u_i}{\partial X_1} \frac{\partial u_i}{\partial X_1}) n_1 - n_j \sigma_{ji} u_{i,1}] ds$$

$$+ \int_{V-V_\varepsilon} (\rho c_1^2 \frac{\partial^2 u_i}{\partial X_1^2} \frac{\partial u_i}{\partial X_1} - \rho c_1^2 \frac{\partial u_i}{\partial X_1} \frac{\partial^2 u_i}{\partial X_1^2}) dV$$

$$= \int_{\Gamma} [(W + \frac{1}{2} \rho c_1^2 \frac{\partial u_i}{\partial X_1} \frac{\partial u_i}{\partial X_1}) n_1 - n_j \sigma_{ji} u_{i,1}] dS \qquad (32)$$

which agrees with the corresponding result of Sih [6].

From the knowledge of the asymptotic strain and stress field near an (arbitrarily) <u>propagating</u> crack-tip the energy release rate G_k can be evaluated directly from (15) and expressed in terms of the mixed mode stress-intensity factors $K_I(t)$, $K_{II}(t)$, and $K_{III}(t)$. From equation (29) it is seen that $J_1^* = G_1$; while from equation (29) it is seen that $J_k' = G_k$. Even though J_k and $\hat{J}_k$ are both not equal to G_K, equations (24 and 25) can be used to relate J_k and $\hat{J}_k$ factors K_I, K_{II} and K_{III}.

The near-tip fields in general mixed-mode crack propagation have recently been succinctly presented in [8]. Using these near-tip fields, one may directly evaluate G_k from equation (15), and this leads to [8]:

$$G_1 = \frac{1}{2\mu} \{K_I^2 A_I(C) + K_{II}^2 A_{II}(C) + K_{III}^2 A_{III}(C)\} \qquad (33)$$

$$G_2 = -\frac{1}{\mu} K_I K_{II} A_{IV}(C) \qquad (34)$$

where μ is the shear modulus, and

$$A_I(C) = [\beta_1(1-\beta_2^2)]/D(C) \quad , \quad A_{II}(C) = [\beta_2(1-\beta_2^2)]/D(C) \quad , \quad A_{III}(C) = 1/\beta_2$$

$$A_{IV}(C) = \frac{(\beta_1-\beta_2)(1-\beta_2^2)}{[D(C)]^2} [\frac{\{4\beta_1\beta_2+(1+\beta_2^2)^2\}(2+\beta_1+\beta_2)}{2[(1+\beta_1)(1+\beta_2)]^{1/2}} - 2(1+\beta_2^2)]$$

where

$$\beta_1^2 = 1 - (C^2/C_d^2) \; ; \; \beta_2^2 = 1 - (C^2/C_s^2) \; ; \; C_d^2 = \frac{\kappa+1}{\kappa-1} \frac{\mu}{\rho} \; ; \; C_s^2 = \frac{\mu}{\rho}$$

$$\kappa = \begin{cases} (3-\nu)/(1+\nu): \text{ plane stress} \\ (3-4\nu): \qquad \text{ plane strain} \end{cases} \; ; \; D(C) = 4\beta_1\beta_2 - (1+\beta_2^2)^2$$

From equation (26) and (29) it is then seen that

$$J_1' \equiv G_1 \; ; \; J_2' = G_2 \; ;$$

and

$$J_1^* = G_1$$

on the other hand, using equation (24) one may derive [8] that:

$$J_1 = \frac{1}{2\mu} \{ K_I^2 F_I(C) + K_{II}^2 F_{II}(C) + K_{III}^2 F_{III}(C) \} \tag{35}$$

$$J_2 = \frac{-K_I K_{II}}{\mu} F_{IV}(C) \tag{36}$$

where,

$$F_I(C) = \frac{\beta_1(1-\beta_2^2)}{[D(C)]^2} \left[4\beta_1 - \frac{1}{\beta_1} (1+\beta_2^2)^2 - 4(\beta_1-\beta_2) \frac{(1+\beta_2^2)}{[(1+\beta_1)(1+\beta_2)]^{1/2}} \right]$$

$$F_{II}(C) = \frac{\beta_2(1-\beta_2)}{[D(C)]^2} \left[4\beta_2 - \frac{1}{\beta_2} (1+\beta_2^2) - 4(\beta_2-\beta_1) \frac{(1+\beta_2^2)}{[(1+\beta_1)(1+\beta_2)]^{1/2}} \right]$$

$$F_{III}(C) = \frac{1}{\beta_2^2}$$

and

$$F_{IV}(C) = \frac{[4\beta_1\beta_2 + (1+\beta_2^2)^2]}{2[D(C)]^2} \frac{(1-\beta_2^2)(\beta_1^2-\beta_2^2)}{[(1+\beta_1)(1+\beta_2)]^{1/2}}$$

Thus, the path-independent integrals J_k' or J_k [which involve space-fixed far-field contours Γ into which the crack grows] are convenient and highly useful in computing dynamic stress-intensity factors for arbitrarily propagating cracks using numerical methods such as the finite element method. As usual, since the integrals involve contours in the far-field, detailed modeling of the crack-tip region (to obtain accurate near-field stress and strain values) may not be necessary. Such applications have been presented [10], and are illustrated below:

A NUMERICAL EXAMPLE

A. Non-Constant-Velocity Crack Propagation in a DCB Specimen:

The problem studied here is the prediction type simulation of fast fracture in a Double-Cantilever-Beam specimen. The same problem has been studied by the present authors in Reference [11] wherein the dynamic stress-intensity factors were directly computed from a "singular element". Figure 2 shows the specimen geometries and the finite element meshes at t = 0 and t = 20 μsec. The plane stress condition is postulated in the analysis because of the relatively thin nature of the specimen.

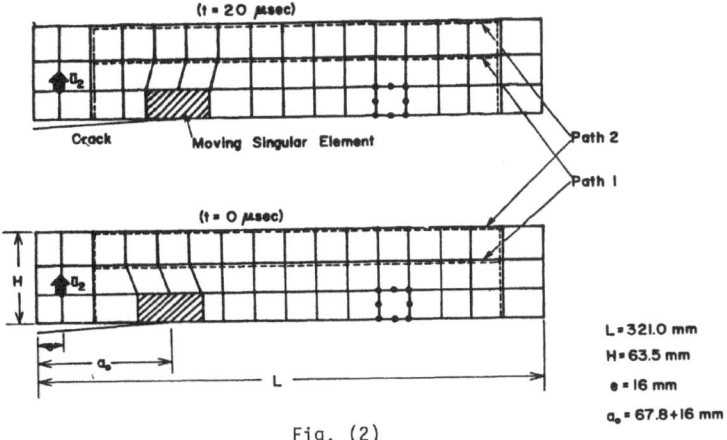

Fig. (2)

Figure 3 shows the fracture toughness K_{ID} versus crack velocity relation determined in the experiment [12] for Araldite B epoxy resin. In the prediction type fracture simulation the instantaneous crack velocity will be determined by using K_{ID} vs C relation. In order to predict the average velocity between the times t_o and $t_o + \Delta t$, the K_I value at time $t_o + \frac{\Delta t}{2}$ is predicted by

$$K_{IP} = K_I(t_o) + \frac{\Delta t}{2} \dot{K}_I(t_o) + \frac{1}{2} \left(\frac{\Delta t}{2}\right)^2 \ddot{K}_I(t_o) + R$$

where R is a corrector of the predicted value. K_I, $\dot{K}_I$ and $\ddot{K}_{II}$ are directly evaluated in the moving singular element at the previous time step. Since the present analysis is based on the standard variational principle, R = 0 was used, whereas the non zero corrector (see Reference [11]) was used for the analysis based on the special variational principle.

In the calculation of the surface integral in equation (29) the necessary quantities were calculated at the element boundaries inside of the paths. The numerical results for the surface integral were found to oscillate with the time

338

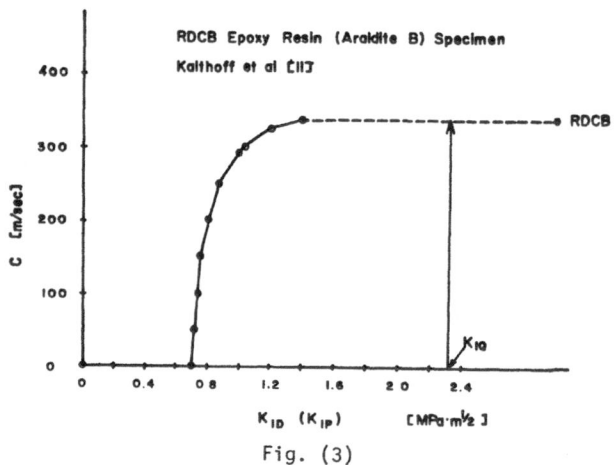

Fig. (3)

step while the volume integral values were in smooth variation. It was found that the source of oscillation is the $t_j u_{j,k}$ term. In the finite element displacement model the traction reciprocity condition $t_j^+ + t_j^- = 0$ at inter-element boundaries, with (+) and (-) denoting the two sides of these boundaries, is satisfied only in an average sense. In this particular modeling of DCB speci-men the traction reciprocity condition was locally not well satisfied under the dynamic condition. To reduce the oscillation in the $t_j u_{j,k}$ term, the average (from the two sides of the path) of the quantity $t_j u_{j,k}$ was calculated at Path 1. For Path 2, $t_j u_{j,k} = 0$ was used at the upper free end of the specimen.

Figure 4 shows the variations of the stress intensity factors and the crack velocities during the time of fast fracture. As seen from the figure the pre-sent fracture simulation gives an excellent prediction of the histories of the stress intensity factors, crack velocity (and the crack length), using the rela-tion of fracture toughness and crack velocity (Figure 3) and the crack initia-tion stress intensity factor for a blunted notch K_{IQ}. The K_I values calculated by the J' integral are seen to be in good agreement with those extracted from the singular element.

ACKNOWLEDGEMENTS

The results herein were obtained during the course of investigations supported by AFOSR under grant 81-0057B. This support as well as the encourage-ment received from Dr. A. Amos are gratefully acknowledged. The authors thank M. Eiteman for her assistance in the preparation of this manuscript.

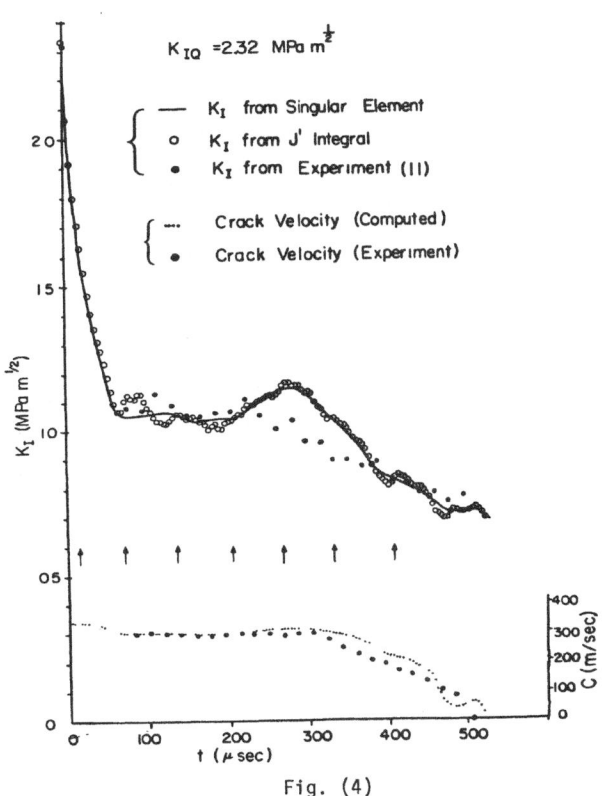

Fig. (4)

REFERENCES

[1] Eshelby, J. D., "The Continuum Theory of Lattice Defects", Solid State Physics, Vol. III, Academic Press, pp. 79-144, 1956.

[2] Rice, J. R., "A Path-Independent Integral and the Approximate Analysis of Strain Concentration by Notches and Cracks", Journal of Applied Mathematics, Vol. 35, pp. 376-386, 1968.

[3] Atluri, S. N., "Path-Independent Integrals in Finite Elasticity and In-Elasticity, with Body Forces, Inertia, and Arbitrary Crack-Face Condi-tions", Engineering Fracture Mechanics, Vol. 16, No., pp. 341-364, 1982.

[4] Kishimoto, K., Aoki, S. and Sakata, M., "On the Path Independent Integral Ĵ, Engineering Fracture Mechanics, Vol. 13, pp. 841-850, 1980.

340

[5] Bui, H. D., "Stress and Crack-Displacement Intensity Factors in Elasto-dynamics", Fracture, Vol. 3, ICF4, Waterloo, pp. 91-95, 1979.

[6] Sih, G. C., "Dynamic Aspects of Crack Propagation:, In Inelastic Behaviour of Solids (Eds. Kanninen, et al.), McGraw-Hill, pp. 607-633, 1970.

[7] Freund, L. B., "Energy Flux into the Tip of an Extending Crack in an Elastic Solid", Journal of Elasticity, Vol. 2, No. 4, pp. 341-349, 1972.

[8] Nishioka, T., and Atluri, S. N., "Path-Independent Integrals, Energy Release Rates, and General Solutions of Near-tip Fields in Mixed-Mode Dynamic Fracture Mechanics, "Engineering Fracture Mechanics" (In Press).

[9] Bui, H. D., Private Communication to S. N. Atluri, 21 April 1982.

[10] Nishioka, T., and Atluri, S. N., "A Numerical Study of the Use of Path-Independent Integrals in Elasto-Dynamic Crack Propagation", Engineering Fracture Mechanics, 1982 (In Press).

[11] Nishioka, T., and Atluri, S. N., "Numerical Analysis of Dynamic Crack Propagation: Generation and Prediction Studies", Engineering Fracture Mechanics, Vol. 16, No. 3, pp. 303-332, 1982.

[12] Kalthoff, J. F., Beinert, J., and Winkler, S., "Measurement of Dynamic Stress Intensity Factors for Fast Running and Arresting Cracks in Double Cantilever-Beam Specimens", Fast Fracture and Crack Arrest, ASTMSTP 627 (Ed. G. T. Hahn et al.), pp. 161-176, 1977.

MODELLING OF SUB-CRITICAL FLAW GROWTH IN THREE DIMENSIONAL CRACK PROBLEMS

C. W. Smith

Virginia Polytechnic Institute and State University
Blacksburg, Virginia 24061

INTRODUCTION

The most frequent chronology of events leading to service fractures in metal parts begins with a defect or tiny flaw, usually on an exposed surface near a re-entrant corner, which enlarges under the action of repeated (vibratory, fatigue, etc.) loading until it reaches a critical size, after which catastrophic fracture results. The problem geometry which exists during this sub-critical flaw growth period is often quite complex, involving irregular body boundaries and often non-planar cracks and/or curved crack fronts. Such problems can readily be identified as three dimensional (3D) cracked, finite body problems where the stress intensity factor (SIF) varies along the flaw border. These problems have resisted efforts of applied mathematicians to render them tractable in modelling such problems numerically [1-4]. A major difficulty has been lack of knowledge of the crack shape during sub-critical flaw growth and its influence upon the SIF distribution.

Over a decade ago, the first author and his associates began to study experimental modelling techniques to use in simulating the growth of natural cracks and from which SIF distributions could be estimated. Originally, the method, involving a marriage between near field equations of linear elastic fracture mechanics with the frozen stress photoelastic technique, was applied only to Mode I problems [5]. Subsequently, it was extended to include mixed mode analysis [6-8]. In the process of modelling nozzle corner cracks in flat plates [9-10], it was noticed that, when cracks were grown from tiny starter flaws in the photoelastic models under monotonic loads, the resulting shapes would, under certain conditions, overlay those produced by tension-tension fatigue in geometrically similar metal models, even though the shapes were not simple.

Such observations suggest that, under the proper conditions, the frozen stress method, when utilizing natural cracks, can provide predictions of both flaw shape and SIF distributions where neither are known a priori. Substantial work has been done (and is continuing) towards the goal of delineating the conditions necessary to produce the desired similarity in model behavior. Important considerations include:

1. Role of model configuration and load orientation.

2. Degree of influence of elastic incompressible response of photoelastic materials above stress freezing temperature.

3. Degree of fatigue enhancement of prototype fatigue data from metals.

After briefly reviewing the experimental technique used, this paper uses experimental results from two finite three-dimensional cracked body problems to qualitatively assess the considerations delineated above.

ANALYTICAL CONSIDERATIONS FOR MODE I LOADING

Before proceeding to the experiments, it would seem appropriate to review the analytical foundations for Mode I loading. We confine ourselves to Mode I here since we are working with growing, stable, natural cracks which, in our experience, exhibit predominantly Mode I deformation fields during flaw growth. Cotterell [11] has discussed this point in detail.

Thus, for the case of Mode I loading, one begins, following the work of Kassir and Sih [12], with equations of the form:

$$\sigma_{ij} = \frac{K_I}{r^{1/2}} f_{ij}(\theta) + \sigma_{ij}^0(r,\theta) \tag{1}$$

for the stresses in a plane mutually orthogonal to the flaw surface and the flaw border referred to a set of local rectangular cartesian coordinates as pictured in Figure 1, where the terms containing K_1, the SIF, are identical to Irwin's equa-

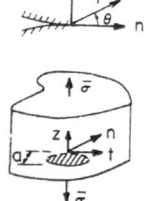

Fig. (1) - General problem and notation

tions for the plane case and σ_{ij}^0 are normally taken to be constant for a given point along the flaw border, within the measurement zone (normally 0.1 to 1.0 mm from the crack plane), but may vary from point to point. Observing that stress fringes tend to spread approximately normal to the flaw surface, Figure 2, equations (1) are evaluated along $\theta = \pi/2$, Figure 1, and the maximum in-plane shearing stress is computed as:

$$\tau_{max}^{nz} = 1/2[(\sigma_{nn}-\sigma_{zz})^2 + 4\sigma_{nz}^2]^{1/1} \tag{2}$$

Fig. (2) - Mode I stress fringes

which, when truncated to the same order as equations (1), leads to the two parameter equation:

$$A = K_1/\sqrt{8\pi}$$

$$\tau_{max}^{nz} = \frac{A}{r^{1/2}} + B \text{ where}$$

$$B = f(\sigma_{ij}^o)$$

$\quad(3)$

which can be rearranged into the normalized form:

$$\frac{K_{AP}}{q(\pi a)^{1/2}} = \frac{K_I}{q(\pi a)^{1/2}} + \frac{f(\sigma_{ij}^o)(8)^{1/2}}{q} \left(\frac{r}{a}\right)^{1/2}$$

$\quad(4)$

where $K_{AP} = \tau_{max}^{nz}(8\pi r)^{1/2}$ and, from the Stress-Optic Law, $\tau_{max} = Nf/2t'$ where N is the stress fringe order, f the material fringe value and t' the slice thickness in the t direction, q is the remote loading parameter (such as uniform stress (σ), or pressure (p)), and a the characteristic flaw depth. Equations (1) with σ_{ij}^o as described above, prescribe a linear relation between the normalized apparent stress intensity factor and the square root of the normalized distance from the crack tip. Thus, one need only locate the linear zone in a set of photoelastic data and extrapolate across a very near field nonlinear zone to the crack tip in order to obtain the SIF. An example of this approach using data from cracked plate tests to be discussed in the sequel is given in Figure 3.

It is important to note that the method is designed for three dimensional problems so that several slices are removed from the same flaw border parallel to the nz plane at intervals along the flaw border. The location of the linear zone cannot always be determined from a single slice and so all slices are surveyed after which a linear zone common to all slices is found. The author has found the approach described here much more reliable in estimating SIF values than attempting to include additional terms in a series expansion of the non-singular stresses about the crack tip in three dimensional problems. Moreover, this technique may be readily extended to mixed mode situations [6] if desired.

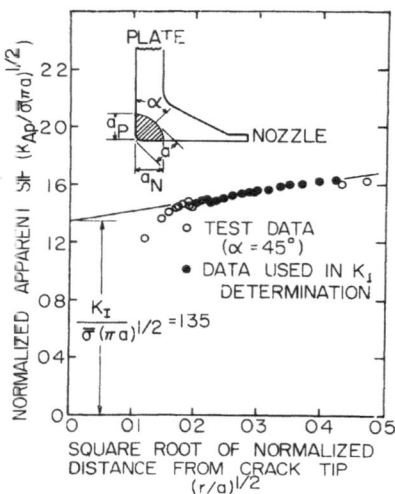

Fig. (3) - Estimating SIF for a slice

EXPERIMENTAL METHODS AND RESULTS

Before proceeding with a discussion of the topic of this section, a few brief remarks appear appropriate relative to the frozen stress method. Materials used for stress freezing, in addition to transparency, exhibit a diphase behavior. At room temperature, the material exhibits time dependent mechanical response with rather high material fringe and modulus of elasticity values. However, when heated to a sufficiently high temperature, called critical temperature, the mechanical response changes rapidly so as to exhibit a material fringe value of about 5% of its room temperature value and a modulus of elasticity of about 0.2% of its room temperature value. Thus, the material is much more sensitive to loads both optically and mechanically (deformation-wise) above critical temperature. Its behavior may be crudely described by the Kelvin model in Figure 4 at room temperature, and when taken above critical, $\mu \to 0$ resulting in purely elastic response. In order to capitalize on this type of behavior to "freeze in" the deformation field produced above critical temperature, the following test procedure is followed:

1. After casting fringe free model parts and inserting cracks at desired locations, the models are glued together with a glue compatible with the model materials. Glue lines are not permitted in the vicinity of crack borders. Natural cracks are inserted by striking a sharp rounded blade held normal to the plate surface at the desired point with a hammer. The starter crack emanates dynamically from the region of impact and arrests after slight enlargement.

2. Place model in a loading rig in an oven with a viewing port. Heat to above critical temperature.

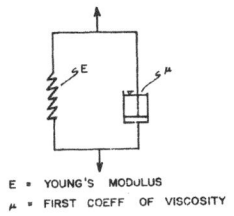

E = YOUNG'S MODULUS
μ = FIRST COEFF OF VISCOSITY

Fig. (4) - Kelvin material

3. Apply sufficient load to grow crack to desired size.

4. Reduce load to stop flaw growth and cool slowly under reduced load to room temperature.

5. Remove load. Since the material is relatively insensitive both mechanically and optically at room temperature, recovery is negligible and stress fringes and deformations produced above critical temperature are retained.

6. Slices approximately 2 mm thick may then be removed parallel to the nz plane for photoelastic analysis without altering fringe or deformation patterns.

7. Photoelastic data are used to construct curves as shown in Figure 3 directly from which SIF values are estimated at point along the flaw border.

From the above procedure, it is clear that we measure the material response above critical temperature, and at this temperature the material is incompressible ($\nu \approx 0.5$). Moreover, our loading technique involves monotonic rather than cyclic loads. We now consider some experiments, the results of which allow some assessments regarding these effects.

A. Program I - Nozzle Corner Cracks in Flat Plates

This study involved tests on photoelastic models of large flat plates, each of which contained a nozzle of varying wall thickness at the center of the plate, Figure 5a, and a single crack emanating from the nozzle corner in a plane normal to

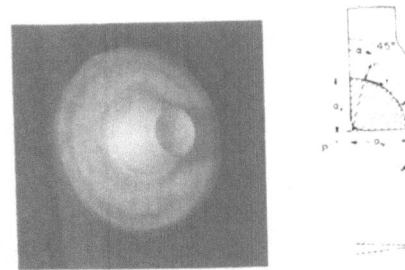
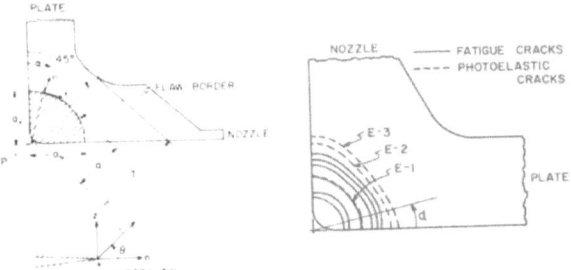

Fig. (5a) - Photo of nozzle, (b) nozzle crack notation, (c) nozzle crack shapes

an applied uniaxial tensile stress. The notation used in describing the crack is given by Figure 5b. Cracks were grown in these models under monotonic uniaxial tension to various depths. In a separate program, researchers at the Delft University of Technology conducted fatigue tests on geometrically similar models of A508 steel at stress ratios (R = $\sigma_{min}/\sigma_{max}$) ≈ 0.1 [13]. Figure 5c shows a composite of the crack shapes produced in the programs. Only in one case (E-1) was the photoelastic crack of exactly the same relative size as the fatigue crack in the steel. However, for this case, the crack shapes were exact overlays. In any event, these tests suggested that crack growth was non-self-similar, accompanied by significant "flattening" of the crack shape in the central region of the flaw border, and that the crack growth characteristics as revealed by the flaw shapes were remarkably similar for the two programs. Figure 6 shows a comparison between

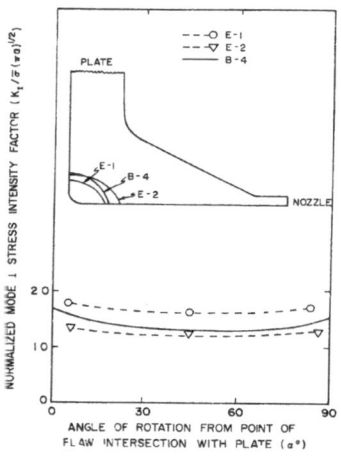

Fig. (6) - SIF distributions when idealized cracks approximate real crack shapes

photoelastic results for two shallow flaws that showed only negligible "flattening" and an analytical result (B4) obtained from the finite element analysis of a quarter-elliptic flaw of about the same relative size as the photoelastic flaws. Here the crack shapes and SIF distributions are quite similar. On the other hand, Figure 7 shows a comparison between an experimental SIF distribution measured photoelastically for a flattened real crack compared with a finite element result obtained by assuming the crack to possess a quarter elliptic shape and using measured values for a_n and a_p for both cases. Since the two cracks were not of identical size, results were normalized with respect to a_4. Figure 7 suggests that a crack will take the shape which minimizes the SIF gradient along the flaw border. It follows that, if one assumes a crack shape in a numerical analysis which is different from the real shape, the SIF gradient along the flaw border will be different (i.e., the nature of the singularity distribution will be different). Taken col-

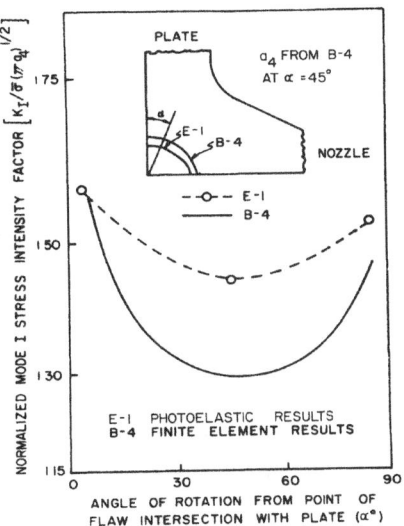

Fig. (7) - SIF distributions when real crack shapes depart
from idealized shapes

lectively, these results demonstrate the importance of the crack shape in the model
configuration influence. They also reveal no indication of the effects of fatigue
enhancement (such as crack closure). Further details of these studies are recorded
in [10] and [13].

B. Program II - Surface Flaws in Flat Plates Under Mode I Loading

Perhaps the best known of the three dimensional cracked body problems involving curved crack fronts is the problem involving a surface flaw in a finite
thickness plate. Since Irwin introduced the notion of a semi-elliptic shape for
such flaws in 1962 [14], much study has been devoted to this class of problems,
both analytically and experimentally, due to the prevalence of such flaws in structural components. In 1975, a group of researchers met at Battelle-Columbus Laboratories and prescribed a set of "benchmark" problem geometries [3] to be used in
verifying problem formulation and computer codes. One such problem was the surface
flaw in a finite thickness plate. Subsequently, a group of numerical analysts provided finite element, alternating technique, boundary integral and hybrid solutions
to the surface flaw "benchmark" geometries [4].

When the author and his colleagues attempted to grow "benchmark" cracks
using monotonic tension and the stress freezing method, they found that the cracks
grew with higher aspect ratios (a/C) than specified as benchmark geometries. Only
by flexing the cracked plates could benchmark aspect ratios be achieved. However,

348

referring to Table 1, when benchmark cracks with a/t ≈ 0.75 were produced by flexing the plate, the crack shapes were no longer semi-elliptic, but developed a
"flattened" region in the central port. This then led to an investigation to study
how natural cracks would grow under uniaxial Mode I loading and what the corresponding SIF distributions and crack shapes would be.

TABLE 1 - SURFACE FLAW BENCHMARK GEOMETRIES [3]

a/C	a/t	C/B	
0.25	0.25	< 0.2	a = flaw depth
0.25	0.75	< 0.2	2C = flaw length
			B = plate width

By selecting one of the numerical solutions which had been extensively
compared favorably with metal fatigue test data and some of this data, it would
be possible to assess the extent to which the cracks grown in photoelastic models
under monotonic loads could predict flaw shapes and SIF distributions resulting
from metal fatigue. The Newman-Raju (N-R) Model [15] was selected for this purpose. It was decided to grow cracks under monotonic tension to depths approaching
benchmark depths, obtain corresponding crack shapes and SIF distributions, and to
compare the crack shapes with test data from metal fatigue tests, and the SIF distributions with the Newman-Raju theory.

Figure 8 presents a summary of the experimental SIF distributions compared
with the N-R Model. All real cracks retained a semi-elliptic shape for all flaw
depths and the N-R Model used semi-elliptic shapes. The N-R Model was used to gen-

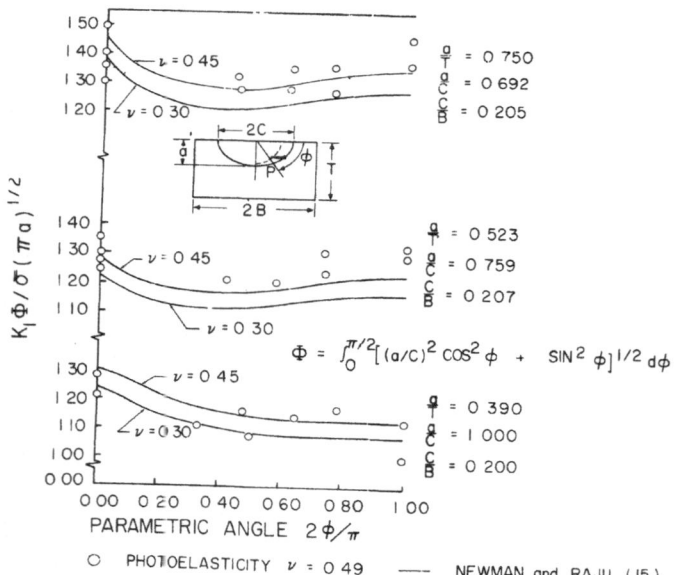

$$\Phi = \int_0^{\pi/2}[(a/C)^2 \cos^2\phi + \sin^2\phi]^{1/2} d\phi$$

O PHOTOELASTICITY ν = 0.49   —— NEWMAN and RAJU (15)

Fig. (8) - Comparison of finite element and photoelastic SIF distributions

erate SIF distributions for ν = 0.3 and ν = 0.45 in order to obtain an assessment of the influence of the elevated value of Poisson's ratio in the stress freezing experiments. From a study of Figure 8, one concludes that the experimental results agree with the N-R Model to within experimental scatter (say ±5%) in general and that the influence of Poisson's ratio is to elevate the SIF distribution by about the same amount (i.e., approximately 5%). However, there does appear to be a slight trend toward higher SIF values at the center for the deeper flaws in the experiments than in the N-R Model. We reiterate that, even for the deep flaws, no deviation from a semi-elliptic shape was observed in the experimental results.

Figure 9 shows a limited comparison between the crack growth characteristics for cracked aluminum plates under fatigue loading (R ≈ 0.1) and cracks grown under monotonic loads in cracked photoelastic plates. Comparisons here are not exact since starter crack ratios were not the same. Nevertheless, some trends may

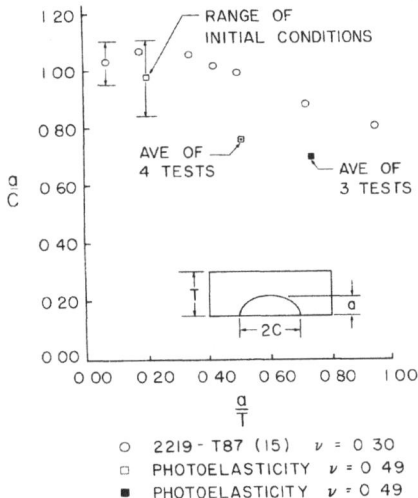

Fig. (9) - Change in crack aspect ratio during crack growth

be noted. These results suggest a divergence in aspect ratios between a/T = 0.2 and 0.5 after which the aspect ratio curves appear to follow parallel paths.

If we note that the bulk of fatigue crack closure will be concentrated near the surface through which the surface flaw enters the plate, we can conjecture that the portion of the flaw growth between a/T = 0.2 and 0.5 was fatigue enhanced by crack closure. Beyond mid-depth, however, the influence of crack closure appears to disappear. In their paper, Newman and Raju integrated a crack growth law with their model, and by adjusting the crack growth law to account for an initial closure effect, were able to generate an aspect ratio curve which closely followed the experimental data.

SUMMARY

After briefly reviewing the analytical foundations and experimental procedures associated with a photoelastic frozen stress technique for estimating flaw shapes and stress intensity distributions in three dimensional cracked body problems where neither are known a priori, results from two experimental programs were described which can be used to assess some of the important considerations (or limitations) of the method. Returning now to three of these considerations cited in the introduction, we offer the following observations:

1. *Role of model configuration and load orientation.* In general, models should be geometrically similar to prototypes. Program I showed that, if this similarity could be maintained, good correlation between photoelastic crack shapes and fatigue crack shapes can result. This implies an absence of fatigue enhancement. On the other hand, Program II showed how fatigue enhancement could influence crack configuration during growth for flaws near a free surface. Program I also showed the importance in maintaining crack shape similarity when predicting SIF distributions analytically. Since Mode I is the dominant mode for growing cracks, load orientation was omitted in the present study.

2. *Degree of influence of elastic incompressible response above critical temperature.* Results of Program II suggest that the incompressibility effect is of the order of the experimental scatter but on the conservative side (i.e., experimental SIF's are elevated).

3. *Degree of fatigue enhancement of prototype fatigue data from metals.* In the present study, results from Program I exhibited no fatigue enhancement but results from Program II suggest the presence of it possibly in the form of crack closure (which did occur in the fatigue tests). It appears that such effects are strongest for flaws which are near the free surface of a material (i.e., shallow surface flaws). For deep flaws, or those not near the entry surface, such effects were not observed.

It is clear from these studies that the frozen stress method is subject to a number of limitations if it is to be used to predict crack shapes and SIF distributions resulting from fatigue crack growth in complex three dimensional problems. Moreover, it is likely that, with increased geometric complexity, the limitations may render results somewhat less exact than the sophisticated mathematician might wish. Nevertheless, the frozen stress method is a powerful tool for addressing three dimensional problems in general, and when coupled with LEFM for cracked bodies, it appears to provide substantial useful information which can be used in the mathematical modelling of such problems. The method is currently being extended to include local displacement field determination through a Moiré interferometric analysis on a slice by slice basis [16].

ACKNOWLEDGEMENTS

The author wishes to acknowledge the assistance of his former students for their considerable help in developing the methods described, especially M. I. Jolles, W. H. Peters, J. J. McGowan and G. C. Kirby. Also, the support of the National Science Foundation Solid Mechanics Program under Grant No. MEA-811-3565 is gratefully acknowledged.

REFERENCES

[1] Swedlow, J. L., ed., The Surface Crack: Physical Problems and Computational Solutions, Applied Mechanics Division of ASME Special Publication, 1972.

[2] Rybicki, E. F. and Benzley, S. K., Computational Fracture Mechanics, Pressure Vessels and Piping Division of ASCE, Special Publication, 1975.

[3] Hulbert, L. E., "Benchmark problems for three dimensional fracture analysis", Int. J. Fracture, Vol. 13, pp. 87-91, 1977.

[4] McGowan, J. J., ed., "A critical evaluation of numerical solutions to the 'Benchmark' surface flaw problem", Fracture Committee of SESA Monograph, 1980.

[5] Smith, C. W., "Use of three dimensional photoelasticity and progress in related areas", Experimental Techniques in Fracture Mechanics, Vol. 2, SESA Monograph, A. S. Kobayashi, ed., pp. 3-58, 1975.

[6] Smith, C. W., Peters, W. H. and Andonian, A. T., "Mixed mode stress intensity distributions for part circular surface flaws", Journal of Eng. Fract. Mech., Vol. 13, pp. 615-629, 1979.

[7] Smith, C. W. and Hardrath, W. T., "Photoelastic determination of stress intensities for Mode III", Recent Advances in Engineering Science, Vol. 15, pp. 195-200, 1978.

[8] Smith, C. W. and Hardrath, W. T., "A method for measuring K_3 in three dimensional cracked body problems", Developments in Mechanics, Vol. 10, pp. 225-230, 1979.

[9] Smith, C. W. and Peters, W. H., "Experimental observations of 3D geometric effects in cracked bodies", Developments in Theoretical and Applied Mechanics, Vol. 9, pp. 225-234, 1978.

[10] Smith, C. W. and Peters, W. H., "Prediction of flaw shapes and stress intensity distributions in 3D problems by the frozen stress method", Preprints from Sixth International Congress on Experimental Stress Analysis, pp. 861-865, 1978.

[11] Cotterell, B., Int. J. of Fracture, Vol. 2, No. 3, pp. 526-533, 1966.

[12] Sih, G. C. and Kassir, M. K., "Three dimensional stress distribution around an elliptical crack under arbitrary loadings", J. of Applied Mechanics, Vol. 33, pp. 601-611, 1966.

[13] Broekhoven, M. J. G., "Fatigue and fracture behavior of cracks at nozzle corners: comparison of theoretical predictions with experimental data", Proceedings of the Third International Conference on Pressure Vessel Technology (Part II), Materials and Fabrication, pp. 839-852, April 1977.

[14] Irwin, G. R., "The crack extension force for a part through crack in a plate", J. of Applied Mechanics, Vol. 29, ASME Transactions, Vol. 79, Series E, pp. 651-654, December 1962.

352

[15] Newman, J. C., Jr. and Raju, I. S., "Analyses of surface cracks in finite plates under tension or bending loads", NASA-TP No. 1578, December 1979. See also J. of Engineering Fracture Mechanics, Vol. 15, No. 1-2, pp. 185-192, 1981.

[16] Smith, C. W., Post, D. and Nicoletto, G., "Prediction of sub-critical flaw growth data from model experiments", (in press), Developments in Theoretical and Applied Mechanics, Vol. 11, 1982.

MATERIAL SELECTION AND STRUCTURAL DESIGN FOR SAFETY AND DURABILITY

R. J. H. Wanhill

National Aerospace Laboratory NLR
Amsterdam, The Netherlands

ABSTRACT

Selection of structural materials is considered in a general way and then specifically with respect to fracture mechanics. There follow descriptions of structural design philosophies and planning for fracture control in order to ensure operational safety and durability. The ways in which deciding to plan for fracture control may affect the material selection and structural design procedures are discussed and illustrated with widely different examples.

MATERIAL SELECTION

A. General Considerations

A schematic of the procedure for selection of structural materials is shown in Figure 1. Performance and design requirements lead to a provisional

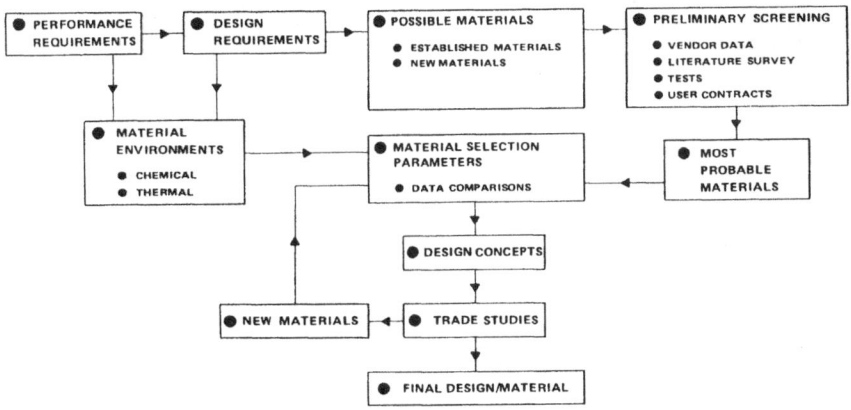

Fig. (1) - Material selection procedure, after [1]

selection of materials, possibly including both established and new materials. Design concepts are then developed and, in the normal way, trade studies result in final choice of design and material. However, a reiteration loop shows the possibility that new materials are considered only after carrying out a "first round" of trade studies.

Consideration of new materials is a significant departure. It implies that past experience with designing and fabricating particular types of structure is no longer adequate. There may be one or two main reasons for this:

(1) Service experience indicates structural or material deficiencies even though current design requirements are met.

(2) Performance and/or design requirements have been radically altered.

The inadequacy of past experience does not necessarily mean that new materials will be used. But at the very least, there must be changes in the relative importance of various parameters used for material selection.

B. Material Selection Parameters

Mechanical properties, aspects of structural application and manufacturing technology, and material and production costs are always involved in selecting candidate materials. Corrosion, with its effects on maintenance and repair costs, is often considered also.

However, and hence the necessity and importance of conferences such as this one, the use of fracture mechanics, both in material selection and structural design, is not universally established and familiar.

C. Applicability of Fracture Mechanics Concepts

In the first instance, the applicability of fracture mechanics to material selection will be determined by the class of most probable materials and whether structural failure is envisaged to be in the yielding-dominant (plastic collapse) or fracture-dominant modes. In turn, these aspects are determined by the types of structure.

Figure 2 exemplifies relations between all these factors assuming, for illustration's sake, that there is a flaw or crack in a load-bearing structural component at the point of *incipient fracture and failure*. Aerospace structures, including airframes, landing gear and parts of engines, are made of high strength, relatively brittle materials. The fracture behaviour is for the most part adequately characterized by Linear Elastic Fracture Mechanics (LEFM), in particular, by the stress intensity factor K. The same is true of low ductility materials such as steel rails and welds at low temperatures.

On the other hand, many important types of structure, including pressure vessels, ships, pipelines and offshore rigs are constructed of materials, which under service conditions, behave in a more ductile manner. LEFM is not applicable but Elastic-Plastic Fracture Mechanics (EPFM) may be. This depends on the *constraint*: in situations of high constraint, e.g., cracks in thick sections, the effective yield stress in the vicinities of crack tips is raised considerably and EPFM can be used to predict fracture behaviour. Otherwise, the fracture behaviour is likely to be controlled by general yielding.

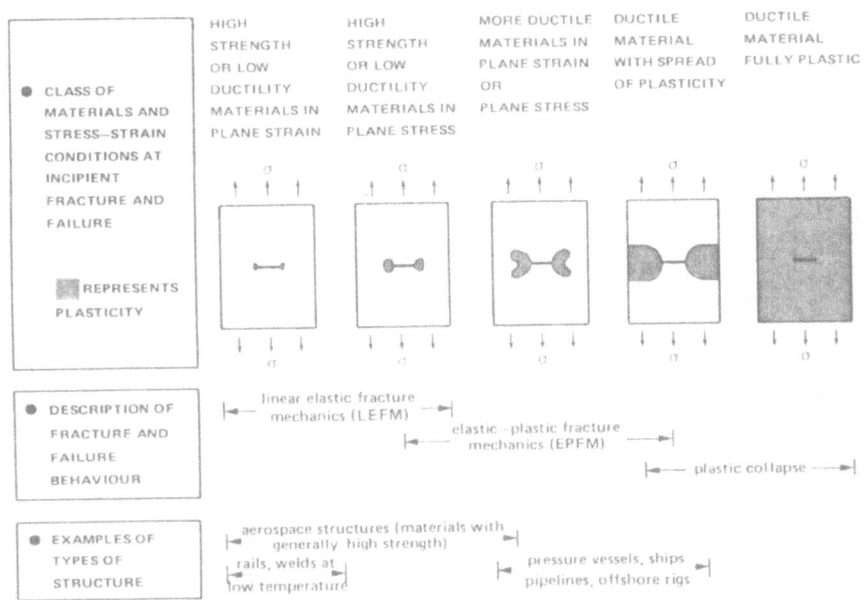

Fig. (2) - Illustration of relations between materials, structures and
fracture and failure

It might be thought that situations of high constraint are rather special cases. In fact, they are of prime importance. In the power generating and
chemical processing industries, most cracks occur in high pressure parts, which
are thick-walled vessels and pipes. Also, the offshore industry must cope with
cracks in very large thick-sectioned welded structures. It is therefore no sur-
prise that most contributions to the development and application of EPFM have come
from these industries.

The characterization of fracture behaviour of more ductile materials by
EPFM is less well-defined than in the case of high strength/low ductility materials with LEFM. Thus, although there are several EPFM concepts for fracture, only
two of these have gained fairly general acceptance: the J integral and Crack
Opening Displacement (COD) approaches.

So far, only fracture leading directly to structural failure has been
considered. However, *subcritical crack growth* may also be relevant. In this
category, the most important is fatigue crack growth, followed (figuratively
speaking) by slow stable tearing, stress corrosion and creep cracking.

Under simple loading conditions, fatigue crack growth in structural materials is generally characterized by the LEFM stress intensity factor range ΔK. Problems arise when the loads are non-stationary (i.e., the load sequence is random and/or not statistically constant with time) and when crack growth is in the EPFM regime. For long cracks, EPFM fatigue crack growth is of little importance since the structure is probably close to failure: in other words, the amount of crack growth life spent under EPFM conditions can probably be neglected in comparison to the total life. At the other extreme, EPFM conditions may be very important to the growth of small fatigue cracks at notches, a problem which is increasingly being recognized as of major significance in structural design. There are several approaches to this problem. One of the most successful is to add a constant to the physical crack size and thereby derive "corrected" ΔK or ΔJ values (depending on whether the notch strains are elastic or elastic-plastic) for correlating the crack growth data [2].

Slow stable tearing in the LEFM regime may be described using the R-curve approach. But treatment of the problem under EPFM conditions is extremely difficult. Currently, the most promising concept appears to be the tearing modulus T [3]. The practical use of this concept is, however, subject to many restrictions. The same is true of so-called J resistance (J-R) curves.

Stress corrosion crack growth is especially associated with high strength in structural materials. The stress intensity factor K can be used to characterize crack growth, at least under well-defined experimental conditions.

Creep crack growth has received much attention over the last few years. Several concepts have been used to try and correlate crack growth rates, but it is evident that none have wide applicability. Correlations with J and the closely related C* integral [4] appear to be more promising for practical applications than other approaches.

D. Use of Fracture Mechanics in Material Selection

Selection of structural materials with the aid of fracture mechanics may be done on three levels of increasing refinement: screening criteria, data comparisons, and design studies. Referring back to Figure 1, these three levels come under the headings of PRELIMINARY SCREENING, MATERIAL SELECTION PARAMETERS and DESIGN CONCEPTS, respectively.

Screening criteria enable a preliminary choice. Their very generality requires that they be applied with caution. At the next level, screened materials may be compared for properties determined by fracture mechanics tests under more or less standard conditions, with the emphasis on those properties considered to be of most importance for the actual structure. At the final level, materials are compared in terms of engineering aspects of cracked structures, Figure 3. This can be a long and complicated process.

The question now arises: which fracture mechanics concepts can be used in material selection and at which levels? Table 1 is an attempt to provide an answer. The choice of concepts is based on the discussion in subsection C. Both this choice and the filling in with ●'s are arguable, especially for EPFM concepts. But this only serves to demonstrate the general observation that use of LEFM concepts is better established.

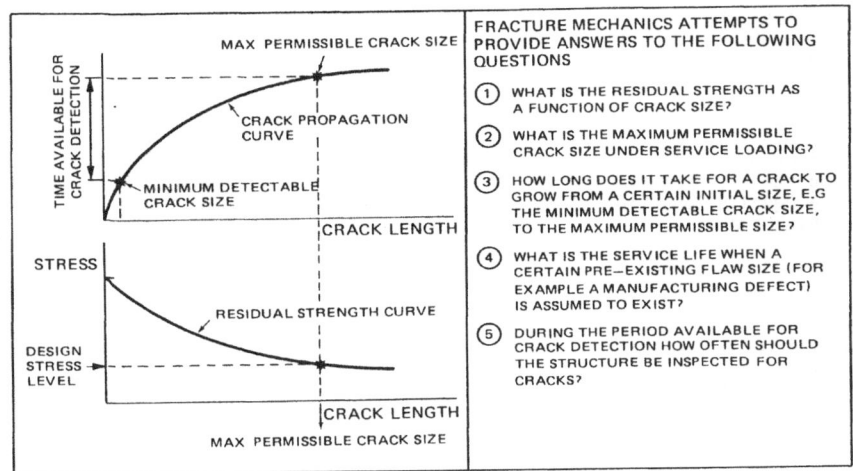

FRACTURE MECHANICS ATTEMPTS TO PROVIDE ANSWERS TO THE FOLLOWING QUESTIONS

1. WHAT IS THE RESIDUAL STRENGTH AS A FUNCTION OF CRACK SIZE?

2. WHAT IS THE MAXIMUM PERMISSIBLE CRACK SIZE UNDER SERVICE LOADING?

3. HOW LONG DOES IT TAKE FOR A CRACK TO GROW FROM A CERTAIN INITIAL SIZE, E.G THE MINIMUM DETECTABLE CRACK SIZE, TO THE MAXIMUM PERMISSIBLE SIZE?

4. WHAT IS THE SERVICE LIFE WHEN A CERTAIN PRE—EXISTING FLAW SIZE (FOR EXAMPLE A MANUFACTURING DEFECT) IS ASSUMED TO EXIST?

5. DURING THE PERIOD AVAILABLE FOR CRACK DETECTION HOW OFTEN SHOULD THE STRUCTURE BE INSPECTED FOR CRACKS?

Fig. (3) - Engineering aspects of cracked structures

TABLE 1 - FRACTURE MECHANICS CONCEPTS FOR MATERIAL SELECTION

OVERALL STRUCTURAL PROBLEM	SPECIFIC ASPECTS OF STRUCTURAL PROBLEMS	CONDITIONS	CONCEPT	PRESENT USEFULNESS		
				SCREENING CRITERIA	DATA COMPARISONS	DESIGN STUDIES
incipient fracture and failure	initiation of crack extension	LEFM	K	● (K_{Ic})	●	●
		EPFM	J COD	● (J_{Ic})	● ●	●
subcritical crack growth	fatigue crack growth	LEFM	ΔK ΔK_{th}		●	● ●
		EPFM	ΔJ			
	slow stable tearing	LEFM	R			●
		EPFM	J-R T			●
	stress corrosion cracking	LEFM	K	● (K_{Iscc})		
	creep crack growth	EPFM	J,C*			

Table 1 shows that use of fracture mechanics concepts as screening criteria for material selection is limited to the initiation of crack extension (note that for stress corrosion cracking, the concept is K_{Iscc}, which is the threshold stress intensity for crack growth).

At the more detailed levels of data comparisons and design studies, there are several possibilities for characterizing incipient fracture and failure. However, characterization of subcritical crack growth, in particular, fatigue, is more the province of LEFM concepts. This provides partial explanation of the aerospace industry's extensive use of fracture mechanics. The other part of the explanation - equally important, if not more so - concerns structural design philosophies, which are the subject of the next section.

STRUCTURAL DESIGN PHILOSOPHIES

A. Safe-Life

The most widespread and initially the only philosophy of structural design is the safe-life approach. This means to design for a finite life during which significant damage (i.e., cracking and/or failure) will not occur. Since it is the intention to avoid cracking altogether, no account is taken of fracture related properties except to use some form of impact notch strength or notch toughness, chiefly as a screening criterion for selection of structural steels and weldments.

B. Fail-Safe

In the 1950's, the aerospace industry-stimulated by the COMET disasters-evolved the fail-safe approach, specifically to take account of fatigue cracking. This approach requires designing for an adequate service life without significant damage but also enabling operation beyond the actual life at which such damage occurs. Operation is thus permitted when the structure is cracked - as might occur prematurely - but it must be shown that cracking will be detected by routine inspection before it propagates to the extent that residual strength falls below a safe level.

There are two very effective ways of designing for fail-safety. One is to provide multiple load paths such that damage can be contained when a load path fails. Alternatively, crack arrestment features, for example, tear straps in airframes and welded-in strips of notch tough material in welded steel ship hulls, can be incorporated. In both cases, safety is to be ensured by allowing partial failure without unsafe reduction in residual strength and by enabling continued operation for a period in which the failure must be found and repaired.

C. Damage Tolerance

In 1969, an F-111 was lost owing to catastrophic fracture in the wing with *only 105 flight hours* accumulated. This accident resulted in development of the damage tolerance approach by the United States Air Force [5].

The damage tolerance philosophy differs from the original fail-safe approach in two major respects:

(1) The possibility of cracks or flaws *already in new structure* must be accounted for.

(2) Structures may be inspectable or non-inspectable in service.

Inspectable structures can be qualified either as fail-safe or as slow crack growth structures, for which initial damage must grow slowly and not reach a size large enough to cause failure between inspections.

Non-inspectable structures must be qualified for slow crack growth, which in this case means that initial damage will not propagate to a size causing failure during the design service life. This is called the safe crack growth requirement. It depends entirely on the establishment of a minimum detectable flaw size for pre-service inspection.

D. Use of Fracture Mechanics in Structural Design

There is a trend to increase application of damage tolerance design to many types of structure. This necessitates providing quantitative answers to some or all of the questions posed in Figure 3. In turn, such answers can only be given or attempted via a thorough knowledge of the fracture mechanics characterizations of incipient fracture and subcritical crack growth.

The aerospace industry and its customers have played major roles in applying the fail-safe and damage tolerance philosophies to structural design. This is the remaining part of the explanation why fracture mechanics is extensively used in the design of aerospace structures.

E. Safety and Durability: Definitions

So far, safety has been discussed without qualification. However, actual design balances performance requirements against economic factors, such that the probability of failure is less than a value deemed to be an acceptable risk. Since all structures deteriorate in service, i.e., the probability of failure increases with time, there is a safety limit.

The safety limit may be defined as the time beyond which the risk of failure is considered to be unacceptable unless preventative actions are taken. For safe-life and non-inspectable slow crack growth (safe crack growth) structures, the necessary preventative action is retirement from service. But for the inspectable fail-safe and slow crack growth structures, retirement is not the only option. In the first instance, preventative action is inspection, followed by repair if it is required and if it is feasible. Only when repair is not feasible or when the frequency of inspection and repair becomes uneconomic need the structure be withdrawn from service.

Consideration of preventative action for ensuring safety leads to a second operational aspect, durability. The necessity for a structure to be durable primarily means that the economic life (including any inspections and repairs) should equal or exceed the design service life.

F. Safety and Durability: Status of Analyses

Whichever structural design philosophy is used, the methods for quantifying the design service lives of structures are fairly well established. This is because intensive effort has gone into introducing the more recent fail-safe and damage tolerance approaches.

Durability analysis techniques to quantify economic life at the design stage are much less developed. Perhaps the most advanced analysis is that concerned with the widespread initiation and growth of small cracks at fastener holes in airframe structures [6]. The upper limit of crack size affecting durability is defined on the basis of economic repair, e.g., the largest radial crack that can be cleaned up by reaming the fastener hole to the next fastener size and installing the appropriate fastener.

FRACTURE CONTROL

A. Planning for Fracture Control

As mentioned earlier, the use of fracture mechanics in material selection and structural design is not universally established and familiar. However, once it is decided, as is increasingly the case, that it is necessary to change the design philosophy from a safe-life to fail-safe or damage tolerance approach, it becomes necessary also to formulate a fracture control plan.

There are four basic parts to a fracture control plan [7]:

(1) Determination of factors that may contribute to fracture of a structural component or to failure of the entire structure, including description of service conditions and loadings.

(2) Establishment of relative contributions of each factor to possible fracture or failure.

(3) Determination of the relative effectiveness of various design methods to minimize the possibility of fracture or failure.

(4) Recommendation of specific design considerations to ensure safety and durability of the structure.

Figure 4 lists various factors and aspects which have to be considered in each part of a comprehensive fracture control plan for a large, complex structure. The list is certainly incomplete, but it is already evident that development of a fracture control plan is a formidable task. Consequently, a detailed treatment is beyond the scope of this paper.

Instead, Figure 4 provides a background to the remaining subsections, where the ways in which deciding to plan for fracture control may affect the material selection and structural design procedures are discussed and illustrated with widely different examples.

B. Fracture Control and Material Selection

A decision to change the design philosophy from a safe-life to a fail-safe or damage tolerance approach and to plan for fracture control can affect several stages in the material selection procedure in the following ways, see also Figure 1:

(1) DESIGN REQUIREMENTS — major alterations.

(2) POSSIBLE MATERIALS — consideration of both established and new materials.

DETERMINATION OF FRACTURE AND FAILURE FACTORS
- SERVICE LOAD SPECTRA AND RATES
- SERVICE ENVIRONMENT
- PROPERTIES RELATED TO SUBCRITICAL CRACK GROWTH AND INCIPIENT FRACTURE AND FAILURE
- FABRICATION QUALITY

RELATIVE CONTRIBUTIONS TO FRACTURE AND FAILURE

SUBCRITICAL CRACK GROWTH
- PRIMARY FACTORS. ENVIRONMENTAL FATIGUE, TENSILE STRESSES
- SECONDARY FACTORS. SLOW STABLE TEARING, STRESS CORROSION, CREEP

INCIPIENT FRACTURE AND/OR FAILURE
- PRIMARY FACTORS: STRENGTH, DUCTILITY, CONSTRAINT, TENSILE STRESSES (FRACTURE) AND TENSILE OR COMPRESSIVE STRESSES (FAILURE)
- SECONDARY FACTORS: TEMPERATURE, LOADING RATE, CORROSION

RELATIVE EFFECTIVENESS OF DESIGN TO MINIMISE FRACTURE AND FAILURE
- MULTIPLE LOAD PATHS: SUBCRITICAL CRACK GROWTH, FRACTURE, FAILURE
- ADEQUATE FRACTURE TOUGHNESS AND CRACK ARRESTERS: FRACTURE
- TENSILE STRESS LEVELS, IMPROVED CORROSION PROTECTION: SUBCRITICAL CRACK GROWTH, FAILURE
- INITIAL FLAW SIZE: FATIGUE
- COMPRESSIVE STRESS LEVELS: FAILURE

RECOMMENDATIONS TO ENSURE SAFETY AND DURABILITY
- PERFORMANCE REQUIREMENTS
- DESIGN REQUIREMENTS
- MATERIAL SELECTION
- DESIGN STRESS LEVELS
- FABRICATION
- INSPECTION

Fig. (4) - Outline of a comprehensive fracture control plan

(3) PRELIMINARY SCREENING - inclusion of fracture mechanics concepts in screening criteria.

(4) MATERIAL SELECTION - inclusion or greater importance of fracture mechanics concepts for data comparisons.

(5) DESIGN CONCEPTS - provision of damage tolerance features and consideration of inspection and repair possibilities.

(6) TRADE STUDIES   - rejection of design concepts in established materials and reiteration to consider new materials.

Table 2 gives some examples of the effects of planning for fracture control on material selection. There is a strong bias to the use of LEFM concepts,

TABLE 2 - EXAMPLES OF EFFECTS OF FRACTURE CONTROL ON MATERIAL SELECTION

TYPE OF STRUCTURE	POSSIBLE MATERIALS CONSIDERED	APPROXIMATE HIERARCHY OF MATERIAL SELECTION PARAMETERS		USE OF FRACTURE MECHANICS				REMARKS
		SAFE-LIFE	DAMAGE TOLERANCE	CONCEPT	PRELIMINARY SCREENING	DATA COMPARISONS	DESIGN STUDIES	
fighter aircraft wings [8-10]	established and new metal alloys	• weight, cost • fatigue strength • static strength	• weight, cost • safe crack growth, fail-safety and fatigue strength • manufacturing technology and structural application • fracture and failure	K_{Ic} K_{Iscc} ΔK K_c	● ●	 ●	 ● ●	combination of new materials, advanced design concepts and manufacturing techniques enables damage tolerance with decreased weight and cost
aircraft gas turbine fan disc [11]	established titanium alloys	• strength/weight • low cycle fatigue strength • creep and rupture strength • cost	• strength/weight • low cycle fatigue strength • slow crack growth • fracture toughness • creep and rupture strength • cost	K_{Ic} ΔK		● ●	● ●	damage tolerance achievable by changing to a lower strength material and thereby incurring a 21% weight increase this is unacceptable
helicopter rotor blades	established metal alloys	• fatigue and static strength • corrosion and erosion resistance • cost and fabricability	• fatigue and static strength • slow crack growth • corrosion and erosion resistance • cost and fabricability	ΔK			●	damage tolerance is hardly achievable with metallic blades hence the trend to use composite blades, which are damage tolerant[12]
offshore and littoral structures	established weldable steels	• cost • mechanical properties, weldability • impact notch strength • fatigue strength • corrosion fatigue strength	• cost • mechanical properties, weldability • fracture toughness • corrosion fatigue crack growth • fatigue strength • corrosion fatigue strength	COD ΔK		● ●	● ●	COD tests and the COD design curve are used to specify weld metal requirements by Det Norske Veritas
train rails [13]	established and new steels	• cost • wear resistance • weldability • corrosion resistance	• cost • wear resistance • dynamic fracture toughness • weldability • corrosion resistance	K_{Ic}, K_{Id}		●		increasing fracture toughness from 30 to 80 MPa√m eliminates current problems however, cost and wear resistance are predominant factors

but the inclusion of COD for offshore structures is notable. In two of the examples, the fan disc and helicopter rotor blades, a damage tolerance approach as described earlier in this paper is unsuccessful. For the fan disc, there is an unacceptable weight increase, and resort has now been made to cryogenic proof testing in spin pits. For metallic rotor blades, the fatigue crack propagation rates from detectable defects are often too high to enable damage tolerance, e.g., [12,14]. Consequently, there is a trend to use composite blades. This provides a good illustration of the reiteration loop in Figure 1.

C. Fracture Control and Structural Design

Historically, most design criteria have been established to prevent failure by either tensile overload or compressive instability. In this conventional approach, a designer starts with given materials and design stress levels and proceeds to detail and proportion structural elements to carry the required static loads without exceeding the allowable stresses. These are usually certain percentages of the yield strength for tension elements, and the buckling stress for compression elements. The presence of stress concentrations and discontinuities is assumed to be accounted for by local yielding and load redistribution, but the structure is not usually considered to contain defects or cracks. If fatigue loading in service is important, this conventional approach is extended by redesigning to meet allowable fatigue stress levels based on the safe-life philosophy.

Introduction of fracture control can result, as Figure 4 indicates, in considerably more effort being required in order to arrive at a final design. More specifically, Figure 5 shows a schematic incorporation of fracture control in the structural design procedure. In this procedure, service loads and environmental conditions are determined first, followed by conventional static and fatigue analyses to obtain a baseline design. The baseline design is then subjected to preliminary fracture control analysis (the same questions as those posed in Figure 3) in order to identify structural areas that may be fracture critical. A detailed assessment of the materials proposed for use in these areas is then carried out. This may involve data from basic fracture mechanics tests; tests on coupons, i.e., relatively simple specimens simulating structural details and expected flaw geometries; and tests on full or reduced scale components. The purpose of this assessment is to obtain fracture and subcritical crack growth allowables pertaining to service conditions. In turn, these allowables are used to develop the fracture control analysis for the baseline design and establish the *safety limit* and inspection intervals.

Not included in Figure 5 is the additional operational aspect of *durability*. This is because durability analysis techniques to quantify economic life at the design stage are less well developed, as stated earlier. The conventional approach to durability is to design for an adequate crack-free fatigue life, as exemplified by both the safe-life and original fail-safe philosophies. In other words, conventional designing for durability occurs at the baseline design stage in Figure 5.

On the other hand, the damage tolerance approach to airframe structures provides a quantitative analysis of durability based on the widespread initiation and growth of small fatigue cracks at fastener holes. This analysis formally takes place at the last stage shown in Figure 5. The general procedure is illustrated in Figure 6. Crack propagation curves for cracks at fastener holes are obtained from visual and fractographic measurements and are extrapolated analytically to initial crack lengths. These fictitious crack lengths are termed Equivalent Initial Flaw Sizes (EIFS). The values and statistical distribution of EIFS define the initial fatigue quality and scatter in fatigue life, and these parameters are used in assessing the economic life of the airframe [6].

SUMMARY

Most engineering structures operate safely and it is not unreasonable to claim that conventional design without using fracture mechanics often is satisfactory. However, the occurrence of infrequent but *totally unexpected* failures, for example, the F-111 wing and the Alexander Kjelland offshore platform, shows that all possible failure modes, including incipient fracture and subcritical crack growth, should be considered. This can be done by changing the design philosophy from a safe-life to fail-safe or damage tolerance approach and by trying to develop comprehensive fracture control plans to ensure operational safety and durability. In this paper, the effects of planning for fracture control on the procedures for material selection and structural design have been discussed and illustrated with widely different examples.

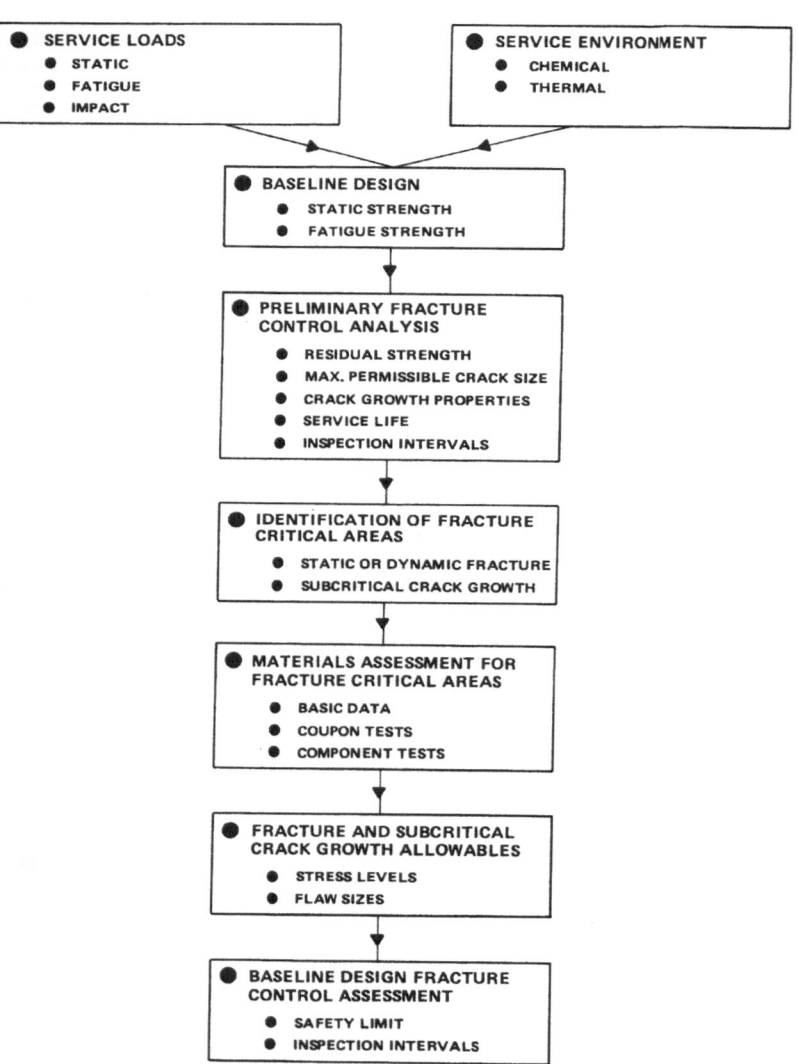

Fig. (5) - Fracture control in the structural design procedure

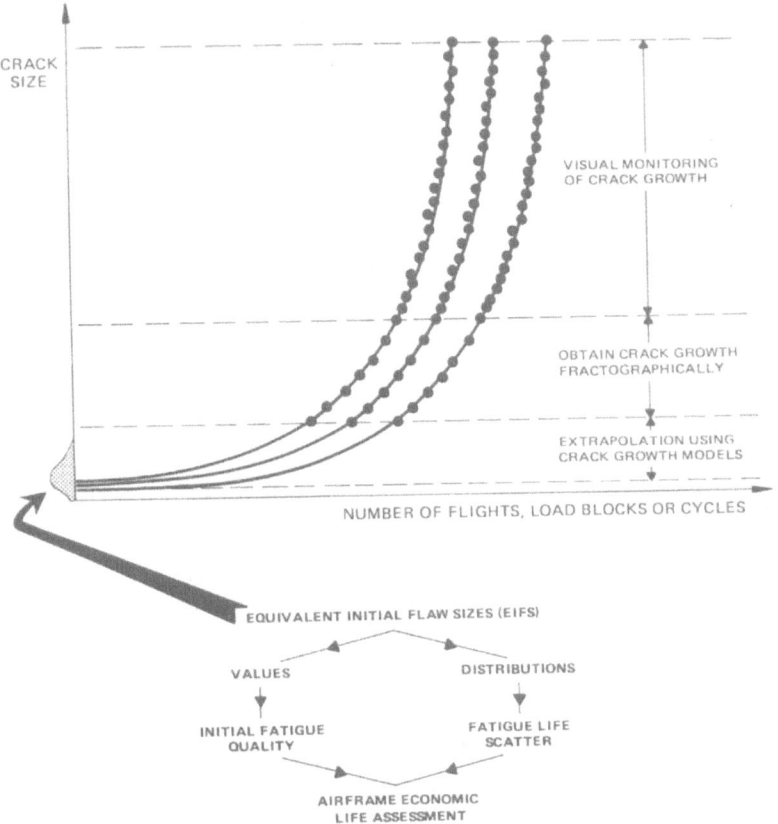

Fig. (6) - Durability analysis for damage tolerant airframe structures

REFERENCES

[1] Shults, J. M., "Material selection procedures for advanced transport air-
     craft", SAE Paper 730884, October 1973.

[2] El Haddad, M. H., Dowling, N. E., Topper, T. N. and Smith, K. N., "J integral
     applications for short fatigue cracks at notches", Int. J. Fract., 16, p.
     15, 1980.

[3] Paris, P. C., Tada, H., Zahoor, A. and Ernst, H. A., "A treatment of the subject of tearing instability", U.S. Nuclear Regulatory Commission Report NUREG-0311, 1977.

[4] Landes, J. D. and Begley, J. A., "A fracture mechanics approach to creep crack growth", Mechanics of Crack Growth, ASTM STP 590, p. 128, 1976.

[5] "Military specification airplane damage tolerance requirements", MIL-A-83444, USAF, July 1974.

[6] Manning, S. D. and Smith, V. D., "Economic life criteria for metallic airframes", AIAA Paper 80-0748, 1980.

[7] Rolfe, S. T. and Barsom, J. M., Fracture and Fatigue Control in Structures, Prentice-Hall, Inc., p. 415, 1977.

[8] Jeans, L. L. and LaRose, R. L., "Application of damage tolerance technology to advanced metallic fighter wing structure", AIAA Paper 74-29, 1974.

[9] Figge, F. A. and Bernhardt, L., "Air superiority fighter wing structure design for improved cost, weight and integrity", AIAA Paper 74-337, 1974.

[10] McAnally, R. W., "Air superiority fighter wing design for cost and weight reduction", AIAA Paper 74-338, 1974.

[11] Meece, C. E. and Spaeth, C. E., "Damage tolerant design: an approach to reduce the life cycle cost of gas turbine engine disks", AIAA/ASME/SAE, Paper 79-1189, 1979.

[12] Grina, K. I., "Helicopter development at Boeing Vertol Company", J. RAe. Soc., 79, p. 401, September 1975.

[13] Cannon, D. F., "A fracture mechanics based assessment of the railway rail failure problem in the U.K.", Case Studies in Fracture Mechanics, Army Materials and Mechanics Research Centre Report AMMRC MS 77-5, Section 4.1, June 1977.

[14] Wanhill, R. J. H., de Graaf, E. A. B. and Delil, A. A. M., "Significance of a rotor blade failure for fleet operation, inspection, maintenance, design and certification", Fifth European Rotorcraft and Powered Lift Aircraft Forum, Paper 38, 1979.

SECTION V
MIXED MODE FRACTURE

STUDY OF THE FRACTURE BEHAVIOUR OF SLAB MILL ROLLS

A. S. Blicblau, C. G. Chipperfield and R. Groenhout

Melbourne Research Laboratories
Clayton, 3168, Victoria, Australia

ABSTRACT

An outline is given of the essential features of the fatigue and fracture behaviour of slab mill rolls. A roll model is proposed which describes the development of crack growth utilizing concepts of fracture mechanics, slab mill operating
data, and allowable safety margins. The results of the model provide guidelines
for the establishment of maintenance intervals and optimum rolling schedules,
whilst allowing the sensitivity of service performance to the roll fatigue and
fracture properties to be determined.

INTRODUCTION

The utilization of rolls for reducing ingots to slabs is of paramount importance to the subsequent manufacture of steel components. Because of the wide use
and importance of rolls, improvements continue to be sought in roll materials,
particularly in terms of their fracture properties and manufacturing processes,
as well as roll handling and maintenance procedures and rolling techniques. In
this regard, the scope of the present study aimed at establishing a model of roll
life which would provide guidelines for the determination of maintenance intervals
and optimum rolling schedules, and assess the sensitivity of service performance
to the roll fatigue and fracture properties.

During the service life of a roll, its surface wears and degenerates, requiring machining or dressing procedures to restore the roll surface to its original
condition. For rolls employed in the early stages of metal processing where the
criteria for the rolled product are not as stringent as for later stages of rolling, the dressing operation is still necessary to remove surface cracks and defects. These cracks, caused by thermal and mechanical stresses, will lead to catastrophic failure of the roll if allowed to grow beyond definable limits in the
roll body whilst reducing ingots.

For the rolling process, a balance is required between the loss in production
stemming from relatively frequent dressings and the increased risk of roll breakage
during extended rolling schedules. To enable rational decisions on optimum rolling schedules to be made, precise knowledge must be available of the stresses occurring in the rolls, and of the fracture behaviour of the rolls under these imposed stresses. To prevent roll failures, remedial action is required to be taken

whilst the fatigue cracks are still of a sufficiently small size that they are
unlikely to cause rapid roll fracture.

Whilst the steel composition and physical properties of the roll are specified
in accordance with the relevant standards, there is a paucity of information concerning the fracture characteristics of the roll steel. The fracture of a roll
is manifested in two distinct ways.

The first, being the appearance of surface "fire-cracks" over that section
of the roll which comes in contact with the ingot and influenced by the thermal
stresses [1], and the second, being the propagation of a crack into the roll body
under the action of mechanical stresses [2], which results in through-the-roll
fracture.

The study detailed in the following sections is aimed at establishing fundamental data on the fracture properties of roll steels, whilst incorporating fatigue fracture characteristics of the roll obtained from the operation of a uni-
versal slabbing mill. It addresses the requirements of specifying maintenance
periods based on crack growths in the roll, thus minimizing the changeout periods
for roll dressing with concomitant savings in maintenance expenditure and roll
costs.

The thermal and mechanical stresses acting on the roll have been analyzed by
Groenhout and Blicblau [3] in a parallel study. The combination of these stresses
with fracture mechanics concepts resulted in the development of a roll life model
based on "fail-safe" design criteria.

THE PROCESS OF ROLL FRACTURE

The failure of a mill roll occurs by propagation of a part circumferential
crack through the roll body. Fracture occurs perpendicular to the roll body axis.
The major stresses acting on the roll are a result of the moments at each end,
the tensile bending stresses resulting from a slab passing through the roll gap,
and the thermal stresses on the surface of the rolls. Crack growth occurs under
a combination of these stresses, shown schematically in Figure 1.

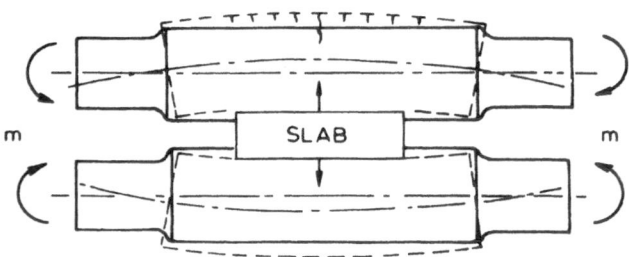

Fig. (1) - The development of cracks under the influence of roll bending
is schematically illustrated in this exaggerated drawing

A characteristic feature of the rolls, whether they break prematurely or are used to their scrap size, is the presence of surface "fire-cracks". The alternating thermal stresses caused by expansion and contraction of the roll surface produce a multiplicity of cracks associated with the initial surface knurling pattern which apparently act as nucleation sites for subsequent crack growth into the roll body. As the crack proceeds into the roll body the thermal effects diminish. For those cracks which have penetrated sufficiently, the influence of mechanical stresses then become dominant. Subsequently, failure of the roll occurs by propagation of one of these cracks through the roll body [4].

For the present analysis, under the influence of mechanical stresses, it is assumed that the roll material is elastic during crack propagation. Consequently, linear elastic fracture mechanics concepts may be applied to the roll fracture situation. Of the three modes of fracture [5] which characterize the fracture process, the Mode I type most closely resembles the roll loading situation, and is employed throughout the ensuing fracture analysis.

CHOICE OF STRESS INTENSITY FACTOR EXPRESSION

A. Initial Fracture Process (Thermal)

A thermal stress intensity factor can be determined as a function of the thermal stresses in a similar manner to the mechanical stress intensity factor. Groenhout and Blicblau [3] established the variation of thermal stresses with roll depth with the utilization of finite element methods. These computed stresses are then used to obtain a stress intensity factor for a surface breaking flaw of depth a, according to Stallybrass [6] as:

$$K_{ITL} = [0.7930A_0 + 0.4829A_1a + 0.3716A_2a^2 + 0.3118A_3a^3$$

$$+ 0.2735A_4a^4 + 0.2464A_5a^5]\sqrt{\pi a} \tag{1}$$

where K_{ITL} - thermal stress intensity factor,

a - maximum crack depth

A_0-A_5 - thermal stress coefficients

Assuming that only positive values of K_{ITL} are effective in promoting crack growth, the thermal stress intensity factor from equation (1) may be interpreted as a ΔK value (i.e., range of thermal stress intensity factor) controlling thermal fatigue growth. Taking a value of ΔK as 8 MPa $\cdot$ m$^{1/2}$ as a relevant threshold value for fatigue crack growth [7], under the influence of thermal stresses, the nucleation of thermal cracks at a multiple level is predicted since:

(i) stress intensity levels K_{ITL} - associated with the knurling pattern depths are greater than the threshold stress intensity factor K_{TH}, as illustrated in Figure 2, and

(ii) stress intensity levels do not reduce to threshold levels until the crack depth approaches 3 millimetres (the knurling pattern is 1.5 millimetres in depth).

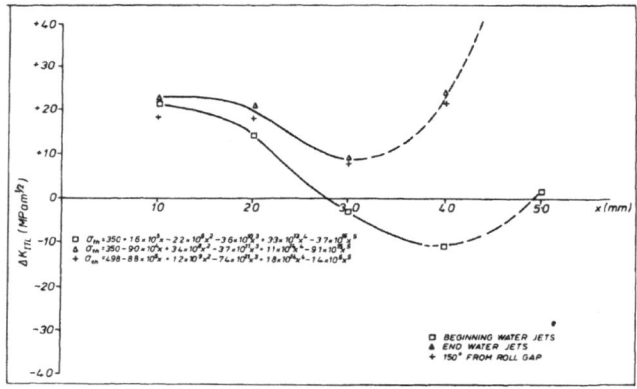

Fig. (2) - Variation of ΔK thermal with roll depth

B. Mechanical Loading

There are a number of stress intensity factor analyses which are approximately relevant to the situation of a slab mill roll containing a surface elliptical flaw. Some of these analyses are more rigorous being determined experimentally [8], or by finite elements [9], whilst others are approximate [10]. The expression advocated is that due to Blackburn [9], being determined by finite element methods for a segment-type edge flaw in a solid cylindrical component, viz:

$$K_I = 0.851 \, \sigma_{max} \sqrt{\pi a} \quad \text{for } a/D < 0.15 \tag{2}$$

where K_I - opening mode stress intensity factor

    σ_{max} - maximum stress at the surface

    a   - maximum crack depth

    D   - roll diameter

FATIGUE FRACTURE

The propagation of a crack under the effect of cyclic loading may be described by the expression:

$$\frac{da}{dN} = C(\Delta K)^m \tag{3}$$

where da/dN is the cyclic crack growth rate, ΔK is the range of the stress intensity factor, and C and m are empirically derived constants [11]. On a log-log plot, the general behaviour of the crack growth rate is sigmoidal in nature, with the majority of growth following the behaviour of equation (3), except for the initiation and rapid growth or incipient fracture regimes of the propagation curve.

Extrapolation of crack growth according to equation (3) into the rapid growth region leads to a non-conservative estimate of growth in that region. Employment of the above equation for an overall description of crack growth behaviour gives an approximate but pertinent description of the life of a crack.

MODELLING ROLL MILL FATIGUE BEHAVIOUR

The principal requirement of fatigue testing is the determination of the expected life of a component. There have been a variety of life models developed; some employing historical data [12], others utilizing the Palmgren-Miner rules [13], whilst a number were based on fracture mechanics principles [14].

The development of the present fatigue life model is based on fracture mechanics principles and utilizes crack growth rates derived from experimental data in a cyclic load programme together with knowledge of the loads acting on the rolls, obtained from the BHP Steelworks.

The first stage of the model determines the cyclic life based on one maximum load acting on the roll, analogous to a constant amplitude fatigue test. However, during a slab rolling schedule, the amplitude and frequency of loads vary greatly, and depend largely on the draft schedule, the slab geometry and material, and the rolling temperatures.

The second stage of the model incorporates the distribution of the loads. The number of cycles experienced by the roll as a function of crack size can then be expressed as [15]:

$$N = \frac{1}{\psi} \cdot [a_o^{\frac{2-m}{2}} - a_i^{\frac{2-m}{2}}] \tag{4}$$

where N   - roll life (cycles)

   ψ - includes the effect of the statistical frequency distribution of stresses, geometric correction factor and the empirical growth law coefficient

   a_o, a_i - final and initial crack depths respectively

   m   - growth law index of equation (3)

EXPERIMENTAL PROGRAMME

   A. Materials

The material examined was taken from a representative steel roll obtained from the Steelworks. The chemical composition of the steel is shown in Table 1, whilst its mechanical properties are indicated in Table 2.

Chemical analysis of the steel, Table 1, indicates that it is a high chromium, manganese steel with over one percent carbon. The accompanying mechanical properties are consistent with this grade of steel. Tensile tests carried out on

TABLE 1 - CHEMICAL COMPOSITION OF ROLL STEEL

Sample:	RX monocast steel						
Composition:	(weight percent)						

C	P	Mn	Si	S	Ni	Cr	Mo
1.25	0.04	1.13	0.39	0.057	0.3	1.15	0.37

TABLE 2 - MECHANICAL PROPERTIES OF THE ROLL STEEL

0.2% Proof Stress MPa	Tensile Strength MPa	Elongation in $5.65\sqrt{S_o}$ %	Reduction of Area %	Hardness Shore C
350	750	6	6	29-31

*S_o - cross sectional area.

similar samples of the steel indicated that roll steel tensile properties vary significantly with depth [16]. Consequently, because of the varied nature of the specimens, the results obtained are taken to be average properties as shown in Table 2.

B. Specimens and Testing

The fracture toughness of the roll steel, K_{IC}, was determined according to BS5547.1977 [17] using a single edge notch specimen subjected to a load in bending.

Compact tension (CT) specimens described by both the British Standards Institution [17] and ASTM [18] were employed for the determination of the fatigue behaviour of the roll steel. Although there are no standard procedures currently available to determine fatigue propagation resistance, a tentative test method recommended by Hudak et al [19a] was adopted (currently the basis of a tentative ASTM Standard [19b]).

The CT specimens were subjected to a sinusoidal tensile loading regime in a Tinius-Olsen servo-controlled hydraulic test machine as illustrated in Figure 3. A fatigue crack starter notch was initiated at the tip of the chevron notch of each sample to a maximum depth of 2.5 mm, whilst experiencing a load range of 1-19 kN at a frequency of 10 Hz. The specimens were subsequently loaded in fatigue using a load range of 1-15 kN at a frequency of 10 Hz. The crack extension was visually measured with a travelling microscope at a X30 magnification to an accuracy of ±0.01 mm. The crack length "a" was measured at a predetermined interval, usually every 10000 cycles. As the crack propagation proceeded and failure became imminent, the interval between crack measurements was reduced.

C. Synthesis of Data

The stress intensity factor for the single edge notched specimen was determined as [17]:

Fig. (3) - Illustration of experimental set-up for fatigue testing showing CT specimen and occular measuring system

$$K_I = \frac{3PL}{BW^{3/2}} \cdot Y_1 \tag{5}$$

where

$$Y_1 = 1.93\left(\frac{a}{W}\right)^{1/2} - 3.07\left(\frac{a}{W}\right)^{3/2} + 14.53\left(\frac{a}{W}\right)^{5/2} - 25.11\left(\frac{a}{W}\right)^{7/2} + 25.80\left(\frac{a}{W}\right)^{9/2}$$

and in a similar manner the stress intensity factor for the CT fatigue specimen was determined as [17]:

$$K_I = \frac{P}{BW^{1/2}} \cdot Y_2 \tag{6}$$

where

$$Y_2 = 29.6\left(\frac{a}{W}\right)^{1/2} - 185.5\left(\frac{a}{W}\right)^{3/2} + 655.7\left(\frac{a}{W}\right)^{5/2} - 1017\left(\frac{a}{W}\right)^{7/2} + 638.9\left(\frac{a}{W}\right)^{9/2}$$

When subjecting the CT specimens to cyclic loads, the alternating stress intensity factor (stress intensity factor range) was expressed as:

$$\Delta K = \frac{\Delta P}{BW^{1/2}} \cdot Y_2 \tag{7}$$

with ΔK - alternating stress intensity factor

ΔP - maximum load minus minimum load in one cycle

P - applied load

W - specimen width

L=2W - half loading span, equation (5)

B - specimen thickness

Y_1Y_2 - geometric correction factors in equations (5), (6) and (7)

a - crack depth

Although the British and U.S. Standards [17,18] recommend that equation (6) be applied to specimens with a/W ratios of between 0.45 and 0.55, Duggan [20] has shown that the errors associated with using equation (6) for a/W ratios of between 0.3 and 0.7 were minimal, thus allowing application of this equation to the present study for which the a/W ranged from 0.41 to 0.71.

The crack growth rate, da/dN, and alternating stress intensity factor, ΔK, were derived from data for the crack growth and elapsed cycles by the application of an incremental polynomial technique fully described by Clark et al [21]. Utilization of this technique requires a second order polynomial to be fitted to groups of seven successive data points of "a" versus "N", commencing with the first pair. The gradient of the parabola at its central point (at crack length, a_i, number of

cycles, N_i) was deemed to be the crack growth rate $(da/dN)_i$. The value of ΔK corresponding to this growth rate was evaluated using an in-house developed computer program for fatigue fracture analysis. The same procedure was subsequently applied to all succeeding data pairs for evaluation of growth rates and alternating stress intensity factors at various stages of the cyclic load programme. Because of the nature of the data reduction, the last three data points were not considered in the analysis.

CRACK GROWTH BEHAVIOUR

Typical crack growth versus elapsed cycles behaviour is shown in Figure 4. The curves show differing behaviour for test pieces from nominally similar material. Although in any experimental programme systematic and random errors are ever present, in this instance they would not be sufficient to cause 100 percent variation in cyclic life. The variation in life is attributed mainly to the variations in material, and to a lesser extent, experimental errors.

Curves of fatigue crack growth rate and corresponding stress intensity factors for the roll samples are shown in Figure 5, where, with the exception of some of the initial and final data points, there exists a relationship between da/dN and ΔK of the form of equation (3).

A linear regression analysis was employed to fit the growth data to equation (3). The results of the analysis are shown in Table 3, where the high index of determination indicates that the fit of the regression lines to the data points is very good.

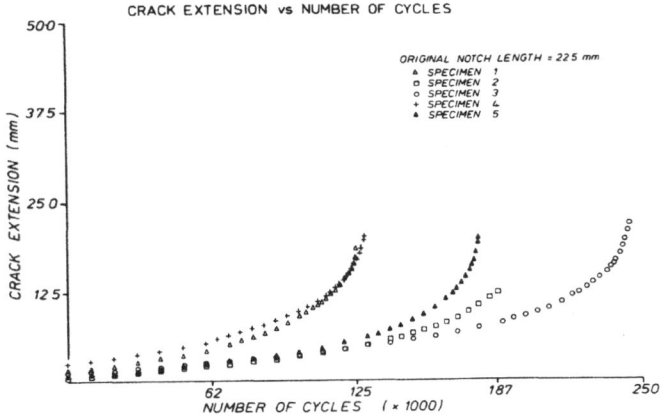

Fig. (4) - Variation of crack extension with elapsed cycles

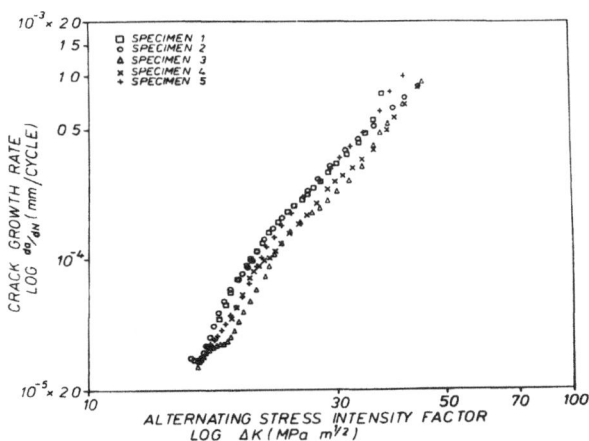

Fig. (5) - Crack growth rate variation with alternating stress intensity factor

From Table 3, the value of the index m for sample V was higher than the other samples. One method for accepting or rejecting outlier data is that due to Dixon [22]. The results of applying this test show that values of m for sample V may be rejected with 98% confidence. The results presented in the following sections are for all samples except V.

378

TABLE 3 - RESULTS OF LINEAR REGRESSION ANALYSIS FOR DETERMINATION
OF LINE OF BEST FIT

Sample	Index of Determination	Coefficient $C \times 10^{-11}$	Index m
I	0.99	1.3	2.93
II	0.99	1.2	2.95
III	0.99	0.99	2.98
IV	0.99	1.2	2.98
V	0.99	0.33	3.32

To incorporate the scatter of results from the crack growth curves, 95 percent
confidence limits of the experimental data points were established and additional
linear regression analyses were applied to these points to obtain upper and lower
limits as shown in Figure 6.

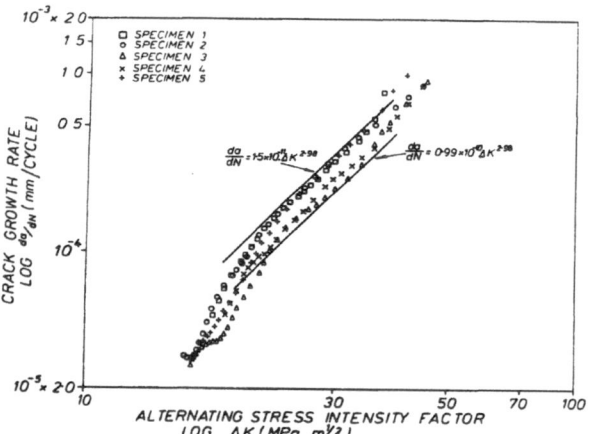

Fig. (6) - Confidence limits for crack growth data

The largest values of the index m and coefficient C, over the interval of constant crack growth for which equation (3) is valid, results in a curve which lies
within the spread of results and encompasses at least 95 percent of all growth
situations. In a similar manner, Broek [23] recommended that as a result of the
large spread of data obtained from fatigue growth tests, scatter bands are required
to obtain a necessarily conservative limit to the crack growth equation: an analogous procedure to that adopted in the current work.

The advocated growth equation was thus defined as:

$$\frac{da}{dN} = 1.5 \times 10^{-11} \; (\Delta K)^{2.98} \tag{8}$$

where

$\frac{da}{dN}$ is m/cycle

and

ΔK is MPa $\cdot \sqrt{m}$

which may be considered as being representative of the fatigue behaviour of the roll steel.

The rapid increase in growth rate towards the end of a component life is typical of fatigue failures has been observed by other workers for similar grades of steel, e.g., [24]. At this stage of crack growth, crack arrest procedures are difficult to implement. Detailed knowledge of the growth behaviour before the onset of rapid fracture allows effective safety measures to be implemented both in the design of the roll and specification of maintenance procedures.

QUANTIFICATION OF ROLL LIFE

Visual inspection of broken rolls together with operational experience have shown that the limiting defect size before the onset of rapid crack propagation is approximately 100 mm in depth, as shown, for example on the fractured roll surfaces in Figure 7, but may be as low as 50 mm in depth. The incidence of cracks in rolls has been recorded by personnel at the Steelworks [25].

There exists a large scatter in results obtained on crack growth data, consequently a crack growth rate was difficult to determine, and required an experimental and analytical programme; as detailed in the present study.

Operational experience in the Steelworks provided the load frequency histogram shown in Figure 8. The stresses resulting from the loads, as determined by finite element methods [3], are displayed in Table 4. The critical defect sizes corresponding to these calculated stress levels, are additionally listed in Table 4. The value of the critical stress intensity factor, K_{IC}, was obtained at 50°C (the average internal temperature of the roll during service), as 45 MPa $\cdot \sqrt{m}$ [25].

Inspection of Table 4 indicates that realistic defect sizes are predicted for the majority of stresses occurring. These defect sizes correspond well to those found in the rolls at the Steelworks [25].

The principle of the roll life model is that of "fail-safe" design [26] frequently utilized in the aircraft industry, i.e., the roll will continue to function effectively in the presence of defects and cracks until the latter reach a specified size, and rectifying action is taken.

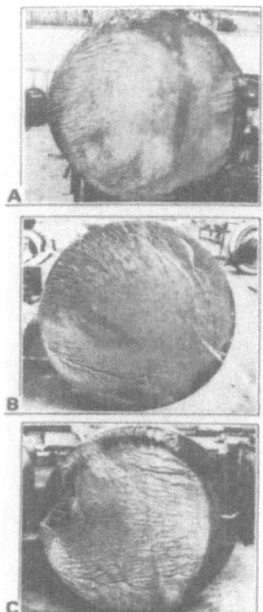

Fig. (7) - Illustrative examples of slab mill roll surfaces of rolls
which failed due to fatigue fracture

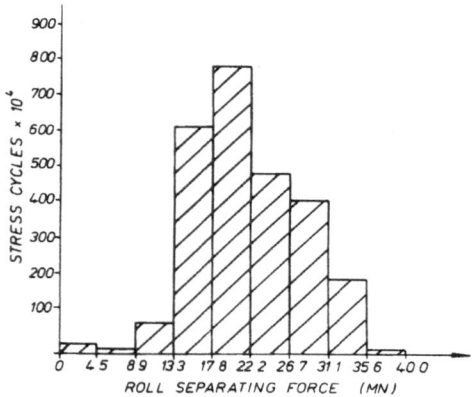

Fig. (8) - Histogram of load frequency

TABLE 4 - CRITICAL DEFECT SIZES AS DETERMINED BY THE ROLL LIFE MODEL

Roll Separating Force (MN)	Finite Element Stresses on Roll Face σ_{max} (MPa)	Critical Defect Size (mm)	Frequency of Occurrence of Force (percent)
2.25	7.3	16695	1.1
6.7	21.8	1880	0.4
11	35.8	698	3.1
15	48.8	375	23.7
20	65.0	211	29.8
24	78.0	147	18.6
29	94.3	100	15.8
33	107.3	78	7.5
38	123.3	58	0.02
40	130.1	53	0.0001

*Calculated defect sizes are for rolling a 1000 mm slab, at a roll temperature of 50°C and K_{IC} = 45 MPa√m.

The preliminary stage of the roll life model assumes that the rate of crack propagation is a function of a constant stress range. Integration of equation (3) for a crack growing from initial size a_i, to final depth a_o, for a range of cycles ΔN, results in:

$$\Delta N = (\frac{1}{\beta}) \cdot {}^{(\frac{2}{2-m})} [a_o^{\frac{2-m}{2}} - a_i^{\frac{2-m}{2}}] \qquad (9)$$

with $\beta = C[Y_2 \Delta\sigma\sqrt{\pi}]^m$

and ΔN = number of cycles if the roll starts from rest.

However, the application of a single value of stress ignores the actual loading situation where a range of stresses act on the roll. The second stage in the development of a roll life model includes the effect of the spectrum of stresses acting on the roll, as detailed below.

For a particular stress of magnitude σ_j, occurring with relative frequency P_j, the growth rate is $(da/dN)_j$. The overall growth rate (assuming discrete stress levels) is expressed as:

$$\frac{da}{dN} = \Sigma(\frac{da}{dN})_j P_j \qquad (11)$$

and substituting equation (3) for $(da/dN)_j$, results in

$$\frac{da}{dN} = C[Y_2\sqrt{\pi a}]^m \cdot \Sigma\Delta\sigma_j^m P_j \tag{12}$$

Integrating equation (12) from the initial crack depth a_i, to the final crack depth a_o, determines the total number of cycles, assuming the roll enters service from rest, as

$$N_\ell = \left(\frac{1}{\psi}\right) \cdot {}^{\left(\frac{2}{2-m}\right)} [a_o^{\frac{2-m}{2}} - a_i^{\frac{2-m}{2}}] \tag{13}$$

with $\psi = C(Y_2\sqrt{\pi})^m \cdot \Sigma\Delta\sigma_j^m P_j$.

When a roll is removed from the mill for maintenance, the surface is ground away until all defects are reduced to a depth of less than 12 mm; the roll is then returned to operation in the mill. The initial depth of a crack in the resulting analysis is thus taken to be 12 mm.

When deriving the life model, many unknowns in the rolling operation are not necessarily taken into account, and may have deleterious effects on the results: a factor of safety on the life cycles is required. Broek and Smith [27] showed that safety factors on loads, stresses, or crack growth data may make some predictions of life more conservative than others; thus, the safety factor should be applied to the crack growth curve itself by dividing the number of cycles for growth of a crack to its final size by a constant factor.

Operational data concerning the mills provided by the Steelworks, indicated that over a 12 month period the maximum and minimum amounts rolled by any roll before being scrapped were 770,000 and 210,000 tonnes respectively. Implications from this data are that safety factor (SF) of 3.7 be applied to the predicted roll life to allow for product variations through the mill, effects of loading sequence, etc.

The growth of an initial defect of depth 12 mm under the influence of the spectrum of stresses is shown in Figure 9, which indicates the theoretical number of cycles for a crack to grow from an initial defect size to an upper defect depth. The maximum load acting on the rolls is 40 MN [25].

The advocated expression for roll life (where approximately 1.2 tonnes of product are rolled for each roll revolution) may be given as:

$$T_{RL} = 1.2 \frac{1}{SF} N_\ell = 168,000 \text{ tonnes} \tag{14}$$

under the influence of a spectrum of stresses with a maximum roll load of 40 MN.

These estimates correspond well with the current guidelines for maintenance, and where necessary, replacement of rolls after approximately 150,000 tonnes of product rolled, as employed by the Steelworks.

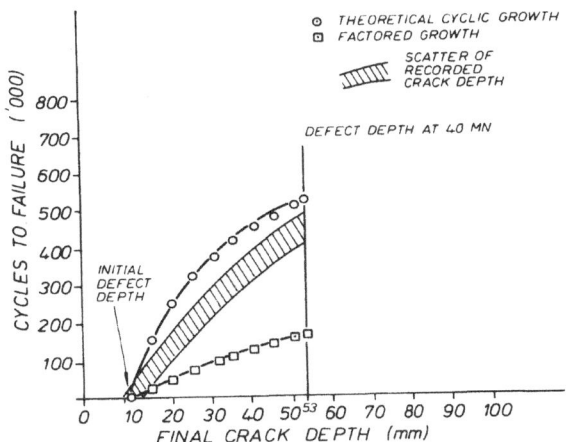

Fig. (9) - Crack growth versus elapsed cycles for spectrum stress conditions
with a maximum force of 40 MN

The usefulness of equation (14) is seen where facilities for monitoring the
number of roll revolutions do not exist and only information concerning the number
of tonnes rolled is available. It should further be emphasized that the conver-
sion factor of 1.2 incorporated in equation (14) is an average value obtained from
an analysis of a 12 month series of roll operations. During this period, a wide
variety of product was rolled under differing operating condition. The versatil-
ity of the roll life expression, equation (14), is such that it may be easily
amended to allow for changes in operating conditions and product material rolled,
hence allowing new roll lives to be established.

The crack growth curve as determined by equation (14) divided by the safety
factor, corresponds favourably with crack data obtained in service conditions as
shown in Figure 9, and, is necessarily conservative. However, it is important to
realize that the roll life model employs general operating data, statistical load
analysis, and experimental fatigue data in its development of life prediction:
there may exist situations when the combination of roll load and operating conditions are not within the bounds of the model, resulting in premature roll fracture.

The emphasis in the development of the roll model has not been on the prevention of crack initiation and propagation, but rather on producing a model which
allows a crack to propagate slowly at a known growth rate whilst satisfactorily reducing ingots to slabs. This technique of "fail-safe" design, which has previously
been widely adopted in the aircraft industry, is now spreading and being applied
to the design of critically dependent components in other industries [28].

The life estimation and prediction techniques described above are adaptable
to the establishment of suitable non-destructive inspection procedures and main-

tenance intervals: which may lead to a lengthening of roll life and concomitant reduction in operating and maintenance costs. Potential usefulness of the approach also extends to assessing the implications of modifying either or both of roll and roll housing geometries, should such modifications be made.

CONCLUSIONS

The significance of thermal and mechanical stresses on the behaviour of slab mill rolls has been assessed by utilization of techniques of linear elastic fracture mechanics. Finite element methods of stress analysis were employed to determine the thermal and bending (tensile) stresses occurring in the roll. Incorporation of experimental fatigue data, measured operational loads and calculated stresses into a fatigue fracture model, gave rise to a life determination model shown in equation (9).

The main conclusions arising from this study were:

1. For safe roll operation, the mill rolls must be damage tolerant. The roll model predicts a planned roll life such that the crack will not grow to critical size before redressing.

2. The approach adopted was shown to be capable of explaining the principal phases of radial cracking in rolls during service; namely the near surface fire cracking and the subsequent fatigue arising from primary mechanical loading.

3. It was shown that the model proposed, adequately predicted the service life of the roll, as evidenced by the agreement of the fracture mechanics theory and service experience.

This agreement strongly suggests that the approach can be used to:

4. Determine the life of the rolls employing finite element methods of stress analysis. It is applicable and adaptable to other operating conditions and rolling procedures to determine the resulting stress and hence predict the fatigue behaviour under the new conditions.

It is of a sufficiently general nature to allow for variations in roll geometry, product properties and operating conditions and yet allow realistic determinations of roll lives.

5. The implications of maintenance procedures are such that regular inspection intervals and changeout periods may be established from knowledge of the crack growth rate as a function of the roll cycle life or number of tonnes of rolled product.

6. It is theoretically possible to establish an economically favourable maintenance procedure for the rolls based on the results of this study. An increase in the maintenance period between roll dressings may be associated with reducing mill downtime whilst improving the productivity.

ACKNOWLEDGEMENTS

The authors wish to acknowledge the assistance of Mr. R. Cornish with the mechanical testing and Mr. D. Barnett AIS, Pt. Kembla, for many useful discussions and helpful suggestions.

This paper is published by the kind permission of The Broken Hill Proprietary Company, Ltd.

REFERENCES

[1] Dugan, J. M., "A study of microstructure of rolls for blooming and slabbing mill applications", Iron and Steel Engineer, p. 93, March 1966.

[2] Hamill, N. C., McLean, L. C. and Ghobarah, A. A., "Why did that roll break?", Iron and Steel Engineer, p. 51, June 1974.

[3] Groenhout, R. and Blicblau, A. S., "Modelling axi-symmetric bodies with applications", in Proceedings of the Fourth International Conference in Australia on Finite Element Methods", Melbourne, August 1982.

[4] Boissenot, J.-M., Dubois, M., Janot, A., Lachat, J-C., Lahlou, N., Marandet, B., Plez, J.-C., Ratte, P. and Sanz, G., "Etude quantitative de la rupture de cylindres de laminoirs a table lisse", Revue de Metallurgie, p. 333, 1977.

[5] Knott, J. F., Fundamentals of Fracture Mechanics, Butterworths, London, 1973.

[6] Stallybrass, M. P., "A crack perpendicular to an elastic half-plane", Int. J. Eng. Sci., 8, p. 351, 1970.

[7] Chipperfield, C. G., Skinner, D. H. and Marich, S., "Influence of inclusions on the fatigue of rail steels", IISI Annual Conference, Sydney, March 1981.

[8] Bush, A. J., "Experimentally determined stress-intensity factors for single-edge-crack, round bars loaded in bending", Experimental Mechanics, 16, p. 249, 1976.

[9] Blackburn, W. S., "Calculation of stress intensity factors for straight cracks in grooved and ungrooved shafts", Engineering Fracture Mechanics, 8, p. 731, 1976.

[10] Harris, D. O., "Stress intensity factors for hollow circumferentially notched round bars", J. of Basic Engineering, Transactions of the ASME, p. 49, 1967.

[11] Paris, P. C., "The fracture mechanics approach to fatigue, fatigue - an interdisciplinary approach", in Proceedings of the 10th Sagamore Army Materials Research Conference, Syracuse University Press, J. J. Burke and V. Weiss, eds., pp. 107-132, 1964.

[12] Frost, N. E., Marsh, K. J. and Pook, L., Metal Fatigue, Clarendon Press, Oxford, 1974.

[13] Hashin, Z., "A reinterpretation of the Palmgren-Miner rule for fatigue life prediction", Journal of Applied Mechanics, Transactions of the ASME, 47, p. 324, 1980.

[14] Coffin, M. D. and Tiffany, C. F., "New Air Force requirements for structural safety, durability, and life management", Journal of Aircraft, 13, p. 93, 1976.

[15] Clark, W. G., Jr., "Fracture mechanics in fatigue", Experimental Mechanics, 11, p. 421, 1971.

[16] Suzuki, K., Takahashi, K., Nishi, T., Kohira, H. and Hori, M., "Mechanical properties of slabbing mill roll materials at room and elevated temperatures", Transactions of the Iron and Steel Institute of Japan, 16, p. 106, 1976.

[17] British Standards Institution, "Methods of test for plane strain fracture toughness (K_{Ic}) of Metallic Materials", BS5447, 1977.

[18] American Society for Testing and Materials, Standards Methods of Tests for Plane Strain Fracture Toughness of Metallic Materials, E399.78a, Philadelphia, Pa.

[19a] Hudak, S. J., Jr., Saxena, A., Buccis, R. J. and Malcolm, R. C., "Development of standard methods of testing and analyzing fatigue crack growth rate data", Technical Report AFML-TR-78-40, Westinghouse Research and Development Center, May 1978.

[19b] American Society for Testing and Materials, Tentative Test Method for Constant-Load-Amplitude Fatigue Crack Growth Rates Above 10^{-8} m/cycle", ASTM Designation E647-78T, Philadelphia, Pa.

[20] Duggan, T. V., Proctor, M. W. and Spence, L. J., "Stress intensity calibrations and compliance functions for fracture toughness and crack propagation test specimens", Int. J. of Fatigue, p. 37, 1979.

[21] Clark, W. G., Jr. and Hudak, S. J., Jr., "The analysis of fatigue crack growth rate data", in 22nd Sagamore Army Materials Research Conference on on Application of Fracture Mechanics to Design, p. 67, 1975.

[22] Dixon, W. J. and Massey, F. J., Jr., Introduction to Statistical Analysis, McGraw-Hill, New York, 2nd edition, 1957.

[23] Broek, D., Elementary Engineering Fracture Mechanics, Sijthoff and Noordhoff International Publishers, Alphen, aan den Rijn, 1978.

[24] Dowling, N. E., "Fatigue crack growth rate testing at high stress intensities", in Flaw Growth and Fracture, ASTM STP 631, p. 139, 1977.

[25] Australian Iron and Steel, Port Kembla, Australia, Personal Communication.

[26] Kaplan, M. P. and Reiman, J. A., "Use of fracture mechanics in estimating structural life and inspection intervals", J. Aircraft, 13, p. 99, 1976.

[27] Broek, D. and Smith, S. H., "The prediction of fatigue crack growth under flight-by-flight loading", Engineering Fracture Mechanics, 11, p. 123, 1979.

[28] Wei, R. P., "Fracture mechanics approach to fatigue analysis in design", Journal of Engineering Materials and Technology, Transactions of the ASME, 100, p. 113, 1978.

THE APPLICATION OF MIXED MODE II/III FRACTURE MECHANICS ANALYSIS TO THE CRACKING OF RAILS

D. J. H. Corderoy, M. B. McGirr and A. K. Hellier

University of New South Wales
Kensington 2033, Australia

ABSTRACT

Experimental mixed mode static and dynamic fracture mechanics data are becoming of increasing demand in the application of fracture mechanics to structural design, particularly in regard to rail geometry and material. A mixed mode loading rig has been designed and is now under construction. The philosophy of the rig design is discussed and currently available test results will be presented at the Conference.

To fully utilize the data from a rig of this type, it is desirable that a finite element analysis of the stress distribution in the test piece be conducted. The approach taken to establish the mixed mode stress field intensity factors K_{II} and K_{III} in the present case is described, and the application of experimental data to the failure of rails is considered.

INTRODUCTION

The vast majority of fracture mechanics data in the literature relates to the opening Mode I, this being the simplest case to handle both experimentally and theoretically. It is becoming increasingly apparent, however, that many problems in engineering design involve the shear Mode II or antiplane strain Mode III or, especially, combinations of the modes. A prime example, which will be discussed in this paper, is the initiation of fatigue cracks at inclusions in the head of a rail, an event dominated by the mixed Modes II and III.

Many specimen configurations have been used for tests in which the stressing is other than pure Mode I. For example, the most commonly used Mode II test piece is the compact shear specimen (a double-cracked compact tension type of specimen) [1], while Mode III testing is usually carried out in torsion [2]. An angled center crack in a plate has been used to produce mixed Mode I/II loading [3], whereas combined Mode I and III has been obtained using a specimen containing an edge crack inclined at an angle to the horizontal in the through-thickness direction [4]. A major problem encountered in such mixed mode tests is that crack growth tends to revert to a single mode, usually Mode I, although some cases of reversion to Mode III are known.

388

Chell and Girvan [5] have proposed an experimental arrangement for fracture testing subject to any combination of Modes I, II and III. However, they encountered practical difficulties when applying the method to steels.

PHILOSOPHY OF TEST RIG DESIGN

A schematic representation of the loading arrangement is shown in Figure 1. The specimen is a bar of square cross-section containing an edge crack whose plane

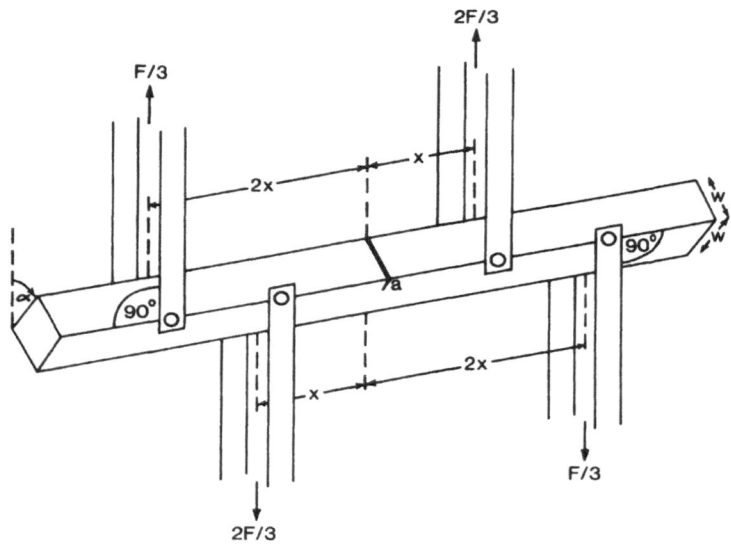

Fig. (1) - Schematic diagram showing mixed mode test geometry

is parallel to, and equidistant from, the ends of the bar, and whose front is parallel to a side of the bar. The loading is similar to that in a four point bend test except that the two loading points on one side of the crack are transposed. These loading points are spaced at distances of x and 2x from the crack plane, and the magnitudes of the respective forces acting at each are in the ratio 2:1. This arrangement is stable under load and produces zero net bending moment about the crack plane. Also, since the applied loads are perpendicular to the length of the specimen, the axial force is zero. Thus, there is theoretically no tensile Mode I contribution to the traction on the crack front but, rather, a system of shears giving a mixture of Modes II and III. By rotating the specimen about its axis, the relative proportions of Mode II and Mode III may be varied. When the angle α in Figure 1 is zero, pure Mode II is obtained, while an angle α of 90° results in pure Mode III.

Chell used pin loading in his mixed mode tests but found that the steel specimens tended to deform around the pin holes [6]. In order to overcome this problem, the essential feature of the present design is that the specimen is held by rotatable inserts within each of four loading arms (refer to Figure 2). Accordingly, no pin holes are required in the specimen, the same specimen being suitable for tests at any angle α. The insert at one end has a locking taper to prevent rotation of the specimen once the desired angle has been set. Also, translation of the specimen is restrained by detachable collars, one at each end.

The basic rig design philosophy has been to make all cross-sections sufficiently large to carry the high loads required in tests on steel without giving rise to fatigue or yielding of the rig. Secondly, the rig has been made as stiff as possible so that the specimen is tested and not the rig. For these reasons, each half of the rig is machined from a single billet of die steel X4150, heat-treated to a hardness of 285 BHN. The test rig has been designed for use with two Instron machines. It is intended to carry out pre-cracking of specimens and static tests on an Instron 8000 series hydraulic machine; if possible, fatigue testing will be done using an Instron 1603 resonant frequency machine. It is anticipated that the test rig will allow cyclic loading to be conducted at both positive and negative stress ratios R ($=\sigma_{min}/\sigma_{max}$), since a close-sliding fit exists between loading arm and insert, and between insert and specimen. The inserts, which are to be heat-treated to a hardness of $50R_c$, are made from a dimensionally stable steel corresponding to AISI 02.

COMPLIANCE FUNCTIONS

The parameters to be measured in the present mixed mode fatigue tests are the threshold stress intensity factor ranges ΔK_{IIth} and ΔK_{IIIth} for various combinations of Modes II and III, and crack growth rate as a function of ΔK. In addition, it is hoped to determine the fracture toughness values K_{IIc} and K_{IIIc} under monotonic loading.

In order to obtain such data, it is necessary to know the compliance functions of these two modes for the specimen. If the crack length is a and the width of the specimen is w, then these are given in [7] as

$$Y_2\left(\frac{a}{w}\right) = \sqrt{\pi}\left[1.122 - 0.561\left(\frac{a}{w}\right) + 0.085\left(\frac{a}{w}\right)^2 + 0.18\left(\frac{a}{w}\right)^3\right]\sqrt{1 - \frac{a}{w}} \tag{1}$$

and

$$Y_3\left(\frac{a}{w}\right) = \sqrt{\frac{2w}{a}\tan\left(\frac{\pi a}{2w}\right)} \tag{2}$$

provided that the distance x in Figure 1 is greater than 1.5w. The stress intensity factors are then obtained as

$$K_{II} = \frac{F\cos\alpha}{3w^2}\sqrt{a}\,Y_2\left(\frac{a}{w}\right) \tag{3}$$

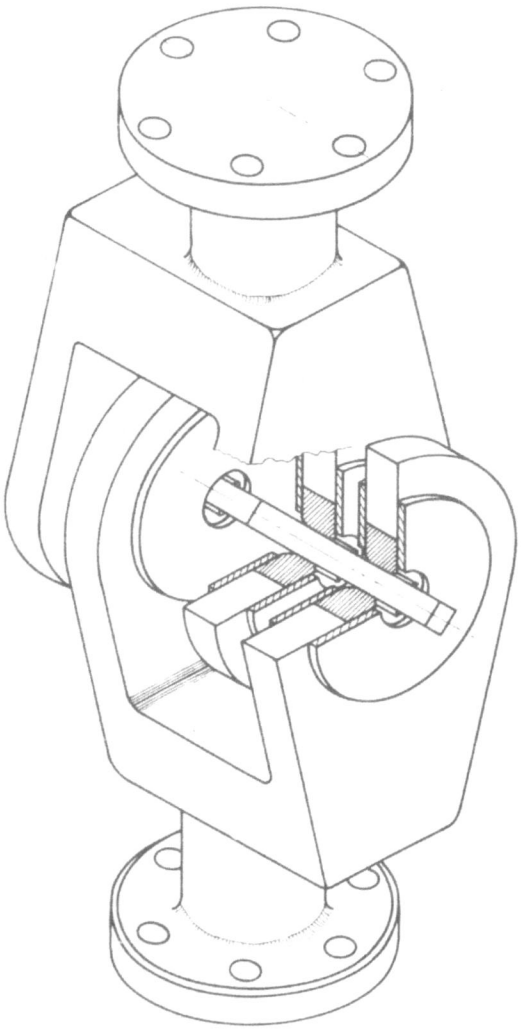

Fig. (2) - Drawing of mixed mode test rig

and

$$K_{III} = \frac{F\sin\alpha}{3w^2} \sqrt{a}\ Y_3\left(\frac{a}{w}\right) \tag{4}$$

where F is the total load applied to the specimen.

FINITE ELEMENT ANALYSIS OF TEST PIECE

It is highly desirable that the compliance functions be confirmed by a finite element analysis of the test specimen, particularly since equations (1) and (2) strictly refer to the pin-loaded case shown in Figure 1. In this regard, a preliminary study of the test piece is currently being made; however, the selection of a suitable set of boundary conditions is somewhat problematic. Whatever boundary conditions are chosen should take into account two factors, namely: the forces acting on the specimen must be in the ratios shown in Figure 1, and the deformed specimen, without the crack, should exhibit a two-fold rotation axis perpendicular to the plane of loading. Unfortunately, it is not possible to simply apply the loads to a finite element mesh of the specimen, since a number of fixities are required to make the solution process numerically feasible.

Figure 3 shows a two-dimensional model of the test piece for the case of pure Mode II loading. Uniformly distributed loads were applied as shown and fixities

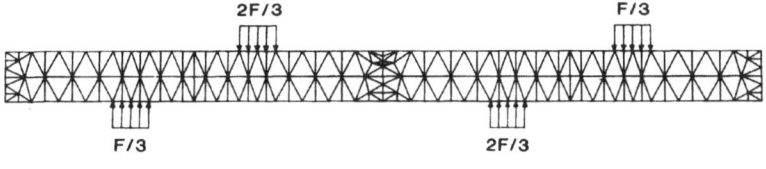

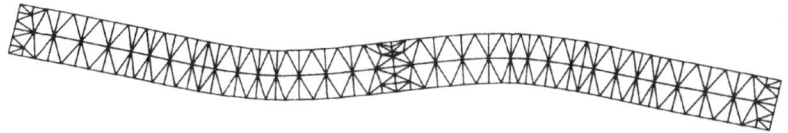

Fig. (3) - Plane strain finite element model of test piece under Mode II loading and deformed mesh with displacements shown magnified

made at three nodes along the mid-plane of the specimen which, in the absence of a crack, would remain along that plane. The result of this analysis for a steel specimen, assuming plane strain conditions, is also shown. The mesh was generated and solved using the authors' own finite element programs [8]. Although not very useful quantitatively, this model does indicate the manner in which the test piece is likely to deform.

Two simple three-dimensional meshes have also been used in this stage of the investigation (refer to Figure 4). All attempts made to model the specimen to

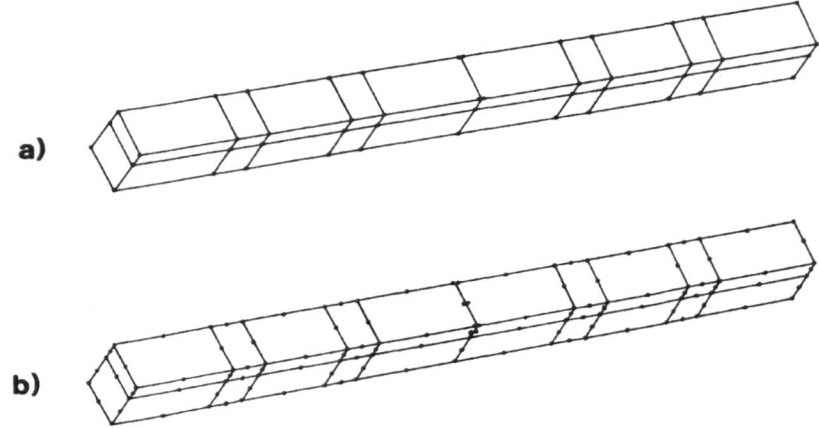

a)

b)

Fig. (4) - Simple three-dimensional finite element meshes of test piece using a) 8 node brick elements and b) 20 node brick elements

date confirm the desired Mode II/III behavior, with little or no Mode I contribution. The three-dimensional analyses are being carried out using ACES, a finite element package available at the University of New South Wales. It is intended that future work to determine stress intensity factors, and hence obtain the compliance functions, be done using a more sophisticated package containing special crack tip elements.

APPLICATION OF EXPERIMENTAL DATA TO THE CRACKING OF RAILS

The trend in recent years towards higher axle loads on railways, particularly in the case of heavy haul systems, has led to an increase in the incidence of "transverse defects" (i.e., those defects which grow approximately in the transverse plane of the rail). The occurrence of these defects necessitates costly, regular ultrasonic inspection of track since, if undetected, they may ultimately lead to rail fracture. A comprehensive metallographic examination of many such defective rails removed from service has been undertaken at the Melbourne Research Laboratories of B.H.P. [9]. This work has shown that transverse defects originate from "shell defects" which grow in a plane containing the longitudinal axis of the rail and which, in turn, initiate beneath the surface at individual inclusions or closely spaced inclusion arrays.

Chipperfield [10] has proposed a model for rail head fatigue, utilizing fracture mechanics concepts together with an analytical stress analysis, which is consistent with these metallographic observations. The model appears to be capable of predicting the inclusion dimensions and sizes which are of importance in the initiation event, and may serve as a basis for the design of rail steels with enhanced fatigue resistance for specific applications. The critical inclusion size, below which fatigue crack initiation does not occur, is sensitive to the threshold stress intensity factor range, being proportional to ΔK_{th}^2. Since threshold data for shear loading are not currently available for rail steel, the usual approach to the application of this model has been in terms of estimates of ΔK_{IIth} and ΔK_{IIIth} based on available Mode I data, and is obviously unsatisfactory. Furthermore, the effect of combined modes on the absolute threshold level is unclear. Since ΔK_{II} and ΔK_{III} are strictly not scalars, they cannot simply be superposed; use of the strain energy density field concept proposed by Sih [11], indicating an elliptical relationship with ΔK, appears to be more appropriate.

The mixed mode testing rig described in this paper offers the means to determine ΔK_{IIth} and ΔK_{IIIth} directly, for any desired combination of Modes II and III, and should permit their relationship with ΔK_{th} to be investigated. Crack growth rate data obtained using the rig may also be used in conjunction with the rail head fatigue model to estimate the lifetime of rails in service.

ACKNOWLEDGEMENTS

The authors wish to acknowledge the financial support given to this work by the Melbourne Research Laboratories of B.H.P.

REFERENCES

[1] Jones, D. L. and Chisholm, D. B., "An investigation of the edge-sliding mode in fracture mechanics", Eng. Fract. Mech., 7, p. 261, 1975.

[2] Pook, L. P. and Sharples, J. K., "The Mode III fatigue crack growth threshold for mild steel", Int. J. Fract., 15, p. R223, 1979.

[3] Erdogan, F. and Sih, G. C., "On the crack extension in plates under plane loading and transverse shear", Trans. ASME, J. Basic Eng., 85, p. 519, 1963.

[4] Pook, L. P., "The effect of crack angle on fracture toughness", Eng. Fract. Mech., 3, p. 205, 1971.

[5] Chell, G. G. and Girvan, E., "An experimental technique for fast fracture testing in mixed mode", Int. J. Fract., 14, p. R81, 1978.

[6] Chell, G. G., Central Electricity Research Laboratories, U.K., Private Communication, 1981.

[7] Tada, H., Paris, P. C. and Irwin, G. R., The Stress Analysis of Cracks Handbook, Del Research Corporation, Hellertown, Pa., 1973.

[8] Corderoy, D. J. H., McGirr, M. B., Easterbrook, P. C. and Hellier, A. K., "A new approach to automatic mesh generation in the continuum", 4th Int. Conf. in Australia on Finite Element Methods, 1982.

[9] Skinner, D. H. and Judd, P. A., "A metallographic study of fatigue defects in rails", Proc. 34th Annual Conf. of the A.I.M., Surfers Paradise, Queensland, p. 68, 1981.

[10] Chipperfield, C. G., "Modelling rail head fatigue using fracture mechanics", Proc. 34th Annual Conf. of the A.I.M., Surfers Paradise, Queensland, p. 63, 1981.

[11] Sih, G. C. and Barthelemy, B. M., "Mixed mode fatigue crack growth predictions", Eng. Fract. Mech., 13, p. 439, 1980.

ON MIXED MODE DUCTILE FRACTURE CRITERIA - A NOTE ABOUT THE CORE REGION

Y.-S. Kao

Nanjing Aeronautical Institute
People's Republic of China

ABSTRACT

A method to estimate the crack tip core region radius was proposed. As an example, the test data of MCIC [9] for 0.079" (2 mm) 2024-T3 plate were used to to evaluate the critical radius, r_o. Fracture angles and critical external loads were calculated by several mixed mode fracture criteria. Exact stress solutions were used. The results were compared with experimental data [4]. It was found that by proper selection of the critical radius, some of the criteria which were derived from linear theory, may be used to estimate the ductile fracture within engineering accuracy.

INTRODUCTION

Since Erdogan and Sih [1] proposed a mixed mode fracture criterion in 1963, many propositions had been published, such as strain energy density theory [2], maximum circumferential strain criterion [5], etc. Certain successes had been achieved specially on brittle fractures. But for ductile fractures, the critical radius of the core region are much larger, it is insufficient to take a singular term of stress function in computing the stress components. It will be much better by taking the exact solutions [3,6,7].

But in some mixed mode criteria, the exact solutions depend upon the critical radius. So there is a problem how to select it. Williams [6] had suggested 0.002" (0.05 mm) for PMMA, which is a brittle material. How about the thin plate of ductile material like aluminium alloy 2024-T3? It is a problem. The author suggests a method to determine it. By this critical radius selected, the mixed mode fracture angles and on set critical loads were calculated by several fracture criteria with exact stress solutions. Comparisons of the calculated results with existing experimental data [4] are presented.

ANALYSIS

A. Critical Radius

The physical meaning of the critical radius, r_o, is not very clear at present; many authors have discussed it [11,10,6]. Recognizing that it is a material constant, it must be able to determine from test. Because critical radius will not depend upon crack angle, many uniaxial fracture tests with loadings normal to the crack surface can be utilized. Assuming that at the point near the crack tip, $r = r_o$, the stress component σ_θ reaches yield stress $\sigma_{0.2}$, the crack will begin to propagate. Then r_o can be solved by the exact equation (A-1), thus

$$\frac{r_o}{a} = -1 + \sqrt{1-1/(1-p^2)}$$

$$p = \sigma_{0.2}/\sigma_i \qquad\qquad (1)$$

σ_i = on set external load (stress).

As an example, the cases of 0.079" (2 mm) 2024-T3 fracture test data were taken from MCIC [9]. There were 79 specimens, 5.9" to 24" wide, with various crack lengths. The critical radii were calculated by equation (1). The results vary from 4 mm to 6 mm, except a few extremities. A typical case is as shown in Table 1. r_o = 5 mm, as an average, is accepted for ductile fracture analysis of 2 mm 2024-T3 plate.

TABLE 1 - CRITICAL RADIUS

(Specimen width = 11.8", Thickness = 0.079", $\sigma_{0.2}$ = 51.8 ksi)

Half Crack Length	On Set External Load σ_i, ksi	$p = \sigma_{0.2}/\sigma_i$	$\frac{r_o}{a}$	r_o
.590	33.4	1.551	.3082	.1818
.590	31.0	1.671	.2482	.1464
.590	32.0	1.619	.2716	.1602
.885	32.3	1.604	.2790	.2469
.885	28.5	1.817	.1977	.1749
.885	30.5	1.698	.2373	.2100
1.770	25.8	2.007	.1534	.2715
1.770	24.9	2.080	.1404	.2485
2.360	17.4	2.977	.0617	.1456
2.360	17.2	3.012	.0601	.1418
			average	.1964" (4.99 mm)

B. Fracture Angles

There are quite a number of mixed mode fracture criteria at present. The purpose of analysis is to see if the selected critical radius fit them well. So, three criteria are chosen to calculate the fracture angles for plane stress condition. The equations to solve them are given as follows:

Maximum circumferential strain criterion:

$$\frac{\partial \sigma_\theta}{\partial \theta} - \nu \frac{\partial \sigma_r}{\partial \theta} = 0 \tag{2}$$

Minimized strain energy density with respect to θ:

$$\sigma_\theta \frac{\partial \sigma_\theta}{\partial \theta} + \sigma_r \frac{\partial \sigma_r}{\partial \theta} - \nu(\sigma_\theta \frac{\partial \sigma_r}{\partial \theta} + \sigma_r \frac{\partial \sigma_\theta}{\partial \theta}) + 2(1+\nu)\tau_{r\theta} \frac{\partial \tau_{r\theta}}{\partial \theta} = 0 \tag{3}$$

Zero in plane shear stress criterion:

$$\tau_{r\theta} = 0 \tag{4}$$

See Appendix for the exact stress components. Poisson's ratio $\nu = 0.33$ for 2024-T3 material. The calculated results were plotted in Figure 1.

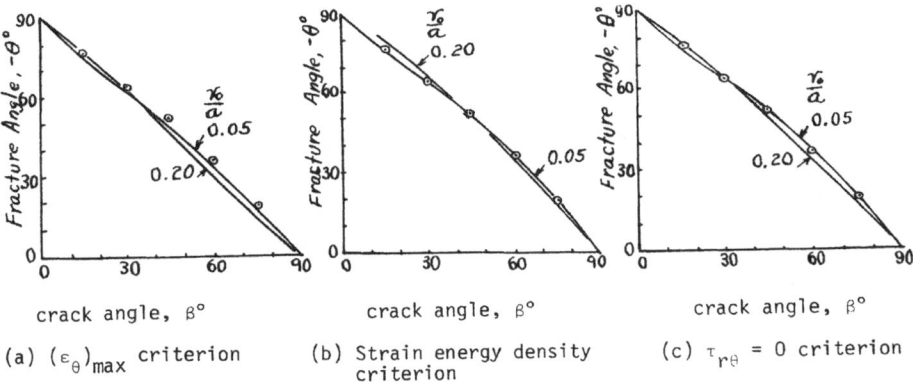

(a) $(\varepsilon_\theta)_{max}$ criterion   (b) Strain energy density criterion   (c) $\tau_{r\theta} = 0$ criterion

Fig. (1) - Fracture angles versus crack angles
(⊙ denotes experimental points [4])

Experiments [4] were made by specimens 300 mm x 830 mm x 2 mm aluminium alloy, very similar to 2024-T3. 2a = 50 mm. ($r_0/a = 0.20$).

C. Critical Loads

The critical loads were determined by the conditions the maximum circumferential strain $\varepsilon_\theta = \varepsilon_{0.2}$ and the maximum circumferential stress $\sigma_\theta = \sigma_{0.2}$.

$\varepsilon_{0.2}$ and $\sigma_{0.2}$ are uniaxial strain and stress of the material with 0.2% permanent set.

Calculated and tested results are shown on Figure 2 and Figure 3. The ordinate σ_i is the on set external load (stress) normalized by uniaxial tensile

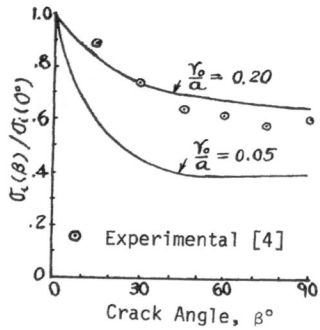

Fig. (2) - Critical load based on
$\varepsilon_\theta = \varepsilon_{0.2}$

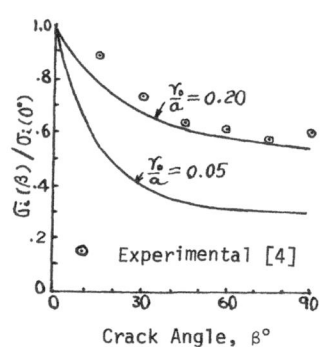

Fig. (3) - Critical load based on
$\sigma_\theta = \sigma_{0.2}$

yield stress $\sigma_i(0°)$. From the figures, it can be seen that there is quite a large difference of σ_i for different values of r_0/a ratios. Although $\sigma_\theta = \sigma_{0.2}$ criterion underestimated the critical load, but it is on the safe side.

RESULTS AND DISCUSSIONS

1. Values of critical radius, r_0/a, are more sensitive to the critical load than to the fracture angle. r_0/a varies from 0.05 to 0.20. The fracture angle can be varied by 3 to 4 degrees but the critical load can be varied more than 30%. It seems that the estimated $r_0 = 5$ mm ($r_0/a = 0.20$) for 2 mm thick 2024-T3 plate fit them both well.

2. The above mentioned criteria all were derived from linear theory of elasticity, but for ductile fracture by properly choosing of r_0/a ratio and using exact stress field equations, the calculated results are within engineering accuracy.

3. Further studies on ductile fracture of thin plates are needed.

APPENDIX - FORMULATIONS

The stress components were calculated by Muskhelishvili's [8] two complex stress functions. The coordinate system (see Figure A-1). For infinite plate with uniaxial tension, the stress components are:

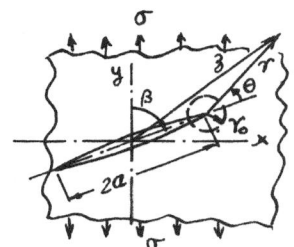

Fig. (A-1) - Coordinates

$$\sigma_\theta = 2\text{Re}[\Phi(z)] + \text{Re}\{[(\bar{z}-z)\Phi'(z) + \bar{\Omega}(z) - \Phi(z)]e^{i2\theta}\} \qquad \text{(A-1)}$$

$$\sigma_r = 2\text{Re}[\Phi(z)] - \text{Re}\{[(\bar{z}-z)\Phi'(z) + \bar{\Omega}(z) - \Phi(z)]e^{i2\theta}\} \qquad \text{(A-2)}$$

$$\tau_{r\theta} = \text{Im}\{[(\bar{z}-z)\Phi'(z) + \bar{\Omega}(z) - \Phi(z)]e^{i2\theta}\} \qquad \text{(A-3)}$$

$$\left.\begin{array}{r}\Phi(z)\\[2ex]\Omega(z)\end{array}\right\} = \frac{\sigma c}{4}\,\frac{z}{\sqrt{z^2-1}} \mp \frac{1}{2}\,\bar{\Gamma}'$$

$$\Phi'(z) = -\frac{\sigma c}{4}\,(z^2-1)^{-3/2}$$

$$\bar{\Omega}(z) = \frac{\sigma \bar{c}}{4}\,\frac{z}{\sqrt{z^2-1}} + \frac{1}{2}\,\Gamma'$$

$$c = 1 - e^{i2\beta} \qquad\qquad \bar{c} = 1 - e^{-i2\beta}$$

$$\frac{1}{2}\,\bar{\Gamma}' = -\frac{\sigma}{4}\,e^{i2\beta} \qquad\qquad \frac{1}{2}\,\Gamma' = -\frac{\sigma}{4}\,e^{-i2\beta}$$

See Figure A-1 for coordinate system. z is normalized with half crack length a, such that

$$z = 1 + \tilde{r}e^{i\theta} \qquad\qquad \bar{z} = 1 + \tilde{r}e^{-i\theta}$$

$$\tilde{r} = r/a$$

400

REFERENCES

[1] Erdogan, F. and Sih, G. C., "On the crack extension in plates under plane loading and transverse shear", J. of Basic Eng., 85, 4, pp. 513-518, 1963.

[2] Sih, G. C., "A special theory of crack propagation", Mech. of Fracture, Vol. 1, Noordhoff Pub., 1973.

[3] Sih, G. C. and Kipp, M. E., Int. J. Fracture 10, pp. 261-265, 1974.

[4] Qi, Y.-X., "Mixed mode ductile fracture criterion - maximum shear theory", MA Thesis, Nanjing Aeronautical Ins., 1981, unpublished.

[5] Chang, K. J., "On the maximum strain criterion - a new approach to the angled problem", Eng. Frac. Mech. 14, p. 107, 1981.

[6] Williams, J. G. and Ewing, P. D., International J. Frac. 8, p. 441, 1972.

[7] Ewing, P. D. and Williams, J. G., International J. Frac. 10, p. 135, 1974.

[8] Muskhelishvili, N. I., "Some Basic Problems of the Mathematic Theory of Elasticity", Noordhoff Pub., 1953.

[9] Damage Tolerance Design Handbook, MCIC-HB-01, Metals and Ceramics Information Center, 1973.

[10] Sih, G. C., Mechanics of Fracture, Vol. II, Page XVIII.

[11] Sih, G. C., Mechanics of Fracture, Vol. III, Page XXVIII.

FURTHER OBSERVATIONS ON MIXED MODE PLANE STRESS DUCTILE FRACTURE

Y. W. Mai and B. Cotterell

Department of Mechanical Engineering, University of Sydney
Sydney, N.S.W. 2006, Australia

ABSTRACT

Mixed mode ductile fractures in thin sheets are shown to be possible for two low carbon steels and two aluminium alloys. The staggered deep edge notch tension specimen allows the mixed mode J_C to be measured and to be partitioned into its component modes J_{IC} and J_{IIC} by extrapolating the specific work of fracture to zero ligament length. For the low carbon steels J_C is shown to be independent of the mode of fracture but for the aluminium alloys J_C is dependent on the stagger angle (θ). It is suggested that if necking is intensive and the strain hardening exponent (n) is low J_C will vary with θ. If however necking is moderate and n is comparatively large J_C is approximately constant. The experimental results obtained from these thin sheet metals seem to support this hypothesis although no theoretical grounds can be offered.

INTRODUCTION

The design of structural components containing defects and flaws under biaxial loading requires a complete knowledge of the physics of mixed mode crack initiation and propagation. Much work has been done in the past on mixed mode fracture of homogeneous isotropic brittle materials and a variety of criteria has been proposed for crack initiation [1-5]. For crack propagation, however, it seems that the fracture path is one that has a mode I stress field at the crack tip [6]. It is not possible therefore to obtain mixed modes I and II crack propagation in brittle solids. However, for ductile materials, localised necks can form that in general are at an angle to the principal stresses so that mixed mode crack propagation is permissible. There have been some attempts to investigate mixed mode fracture in ductile solids but they are not so successful. This is either because the fracture propagates in essentially a mode I direction [7] or the linear elastic fracture mechanics analysis employed is inappropriate [8]. True mixed mode ductile fractures can only occur in specimens that are completely yielded and where post yield fracture mechanics is necessary. The most appropriate ductile fracture criterion is the critical J-integral (J_C) which for mixed mode problems can be separated into its component modes I (J_{IC}) and II (J_{IIC}) so that $J_C = J_{IC} + J_{IIC}$ [9,10].

In a previous paper [10] we have shown that it is possible to obtain true mixed mode ductile fractures in thin sheets using staggered deep edge notch tension specimens, Figure 1, similar to those suggested by Hill [11] for determining the yield criterion of ductile metals. In these specimens localised necks form along the line joining the notches which develop into fractures when the opening at the tip of the notch reaches its critical opening displacement δ_c. The width of the process zone d in which the neck develops is roughly equal to the sheet thickness for zero stagger angle [12] and becomes narrower for larger angles. If we assume that the flow stress curve for the material can be represented by the power law

$$\bar{\sigma} = A\bar{\epsilon}^n \tag{1}$$

where $\bar{\sigma}$ and $\bar{\epsilon}$ are the equivalent stress and strain, then the normal strain across the process zone at which necking commences is n. The velocity discontinuity across the process zone makes an angle ψ with the neck which is given by [13]

$$\tan \psi = \frac{1}{4} \cot \theta . \tag{2}$$

The J integral is given by [14]

$$J = \oint_\Gamma \left[Wdy - T_j \frac{\partial u_j}{\partial x} ds \right] \tag{3}$$

where W is the strain energy density function, T is the stress vector and u the displacement on the contour Γ that can take any path from one crack surface to the other outside the fracture process zone where the constitutive equations are not held. In particular we can evaluate J around the boundary of the process zone to give

$$J_C = d \int_0^{\bar{\epsilon}_n} \bar{\sigma}d\bar{\epsilon} + \int_{nd}^{\delta_{IC}} \sigma d\Delta_I + \int_{ndcot\psi}^{\delta_{IIC}} \tau d\Delta_{II} \tag{4}$$

where σ, τ are the stresses on the edge of the necked zone, Δ_I, Δ_{II} are the normal and tangential displacements of the necked zone which on entering the zone are nd and ndcotψ respectively, and $\bar{\epsilon}_n$ is the equivalent strain at the point of necking. Thus J_C is the fracture work per unit area in the fracture process zone at the tip of the crack necessary to cause tearing. We have previously [12,15-17] called this work the specific essential work of fracture, w_e, to distinguish it from the plastic work performed outside the process zone. We can separate J_C into its mode I and mode II components

$$J_{IC} = \frac{\sin\psi \cos\theta}{\sin(\theta + \psi)} d \int_0^{\bar{\epsilon}_n} \bar{\sigma}d\bar{\epsilon} + \int_{nd}^{\delta_{IC}} \sigma d\Delta_I \tag{5}$$

$$J_{IIC} = \frac{\cos\psi \sin\theta}{\sin(\theta + \psi)} d \int_0^{\bar{\epsilon}_n} \bar{\sigma}d\bar{\epsilon} + \int_{ndcot\psi}^{\delta_{IIC}} \tau d\Delta_{II} . \tag{6}$$

Both of these J-integrals are impossible to compute since δ_{IC} and δ_{IIC} are not material constants and they depend on the ratio τ/σ. However, experimental evaluations for J_{IC} and J_{IIC} are possible using staggered deep edge notch tension specimens with varying ligament lengths (ℓ) and extrapolating the specific work of fracture for the respective fracture modes to zero ligament length [10]. In our previous investigation [10] on a particular low alloy steel (Lyten) J_C is shown to be independent of the mode of fracture, Figure 2. If this result was general it would be a most useful mixed mode ductile fracture criterion. In order to examine this possibility further work on mixed mode ductile fracture of three other sheet metals, a low carbon steel and two aluminium alloys, B1200-H14 and 5251, has been carried out. The major results of this investigation are given in the present paper.

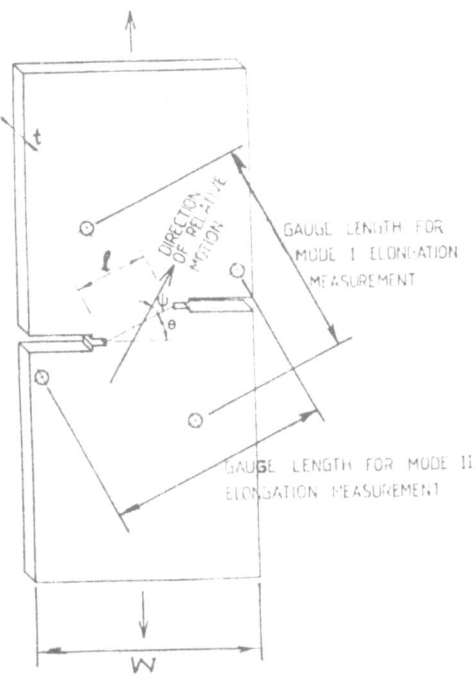

Fig. (1) - Staggered deep edge notch tension specimens

404

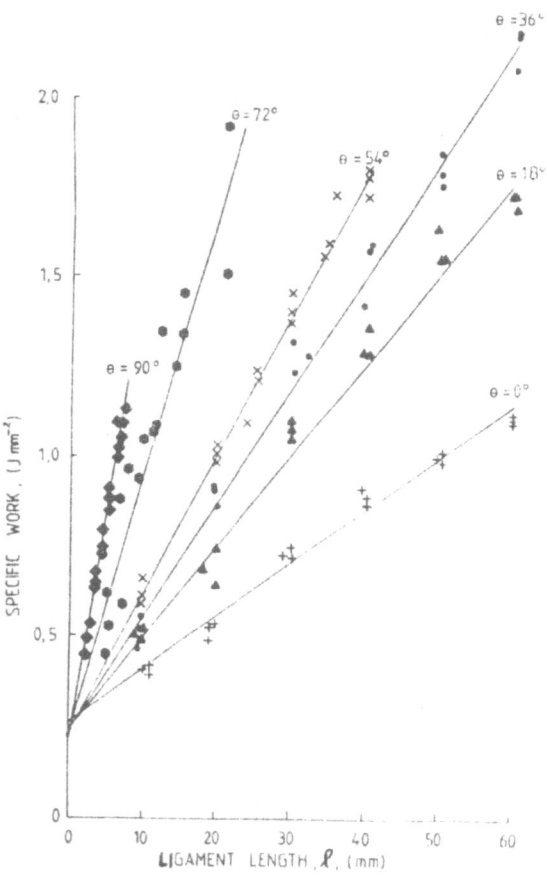

Fig. (2) - Variation of specific work with ligament
length for Lyten staggered
notch specimens

FURTHER EXPERIMENTS ON MIXED MODE PLANE STRESS DUCTILE FRACTURE

Three sheet metals, a temper-rolled low carbon (0.1% C) 16 gauge steel which has been used for similar mode I fracture work [16] and two other aluminium alloys, 5251 and B1200-H14, of nominal thicknesses 2 mm and 1.6 mm respectively were investigated in this work. The chemical compositions and mechanical properties in transverse to rolling are given in Tables 1 and 2. For comparison purposes similar properties for Lyten [10] are also given. It is obvious from Table 2 that the steels have much larger elongations to break, higher strain hardening exponent (n), more width but less thickness and area reductions of the fractured neck. There is however little plastic anisotropy in these metals.

Except for the low carbon steel where fracture anisotropic effects had also been studied all specimens were cut so that the rolling direction was aligned to the line joining the staggered notches. The specimen dimensions were designed to avoid yielding spreading to the free edges and this was confirmed using the brittle lacquer coating technique. The ligament lengths were also chosen so that there was no plane stress-plane strain transition for all staggered angles (θ). However, it was not always possible to obtain true mixed mode fractures for large θ with large ligament lengths (ℓ) because the cracks growing from the two notches did not join up. The last few millimeters of the notches were finished with a fine saw blade to give a sharp initiating notch. All experiments were carried out in a hydraulic controlled Shimadzu testing machine. The elongations normal to and along the ligaments were measured with clip gauges mounted outside the plastic region in pairs on each side of the specimen, Figure 1.

TABLE 1 - COMPOSITION OF MATERIALS

Material	C	Si	Mn	Cr	Ni	Cu	P	S
Lyten	0.10	0.40	0.85	0.80	0.85	0.25	0.09	0.03
Temper-rolled	0.04-0.09	trace	0.25-0.40	-	-	-	0.005-0.020	0.01-0.03

Material	Si	Fe	Cu	Mn	Mg	Cr	Zn	Ti
5251	0.40	0.50	0.15	0.1-0.5	1.7-2.4	0.15	0.15	0.15
B1200-H14	0.40	0.60	0.05	0.05	-	-	0.10	-

TABLE 2 - MECHANICAL PROPERTIES IN TRANSVERSE TO ROLLING

Material	Lyten	TRLCS	5251	B1200-H14
0.2% proof stress (MPa)	360	320	132´	112
UTS (MPa)	510	400	159	121
Elongation (%)	32	40	7	6
Strain hardening index (n)	0.25	0.10	0.05	0.05
Reduction in width (%) (a) Necked section (b) Outside necked section	23 10	26 11	4 3	9 2
Reduction in thickness (%) (a) Necked section (b) Outside necked section	28 6	24 10	57 3	75 2
Reduction in area (%) (a) Necked section (b) Outside necked section	45 17	44 20	58 5	77 4

MIXED MODE FRACTURE RESULTS

A. Temper-Rolled Low Carbon Steel

The mixed mode ductile fracture results for crack propagation in the direction of rolling is shown in Figure 3. Experiments were only performed for $\theta \leqslant 45°$ due to the limited amount of this steel left over from previous work [17]. Nevertheless, it is clear that the specific essential work of fracture (w_e) which is identified as J_C and given by the intercept at $\ell = 0$ is invariant with the mode of fracture. These results seem to confirm the previous finding on Lyten which also shows J_C is constant and independent of θ [10]. However, both Lyten and this temper-rolled low carbon steel have very similar mechanical properties in terms of strain hardening capability and geometric dimension changes of the fractured neck, Table 2. It was therefore necessary to perform further experiments on the aluminium alloys having different mechanical characteristics to confirm if J_C is constant is a generalised result.

Anisotropic fracture effect on the specific essential work of fracture (w_e) or equivalently J_C, is shown in Figure 4 as a function of the angle between the fractured ligament and the rolling direction (β). It is interesting to see that w_e is minimum at $\beta = 0°$ (i.e. fractured ligament is along the direction of rolling) and reaches a maximum at $\beta = 30°$ before it drops monotonically again to $\beta = 90°$. Because of this anisotropic fracture effect on w_e and J_C we have kept $\beta = 0$ for all further mixed mode ductile fracture experiments on the aluminium alloys.

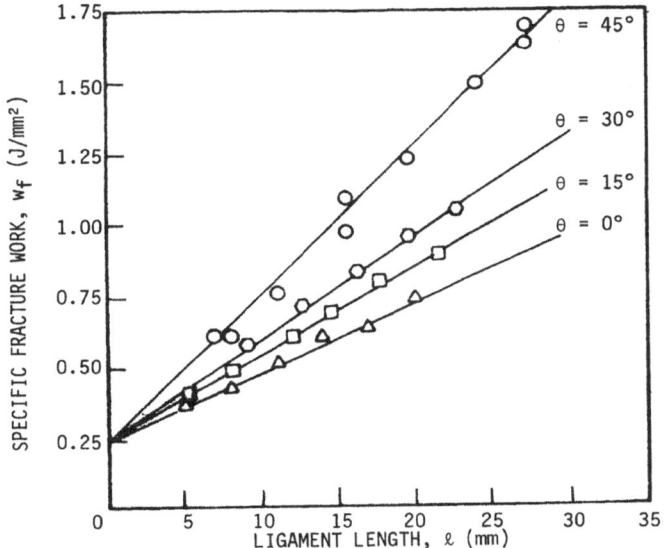

Fig. (3) - Variation of specific fracture work with
ligament length for low carbon steel staggered notch specimens

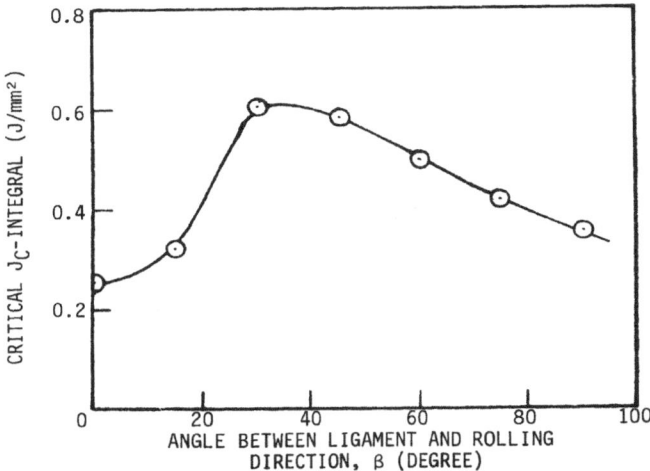

Fig. (4) - Anisotropic fracture effect in low carbon steel

B. Aluminium Alloys 5251 and B1200-H14

Figures 5 and 6 show the variation of the specific work of fracture (w_f) as a function of ligament length (ℓ) for staggered angles between 0° and 90° for both aluminium alloys. It was not possible to obtain a wider range of data for $\theta = 90°$ as the fractures did not join up for large ligaments. However, sufficient information is presented in these results to conclude that the intercept value, i.e. w_e or J_C, is not constant and is dependent on the mode of fracture. J_C seems to increase with θ until a maximum is reached and then it decreases with further increase in θ, Figure 7. Thus the possibility of a constant J_C fracture criterion for mixed mode cracks in ductile solids is not upheld although for some materials such as Lyten and the temper-rolled low carbon steel the constant J_C criterion is apparently valid.

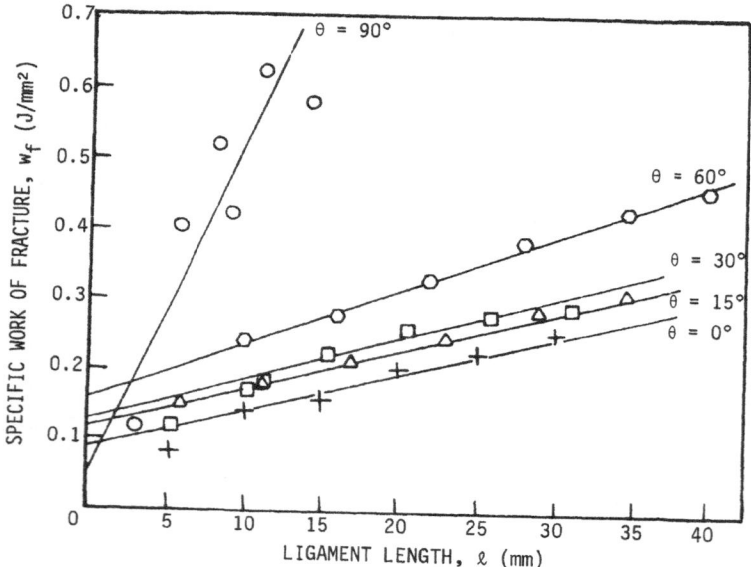

Fig. (5) - Variation of specific fracture work with ligament length for 5251 staggered notch specimens

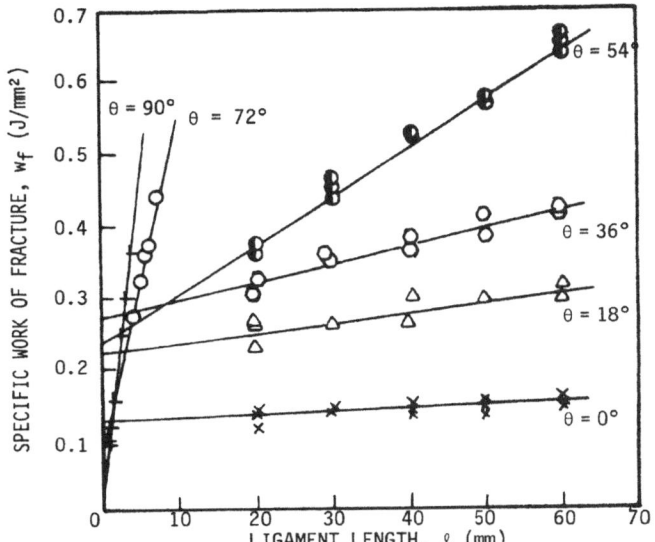

Fig. (6) - Variation of specific work of fracture with
ligament length for B1200-H14 staggered notch
specimens

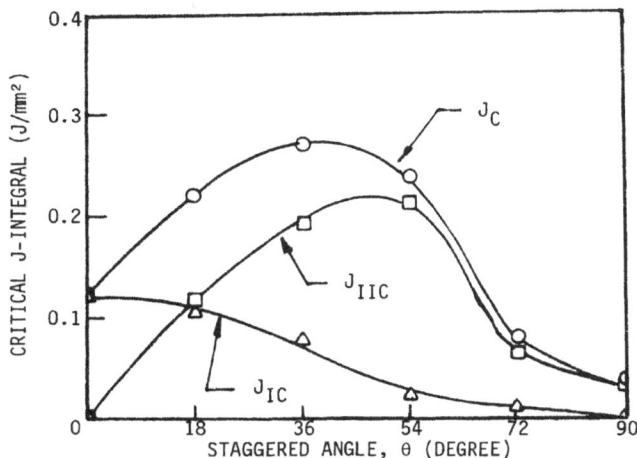

Fig. (7) - Critical J-integral versus staggered angle for B1200-H14

There is a substantial difference on the reduction of thickness in the fractured neck between the aluminium alloys and the low carbon steels. As may be expected from the tensile results in Table 2 the necking is far more intense for the aluminium alloys, Figure 8, than it is for the steels, cf. Figure 4 of [10]. The thickness reduction for the aluminium alloys decreases monotonically as θ increases. This behaviour is different to the Lyten steel which has a maximum thickness reduction at θ = 36° and a stabilized neck within a longer distance of crack extension [10].

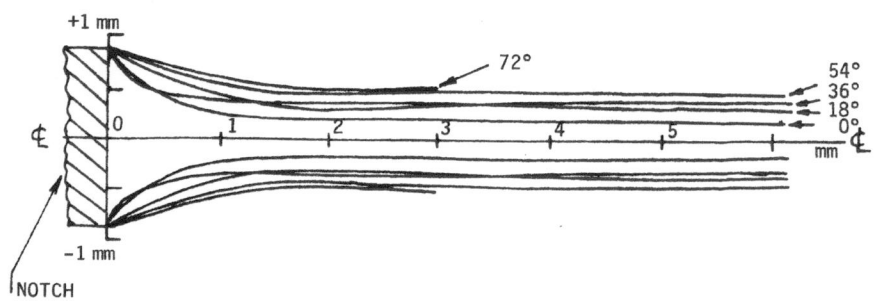

Fig. (8) - Thickness reduction of fractured neck for B1200-H14 staggered notch specimens

C. Evaluation of J_C and Comparison with Experimental Results

In order to evaluate the J_C-integral of equation (4) we need to measure experimentally the process zone width (d), the normal and shear stresses (σ, τ) and the associated crack tip opening displacements (δ_{IC}, δ_{IIC}) as a function of the stagger angle (θ). Theoretical predictions of these physical quantities are difficult to make and experimental measurements are therefore sought. The equivalent stress-strain relationship is given by equation (1) and the equivalent strain for localised necking for all angles except θ = 90° is

$$\bar{\varepsilon}_n = \left[\frac{1 + 3\,\sin^2\psi}{3}\right]^{\frac{1}{2}} \frac{n}{\sin \psi} . \tag{7}$$

Since ψ is related to θ by equation (2) this means that the equivalent necking strain depends on the stagger angle. In general σ and τ are functions of displacements Δ_I and Δ_{II} after necking has occurred. However, for simplicity we have assumed a parabolic relation between these variables so that the second and third integrals of equation (4) become

$$I_2 = \frac{2}{3}\,\sigma(\theta)[\delta_{IC} - nd] . \tag{8}$$

$$I_3 = \frac{2}{3}\,\tau(\theta)[\delta_{IIC} - nd\cot\psi] . \tag{9}$$

The first integral is also given by

$$I_1 = \frac{A\,d(\theta)}{(n+1)} \left[\frac{n}{\sin\psi}\right]^{(n+1)} \left[\frac{1+3\sin^2\psi}{3}\right]^{\left(\frac{n+1}{2}\right)} \tag{10}$$

using equations (6) and (7). For the Lyten steel, A = 992 MPa and n = 0.25; for the aluminium B1200-H14, A = 138 MPa and n = 0.05. Tables 3 and 4 show the predicted $J_C(\theta)$ values and their comparisons with experimental results for these two sheet metals.

TABLE 3 - EVALUATION OF J_C FOR B1200-H14

θ	=	0°	18°	36°	54°	72°	90°
ψ	=	90°	37.6°	19°	10.3°	4.64°	0°
$\bar{\varepsilon}_n$	=	0.06	0.06	0.06	0.07	0.11	-
d (mm)	=	2	1.95	1.65	1.40	0.50	0.30
σ (MPa)	=	128	109	77	47	20	0
τ (MPa)	=	0	35	56	65	68	70
δ_{IC} (mm)	=	1.25	1.20	1.10	0.45	0.51	0
δ_{IIC} (mm)	=	0	1.35	2.75	3.40	1.55	0.10
I_1 (J/mm²)	=	13.25	15.61	19.81	28.4	22.49	0
I_2 (J/mm²)	=	98.62	80.52	52.49	11.1	6.10	0
I_3 (J/mm²)	=	0	28.69	94.20	124.5	39.74	5
J_C $(I_1+I_2+I_3)$ (J/mm²)	=	119	125	167	164	68	5
J_C (Experimental) (J/mm²)	=	125	220	270	235	70	25

A few specific comments may be made with respect to these computed results given in the tables. Firstly, for θ = 90°, J_C is grossly underestimated for both materials because in theory d = 0 so that I_1 is zero but in practice there is a finite process zone width and plastic shearing work must be done in the fracture process. Secondly, for the aluminium specimens I_1 is only a small to moderate fraction of J_C but for the Lyten this is almost all the fracture work J_C. The very small crack tip opening displacements for the steel indicate that both I_2 and I_3 are insignificant whereas for the aluminium these two integrals are major contributors to J_C because of their large critical crack opening displacements. Thirdly, the trends of the predicted results are in good agreement with those of the experimental measurements although the absolute values do not entirely agree with each other, i.e. J_C is invariant with θ for Lyten but not for B1200-H14.

TABLE 4 - EVALUATION OF J_C FOR LYTEN

θ		$0°$	$18°$	$36°$	$54°$	$72°$	$90°$
$\bar{\varepsilon}_n$	=	0.29	0.29	0.31	0.35	0.53	-
d (mm)	=	1.70	1.50	1.00	0.50	0.20	0.10
σ (MPa)	=	486	386	235	178	84	0
τ (MPa)	=	0	125	207	245	260	270
δ_{IC} (mm)	=	0.60	0.46	0.12	-0.09(?)	0	0
δ_{IIC} (mm)	=	0	0.51	0.68	0.55	0.74	0.57
I_1 (J/mm²)	=	285	314	341	321	331	0
I_2 (J/mm²)	=	34.8	6.47	- *	- *	- *	0
I_3 (J/mm²)	=	0	- *	- *	- *	- *	154
$J_C = I_1 + I_2 + I_3$ (J/mm²)	=	320	320	341	321	331	154
J_C (Experimental) (J/mm²)	=	270	250	250	250	250	250

*Not evaluable because integrals become negative which are not physically meaningful.

Based on these experimental results and the J_C-integral computed in Tables 3 and 4 we propose that the strain hardening exponent (n) and the dimensional changes of the fractured neck in a tensile specimen will largely determine whether J_C is invariant with the mode of fracture. If the necking is intensive such as indicated by the large crack tip opening displacements and n is low, e.g. the aluminium alloys 5251 and B1200-H14, J_C varies with θ. If however necking is moderate such as typified by small crack opening displacements and n is comparatively large, e.g. low carbon steels, J_C is approximately constant. We do not claim that we have proven this hypothesis since the computed results given in Tables 3 and 4 rely on experimental measurements of certain physical quantities. Indeed we do not have any theoretical grounds to explain these experimental findings but our suggestion seems to be qualitatively reasonable.

CONCLUSIONS

In contrast to brittle fractures in isotropic materials which are always mode I it is possible to obtain true mixed mode fractures in ductile solids. The critical J_C-integral can be evaluated using the staggered deep edge notch tension specimens. It seems that for the aluminium alloys with low strain hardening exponent but intensive necking J_C is dependent on the staggered angle. For the steels with large n but moderate necking J_C is invariant with θ. There is however no theoretical proof for these experimental results.

ACKNOWLEDGEMENT

The authors wish to thank the Australian Research Grants Commission for the financial support of this research work.

REFERENCES

[1] Erdogan, F. and Sih, G.C., J. Basic Engineering, Vol. 85, p. 519, 1963.

[2] Williams, J.G. and Ewing, P.D., Int. J. Fract., Vol. 8, p. 441, 1972.

[3] Ewing, P.D., Swedlow, J.L. and Williams, J.G., Int. J. Fract., Vol. 12, p. 85, 1976.

[4] Cotterell, B., Int. J. Fract., Vol. 1, p. 96, 1965.

[5] Sih, G.C., Int. J. Fract., Vol. 10, p. 305, 1974.

[6] Cotterell, B. and Rice, J.R., Int. J. Fract., Vol. 16, p. 155, 1980.

[7] Pook, L.P., Engr. Fract. Mech., Vol. 3, p. 205, 1971.

[8] Jones, D.L. and Chisholm, D.B., Engr. Fract. Mech., Vol. 7, p. 261, 1975.

[9] Ishikawa, H., Kitigawa, H. and Okamura, H., "J-integral of a mixed mode crack and its application", Proc. ICM3, Cambridge, Vol. 3, K.J. Miller and R.F. Smith, eds., Pergamon Press, London, p. 447, 1979.

[10] Cotterell, B., Lee, E. and Mai, Y.W., accepted for Int. J. Fract.

[11] Hill, R., J. Mech. Phys. Solids, Vol. 1, p. 271, 1953.

[12] Cotterell, B. and Mai, Y.W., "Plane stress ductile fracture", Advances in Fracture Research, Vol. 4, D. Francois et al., eds., Pergamon Press, London, p. 1683, 1981.

[13] Hill, R., J. Mech. Phys. Solids, Vol. 1, p. 19, 1953.

[14] Begley, J.A. and Landes, J.D., "The J-integral as a Fracture Criterion", in "Fracture Toughness", ASTM STP 514, p. 1, 1972.

[15] Cotterell, B. and Reddel, J.K., Int. J. Fracture, Vol. 13, p. 267, 1977.

[16] Mai, Y.W. and B. Cotterell, J. Mater. Sci., Vol. 15, p. 2296, 1980.

[17] Mai, Y.W. and Pilko, K.M., J. Mater. Sci., Vol. 14, p. 386, 1979.

SECTION VI
FATIGUE CRACK GROWTH

FATIGUE CRACK GROWTH IN AUTOFRETTAGED THICK-WALLED CYLINDERS

G. Clark

Materials Research Laboratories, Department of Defence
Maribyrnong, Victoria 3032, Australia

ABSTRACT

This paper describes the use of cold-working processes to introduce beneficial residual stresses in highly stressed components, with particular reference
to the autofrettage of thick-walled cylindrical pressure vessels for use as large-
calibre gun barrels. The effects of residual stresses on the stress intensity
range controlling the growth of radial fatigue cracks from the bore surface are
estimated using a weight function approach. A parallel experimental determination
of these effects, based on fatigue crack growth rate measurement in autofrettaged
barrel sections, is described. The forging and manufacturing procedures for a
service gun barrel were considered in both investigations; comparison of the predicted and experimentally-determined effects shows satisfactory agreement, indicating that a theoretical approach may be used to predict the fatigue crack growth
rate reductions associated with practical autofrettage conditions.

INTRODUCTION

The introduction of residual stresses by cold work has been recognized for
many years as a means of improving the elastic strength of components which are
required to remain elastic in service; well-known examples include the presetting
of springs and the overstrain (or autofrettage) of thick-walled cylindrical pressure vessels for use as gun barrels. The overstrain procedure involves deliberately producing permanent deformation of a component during manufacture using an
exaggerated form of the anticipated service loading; when this plastic deformation
is non-uniform, as is usually the case, removal of the over-strain load establishes
a residual stress distribution which opposes the normal service stresses. The effect of this process is to prevent further plastic deformation in service unless
conditions are such that the service loading exceeds that used during the original
overstrain. Superficially, this is similar to the effect of strain-hardening, but
the residual stresses arising from the cold-working operation are capable of providing a much more pronounced strengthening effect than would be possible using
only the material's strain-hardening characteristics, this can be done without in-
curring the penalty of excessive reductions in ductility and toughness. For example, the elastic strength of a thick-walled cylindrical pressure vessel with a
wall ratio (od/id) of two may be increased by some 85% by autofrettage even if the
cylinder material displays a strain-hardening capacity (UTS/σ_y) close to unity.

The principal reason for the use of cold-working procedures such as autofrettage is to improve the load-bearing capacity of lightweight components by increasing their effective range of elastic stability, but an important secondary benefit may be observed when a fatigue crack in the component extends into material containing residual compressive stresses. More specifically, if the residual stresses act to bring the faces of the crack into contact during part of the applied loading cycle, the reduction in stress-intensity range which results can lead to a decrease in crack growth rate. This effect, which may be substantial, has been recognized in the aircraft manufacturing industry where the cold working of fastener holes is an accepted means of minimizing the growth rate of any radial fatigue cracks which develop around the circumference of the hole [1].

Application of these cold working procedures is encouraged by the use of increasingly high service stresses in the high-strength components used for lightweight military equipment. Such changes in design and materials are largely responsible for the increasing number of components having service lives which are limited by fatigue crack growth. For example, many large calibre guns are now retired from service when the number of rounds fired exceeds a safe proportion of an experimentally-determined fatigue life, whereas older designs which were manufactured from lower-strength steels could remain in service until wear of the bore surface produced an unacceptable loss of accuracy. The need to incorporate a relatively large safety factor in service fatigue lives leads to retirement of perfectly usable equipment, and in the case of large calibre gun barrels has generated considerable interest in fracture mechanics analyses of fatigue and fracture in thick-walled cylinders. Such analyses are intended to improve initial estimates of gun life and to assist in the development of non-destructive inspection programs which could greatly extend the barrel service life. In addition, the results of investigations of this kind are of course applicable to most thick-walled pressure vessels and find widespread use in other applications of fracture mechanics to highly stressed components.

The initiation stage of fatigue cracking does not contribute significantly to barrel life since a thermal craze-crack network forms on the bore surface within a few rounds of the barrel entering service, and any investigation of fatigue in gun barrels consists of analyzing the factors which control fatigue crack propagation. As a result of intensive work (much of it conducted in the last ten years) the effects of many of these factors are now well understood [2-4]; unfortunately, their use in life prediction exercises, despite being a relatively straightforward procedure, has little practical value until the remaining factors can be incorporated in the model. Furthermore, much of the theoretical work carried out to date has not yet been adequately tested by experiment, primarily because of the difficulties involved in isolating individual factors. The investigation of the effects of autofrettage residual stresses on fatigue crack growth, in particular, was neglected for many years in spite of the fact that the use of autofrettage can increase barrel fatigue life by a factor of approximately 2.5 [5,6]. The techniques for predicting residual stresses in autofrettaged thick-walled cylinders have been available for over thirty years [7-9] and the remaining problem - that of estimating the stress intensity produced by complex crack face loadings - has recently become amenable to analysis by various methods, including superposition of standard results [4], numerical analysis [8] and weight function (Green's function) techniques [10,11]. This paper describes the results of an investigation which combines a theoretical analysis with an experimental procedure in order to determine how different levels of autofrettage affect the growth of fatigue cracks in sections of a service gun barrel. The approach adopted was to conduct parallel investigations (one predictive, the other experimental) to estimate the reduction in crack tip stress intensity associated with three different levels of autofrettage; in this way, the accuracy of the theoretical techniques could be tested, and in addition, the choice of a service gun barrel and use of normal manufacturing conditions would ensure that all of the more practical aspects involved in applying the theoretical procedure were given adequate attention.

MANUFACTURING AND SERVICE CONDITIONS

The gun barrel considered is that from a 76 mm gun manufactured in Australia, and is usually autofrettaged using a swage technique, in which an oversize mandrel is forced through the bore of the hollow cylindrical forging. The forging is restrained at the entry point of the tapered mandrel, and is provided with various lubricating films on the bore surface. For this investigation, a production forging was cut into sections, each of which was machined to give a different swage/bore interference and then autofrettaged using the normal manufacturing procedure. Barrel sections A, B, C and D were machined to give diametral interferences of 0.363, 0.605, 0.848 and 1.097 mm; condition A corresponded to purely elastic deformation, in order to provide a stress-free control for determination of material properties. Sections C and D corresponded approximately to the range of conditions normally used in the manufacture of the service barrel and were therefore of major interest. The wall ratio was approximately 1.8.

After autofrettage, each section was cut into rings approximately 30 mm thick; a number of these specimens were taken from the central part of each section, to avoid end effects, and were ground to a thickness of 25.4 mm.

STRESS INTENSITY REDUCTION - THEORETICAL

A. Residual Stress Distribution

The effects of applying an internal pressure to a thick-walled cylinder are illustrated in Figure 1; as the radial pressure increases, yielding initiates at the bore surface and the boundary of the plastically-deformed region then moves into the wall. The extent to which this occurs depends on the maximum pressure applied; most autofrettage operations rely upon the yield zone extending through more than half the wall thickness of the cylinder, and some involve producing plasticity throughout the wall. When the autofrettage pressure is removed, the inner wall region (which has been subjected to the maximum plastic strain) is compressed by the outer "layer" of the cylinder, leading to the establishment of residual hoop compressive stresses near the bore, balanced by residual tensile hoop stress nearer the external surface. For some combinations of cylinder geometry and autofrettage pressure, the magnitude of the compressive stress may be sufficient to cause reversed yielding at the bore [7].

The calculation of residual stresses for the expansion of a cylinder by internal pressure is relatively straightforward, and was described in detail by Hill [8], Prager and Hodge [7] and others some thirty years ago. The approach used here (described in more detail in [12]) is that of [7], the residual hoop stress σ_r at radius r in a cylinder with internal and external radii R_i, R_o and displaying monotonic and reversed yield to radii R and ρ produced by internal pressure P is given by

420

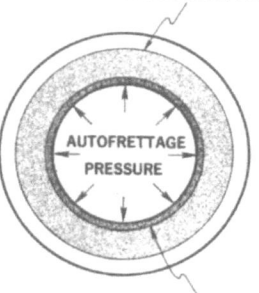

LIMIT OF YIELD ZONE
AT AUTOFRETTAGE PRESSURE

AUTOFRETTAGE
PRESSURE

REVERSED YIELD ZONE
AFTER REMOVAL OF PRESSURE

Fig. (1) - Yield zone formation during autofrettage of a thick-walled cylinder

$$\sigma_r = - \beta Y^* [1 + \ln(r/R_i)] \text{ where } R_i < r < \rho$$

$$= Y^* + Y^*\ln(r/R_i) - P - [P - (1+\beta)Y^*\ln(\rho/R_i)][(R_o/r)^2 + 1]/[(R_o/\rho)^2 - 1]$$

$$\text{where } \rho < r < R$$

$$= (Y^*/2)[(R_o/r)^2 + 1](R/R_o)^2 - [P - (1+\beta)Y^*\ln(\rho/R_i)] \times$$

$$\times [(R_o/r)^2 + 1]/[(R_o/\rho)^2 - 1]$$

$$\text{where } R < r < R_o \tag{1}$$

The relationship between the internal pressure P and the monotonic yield radius is

$$P = (Y^*/2)[1 - (R/R_o)^2 + \ln(R/R_i)^2] \tag{2}$$

While a Tresca yield criterion is assumed for ease of calculation, it has been shown [13] that the yielding of steel cylinders for gun barrels is more closely represented by von Mises criterion. However, it has been observed [8,14] that the use of a suitably modified yield stress in the Tresca condition produces a criterion which is a satisfactory approximation to that of von Mises. The modified yield stress Y* used in the present work is 8% greater than the measured 0.2% offset proof stress of 1050 MPa, this value being selected on the basis of an experimental investigation of yielding in large-calibre gun barrels [14]. Throughout the calculation, elastic-perfectly plastic behavior is assumed; since the steels used in gun barrel manufacture have low UTS/yield strength ratios, this assumption is believed to be a reasonable one.

The parameter β in equation (1) provides a means of allowing for reversed yielding of the cylinder material. This behaviour may involve the Bauschinger effect [15] which describes yielding in compression at a stress which may be considerably lower than that reached during prior plastic deformation in tension. The Bauschinger effect parameter β as used here, after [16,17], indicates the ratio of compressive yield stress to the yield stress in tension and since this can have a significant effect on the residual stress distribution near the bore [17], an experimental determination of this parameter was made using material from barrel section A [12]; a value of β = 0.40 was used in all calculations.

Equation (1) was used to calculate the residual stress distribution expected for autofrettage conditions B, C and D; the calculation and assumptions made are discussed in greater detail in [12], and the predicted distributions are shown in Figure 2. Some of the more important aspects of this procedure are as follows:

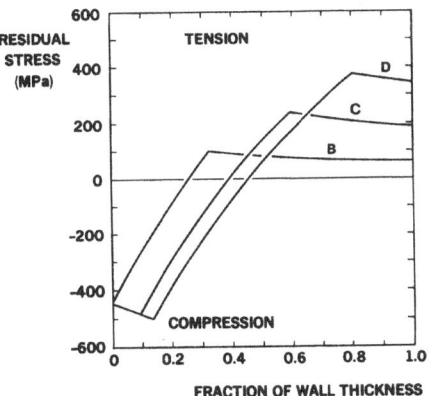

Fig. (2) - Results of residual stress prediction procedure for autofrettage conditions B, C and D

1. The calculation is based on theory which assumes that hydraulic pressure is applied uniformly to the bore. However, the mechanical autofrettage technique used in this case involves a complex stress distribution in the material near the swage, and the use of a technique capable of incorporating any variations between the two procedures would have been preferable. Such an analysis is not yet available, and the assumption was made that the radial pressure produced by the swage on the bore behaves exactly as hydraulic pressure; the use of this assumption has been reported [18] to provide satisfactory results in predicting swage autofrettage behaviour. The swage/bore pressure was estimated using an iterative technique which made allowance for elastic deformation of the swage under pressure.

2. A major effect of increasing the autofrettage pressure (or swage interference) is to produce marked increases in the hoop tensile stress at the external surface of the cylinder as shown in Figure 2. This is most undesirable in that such increases act to reduce the critical length of any fatigue crack growing

from an external stress-concentrator, and limit the extent to which higher autofrettage pressures may be used to improve the elastic strength of the cylinder.

3.   The effect of reversed yielding is to limit the compressive stress at
the bore; increases in autofrettage pressure beyond that required to produce reversed yielding on unloading simply extend a zone of approximately constant residual stress into the wall.   This behaviour suggests that for very short cracks
on the bore surface, little or no benefit in terms of a reduction in crack growth
rate may be associated with an increase in the autofrettage pressure.   However,
the calculation of residual stresses in the reversed yield zone is based on an assumption that the material behaves in an elastic-perfectly plastic manner, whereas
the experimental determination of the Bauschinger effect parameter β referred to
above was complicated by the fact that this steel exhibited pronounced strain hardening in the compressive part of the test [12].   This difficulty suggests that a
different approach to calculation of reversed yielding effects is required in order
to improve the accuracy of residual stress predictions for the bore surface region.

4.   The predicted residual stress distributions shown in Figure 2 have been
compared [12] with measurements of residual hoop stress obtained by X-ray diffraction from the surface of ring specimens in conditions A, B, C and D; the experimental measurements available were limited in number, but corresponded closely with
the predicted distributions, the agreement for sections C and D being better than
for condition B.   Section A was confirmed to be essentially free of hoop stresses
within the limits of experimental accuracy.

B.   Calculation of Stress Intensity Reduction

Weight function (Green's function) techniques [10,19] are now a well-established means of determining the stress intensity K which is generated at the
tip of a crack (of length a) when that crack is introduced into an uncracked body
[11].   If the body is assumed to contain stresses $\sigma(t/a)$ normal to the plane on
which the crack is introduced, the stress intensity is usually determined from an
expression of the form

$$K = A \int_0^1 \sigma(t/a)F(t/a)d(t/a)$$

$\qquad\qquad\qquad\qquad\qquad\qquad\qquad\qquad\qquad\qquad\qquad\qquad\qquad\qquad$ (3)

where A is a geometry-related factor and t is a distance along the crack face.

The principal benefit offered by this technique, namely that once the
weight function F(t/a) is known, stress intensities may be determined for any
stress distribution $\sigma(t/a)$ simply by evaluating equation (3), makes it particularly suitable for use in cases like the present one in which various residual
stress distributions need to be considered.

In practice, F(t/a) must usually be determined for a range of crack length/
specimen wall thickness ratios (a/T), and may be derived numerically (or otherwise) from crack surface displacement estimates [19,21] or more directly as was
done here, by a numerical determination of the relationship between crack face
loading and the crack tip stress intensity.   Using a finite element analysis of
ring specimens containing radial bore cracks of various lengths [20], the weight
function G(a/T, t/a), in a slightly modified form compared with that in equation
(3), was derived [21] by determining the stress intensity produced by applying
known loads to the crack faces

$$K(a/t) = (2a/\sqrt{\pi}) \int_0^1 \sigma(a/T,t/a)G(a/T,t/a) \, \sqrt{1/(1-t^2/a^2)} \, d(t/a) \qquad (4)$$

The function G(a/T,t/a) determined using this procedure is shown in Figure 3 in the form of a bicubic spline surface which represents the numerical data and which

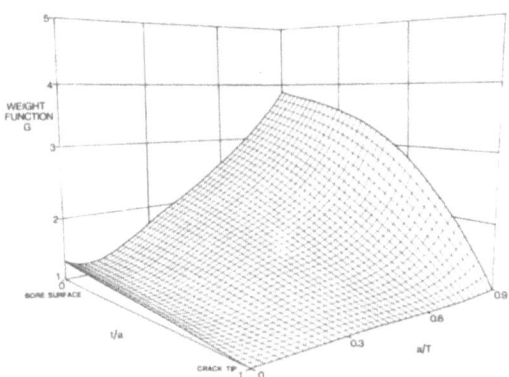

Fig. (3) - The weight function G(a/T,t/a) in equation (4) for a single crack in a ring loaded in compression across a diameter (see Figure 6); 't' represents a distance along the crack (of length 'a'). The wall thickness of the ring is T

facilitates evaluation of equation (4) for other stress distributions. The surface is described in more detail in [12].

The ease with which stress intensities may be determined using a known weight function and known stress distributions makes testing the accuracy of the weight function a relatively simple procedure; in this case, the two configurations for which stress intensities were estimated (and for which published solutions are available) were uniform internal pressure on the bore and crack faces, and compression of a radially-cracked ring along a diameter. In both cases, use of the weight function produced estimates of stress intensity which were within 5% of the values published in [22-24], respectively.

To apply the weight function technique to the case of residual stresses in a cracked ring, equation (4) was integrated numerically using the appropriate stress distribution, equation (1), as σ(a/T,t/a), and the weight function shown in Figure 3. The stress intensities calculated using this method for conditions B, C and D are shown as solid lines in Figure 4 and represent the change in stress intensity associated with the presence of the residual stresses; note that all values are negative, showing that over the range of crack lengths considered, the residual stress acts to close the crack. Two features of these curves are worth emphasizing; firstly, as would be expected from the prediction that reversed yield-

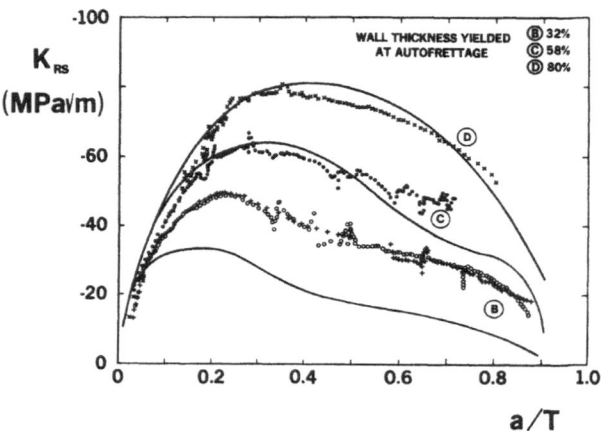

Fig. (4) - Predicted (solid lines) and experimentally-determined stress
intensity changes associated with three different levels of
autofrettage B, C and D. The crack length/wall thickness
ratio is a/T. Results from two specimens are shown for condition B

ing results in the residual compressive stresses near the bore being maintained at
an approximately constant level, the stress intensities for short cracks are similar for all three autofrettage conditions. Secondly, the effects of the compressive residual stresses are considerable, with changes in stress-intensity of up to
80 MPa$\sqrt{m}$ being predicted for condition D, and reach a maximum near the transition
from compressive to tensile residual stress for each autofrettage level.

STRESS INTENSITY REDUCTION - EXPERIMENTAL DETERMINATION

The effect of the residual stresses in an autofrettaged cracked cylinder is
to produce crack face contact over part of the applied loading cycle, the crack
growth rate being controlled by the stress intensity range ΔK_{eff} for which the
crack is effectively open, as illustrated in Figure 5. For life prediction purposes, however, the apparent stress intensity K_o which must be applied to open
the crack is of more direct interest since this should be equal to $-(K_{RS})$. In
practice, K_o may be determined simply by subtracting ΔK_{eff} from the maximum "applied" stress intensity K_{max}.

In order to determine ΔK_{eff} experimentally, use was made of the sensitivity of
fatigue crack growth rate to stress intensity range; once the characteristic relationship between these parameters has been established by a suitable testing
procedure, the value of ΔK_{eff} for any observed crack growth rate may be determined
directly from this relationship.

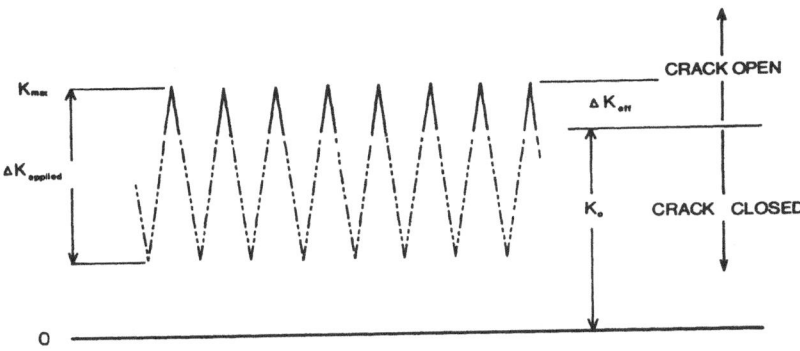

Fig. (5) - Closure of the crack (at an externally applied load which acting
alone would correspond to a stress intensity of K_o) reduces the
stress intensity range for ΔK_{appl} to ΔK_{eff}, which may be determined from measurements of fatigue crack growth rate

This technique, when used with a suitably reliable crack growth rate measurement system has been shown [25,26] to be capable of providing accurate stress intensity calibrations for laboratory fatigue specimens. In the present investigation, it was applied to ring specimens containing radial fatigue cracks which were
grown from a notch on the inner surface by loading the ring in compression across
a diameter, as shown in Figure 6. In order to ensure that variations in material
properties between specimens were not a significant source of error, the material's
fatigue crack growth rate characteristics were established from tests on four
rings cut from barrel section A. Testing involved growing the crack using various values of stress intensity range ($K_{max}-K_{min}$), the loads being changed periodically as the crack extended. A different sequence of load levels was used for each
specimen in order to ensure the detection of any significant errors arising from
the experimental procedure and, more particularly, the crack measurement system.
Since this particular specimen geometry provides almost constant stress intensity
conditions when the crack extends under constant cyclic load range conditions [26],
the material's fatigue characteristics are displayed, Figure 7, as discrete points
rather than a continuous curve; a third-order polynomial fitted to these data was
used to represent the relationship between crack growth rate and alternating stress
intensity.

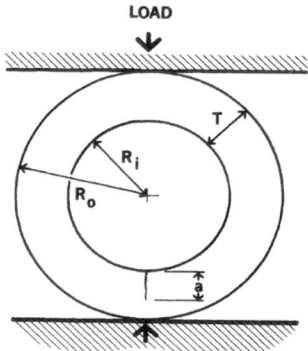

Fig. (6) - Loading configuration for experimental determination of
fatigue crack growth rate in both autofrettaged and non-
autofrettaged specimens

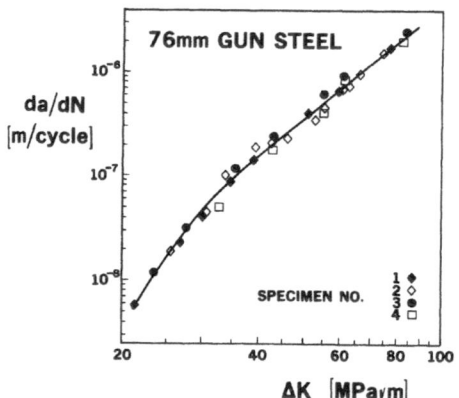

Fig. (7) - Fatigue crack growth rate properties determined using
four specimens of the kind shown in Figure 6; loads were
varied during the test as this specimen provides almost
constant stress intensity conditions when the crack grows
under constant load range. The solid line is a least-
squares polynomial fit to the data

An identical technique was used for testing autofrettaged specimens in conditions B, C and D. However, considerable increases in maximum load during fatigue cycling were necessary to produce crack extension, and it was also necessary to change this load repeatedly as the crack length changed in order to maintain a suitable growth rate. For rings B, C and D, growth rates were measured continuously as the crack extended, and the computer program which processed this data then made use of the polynomial representation of the material's fatigue properties to estimate the true stress intensity range ΔK_{eff} which produced each growth rate. A comparison of ΔK_{eff} with the stress intensity expected from the applied loading could then be used (as shown in Figure 5) to estimate the change in stress intensity resulting from autofrettage, K_{RS} (= $\Delta K_{eff} - K_{max}$).

Values of K_{RS} determined from duplicate tests on rings in condition B showed that the experimental techniques used in the investigation provide excellent reproducibility of results. The experimentally determined stress intensity reductions K_{RS} for conditions B, C and D are compared in Figure 4 with values predicted using the theoretical approach described earlier; for the autofrettage conditions normally used for this particular barrel, conditions C and D, the experimental and theoretical results correspond closely, while for condition B, the stress intensity effects observed experimentally are greater than those predicted.

DISCUSSION

The magnitude of the experimentally-determined stress intensity reduction shown in Figure 4 indicates that the residual stresses associated with cold-work treatments can have significant effects on crack growth rates. For example, from Figure 7, the greatest reduction observed (approximately 80 MPa√m) corresponds to a reduction in fatigue crack growth rate for this material of at least a factor of 400, and it is therefore clear that any fracture safety assessment procedure for autofrettaged thick-walled cylinders must incorporate a suitable means of predicting these effects.

Since the experimental techniques used in this investigation are based on measuring the crack growth rate reductions produced by autofrettage, the stress-intensity effects shown in Figure 4 may be used with confidence with the materials data in Figure 6 as part of any prediction procedure for crack growth rates and critical crack length. In addition, the close correspondence between predicted and measured stress intensity reductions for conditions C and D indicates that prediction of stress intensity and crack growth rate effects from knowledge of autofrettage conditions and materials characteristics is feasible.

For the low level of autofrettage used in condition B, agreement between predicted and measured stress intensity changes is not as good as that for the more usual manufacturing conditions. In this particular case, the difference between the predicted and measured residual stresses is believed to arise from the fact that the swage/bore interference in this example is largely elastic, and any errors introduced in measurement and in calculating elastic deformations produce relatively large errors in the plastic component of bore expansion, and hence in the residual stress. For example, an error of 0.025 mm in the measured diameter of the swage or bore (equivalent to approximately 30°C temperature differential between the two components, both of which are heated in practice) can produce a variation of more than 20% in the predicted permanent bore expansion for condition B, and since residual stresses are particularly sensitive to changes in permanent expansion at lower levels of autofrettage, the accurate prediction of residual stresses from low strain swage autofrettage data can be seen to be particularly difficult. In contrast, the same initial error would lead to only 8% and 4% variation in residual bore expansion for conditions C and D respectively, and these variations would themselves have relatively small effect on residual stress.

The variation in residual hoop stress through the wall produces distortion of the crack faces, reducing the crack opening which would otherwise be observed near the bore surface. The effect of this distortion is to wedge open the crack and to raise the applied load necessary to achieve full crack opening, effectively increasing the magnitude of the stress intensity reductions shown in Figure 4. While this effect has been studied numerically for a ring loaded in diametral compression by Jones and Callabresi [27], the $R_i/R_o = 0.9$ geometry they considered is not comparable to that used here, and the magnitude of any wedging contribution to the normal closure conditions is not known for the present case. However, the close agreement between predicted and measured stress intensity reductions suggests that any contribution of crack face distortion to the normal crack closure conditions encountered in autofrettage is relatively small.

CONCLUSIONS

1. Measurements of fatigue crack growth rates show that autofrettage of a thick walled cylindrical pressure vessel can provide considerable reductions in the stress intensity at the tip of a radial bore crack. This data was obtained for various levels of autofrettage of a large calibre gun barrel, and is directly relevant to the service conditions of this component.

2. Theoretical predictions, using established techniques for estimating residual stresses, and a weight function approach to determine the effect of these stresses on crack tip stress intensity, compare favourably with measured values, where initial conditions are known to sufficient accuracy.

3. The influence of the Bauschinger effect on the residual stress distributions generated by overstrain of cylindrical pressure vessels cannot yet be satisfactorily incorporated in models which rely on elastic-perfectly plastic material behaviour.

4. While the effect of crack face distortion by residual stresses is expected to provide an additional contribution to the crack closure which retards fatigue crack growth, this contribution does not appear to be large.

ACKNOWLEDGEMENTS

The author would like to express appreciation for the assistance provided by Mr. T. V. Rose in this program.

REFERENCES

[1] Grandt, A. G. and Gallagher, J. P., "Proposed Fracture Mechanics Criteria to Select Mechanical Fasteners for Long Service Lives", STP 559 ASTM, pp. 283-297, 1974.

[2] Clark, G., "Fatigue in Gun Barrels", Proc. Symp. Fracture Mechanics, Australasian Institute of Metals, Krisbane, 1977.

[3] Davidson, T. E. and Throop, J. F., "Practical Fracture Mechanics Applica-
     tions to Design of High Pressure Vessels", Report WVT-TR-76047, Watervliet
     Arsenal, N.Y., 1976.

[4] Underwood, J. H. and Throop, J. F., "Surface Crack K-Estimates and Fatigue
     Life Predictions in Cannon Tubes", STP 687, ASTM, pp. 195-210, 1979.

[5] Davidson, T. E., Throop, J. F., Austin, B. A. and Reiner, A. N., "Analysis
     of the Effect of Autofrettage on the Fatigue Life Characteristics of the
     175 mm M113 Gun Tube", Report WVT-6901, Watervliet Arsenal, N.Y., 1969.

[6] Nishioka, K. and Hirakawa, K., "Effect of Internal Surface Flaw on the Fa-
     tigue Strength of a Thick Walled Cylinder under Cyclic Internal Pressure",
     Proc. 2nd Intl. Conf. High Pressure Engng., Brighton, July 1975, I. Mech. E.,
     1977.

[7] Prager, W. and Hodge, P. G., Jr., Theory of Perfectly Plastic Solids, John
     Wiley, 1951.

[8] Hill, R., "The Mathematical Theory of Plasticity", Oxford University Press,
     1950.

[9] Nadai, A., Theory of Flow and Fracture of Solids, McGraw-Hill, 1950.

[10] Bueckner, H. F., "A Novel Principle for the Computation of Stress Intensity
     Factors", Z. Angew. Maths. Mech. 50, pp. 529-546, 1970.

[11] Cartwright, D. J. and Rooke, D. P., "Green's Functions in Fracture Mechanics",
     Proc. Conf. Fracture Mechanics: Current Status, Future Prospects, Cambridge,
     1979.

[12] Clark, G., "Residual Stresses in Swage-Autofrettaged Thick-Walled Cylinders",
     Report MRL-R-847, Materials Research Laboratories, Melbourne, 1982.

[13] Davidson, T. E., Barton, C. S., Reiner, A. N. and Kendall, D. P., "Overstrain
     of High-Strength Open-End Cylinders of Intermediate Diameter Ratio", Proc.
     1st Intl. Congress Exp. Mechanics, December 1961.

[14] Warren, A. G., "Autofrettage", Proc. Symp. Internal Stresses in Metals and
     Alloys, Inst. of Metals, pp. 209-218, 1947.

[15] Bauschinger, J., Civilingenieur, 27, p. 289, 1881.

[16] Welter, G., "Micro- and Macro-Deformations of Metals and Alloys under Longi-
     tudinal Impact Loads", Metallurgica, Part I, pp. 287-292, Part II, pp. 328-
     330, Part II, pp. 13-17, 1948.

[17] Milligan, R. V., "The Influence of the Bauschinger Effect on Reverse Yield-
     ing of Thick-Walled Cylinders", Report WVT-7036, Watervliet Arsenal, N.Y.,
     1970.

[18] Davidson, T. E., Barton, C. S., Reiner, A. N. and Kendall, D. P., "A New
     Approach to the Autofrettage of High Strength Gun Tubes", IPM Project Report,
     Watervliet Arsenal, April 1959.

430

[19] Rice, J. R., "Some Remarks on Elastic Crack Tip Stress Fields", Int. J. Solids Struct. 8, p. 751, 1951.

[20] Clark, G. and Rose, T. V., "Stress Intensity Calibration of a Ring Specimen using a Finite Element Technique", Report MD 80-4, Materials Research Laboratories, Melbourne, 1980.

[21] Clark, G., "Use of Weight Functions to Determine the Stress Intensity for a Cracked Thick-Walled Cylinder", Report MRL-R-814, Materials Research Laboratories, Melbourne, 1981.

[22] Bowie, O. L. and Freese, C. E., "Elastic Analysis for a Radial Crack in a Circular Ring", Engineering Fracture Mechanics, 4, pp. 315-321, 1972.

[23] Jones, A. T., "A Radially-Cracked Cylindrical Fracture Toughness Specimen", Engineering Fracture Mechanics 6, pp. 435-446, 1974.

[24] Grandt, A. F., Jr., "Evaluation of a Cracked Ring Specimen for Fatigue Testing under Constant Range in Stress Intensity Factor", Proc. Conf. Fracture Mechanics and Technology, Hong Kong, March 1978.

[25] Clark, G., "Using Fatigue Crack Growth Rates to Determine Stress-Intensity Calibrations", Proc. Conf. Australian Fracture Group, Melbourne, November 1980.

[26] Clark, G., "Experimental Determination of Stress-Intensity in a Cracked Cylindrical Specimen", Report MRL-R-774, Materials Research Laboratories, Melbourne, 1980.

[27] Jones, A. T. and Callabresi, M. L., "Numerical Analysis of the Influence of Residual Stresses on Crack Closure in Rings", Engineering Fracture Mechanics, 11, pp. 675-688, 1979.

USE OF DAMAGE CONCEPT ON PREDICTING FATIGUE LIFE IN TWO-LEVEL STRAIN
CONTROLLED TESTS

Z. Azari, M. Lebienvenu and G. Pluvinage

Laboratoire de Fiabilité Mécanique, Université de Metz
Ile du Saulcy - 57000 Metz, France

ABSTRACT

Damage concept is discussed through several definitions. Damage is associated through physical, mechanical or terotechnological properties. Experimental
verifications show that damage evolution curves strongly depend on properties
used as indicator.

Predicting fatigue life of two-level strain controlled tests can be done by
using damage concept. Special attention is focused on damage transition to be
used for prediction.

ON DAMAGE CONCEPT

Damage concept is a useful medium to study evolution of mechanical properties
and particular cyclic properties. Damage definition is controversial because the
word "DAMAGE" has at least two different meanings: creation of defects at the macro or microscale and alteration of material properties. Measurement of defects
is not directly useable for mechanical analysis but leads to physical modeliza-
tion. Another way is to consider damage D by assuming that D is a function of the
alteration of a property X:

$$D = f(X) \tag{1}$$

The damage parameter D is function of the variables stress σ, strain ε, time
t, hardening variable α, etc.

$$\frac{dD}{dt} = f(\sigma, \varepsilon, t, D, \alpha, ..) \tag{2}$$

There is no restriction for properties which can be used for damage evolution.
The following properties have been chosen:

- physical properties (resistivity, acoustic emission, density);

- mechanical properties (ductility, fatigue limit, internal damping, yield
  stress, etc.);

- terotechnological properties (remaining life, reduced creep life).

Damage representation comprises of different forms:

(1) tensorial representation of effective stress according to the Katchanov's basic idea [1];

(2) description lying within the thermodynamic framework of materials with internal variables [2];

(3) homogenization which allows to derive macroscopic properties of highly heterogeneous quasi-periodic medium from their microscopic properties [3].

According to this definition

$$D_X = f^{-1}(X)$$ (3)

Damage is dependent on the support properties used for definition. For this reason, we use the subscript X after capital letter D_X to indicate these properties. The relation between damage and change of properties is assumed to be linear as that used in the effective stress concept ($\hat{\sigma}$):

$$\hat{\sigma} = \frac{\sigma}{1-D_\sigma}$$ (4)

The Miner definition used in low cycle fatigue or creep test leads to the expressions

$$D_\beta = \frac{n}{N_r} \text{ or } D_\beta = \frac{t}{t_r}$$ (5)

where N_r and t_r are the number of cycles and time to rupture. The quantities r and t are the applied cycle number and creep time while β is the fraction of life.

The damage function $D = k(\Phi)$ shall be referred to as the elementary curve of damage with Φ being a forcing variable and other variables are kept constant. Damage evolution is generally divided into four possible processes, Figure 1:

(1) damage increases linearly (Miner - process A):

(2) nonlinear increasing of damage (Gatts [4], Buiquoc [5]) with eventually acceleration or deceleration at the end of life of materials (Chaboche [6]): processes B and C;

(3) damage accumulation at the beginning and at the end of life of materials: process D and;

(4) damage accumulation at the beginning, quasi disappears after a maximum and develops at the end of life: process E.

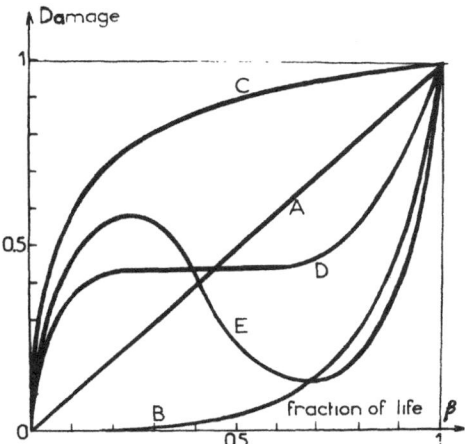

Fig. (1) - Damage accumulation process

Elementary curves of damage are considered as material constants. Predicting fatigue or creep life requires a knowledge of the damage accumulation rule when a second forcing variable is modified during life, and two elementary curves of damage. For schematization, we examine the case of low-cycle fatigue with a change of level of applied strain ε_1 to ε_2 at life fraction β_1.

Two possibilities are presented:

(1) a transition at non constant damage level with interaction between the first and second levels, Figure 2, to explain gap with Miner's rule;

(2) a constant damage transition with interaction or non interaction at the second level, Figure 3.

EXPERIMENTAL METHODS TO MEASURE DAMAGE

In order to value damage through different properties, we have tested 2 bainitic steels with chromium and molybdenum called PM20 and NF30. Tests were conducted on specimen with a closed loop fatigue machine. Total applied strains ε_t were controlled by a longitudinal extensometer at two different levels, 0.64%, 0.73%. Specimens were heated at two temperatures, 560° and 610°. During mechanical cycling, many defects can appear: point, line, surface and volume defects. They do not appear in the same time, or period of stress-number of cycle diagram. On such a diagram for NF30, Figure 4, three stages are observed:

(1) stage I: accommodation;

(2) stage II: stabilization of cyclic material properties;

434

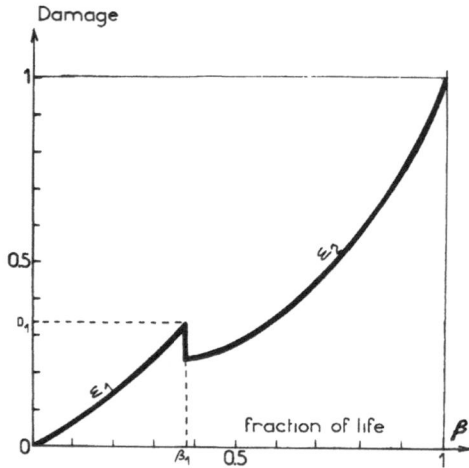

Fig. (2) - Damage accumulation at non constant damage transition

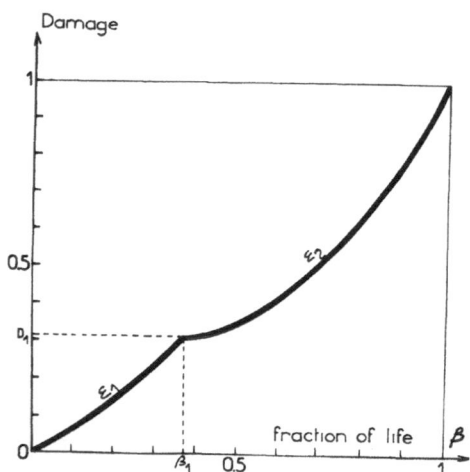

Fig. (3) - Damage accumulation at constant damage transition

(3) stage III: growth of principal(s) crack(s) before failure.

Examination of metals with a transmission electron microscope shows multiplication of dislocations during stage I and formation of cells which are completely

achieved at the beginning of Stage II.

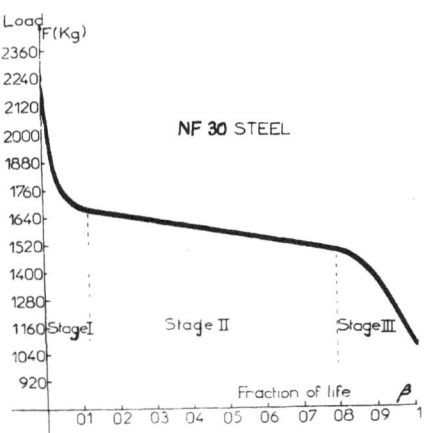

Fig. (4) - Evolution of applied stress during mechanical cycling

In the same time, we have measured the normalized cumulative crack length $\frac{\ell}{\ell_r}$ (ℓ_r, cumulative crack length at rupture) on a gauge length of 5 mm. These quantities increase with life fraction β after an incubation time of $\beta = 0.4$-0.5. Principal cracks appear for $\beta = 0.7$-0.8, Figure 5.

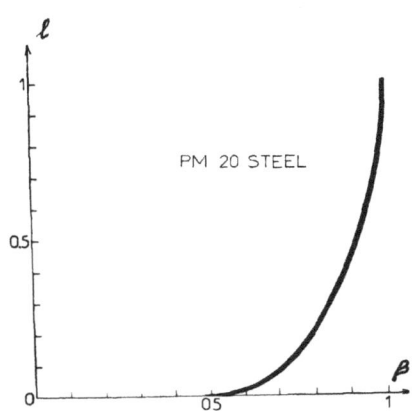

Fig. (5) - Normalized cumulative crack length versus fraction of life

We have studied the influence of these defects on physical, mechanical and terotechnological properties alteration during mechanical cycling. In order to compare every damage parameter, a standard relationship is used:

$$D_X = \frac{X_N - X_o}{X_R - X_o} \qquad (6)$$

where X_o is the value of the considered properties for the virgin material, X_R is the value of the considered property when failure occurs, X_N is the value of the considered property after N cycles of strain. These properties are:

Electrical Resistance. Using the same formalism as the effective stress concept, it is possible to define effective current density $\tilde{i}$ given by the following relationship:

$$\tilde{i} = \frac{i}{1 - D_i} \qquad (7)$$

where D_i is damage associated with change of resistance. We consider that damage effect resistivity by multiplication of point and line defects in the most part of material life. But when cracks develop damage, affect principally effective section, this influence increases rapidly damage at the end of life. Electrical measurements are done unloaded to take off elastic stress effects on electrical resistance. A constant direct current is applied to the specimen. Difference of electrical potential is measured with a high resolution voltmeter (10^{-7} volt). Average resistance of virgin specimen is about $2,6\ 10^{-4}$ ohm and resistance variation after failure is about 7%.

Density. By using this property, we consider that damage is principally correlated with volumic defects. Density measurements are done according to Ratcliffe's method [7]. Electronic pair of scales with a 10^{-7} sensibility, in isothermal precinct allow to measure density variation better than 5×10^{-5}. This method shows microvoids creation during first loading and it is not accurate for damage resulting to microcracks.

Hardness. Brinell hardness (ball diameter of 2.5 mm and load of 187,5 kg) was used to see damage due to principally linear defects.

Reversibility Limit. Use of reversibility limit was recently used by Tamhan [8] as damage indicator. We have used this method not in torsion but in compression. It is defined as an applied stress giving a permanent strain ε_r of 10^{-5}. This strain is determined by incremental method. If applied strain is lower than ε_r, dislocation displacement is great and some are trapped. Multiplication of defects which trapped dislocations tends to decrease reversibility limit.

Young's Modulus. Effective stress $\tilde{\sigma}$ is proportional to elastic strain ε_e according to the following relationship

$$\tilde{\sigma} = E_o \epsilon_e \tag{8}$$

E_o is initial value of Young's modulus; introducing definition of effective stress in relationship number (8), we obtain

$$\sigma = E_o (1-D_E)\epsilon_e \tag{9}$$

when uniaxial case is only considered. It is possible to give relationship between damage and Young's modulus E.

$$E = E_o (1-D_E) \tag{10}$$

This method was first used by Lemaitre [9]. Experimental measurements are done by measuring the slope of the loading part of hysteresis loop between $\pm \frac{1}{2} F_{max}$ (F_{max} = maximum load). An important evolution appears for the first cycles and when microcracks appear.

Residual Ductility. By extracting small cylinders (ϕ 8 mm, height 6 mm), it is possible to measure residual conventional ductility in compression by Polhemus' method [10].

Residual Fatigue Life. Residual fatigue life at second or last level is considered as a damage parameter by many authors. Damage is given by

$$D_\beta = 1-\beta_2 \tag{11}$$

with $\beta_2 = n_2/N_{r_2}$, life fraction at second level. This definition of damage is used in Palgreen-Miner's linear cumulative law.

Load Fall During Strain Controlled Tests. Stress evolution is sensitive to all defects produced during mechanical cycling and can be considered as a damage parameter. Conventionally, failure is taken as a 50% fallen load.

Plastic Strain on Hysteresis Loop. Possibilities to measure damage with plastic strain of hysteresis loop was introduced by Chaboche [6]. This author proposes to exprime a strong nonlinear damage by the equation

$$D = 1 - (\epsilon_p/\epsilon_o)^{1/m_c}$$

with m_c = b/c where b and c are respectively Basquin and Coffin exponents; ϵ_{p_o} represents plastic strain measurement on the stabilized cycle. These elementary curves of damage $D_\chi = f(\beta)$ are plotted on Figure 6.

This figure shows clearly that all kinds of nonlinear damage process described earlier is represented and depends of chosen properties. Some properties are sensitive to linear and volumic defects which can appear principally during stage I (density, hardness); they give a rapid increasing of damage at beginning of life. Some others are sensitive to microcracks (electrical resistance, Young's modulus) and give a strong increasing of damage during stage III. Cycling mechanical prop-

438

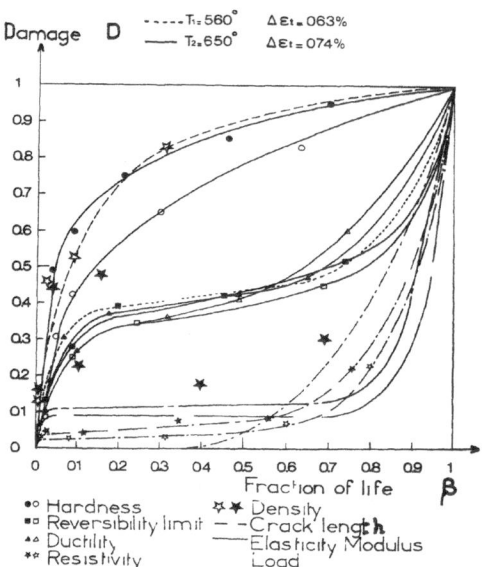

Fig. (6) - Damage evolution versus fraction of life for different experimental methods

erties give stable value of damage during a large fraction life (75%). All these methods give the same limited value of stages I, II and III with a small scattering. This result is according to invariable process postulate of Erismann [11] which assumes that materials have an invariable modification process during fatigue life. This postulate predicts also that there is a correlation between different damage elementary curves; correspondingequal damage points are tied with equal value of fraction life β. Yet, this postulate is not always verified, Figure 7.

To choose one property as a damage parameter needs three remarks: some properties give an important evolution (50% fall of load, 20% reversibility limit, 12% ductility); density (0,1%) or electrical resistance (4%) have a poor sensitivity. The possibility of evaluation damage continuously is an important element of choice. According to these three remarks, we have given special attention to damage evolution by fall of load and plastic strain evolution. These two properties show clearly the three stages of damage process and give a reasonable value of damage at the end of stage I, according to interaction effect during this period. The type of damage curve shows that damage is coupling with elastic strain energy and strain hardening in formulation of material free energy. This assumption is different from Chaboche's assumption [6] which gives a strong nonlinear damage curve, Figure 7.

USE OF DAMAGE CONCEPT IN TWO LEVEL STRAIN CONTROLLED TESTS

The principal interest of damage concept in fatigue is to predict fatigue life at non uniform strain or temperature level. Predicting fatigue life methodology requires a knowledge of

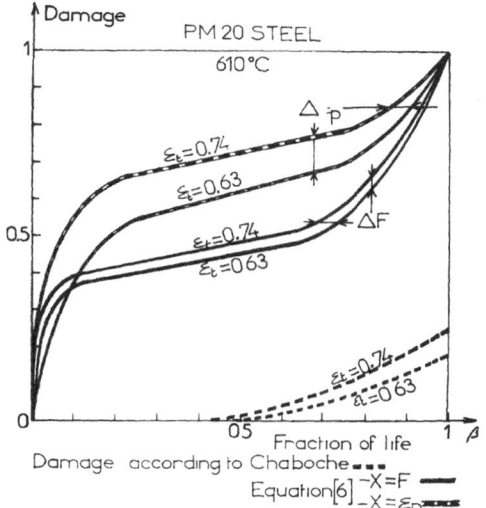

Fig. (7) - Damage elementary curves at two levels of strain
and invariable succession postulate

- damage elementary curves $D_x = f(\beta)$ for each variable chosen (strain, temperature);

- damage transition rule;

- second level interaction law.

In order to precise these three points we have performed tests at two different levels of strain, temperature or mixed on two bainitic steels PM20 and NF30. Cylindrical specimens with collars used for fixing extensometer are put in electrical furnace and cycled mechanically on servohydraulic machine; experimental parameters are as following:

- total strain amplitude: 0.63%   0.70%   0.74%   0.80%

- testing temperature: 560°C - 610°C

- strain rate: $3.2 \ 10^{-3} \ s^{-1}$

Six kinds of tests are done: "low-high" transition of strain and temperature, "high-low" transition strain and temperature and mixed "low-high" "high-low" temperature strain. Results from PM20 steel and for low-high transition are presented in Figure 8. Results are plotted on a graph fraction life β_2 at second level versus fraction life at first level β_1. Miner's laws ($\beta_1 + \beta_2 = 1$) are plotted also on the same graph. For the two steels, we can remark that maximum interaction are present for small value of β_1, i.e., in stage I. A difference of 30% with Miner's law can occur.

Possibilities of interaction have been suggested by Corten and Dolan [12]. They give a differential interaction law as follows:

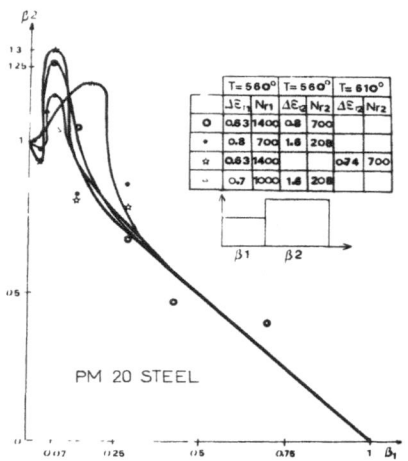

The table within the figure:

	T=560°		T=560°		T=610°	
	$\Delta\varepsilon_1$	N_{r1}	$\Delta\varepsilon_2$	N_{r2}	$\Delta\varepsilon_2$	N_{r2}
O	0.63	1400	0.8	700		
•	0.8	700	1.6	208		
☆	0.63	1400			0.74	700
..	0.7	1000	1.6	208		

PM 20 STEEL

Fig. (8) - Fraction of life at second level β_2 versus fraction of life
at first level and representation of Miner's law

$$dD = arN^{a-1}dN \qquad (12)$$

in which r is a function of stress at the first level and a is a constant material.

In practice, interaction is seen by modification of damage curve at second
level. An example of damage evolution during tests at two levels of deformation
on NF30 steel type "high-low" and "low-high" at 560°C is given in Figure 9.

Elementary damage curves for second level are plotted on the same curves and
give possibilities of comparison. We see that first level interaction gives an
important modification. Other results show that this interaction is not only dependent of strain, at second level, but also of properties used to define damage.
For example, damage through curves at second level, fall of load is little affected.

A modification of strain level during mechanical cycling is characterized by
a change of load or plastic deformation. But generally, it is not possible to re-
cover this level of load or plastic deformation at the continuous second level.
The difference between the two values characterizes also the transition level. A
schema of this is given in Figure 10 for the particular case of fall of load. If
elementary curve at second level is above two level curves, this difference is
conventionally taken as negative; otherwise, it is taken as positive. These
two situations correspond to "low-high" and "high-low" transitions. Experimental
data show that in this case, Palmgreen-Miner's law is not followed. In most cases,
"low-high" transition gives a Miner parameter value more than unity and "high-low"
less than unity:

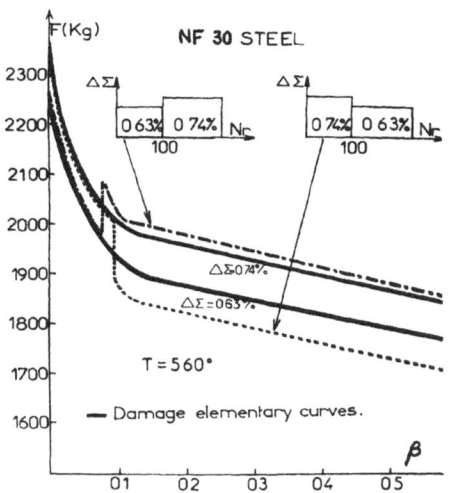

Fig. (9) - Damage evolution during a two-level strain controlled test
        − − Damage elementary curves

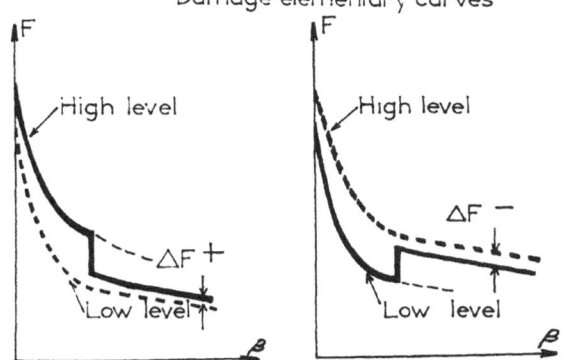

Fig. (10) - Definition of transition characteristics

"low-high" $\Sigma \dfrac{n_i}{N_i} \geq 1$

"high-low" $\Sigma \dfrac{n_i}{N_i} \leq 1$

This particular attention to transition was extended to another kind of test.

Data for two hundred overloads block at 610°C with total strain amplitude 0.74% were collated with those taken at 20, 50, 200, 400 cycles at 560°C and 0.63% strain amplitude. All the results for each steel were plotted in a diagram: difference of load ΔF or difference of plastic deformation $\Delta\varepsilon_{tr}$ as a function of Miner's parameter. A linear relationship was obtained in each case as it can be seen in Figure 11 with a good correlation factor.

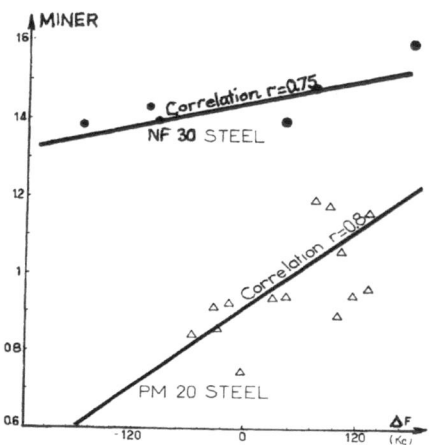

Fig. (11) - Linear correlation with transition parameter ΔF
and Miner's parameter

These results indicate that the transition parameter is a good indicator for predicting fatigue life. During mechanical cycling, energy is dissipated in hysteresis loop. Halford [13] gives a relationship between stress, strain and energy per cycle ΔW_e

$$\Delta W_c = \Delta\varepsilon_p \Delta\sigma \left(\frac{1-n'}{1+n'}\right) \tag{13}$$

where n' is cyclic strain hardening exponent. This energy per cycle is function of number of cycle to failure N_2

$$\Delta W_c = BN_2^{b-1} \tag{14}$$

and total dissipated energy W_T is given by

$$\Delta W_T = BN_2^b \tag{15}$$

b and B are material constants.

It is possible to extend Katchanov's concept to energy by replacing elastic energy ΔW_e by $\Delta \tilde{W}_e$ effective elastic energy with

$$\Delta \tilde{W}_e = \frac{\Delta W_e}{1-D} \tag{16}$$

In this particular case, effective stress is given by

$$\tilde{\sigma} = \frac{\sigma}{\sqrt{1-D}} \tag{17}$$

and effective strain by

$$\tilde{\varepsilon} = \frac{\varepsilon_e}{\sqrt{1-D}} \tag{18}$$

For a two level test, second level hysteresis loop is modified in comparison of elementary hysteresis loop. Experimental data show that energy dissipated at second level and Miner's parameter have a good linear correlation, Figure 12.

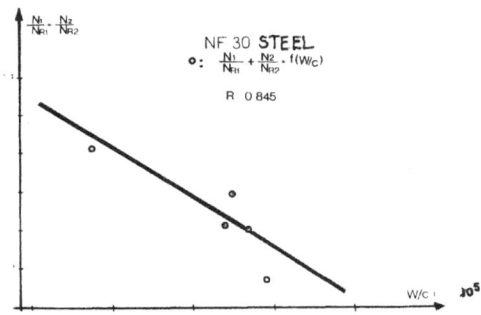

Fig. (12) - Energy dissipated in hysteresis loop at second level versus Miner's parameter

DISCUSSION

Many definitions of damage exist as it was discussed in EUROMECH Colloquium 147 in 1981 [14]. Our definition is the alteration of a property x which can be written $\dot{D} = f(x, \dot{x}, D)$ of the property and depends on easy meaning or interpretation of the chosen property like elasticity, as a consequence of phenomenological uniaxial theory due to Katchanov on the assumption of coupling elasticity-damage [1]. Damage evolution is given by

$$\dot{D} = f(\hat{\sigma}, \dot{\hat{\sigma}}, D) \tag{19}$$

where $\hat{\sigma}$ is the effective stress according to Katchanov's postulate.

The effective stress tensor $\hat{\underset{\sim}{\sigma}}$ is defined by

$$\hat{\underset{\sim}{\sigma}} = \frac{\sigma}{1-D} \tag{20}$$

A geometrical approach of damage makes the assumption that the elasticity is due to cracks and cavities occurring during plastic deformation (cyclic deformation, monotonic deformation or time dependent). In the case of isotropic damage, defects are plastically random; anisotropic case is given, for example, by a distribution of parallel microcracks.

A   representation is used to describe anisotropic character of damage.

$\hat{\underset{\sim}{\sigma}}$ represents the stress applied to the damaged material to obtain the same strain ε_e that is given by undamaged material under $\underset{\sim}{\sigma}$ stress.

$$\underset{\sim}{\sigma} = \varepsilon_e \hat{\underset{\sim}{H}}(D)$$

$$\hat{\underset{\sim}{\sigma}} = \varepsilon_e H_o$$

This representation makes the assumption that damage and elasticity are coupled.

Hooke's Law. Damage tensor has the relation between H_o and $\underset{\sim}{H}(D)$ as

$$\hat{\underset{\sim}{H}}(\underset{\sim}{D}) = (1-\underset{\sim}{D}):\underset{\sim}{H}_o$$

with $\underset{\sim}{D}$ being a four order tensor. Damage evolution is then given by:

$$\dot{\underset{\sim}{D}} = \underset{\sim}{G}(\tilde{\sigma})\dot{\underset{\sim}{D}}$$

with $\underset{\sim}{G}$ also a fourth order tensor.

This approach is very convenient because it only requires the measurement of elastic coefficient (Young's modulus, Poisson's coefficient, shear modulus) to build the damage tensor [14].

Thermodynamical Approach. In low cycle fatigue, a coupling between damage, defects on strain hardening exists and it is more convenient to use thermodynamic framework for this problem. Damage can be represented by an internal variable.

Free energy E of a mechanically loading system can be written in term of stress tensor $\underset{\sim}{\sigma}$, damage tensor $\underset{\sim}{D}$, strain hardening internal variables α_i.

Damage can be separated into two damages due to defect damage and damage due to strain hardening D_α

$$E(\underset{\sim}{\sigma}, \underset{\sim}{D}_d, \underset{\sim}{D}_{\alpha_i}) = W_e(\underset{\sim}{\sigma}, \underset{\sim}{D}_d, \underset{\sim}{D}_{\alpha_i}) + \Phi(\alpha_i) \tag{21}$$

where W_e is elastic energy and Φ a function of strain hardening internal variables. Enthalpy V is given by:

$$V(\underset{\sim}{\sigma}, \underset{\sim}{D}_d, \underset{\sim}{D}_{\alpha_i}, T) = W_e(\underset{\sim}{\sigma}, T, \underset{\sim}{D}_d, \underset{\sim}{D}_{\alpha_i}) - \Phi(\alpha_i, T)$$

or

$$V(\underset{\sim}{\sigma}, \underset{\sim}{D}_d, \underset{\sim}{D}_{\alpha_i}) = \frac{1}{2} \alpha_i \underset{\sim}{\sigma} : (D_d, D_{\alpha_i}) \tag{22}$$

The dissipativity can be written as

$$\underset{\sim}{\sigma} : \dot{\varepsilon}_p + A_i \dot{,}_i + \underset{\sim}{y} \dot{D}_d + \underset{\sim}{x} \dot{D}_{\alpha_i} \geq 0$$

where $A_i = \partial V / \partial \alpha_i$, $\underset{\sim}{y} = \partial V_d / \partial D_d$ and $\underset{\sim}{x} = \partial V / \partial D_{\alpha_i}$ are the thermodynamical forces associated with D_d, α_i, D_{α_i} internal variables and derived from a pseudo-potential of dissipation Ω [15]:

$$\alpha_i = \frac{\partial \Omega}{\partial A_i}, \quad \dot{D}_d = \frac{\partial \Omega}{\partial y}, \quad \dot{D}_{\alpha_i} = \frac{\partial \Omega}{\partial x}$$

For simplification, it is easy to consider isotropic damage: damage due to strain hardening appears at the beginning of life, damage due to cracks is significant until stage III has been reached. These considerations give a low damage with three constitutive laws:

$$D_F = A_i + B_i (\beta + C_i) - K_i (D_i - \beta)^{n_i} \tag{23}$$

where A_i, B_i, C_i, K_i, D_i and n_i (i = 1,2,3) are material constants. These coefficients can be simplified by appropriate tests.

Prediction of two level strain controlled tests is possible by integrating fatigue damage equation in two steps. By this way, two kinds of problems appear: possibilities of interaction of finite level on the second and non constant damage level through the definition and the properties used as damage indicator.

Remember that the property is function of observable variables x and internal variables i

$$X = f(D, x, i)$$

A change of observable variables induced a change of damage due to this anisotropic character, and a change of the value of X properties. Another way for predicting fatigue life is to use invariable process postulate of Erismann [11] and consider that according to this postulate, a change of damage through a property X or (fall of load) ΔD_F is represented on a change of another kind of damage and particular fraction of life D. Linear relationship between ΔD_F and D_β indicates that the material follows the Erismann's postulate.

This correlation can be obtained by experience or kinematic hardening model. An example of this kind of experimental correlation is given in Figure 13 where

$$dD_F = f(\epsilon\beta \frac{N_{r1}}{N_{r2}})$$

N_{r1}, N_{r2} are numbers of cycle to failure for level 1 and level 2.

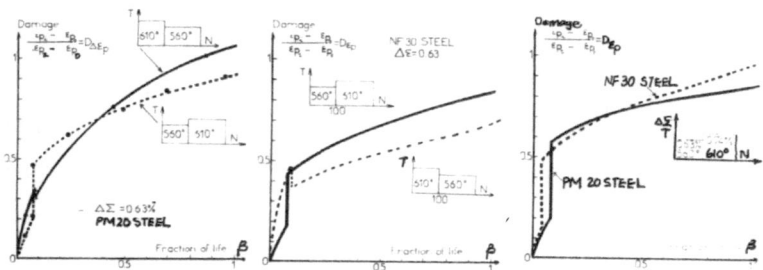

Fig. (13) - Damage evolution at two level temperature, strain or mixed strain temperature controlled tests

All these remarks indicated that damage concept occurs to be an interesting way for predicting fatigue life in low cycle but needs further research on the following points:

- definition of damage

- choice of properties used as damage indicator

- anisotropic character of damage

- choice of formation (geometrical or thermodynamical formations)

- interaction at second level

- invariable process of damage

CONCLUSION

Damage definition is always associated with the properties which use an indicator of damage. Elasticity is a particular property which is often used.

Experimental verifications of damage evolution law are done by methods in low-cycle fatigue tests at high temperature. Damage evolution depends strongly of damage indicator but seems to verify invariable process of damage postulate.

Damage concept can be used for predicting fatigue life for two level strain controlled tests:

(1) by integrating method which needs to precise damage transition and interaction law;

(2) by correlation between damage transition through fall of load or plastic deformation and fraction of life "damage".

REFERENCES

[1] Katchanov, L. M., "Time of the rupture process under creep conditions", Izk. Akad. Nauk, No. 8, pp. 26-31, 1958.

[2] Sidoroff, F., "Anisotropic damage and thermodynamics", "Damage Mechanics", EUROMECH 147, Cachan, France, 1981.

[3] Marigo, J. J. and Suquet, P. M., "An approach by homogenization of a form of damage for elastic bodies", "Damage Mechanics", EUROMECH 147, Cachan, France, 1981.

[4] Gatts, R. R., "Application of cumulative damage concept to fatigue", Journal of Basic Engineering, Trans. ASME, Series D, Vol. 83, No. 4, pp. 529-540, 1961.

[5] Buiquoc, T., "Damage cumulatif in fatigue", Ecole d'été sur la fatigue, Sherhooke, 1981.

[6] Chaboche, J. L., "On the interaction of hardening and fatigue damage in the 316 steel", ICF 5, Vol. 3, pp. 1381-1393, 1981.

[7] Ratcliffe, R. T., Britt. J., Applied Physics 16, p. 1193, 1965.

[8] Tamhan, Z., "Essais de fatigue par la méthode des blocs programmés; constructio et exploitation d'une machine de flexion rotative", Thesis, University Lyon I, 1979.

[9] Lemaire, J., Cordebois, J. P., Dufailly, J. and Cras, T., 238, Série B, pp. 391-394, 1979.

[10] Polhemus, J. F., Spaeth, C. E. and Vogel, W. H., ASTM STP 560, pp. 625-629, 1979.

[11] Erismann, T. M., "Unconventional reflections on damage accumulation", Engineering Fracture Mechanics, Vol. B, pp. 115-121, 1976.

[12] Corten, H. T. and Dolan, T. J., "Cumulative fatigue process", International Conference on Fatigue of Metals, Inst. of Mechanical Engineering, ASTM, pp. 235-246, 1956.

[13] Halford, G. R., "The energy required for fatigue", Journal of Materials, Vol. 1, No. 1, pp. 3-17, 1966.

[14] Nouailhas, D., "Etude expérimentale de l'endommagement de plasticité ductile anisotrope", Thesis, University Paris 6, 1980.

[15] Mandel, J. and Cras, T., 284, 1977.

FATIGUE BEHAVIOUR OF STEEL RECTANGULAR HOLLOW SECTIONS USED IN BUS STRUCTURES

C. Moura Branco

University of Minho
Largo do Paco, 4719 Braga Codex, Portugal

and

A. Augusto Fernandes

Oporto University
Rua dos Bragas, 4000 Porto, Portugal

ABSTRACT

The fatigue behaviour of steel rectangular hollow sections used in welded bus structures was studied. The S-N curves obtained show a significant improvement in the fatigue strength when the section thickness is increased and the toe of fillet welds ground. A significant crack initiation phase was observed specially for low stresses.

A Linear Elastic Fracture Mechanics model used to predict the fatigue crack propagation behaviour of the welded hollow sections only produced a reasonable correlation with the experimental test results due to the crack initiation phase and the approximate solution used for the stress intensity factor. The model considers the effect of flaw size and shape, weld location and geometry. Work is in progress to obtain fatigue crack growth data in the hollow sections, compute more precise solutions of the stress intensity factor and predict the crack initiation phase.

INTRODUCTION

Thin walled rectangular hollow sections are extensively used in welded structures where the weight is a key factor of its design. The city bus structures fall into this category. It is known that a weight reduction produces a reduction in material cost, fuel consumption and also increases the loading capacity of the vehicle. The thin walled rectangular hollow sections combine a reduced weight with high section modulus and torsional strength which makes them very attractive for this kind of application.

Since the vehicle structure is subjected in service to dynamic loads, it is essential to know the fatigue strength of the more significant weld details in order to establish a correct design for the structure. The fillet welded joints, which show poor fatigue strength [1], are extensively used in this type of structure.

To establish fatigue design stresses taking into account the fatigue behaviour of the welded hollow sections, it is necessary to obtain the appropriate S-N curves. To the authors' knowledge, there are no published fatigue test results which allow those stresses to be obtained. Experimental stress analysis data obtained using strain gauges in a bus structure in service showed that the load spectra was mainly constant amplitude bending. In order to investigate fatigue failures occurring in service in these structures, it was decided to study the fatigue behaviour of the hollow sections with three main objectives:

- Determination of basic S-N curves for the more critical welded joints of the bus structure subjected to constant amplitude bending loads.

- Study the influence of tube thickness and weld surface finish on the fatigue strength of the hollow sections.

- Correlation of the experimental test results using a fracture mechanics crack propagation model in order to assess fatigue life.

MATERIALS, SPECIMENS AND TESTING EQUIPMENT

The hollow sections used in the experimental program were full scale tubes with a longitudinal weld seam having the following dimensions: 82 x 38 x 1.5 and 82 x 38 x 2.0 mm. The material was an St 37 - 2 steel DIN 17100. The mechanical properties obtained in tensile tests for various types of specimens taken from the tubes are presented in Table 1.

TABLE 1 - MECHANICAL PROPERTIES OF THE HOLLOW SECTIONS MATERIAL

Specimen	Mechanical Properties		
	Ultimate Stress Kg/mm^2	Yield Stress Kg/mm^2	Elongation
Hollow section	44.0	39.8	9.3
Plate without weld seam	50.4	45.0	29.1
Plate with weld seam	54.0	48.8	13.4
Plate without weld seam (annealed)	41.0	32.2	43.2
Plate with weld seam (annealed)	35.2	25.7	31.5

The alternating bending fatigue testing machine used in the experimental program was designed and built specially for these tests and is shown in Figure 1.

Fig. (1) - Alternating bending fatigue test machine
used in the experimental program

The machine imposes a constant alternating deflection to the specimen, through a manually adjustable eccentric and lever system fitted to the shaft of the electric motor.

The specimens had one edge clamped, the bending load being applied to the free end, Figure 1, with a loading frequency of 1410 cycles/min. Specimen failure was defined when the fatigue crack, initiated at the fillet weld, propagated till the specimen middle height. At this position, the crack broke an electrical wire connected to a D.C. circuit thus stopping the motor and the cycle counter. All the tests were carried out at a stress ratio R = -1.

The specimen geometry is shown in Figure 2. Three types of welded joints were tested using a manual MIG welding process with the same specifications as in service. Those joints were:

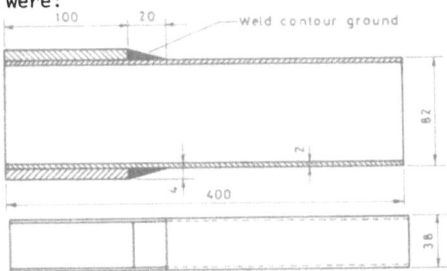

Fig. (2) - Tubular specimen (82 x 38 x 1.5 and 82 x 38 x 2)

- as welded joints

- annealed (650°C for one hour)

452

- joints where the contour of the fillet weld was ground with a manually
operated grinding tool fitted with a small grinding wheel

The weld profiles were checked in all the specimens using a low power travelling microscope fitted with a dial gauge.

The nominal stresses near the fillet weld area were measured with strain
gauges bonded as close as possible to the clamped edges of exactly similar specimens without welds. Thus the stress concentration effect of the weld profile was
not considered and only the nominal stresses were taken as if the weld was not
present. Calibration curves were obtained relating the static deflection at the
free end with the strain readings, Figure 3. From these curves, the nominal
stress values were selected in the tests.

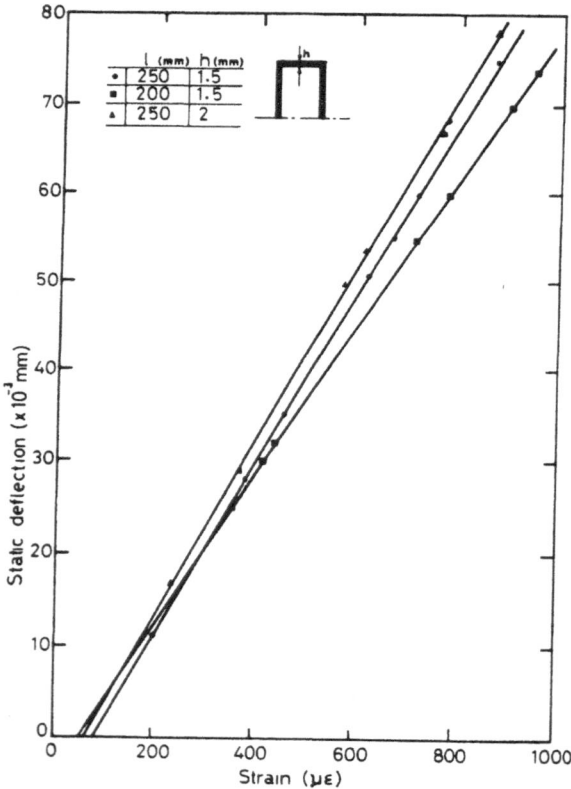

Fig. (3) - Calibration curves for the nominal stresses in the weld zone

EXPERIMENTAL RESULTS

A. S-N Curves for Alternating Fatigue Loading

The fatigue tests results obtained with 1.5 mm thickness tubes both in the as welded and annealed conditions are given in Table 2. The annealing treatment produced a softer material structure with lower yield strength values.

TABLE 2 - ALTERNATING BENDING FATIGUE TEST RESULTS FOR WELDED HOLLOW SECTIONS 82 x 38 x 1.5 IN THE AS RECEIVED AND ANNEALED CONDITION

Specimen No.	Nominal Stress N/mm^2	Fatigue Life Cycles
1	264.6	3,917
2	"	4,600
3	221.5	3,341
7	230.3	5,216
4	200.9	5,197
5	"	5,185
6	176.4	13,030
8	"	15,469
9	166.6	17,052
10	"	17,154
11	147	89,184
12	"	48,466
13	"	62,856
14	127.4	154,244
15	"	625,638
16	"	124,089
17	"	417,464
18	112.7	253,167
19	93.1	483,951
20	"	779,521
21	78.4	3,247,908
22	"	> 4 x 10^6
Annealed Specimens		
1	264.6	5,500
2	225.4	17,588
3	176.4	28,515
4	147.0	44,350
5	127.4	93,194

Table 3 shows the results obtained with thicker hollow sections (82 x 38 x 2) and Table 4 presents the fatigue test results for specimens with the fillet weld contour ground in order to reduce the stress concentration factor.

The experimental test results were analyzed, using a linear regression method. The equations for the mean curves obtained in each series of tests are presented in Table 5, σ being the nominal stress and N the fatigue life of the specimen: the correlation coefficients are also given. S-N curves for each test series are plotted in Figures 4 to 6. A statistical analysis of the test results was carried out in order to calculate the 95% confidence limits which are also shown in Figures 4 to 6.

TABLE 3 - ALTERNATING BENDING FATIGUE TEST RESULTS FOR WELDED HOLLOW SECTIONS
(82 x 38 x 2)

Specimen No.	Nominal Stress N/mm^2	Fatigue Life Cycles
1	264.6	21,977
2	225.4	47,341
3	205.8	85,576
4	176.4	438,521
5	147.0	133,027
6	147.0	235,182
7	127.4	261,697
8	88.2	2,386,504

TABLE 4 - ALTERNATING BENDING FATIGUE TEST RESULTS FOR HOLLOW SECTIONS
(82 x 38 x 2) WITH GROUND WELDS

Specimen No.	Nominal Stress N/mm^2	Fatigue Life Cycles
1	264.6	15,679
2	225.4	46,237
3	205.8	47,193
4	176.4	67,642
5	176.4	48,099
6	176.4	141,603
7	147.0	537,268
8	127.4	877,543
9	104.9	1,235,762
10	98.0	5,030,000
11	93.1	> 1.1 x 10^7
12	88.2	> 2 x 10^6

TABLE 5 - EQUATIONS OF THE MEAN S-N CURVES

Test Series	Mean Curve Equation	Correlation Coefficient
Hollow section 82 x 38 x 1.5 as received	Logσ = - 0.1554 LogN + 29135	r = 0.90
Hollow section 82 x 38 x 1.5 annealed	-	-
Hollow section 82 x 38 x 2	Logσ = - 0.2229 LogN + 33828	r = 0.92
Hollow section 82 x 38 x 2 with weld toe ground	Logσ = - 0.177414 LogN + 31418	r = 0.85

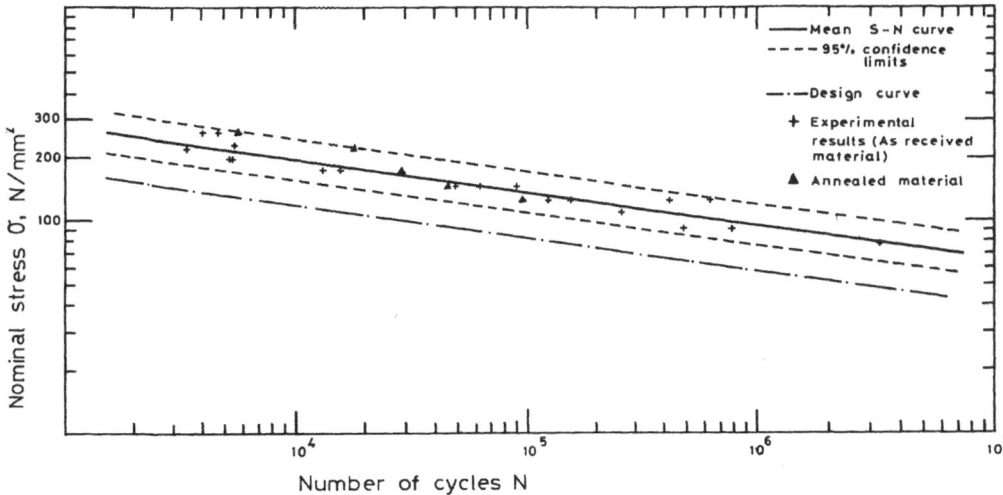

Fig. (4) - S-N fatigue curves for rectangular hollow sections with 1.5 mm
thickness. St 37-2 steel. Tubes 82 x 38 x 1.5. R = -1

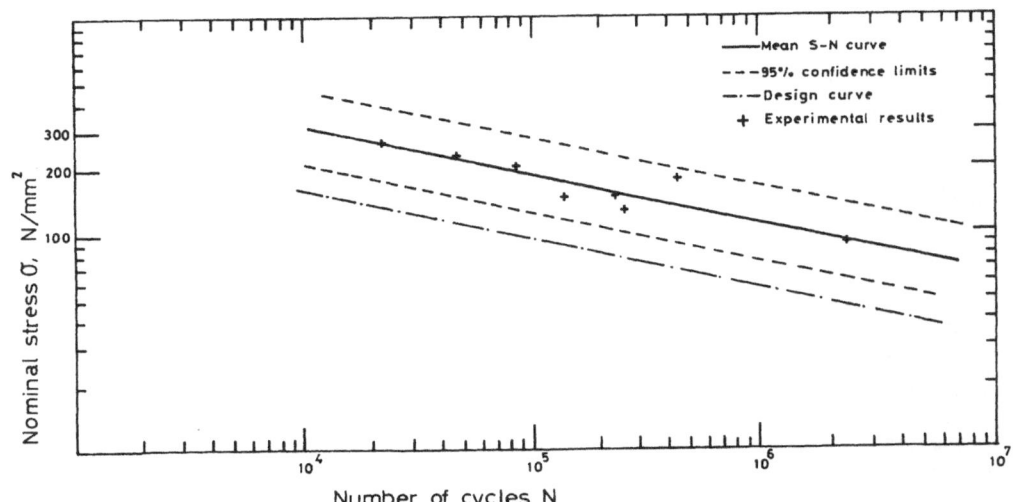

Fig. (5) - S-N fatigue curves for rectangular hollow sections with 2 mm
thickness. St 37-2 steel. Tubes 82 x 38 x 2. R = -1

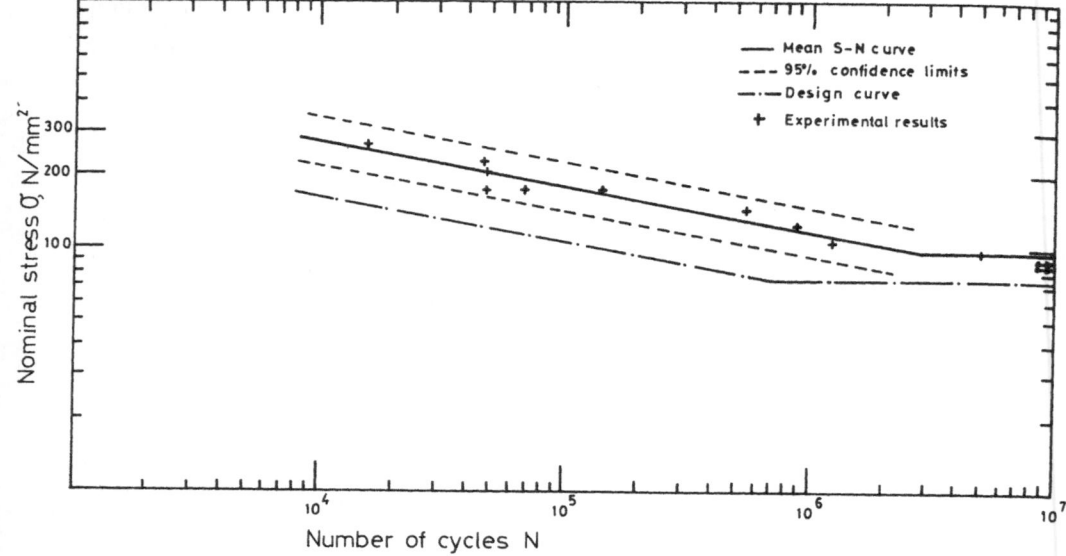

Fig. (6) - S-N fatigue curves for rectangular hollow sections with 2 mm thickness and ground welds. St 37-2 steel. Tubes 82 x 38 x 2. R = -1

The design curves shown in these curves were calculated applying a 4/3 safety factor to the lower bound confidence limits.

The design fatigue stress was taken from the design curves and for a fatigue life of 2 x 10⁶ cycles. The S-N curve defined for hollow sections with 2 mm thickness and fillet weld toe ground exhibits a fatigue limit, Figure 6.

B. Fracture Mechanics Analysis of Data

In welded joints subjected to fatigue loading, fatigue strength depends considerably on the crack propagation stage. The crack grows from an initial size (flaw size) till a critical crack length for the final failure. Therefore, and not taking into account a small crack initiation phase, it is possible to predict the life of the welded joint using linear elastic fracture mechanics methods (LEFM) [3]. If the stress level is sufficiently low so that the stress field near the crack tip is characterized by the stress intensity factor K [3], it is possible to correlate the fatigue crack propagation rate da/dN with the amplitude ΔK of the stress intensity factor in the loading cycle. The Paris law [4] is a classical example of one of the more widely used equations

$$\frac{da}{dN} = C(\Delta K)^m \tag{1}$$

where C and m are experimental constants and $\Delta K = K_{max} - K_{min}$ where K_{max} and K_{min} are the maximum values of K in the loading cycle. From LEFM, it is known that

$$K_{max} = Y\sigma_{max}\sqrt{\pi a} \tag{2a}$$

$$K_{min} = Y\sigma_{min}\sqrt{\pi a} \tag{2b}$$

where Y is the geometrical dimensionless factor. a is the crack length and σ_{max} and σ_{min} are the maximum and minimum values of stress in the loading cycle. In the present study, the loading cycle is sinusoidal with constant amplitude (R = -1). The compressive part of the cycle was neglected because the stress intensity factor was herein only defined for Mode I in tension.

In the fatigue tests, the cracks initiated in the weld toe and propagated first through the top and bottom wall of the tube and then in the lateral sides in an approximately vertical direction. It was observed that the crack took considerably more time to propagate through the top and bottom parts of the tubes than in the lateral sides. Based on this experimental observation, it was decided to use the Maddox [5] stress intensity factor solution for a plate specimen in tension with a fillet weld with the plate thickness equal to the tube thickness. The geometric factor Y is given by the equation:

$$Y = M_{Ka} \left[1.112 - 0.231\left(\frac{a}{B}\right) + 10.55\left(\frac{a}{B}\right)^2 - 21.7\left(\frac{a}{B}\right)^3 + 33.19\left(\frac{a}{B}\right)^4 \right] \tag{3}$$

where B is the hollow section thickness and M_{Ka} is a magnification factor which takes into account the stress concentration at the weld toe. M_{Ka} is a function of a/B and was derived in tabular form.

In equation (3), it is assumed that the stress distribution was constant through the tube thickness. This is a reasonable approximation considering the small values of thickness. Also, the crack front was assumed to be straight instead of a semi-elliptical shape. The integration of Paris' law gives

$$\frac{1}{\sigma_{max}} \int_{a_i}^{B} \frac{da}{(Y\sqrt{\pi a})^m} = CN_f \tag{4}$$

where N_f is the number of cycles to failure and a_i is the initial crack length. Equation (4) gives an S-N type curve

$$(\sigma^*)^m N_f = \frac{1}{C} \tag{5}$$

where

$$\sigma^* = \sigma_{max}(I)^{-1/m} \quad \text{and} \quad I = \int_{a_i}^{B} \frac{da}{(Y\sqrt{\pi a})^m} \tag{6}$$

Equation (5) may predict the fatigue life in the welded joint as a function of flaw size and location. These important variables are not included in the simple

S-N curves. In order to assess the influence of flaw size, equation (5) was integrated. Hence, the integral I was evaluated numerically using Simpson's rule with
an accuracy better than 10^{-2}. The values of C and m were taken from the literature [6] and refer to region II of the crack propagation curve (da/dN) for St-37
-2 type steels (m = 3.07 and C = 2.16×10^{-13}, for da/dN in mm/cycle and ΔK in
$Nmm^{-3/2}$). The best correlation was obtained with a_i values of 0.2 mm for the fillet weld in the as welded specimens and 0 for the specimens with ground welds. A
plot of log σ^* against log N is shown in Figure 7 for all test data taken from
Tables 2 to 4. The straight line corresponding to equation (5) is also plotted
in Figure 7 for comparison with the experimental results.

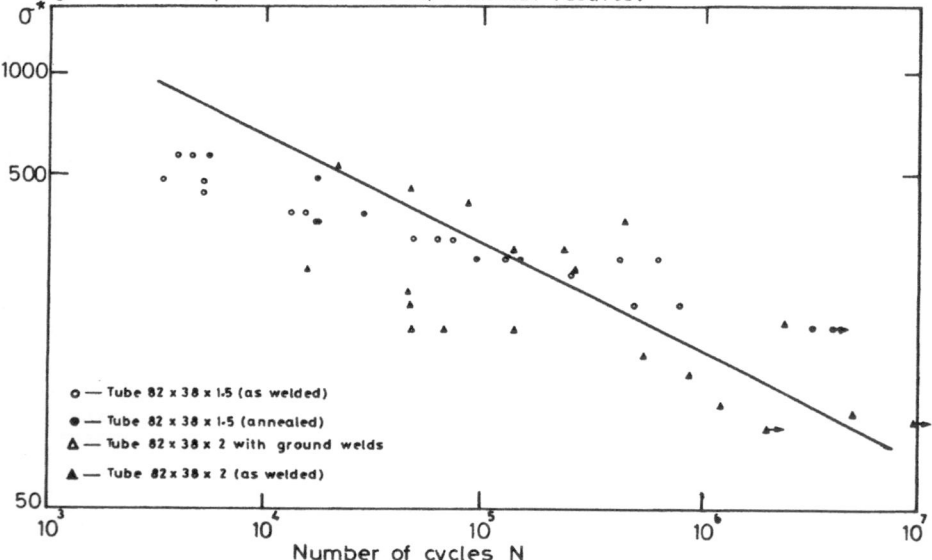

Fig. (7) - Logarithmic plot σ^* against N for rectangular hollow sections.
St 37-2 steel. R = -1

DISCUSSION OF RESULTS

A. Analysis of S-N Data and Failure Modes

Firstly, it should be pointed out that the fatigue failures, simulated in
the alternating bending tests reproduced rather closely the actual fatigue failures occurred in the bus structure as shown in Figures 8 to 10. Thus, Figure 8
shows typical fatigue cracks in specimens where the weld toes were and were not
subjected to the grinding treatment. In all the specimens, the fatigue cracks
initiated from the weld toes, and the similarity between the specimen failures,
Figures 9 and 10, and the failures in the structure, Figure 8, is clear.

Fig. (8) - Typical failure in one of the welded joints of a bus structure

Fig. (9) - Typical failure obtained in one as welded tubular specimen

A more detailed fractographic analysis using both optical and scanning microscope was carried out and the results were reported in the literature [7]. Similar failure modes were observed and discussed by Stephens et al [8] in square tubes under static torsion and Watanabe et al [9] in fatigue tests of welded round tubes in bending.

The mean S-N curves taken from Figures 4 to 6 are plotted in Figure 11. These curves show that fatigue life increases when the hollow section thickness increases from 1.5 to 2 mm. In the high stress region (life below 10^6 cycles) there is not a significant influence of the grinding treatment in the fatigue strength. However, this influence increases for higher fatigue lives and these are of greater interest in terms of the expected life for the structure.

The annealing treatment did not produce any effect on the fatigue strength as shown by the S-N curve presented in Figure 4. The design stress for 2×10^6

460

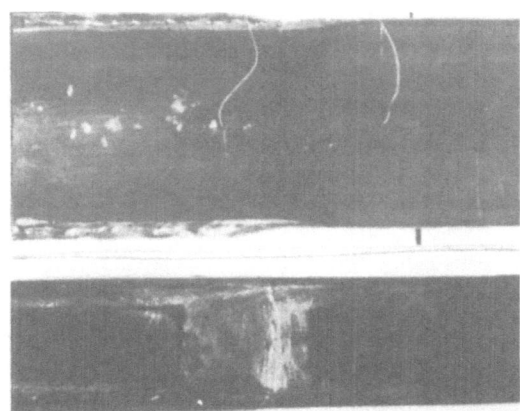

Fig. (10) - Typical failure obtained in one tubular specimen with the weld ground

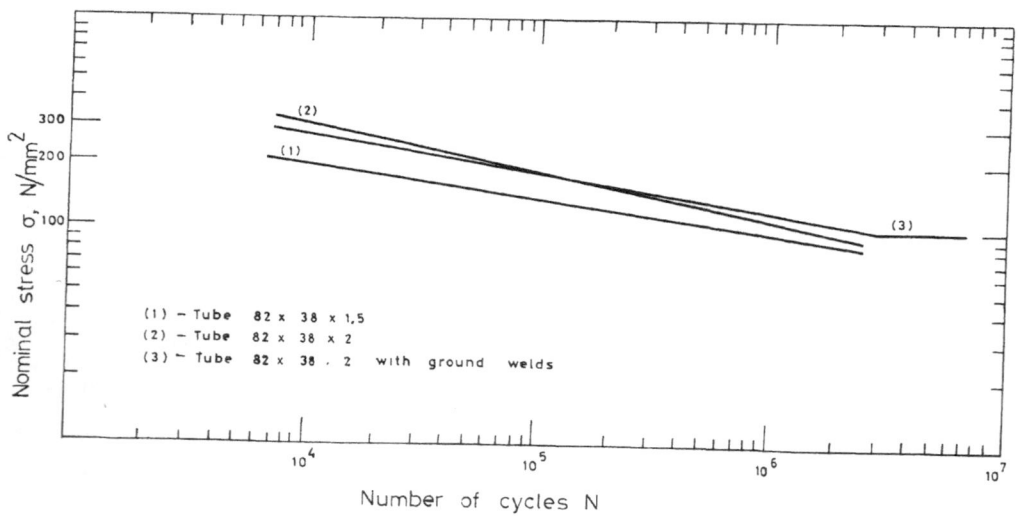

(1) — Tube 82 x 38 x 1,5
(2) — Tube 82 x 38 x 2
(3) — Tube 82 x 38 . 2 with ground welds

Fig. (11) - Mean S-N curves for the four series of tests

cycles increased from 52 N/mm² to 75 N/mm² when the hollow section thickness was
increased from 1.5 to 2 mm. The greater life obtained in the hollow sections
with 2 mm thickness may be due to the fact that the crack path is longer. The

highest design stress occurs in the hollow sections with ground welds and greater weld length due to the lower stress concentration. When the stress is sufficiently low, crack initiation does not occur in the ground welds and a fatigue limit was observed as shown in Figures 6 and 11.

B. Analysis of LEFM Results

The plot σ^* against N in Figure 7 shows that the trend of the experimental results does not correlate very well with the theoretical relation given by equation (5). The LEFM approach in welded joints means that provided only crack propagation takes place, the logarithmic plot σ^* against N should give one straight line irrespective of the weld geometry and loading conditions, when the values of m and C are constant. Also, the shape of the experimental curve should be equal to the exponent m.

There are several factors which may have contributed for not obtaining a better correlation of results; namely,

1. The approximate solution used to compute the stress intensity factor. The stress intensity factor should be obtained more accurately taking into account the whole geometry, crack front and loading conditions. The strain energy density factor is also being considered in the analysis [10].

2. Lack of fatigue crack propagation data for these specimens. This data is necessary to obtain the appropriate function $da/dN = f(\Delta K)$. The similarity approach may also be applied with crack propagation data obtained in specimens with a known stress intensity factor solution. Also, the values of C and m were taken for region II crack propagation only and for thicker specimens.

3. The crack initiation stage should be a significant fraction of the fatigue life specially for long endurances and therefore must be taken into account in any prediction method.

CONCLUSIONS

This preliminary study reported the fatigue behaviour of welded joints of steel rectangular hollow sections used in bus structures. The following conclusions are important:

1. Fatigue strength in these sections increases considerably when the fillet weld contour is ground, thus obtaining a better surface finish and smaller stress concentration factor.

2. Fatigue strength increases when the section thickness is increased from 1.5 to 2 mm due to a longer crack path.

3. The LEFM approach used was not sufficiently accurate to correlate the fatigue crack propagation results due to the approximate solution used to compute the stress intensity factor.

4. There is a significant crack initiation phase in the long endurance region ($N_f > 10^5$ cycles) which should be measured and taken into account in any life prediction method.

Work is in progress to assess the influence on the fatigue life of such variables as the type of steel, heat treatment, section thickness, weld detail and geometry, etc. Moreover, the crack initiation and crack propagation stages are being studied obtaining crack propagation data in the hollow sections and deriving more accurate methods to calculate the stress intensity factor or other suitable fracture mechanics parameters.

REFERENCES

[1] Gurney, T. R. and Maddox, S. J., "A reanalysis of fatigue data for welded joints in steel", WI Research Report E 144/72, 1972.

[2] Gurney, T. R., "Fatigue of Welded Structures", Ed. by Cambridge University Press, 1979.

[3] Branco, C. M., "Fracture Mechanics Applications", Lecture Notes Short Course in Lisbon, CEMUL, 1979.

[4] Paris, P. C. and Erdogan, F., "A critical analysis of crack propagation laws", Trans. ASME, J. Bas. Eng., p. 528, 1963.

[5] Maddox, S. J., "Assessing the significance of flaws in welds subject to fatigue", Weld. Res. Suppl., p. 401, 1974.

[6] Maddox, S. J., "Fatigue crack propagation data obtained from parent plate, weld metal and HAZ in structural steels", WI Research Report, E/48/72, 1972.

[7] Ferreira, J. A., Branco, C. M. and Radon, J. C., "Crack initiation and propagation in welded rectangular hollow sections", submitted and accepted in the Int. Conf. Fracture Mechanics Technology Applied to Material Evaluation and Structure Design, Melbourne, Australia, 1982.

[8] Stephens, R. I. and Glinka, G., "Experimental determination of K_I for surface cracks in a square tube under torsion", Exp. Mech., p. 24, 1980.

[9] Watanabe, T., Husegawa, Y. and Matsumoto, W., "Fatigue tests on defective circumferential welded pipe joints", Proc. 5th Int. Conf. on Fracture, Cannes, France, p. 41, 1981.

[10] Sih, G. C., "The strain energy density criterion", Stress Analysis of Notch Problems, Sijthoff and Noordhoff, Netherlands, 1978.

ACKNOWLEDGEMENTS

The authors wish to acknowledge the financial support given by the Portuguese Science and Technology Research Council (JNICT, Research Contract No. 101.79.07) and by the Oporto Transport Company (Research Contract No. RS 4853). The fatigue tests were conducted with great care and efficiency by Mr. José Francisco Ferreira (Dipl. Eng.) to whom the authors acknowledge his collaboration.

FATIGUE CRACK GROWTH IN WELDED STEEL RECTANGULAR HOLLOW SECTIONS

J. A. Ferreira

Coimbra University
Portugal

C. M. Branco

University of Minho
Braga, Codex, Portugal

and

J. C. Radon

Imperial College
London SW7 2AZ, England

ABSTRACT

The results of fatigue crack propagation tests carried out on steel rectangular hollow sections with fillet welded gusset plates are reported. The crack propagation data was obtained on the lateral sides of the tubes subjected to plane bending. Crack initiation was studied with the scanning electron microscope and a fracture surface analysis was made using the same technique. It was found that crack initiation in thin walled tubes depends basically on three factors: (i) The stress concentration in the tube corners, (ii) The surface ripple in the weldment and (iii) The level of porosity in the weld. Further work is necessary to draw more definitive conclusions on the influence of the factors.

INTRODUCTION

Fatigue failures are known to occur frequently in transport structures such as buses and aircraft, railway axles, etc. These fractures are very varied and can sometimes become rather complex. McLester [1] analyzed some fatigue problems in land transport, mainly the fatigue behaviour of welded joints in boggies and also crack propagation in shafts. Welded rectangular hollow sections are normally used for bus structures and these are the main areas of application for the results presented in this paper. These sections are lightweight and provide high values of both bending and torsion moduli when compared with other sections having the same cross-sectional area. In these structures, it is very important to combine low weight with high strength and stiffness values that can be achieved with hollow sections.

There are some results published in the literature on the fatigue behaviour of rectangular hollow sections. Stephens [2,3] studied fatigue crack propagation for surface cracks in square tubes subjected to torsion [2] and also corner cracks in rectangular tubes in three and four point bending [3]. The stress intensity factor K_I was obtained in these geometries using the similarity approach based on data obtained in the same material and thickness with compact tension specimens. Branco et al [4] reported S-N data obtained in full scale rectangular steel hollow sections with two different thicknesses (1.5 and 2 mm). The loading applied was alternating bending and the fillet weld at one end of the specimen simulated the welded detail in the structure. The objective of the investigation was to study the effect of thickness and the weld surface finish on the fatigue life. It was found that the fatigue life increased by about 10% when the weld toes were ground with an appropriate wheel. Another fatigue life improvement was achieved by varying the weld details. A simplified fracture mechanics model was applied in that work to predict the fatigue life. The model considered the crack initiating from the weld toe subjected to a constantly applied tensile stress and propagated in the top wall of the tube. A straight crack front was assumed and the stress intensity factor solution proposed by Maddox [5] for welded joints was applied. This analysis was followed up by Ferreira [6] who obtained experimental values of K_I for these geometries in plane bending (R=0) using the similarity approach with compact type specimen data in the same material thickness and crack orientation. The results were compared with several fracture mechanics models applicable to crack propagation in the top and lateral walls of the tubes. It was also observed that in the ground welds, the crack initiation plane covered a significant proportion of the fatigue life. It was therefore decided to study the crack initiation mechanisms in more detail, using appropriate fractographic techniques.

In this paper, fatigue crack propagation data in the lateral walls of the tubes are discussed and fatigue crack initiation and propagation are studied by means of both optical and scanning microscopy.

EXPERIMENTAL

The material used in the manufacture of hollow sections was St 37-2 steel in plate form conforming with DIN 17100 specifications. The plates were cold worked to produce a rectangular hollow tube with a longitudinal weld seam. No annealing heat treatment was applied after the forming operation. The nominal dimensions of the specimens were 82 x 38 x 1.5 and 82 x 38 x 2 (mm) (height x width x wall thickness) those being the same dimensions as the tubes used in the bus structure. Table 1 shows the mechanical properties recorded from the tensile tests carried out on sheet specimens made from the tubes and also four other tests were performed on the actual tubes. The mean value of the surface hardness was 88 HRB 15T. It will be seen that due to the cold work operation, the plate steel hardened considerably giving higher values of yield stress and ultimate tensile strength than those expected for a St 37-2 steel.

The specimen used in the fatigue tests is shown in Figure 1. It is a tube with the nominal dimensions given in Figure 1 and a 3 mm thick gusset plate fillet welded along its contour. The weld detail was designed to simulate the detail more frequently found in the bus structures. The weld contour was ground with an appropriate grinding wheel. A fatigue machine applying bending loads in both alternating and plane bending was used for all the fatigue tests. The machine especially designed and built for this project is shown in Figure 2. The load was applied

TABLE 1 - MECHANICAL PROPERTIES OF ST 37-2 STEEL TUBES AND SHEET MATERIAL

	Tensile Strength (MPa)	Yield Strength (MPa)	Elongation at Rupture (%)
Tube	44.0	39.8	9.8
Sheet without the weld[*]	50.4	45.0	29.1
Sheet with the weld[*]	54	48.8	18.4

[*]This is the longitudinal weld seam in the tubes.

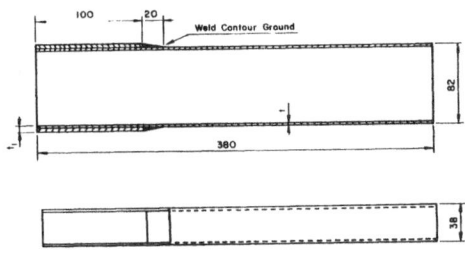

Fig. (1) - Tubular specimen used in the fatigue tests in plane bending

through a constant displacement to the free end of the specimen via an adjustable eccentre and a lever system. All the tests were carried out in air at room temperature with a loading frequency of approximately 24 Hz. The specimens were clamped at one end and provided with gusset plates, Figure 1. The nominal stress close to the weld toe was measured using strain calibration specimens (without the weld). The dynamic stress was related to the static deflection at the free end and a linear calibration curve was obtained. Calibration curves were obtained for each specimen thickness and mean stress (R=0 and R=1). These are shown in Figures 3 and 4 for the specimens having 1.5 and 2 mm thick walls respectively.

Crack propagation was measured on the lateral walls of the specimens as shown schematically in Figure 5 using a low power travelling microscope, a stroboscopic light and a dial gauge with an accuracy better than 0.01 mm.

RESULTS AND DISCUSSION

A. Crack Propagation Curves

These tests were carried out using several specimens with a constant stress ratio value R=0. The results obtained on two of these specimens are included here.

Fig. (2) - Alternating bending fatigue testing machine

Figure 6 is the crack propagation curve for specimen G1 with 1.8 mm thickness. The nominal stress value at the weld toe was 272 MPa. The same stress value was applied to the specimen F2, for which the equivalent crack propagation curve is shown in Figure 7. In specimen G1, a semi-elliptical notch, 6.5 mm long and 0.5 mm deep was cut in one of the top corners of the tube. Both crack propagation curves presented in Figures 6 and 7 were obtained using a polynomial regression program with the least square method.

It will be seen that the crack propagation rate increases considerably with the number of cycles. The number of cycles spent in propagating the crack in the upper horizontal part of the tube, Figure 5, is greater than the number of cycles in the lateral sides. Thus, Figure 6 indicates that even with an initial crack, the crack only starts to propagate in the vertical sides after about 2.5×10^4 cycles and takes only 10,000 cycles to propagate in the lateral walls. In Figure 7, the crack propagation plane in the lateral walls is smaller (approximately 1,000 cycles in a total of 23,800). Since the stress level is the same in both specimens, the crack then propagates faster in the thinner specimen (F2). These results suggest that crack propagation takes place predominantly in the top walls, where a significant crack initiation plane exists. Further tests are in progress to obtain additional data using more accurate experimental methods to detect both crack initiation and propagation in the top wall of the tubes.

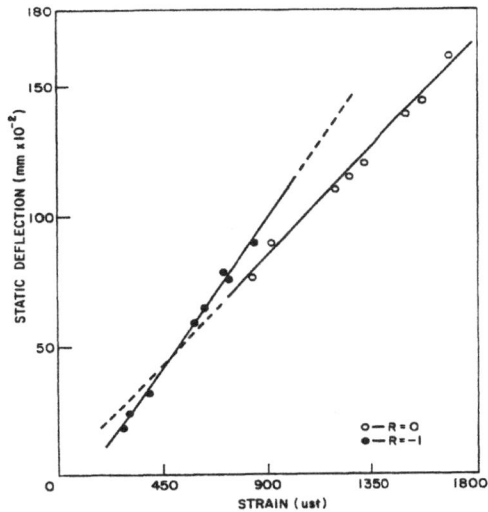

Fig. (3) - Calibration curves static deflection against dynamic strain.
Rectangular tubes 82 x 38 x 1.5

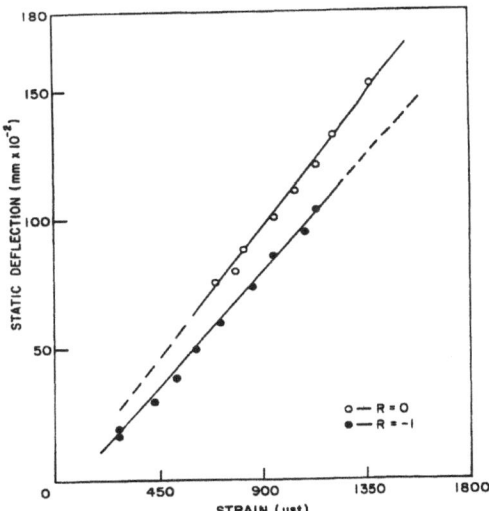

Fig. (4) - Calibration curves static deflection against dynamic strain.
Rectangular tubes 82 x 38 x 2

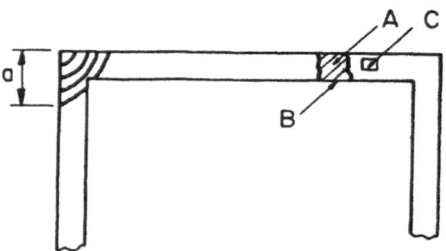

Fig. (5) - Schematic view of crack propagation measurements and regions A, B and C

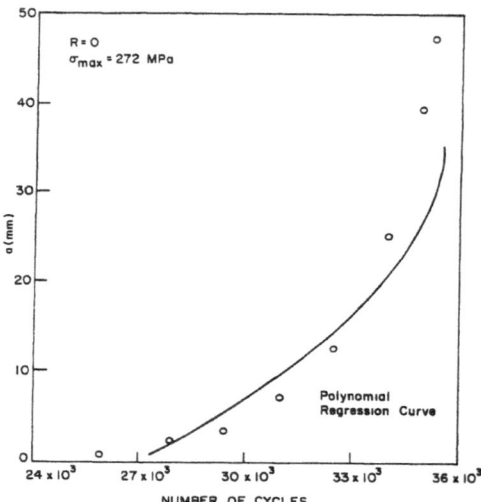

Fig. (6) - Crack propagation curve (specimen G1, Tube 82 x 38 x 2 with an initial crack)

B. Fractographic Analysis

Experimental observation during the fatigue tests and the fractographic study taken at a low magnification seem to indicate that in all the tests performed, the crack initiated in one of the top corners of the tube. Figure 8 shows the fracture surface of one of the specimens where the crack initiation area in the top corner of the tube may be observed. The crack then propagated through the top and lateral sides as schematically shown in Figure 5 and in a plane stress state with the fracture plane at approximately 45° to the longitudinal axis of the tube.

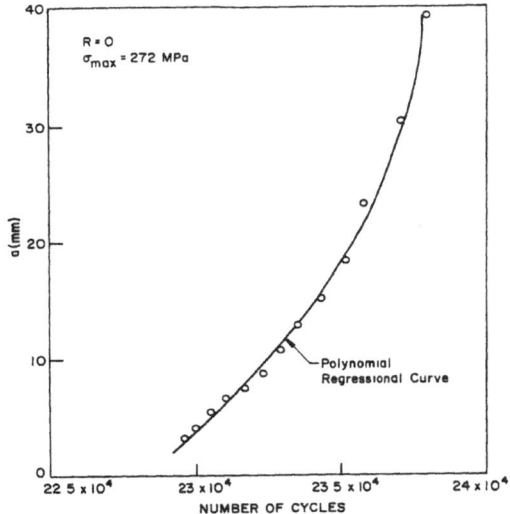

Fig. (7) - Crack propagation curve (specimen F2, Tube 82 x 38 x 1.5)

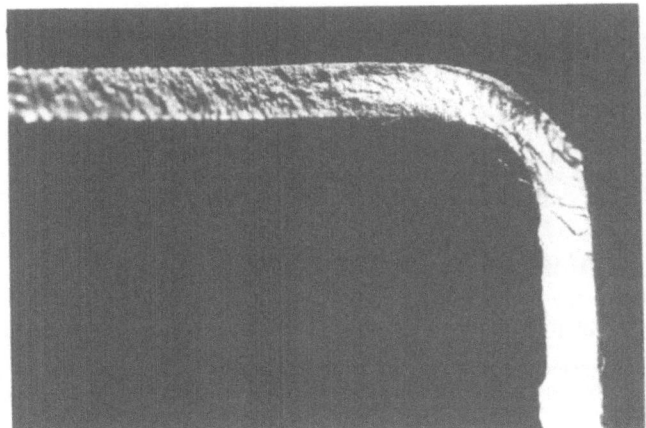

Fig. (8) - Crack initiation in specimen F3 (Magnification 8X)

A similar failure mode was observed by McDermott [3] in a thicker tube where a predominantly plane strain fracture took place. It was observed that the crack propagated till it became visible in the lateral sides. It was decided to carry out some scanning electron microscope analysis to detect any other crack initiation sites. Specimen G1 with the crack propagation given in Figure 6 shows a 45X magnification view of the area marked A in Figure 9, where a high density

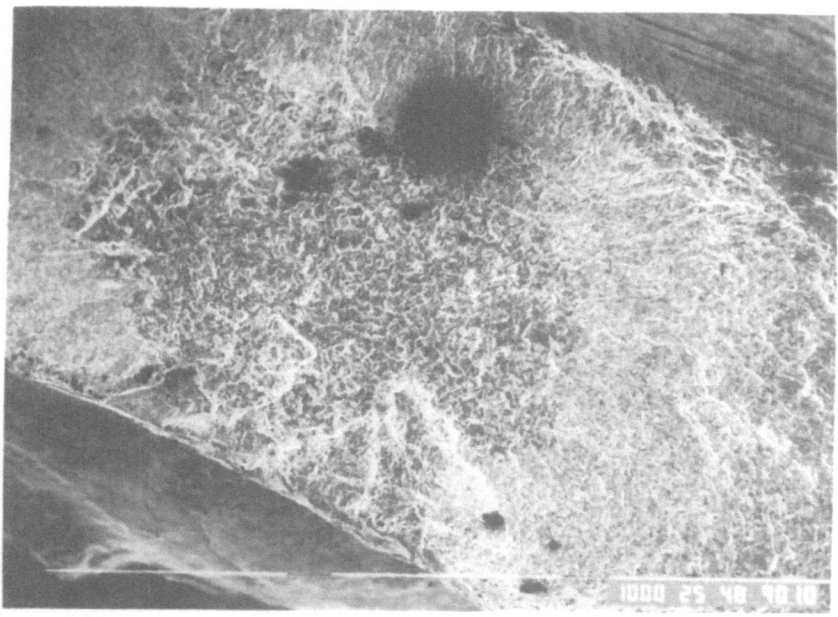

Fig. (9) - Porosity in specimen G1 (Tube 82 x 38 x 2) 45X Magnification

of porosity was seen, some pores being of a considerable size. The crack was initiated from those pores and also from the inner surface of the tube, region B, Figure 5. In this case, the stress at the inner surface was greater than at the weld toe. In thin tubes, the stress distribution can be assumed to be approximately constant and hence a deep surface roughness at the inner surface can give a stress concentration value higher than the stress concentration at the ground weld toe, the latter reduced by the grinding operation. Also, areas in the weldment where surface ripple is pronounced can lead to high stress concentration values and become crack initiation points. A similar crack initiation at the inner surface of circumferential welds in tubes subjected to bending and of thickness varying from 5.5 to 11 mm was reported by Watanabe [7].

Figure 10 shows fatigue cracks initiated in the region of heavy porosity. It was possible to distinguish the fatigue striations starting in the porosity re-

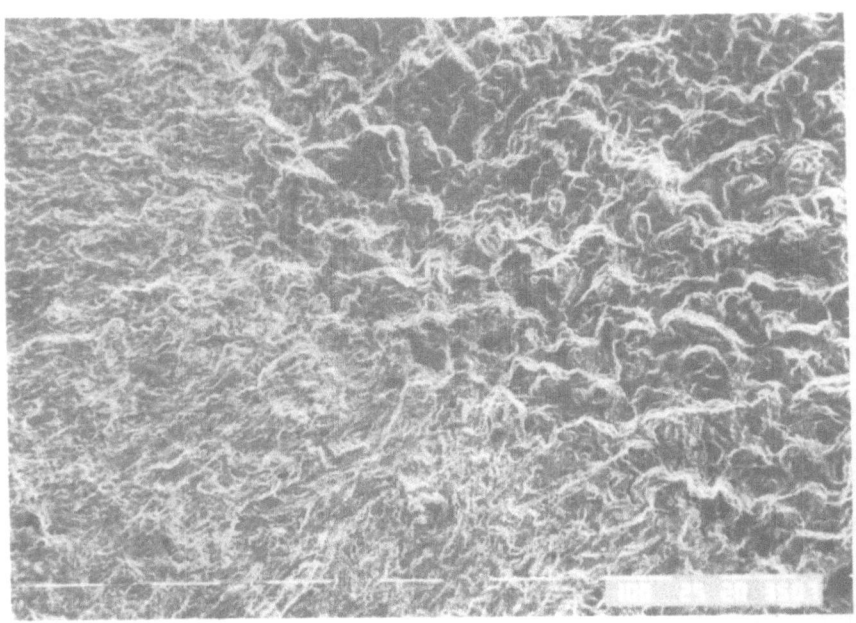

Fig. (10) - Fatigue striations near a porosity area (specimen G1)
120X Magnification

gion C, Figure 5. Crack propagation and the fatigue striations in that area can easily be identified in Figure 11 with a 10,000X magnification. The interval between striations was recorded and found to be dependent on the mean and maximum values of the stress intensity factor [8]. No striations were observed on the top wall of the tubes.

CONCLUSIONS

1. Fatigue crack propagation tests in steel rectangular hollow sections with fillet welded gusset plates indicated that the major part of fatigue life is spent in propagating the crack in the top wall of the tubes.

2. In some specimens where the weld porosity is substantial, the crack does not start at the top corner, but will initiate at the inner surface of the tube or

472

from the inside of the tube wall. More data is necessary to assess this phenomenon and should include further crack initiation and propagation tests for other thicknesses and stress levels.

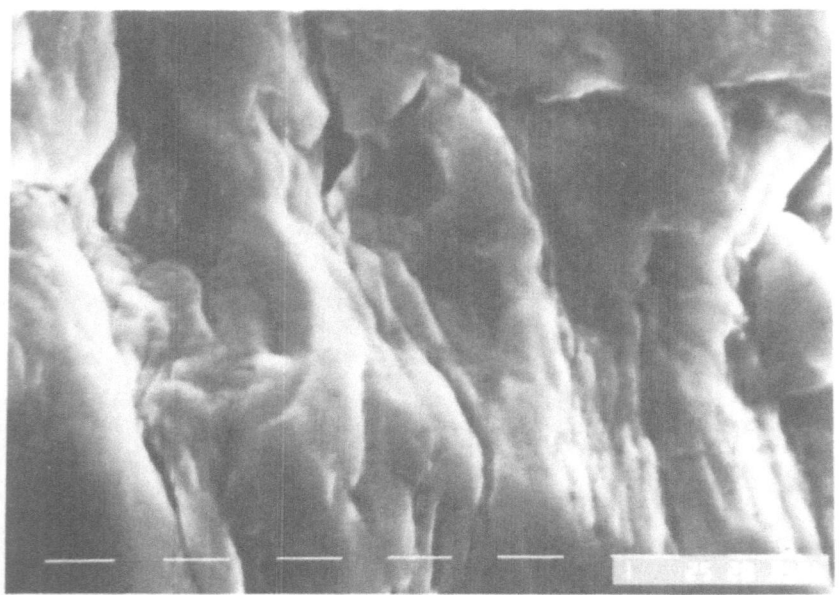

Fig. (11) - Fatigue striations in the crack propagation through the lateral sides of the tubes (specimen G1) 10000X Magnification

ACKNOWLEDGEMENTS

The author wishes to acknowledge the help of Dr. P. S. J. Crofton of Imperial College in the scanning electron microscope work and of Miss Daniela Cuez and Mr. Pedro Matos in the fatigue tests at the University of Minho.

This work was sponsored by the Portuguese National Council for Scientific and Technological Research (JNICT) under contract number 101.79.07. The financial support from the City of Oporto Transport Company is also acknowledged.

REFERENCES

[1] McLester, R., "Fatigue problems in land transport", Metal Science, pp. 303-307, August/September 1977.

[2] Stephens, R. I. and Glinella, G., "Experimental determination of U_I for surface cracks in a square tube under torsion", Experimental Mechanics, pp. 24-30, June 1980.

[3] McDermott, M. E. and Glinella, G., "Experimental determination of K_I for hollow rectangular tubes containing corner cracks", Fracture Mechanics, ASTM STP 677, pp. 719-733, 1979.

[4] Branco, C. M. and Fernando, A. A., "Fatigue behaviour of steel rectangular hollow sections found in bus structures", IIW, Doc. XIII, pp. 1030-1081, Oporto, 1981.

[5] Ferreira, J. A., "Fracture mechanics analysis of fatigue crack propagation in rectangular hollow sections", MS.c. Thesis, Oporto University, 1982.

[6] Maddox, S. J., "Assessing the significance of flaws in welds subject to fatigue", Weld. Res. Suppl., p. 401, 1974.

[7] Watanabe, T., Haugawa, Y. and Matsumoto, K., "Fatigue tests on defective circumferential welded pipe joints", Proc. 5th Int. Conf. Fracture, Cannes, France, pp. 41-48, 1981.

[8] Hertzberg, R. W., "Fatigue fracture surface appearance", Fatigue Crack Propagation, ASTM STP 415, pp. 205-225, 1967.

AN EFFECT OF SPECIMEN SIZE ON FATIGUE CRACK GROWTH AND PLASTIC ZONE SIZE

K. U. Snowden, P. D. Smith and P. A. Stathers

Australian Atomic Energy Commission
Sutherland, New South Wales, Australia, 2232

ABSTRACT

A comparison was made of the fatigue crack growth rates in type 321 stainless steel in the form of 1 mm thick SEN tension specimens and 10 mm thick SEN bending specimens at room temperature. Comparisons were also made of the fatigue crack growth rates in 1 mm thick SEN tension specimens of type 310 stainless steel and type 533B pressure vessel steel and the rates reported in the literature for thicker specimens of these materials. These comparisons showed that, at low values of ΔK, there was little difference between the rates in thick and thin specimens. However, at high ΔK, the fatigue crack growth rates were <u>lower</u> in the thin specimens by up to an order of magnitude compared to those in the thick specimens.

Measurements of cyclic plastic zone sizes using micro-hardness surveys around crack tips in specimens of type 321 stainless steel, revealed that the cyclic plastic zone size was proportional to $(\Delta K)^n$ where $n \doteq 2$ for the SEN bending (B = 10 mm) specimens and $\doteq 0.5$ for the SEN tension (B = 1 mm) specimens. The fatigue crack growth rate behaviour of the thick and thin specimens was also reflected in the size of the cyclic plastic zone; at $\Delta K > 17$ MPa$\sqrt{m}$, the cyclic plastic zone size in the thin specimens was less than that in the thick specimens for the same ΔK. Further, da/dN was approximately proportional to (cyclic plastic zone size)$^{3.5}$ and this relationship appeared to be independent of thickness.

Scanning electron microscopy of the cyclic plastic zone at the crack tip revealed a grid of coarse slip lines which were consistent with those expected from a shear-sliding mechanism for crack growth. This model predicts crack growth rates in agreement with those in the SEN tension specimens of type 321 stainless steel at $\Delta K \geq 17$ MPa$\sqrt{m}$.

The values of ΔK at which deviations in the da/dN - ΔK curves for thick and thin specimens occurred were compared with the limitations imposed by three different size criteria. The criterion which gave the best agreement limited the cyclic plastic zone size to ~3% of the uncracked ligament.

INTRODUCTION

The possibility that material thickness may influence fatigue crack growth is of fundamental importance in the application of fatigue data to the design and performance of real engineering structures. It is reasonable to expect that fatigue crack growth behaviour in thin and thick sections may be different because of possible differences in the state of stress at the crack tip and because through-thickness constraint variations could affect the onset of yielding in the uncracked ligament. A recent review of the effects of thickness on fatigue crack growth has shown that no clear picture has yet emerged; for example, increasing thickness (B) has been found to increase, decrease, or have no effect at all on the fatigue crack growth rate (da/dN) [1].

The present study began as an investigation to determine whether small side-notched tension (SEN tension) specimens (B = 1 mm) of type 310 and 321 austenitic stainless steels and of type 533B ferritic pressure vessel steel could provide meaningful data on the dependence of da/dN on the stress intensity factor range (ΔK). Specimens of this size have a number of significant advantages over thicker or more massive specimens for microstructural or neutron irradiation studies, e.g., the small specimens can be more easily thinned to produce transmission electron microscopy samples, and small irradiated specimens are more readily handled and tested because of lower levels of activity. The da/dN versus ΔK relationships determined for these small specimens were compared with published data for larger specimens of nominally the same material and, in the case of type 321 stainless steel, compared with the da/dN versus ΔK results for side-notched bend (SEN bend) specimens with B = 10 mm of the same steel. These tests indicated that there is a definite influence of specimen thickness on da/dN above a particular value of ΔK, and that this influence is similar for each of the three different materials studied.

In view of the conflicting reports of effect of thickness on da/dN [1], the present type 321 stainless steel specimens were examined in more detail by making micro-hardness surveys around crack tips to indicate plastic zone size, and using scanning electron microscopy to study fracture surfaces and the nature of the deformation around crack tips. The differences in dependence of da/dN on ΔK due to thickness for the three materials were also examined for three different size criteria; the first criterion is being considered for adoption by ASTM and takes the form $(W-a) \geq \frac{4}{\pi} (K_{max}/\sigma_{ys})^2$, where (W-a) is the uncracked ligament length, σ_{ys} is the yield stress, and K_{max} is the maximum stress intensity in a loading cycle [2]; the second gives weight to the strain hardening capacity of the material [2]; and the third is based on the net section stress ≤ 0.8 times the gross yield stress [3].

MATERIALS AND EXPERIMENTAL PROCEDURES

The two stainless steels were supplied in the solution-treated condition and type 533B steel had been given a multistage austenitising, tempering and stress-relieving treatment which is described in detail in [4]. The grain sizes of the type 310 and 321 stainless steel specimens were 20 and 10 to 15 μm respectively. The side-notched tension specimens with B = 1 mm were machined from the three materials in the as-received condition to the dimensions shown in Figure 1. The SEN tension and SEN bending specimens of type 321 stainless steel were given a stress-relieving treatment of 1 hour at 860°C in vacuum to remove any possible effect of

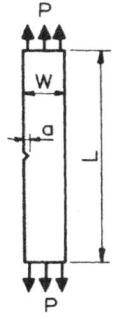

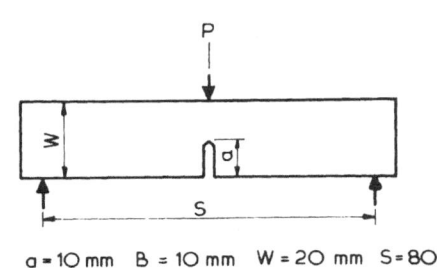

a = 10 mm   B = 10 mm   W = 20 mm   S = 80 mm

a = 2 mm
B = 1 ±01mm
W = 10 mm
L = 55 mm

Fig. (1) - a) SEN tension fatigue specimen;
        b) SEN bending fatigue specimen

surface machining in the micro-hardness tests (see later). This treatment produced no detectable effect on the da/dN - ΔK relationships.

The SEN tension and SEN bending specimens were tested in sinusoidal loading in laboratory air, at ambient temperature and at a frequency of 10 Hz; the ratio of the minimum to maximum stresses was zero. The SEN tension specimens were tested in a Schenck constant-load fatigue machine with peak load constant to within ±2%. The SEN bending specimens were fatigued in three-point bending in an Instron servo-hydraulic testing machine operated in the constant displacement mode in which the load was constantly monitored. The tests were interrupted at various stages to measure crack length using a low-powered microscope (X20) with a long-distance objective. In both SEN tension and SEN bending specimens, a short starter crack, ~1 mm long, was initiated at the notch prior to undertaking the growth measurements. Micro-hardness was measured on a Vickers micro-hardness tester, using a 15 or 25 g load. The typical scatter of measured hardness values was approximately ±5 hardness units.

The fatigue crack growth data were analysed in terms of the Paris-Erdogan relationship [5]

$$da/dN = C(\Delta K)^m \tag{1}$$

where C and m are constants over limited ranges of ΔK. The stress intensity factor range (ΔK) for the SEN tension specimens was derived from [6]

$$\Delta K = (Pa^{\frac{1}{2}}/BW) \cdot Y \tag{2}$$

where P is the maximum cycle load, B is the thickness, a is the total fatigue crack length plus notch, W is the specimen width and Y is given by [6] as

$$Y = 1.99 - 0.41(a/W) + 18.70(a/W)^2 - 38.48(a/W)^3 + 53.85(a/W)^4 \tag{3}$$

The stress intensity values for the SEN bending specimens were obtained from [7]

$$\Delta K = (3PS/2\ BW^{3/2}) \cdot Y \tag{4}$$

for $0 \leq a/W \leq 1$ and where S is the support span (= 4W ± 0.01 W), and

$$Y = \frac{(a/W)^{\frac{1}{2}}\{1.99-(a/W)[1-(a/W)][2.15-3.93(a/W)+2.7(a/W)^2]\}}{[1+2(a/W)][2-(a/W)]^{3/2}} \tag{5}$$

The growth rate measurements were made on at least five to six specimens of each material and specimen size.

Tables 1 and 2 give the compositions and the mechanical properties of the three materials.

TABLE 1 - COMPOSITION (wt %)

Material	C	Ni	Cr	Ti	Si	Mn	S	P	Mo	Fe
Type 321 ss	0.04	10.3	17.0	0.49	0.54	0.68	0.024	0.041		balance
Type 310 ss	0.06	21.2	23.1	0.10	0.68	1.04	0.03	0.02		balance
Type 309s ss [9]	0.06	14.56	22.5	-	0.37	1.72	0.015	0.025		balance
Type 533B Cℓ 1	0.18	0.56	0.06	-	0.20	1.38	0.006	0.006	0.53	

TABLE 2 - MECHANICAL PROPERTIES

	0.2% σ_{ys}, MPa	UTS MPa	Elongation %
Type 321 ss	293	616	67
Type 310 ss	385	657	47
Type 533B Cℓ 1	531	633	19

RESULTS

A. Fatigue Crack Growth

The fatigue crack growth measurements and their dependence on ΔK for the SEN tension (B = 1 mm) and SEN bending (B = 10 mm) type 321 stainless steel specimens are given in Figure 2(a). The results show that there was agreement between the growth rates in the thick and thin specimens for $\Delta K \leq 17$ MPa√m and that, above this value of ΔK, the growth rates in the thin specimens were significantly less than those in the thicker specimens. A single straight line could be used to repre-

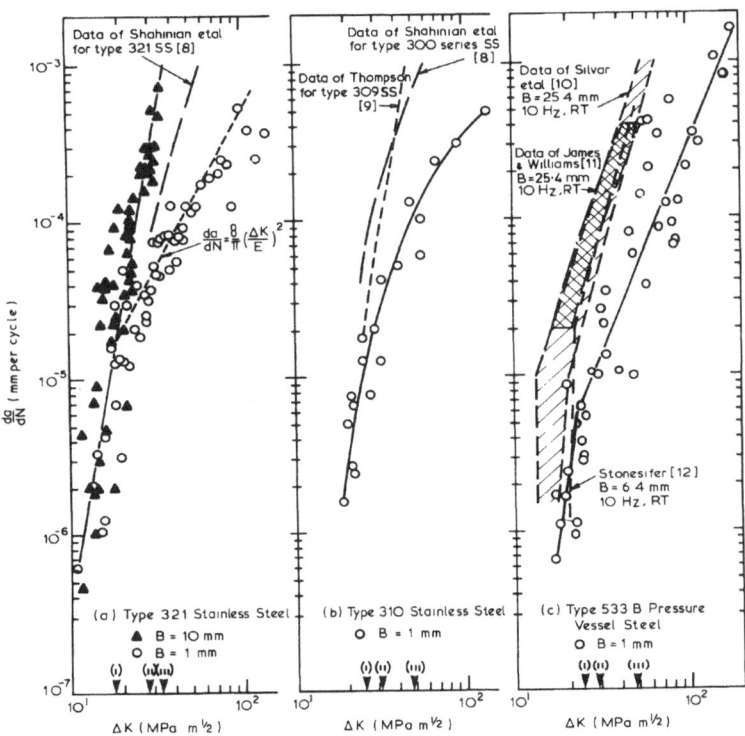

Fig. (2) - a) Crack growth rates for type 321 stainless steel SEN tension
(B = 1 mm) specimens, open points; and SEN bending (B = 10 mm)
specimens, closed points. Dotted lines represent data of Shahinian et al [8] for type 321 stainless steel specimens with B
= 12.7 mm. Dashed line represents equation (10), da/dN = (8/π)
(ΔK/E)², derived by Pook and Frost [22];
b) Crack growth rates for type 310 stainless steel SEN tension
(B = 1 mm) specimens. The dashed curve represents Thompson's
data for type 309S stainless steel specimens with B = 7.5 mm
[9], and the dotted curve represents the composite data of Shahinian et al [8] for type 300 series stainless steels with B
= 12.7 mm;
c) Crack growth rates for type 533B pressure vessel steel SEN
tension (B = 1 mm) specimens. The dashed and dotted curves are
taken from [10-12] for the same steel with B = 5.08 to 25.4 mm.
Arrows (▼) indicate values of ΔK at which violations of inequalities (6-8) occurred

sent the SEN bending data over the entire stress intensity range investigated (ΔK = 11 to 34 MPa$\sqrt{m}$) and this indicated that the exponent of ΔK in equation (1) had a value of $m \cong 6$. The exponent of ΔK for the SEN tension specimens at $\Delta K \geq 17$ MPa$\sqrt{m}$ was significantly less than for the thicker specimen, i.e., $m \cong 2$. The results for the SEN bending specimens are in reasonable agreement with those reported by Shahinian et al [8] for bending specimens of type 321 stainless steel with B = 12.7 mm at $\Delta K \geq 30$ MPa$\sqrt{m}$.

The SEN tension (B = 1 mm) specimens of type 310 stainless steel exhibited a dependence of da/dN on ΔK similar to that shown by the thin type 321 stainless steel specimens, Figure 2(b). At values of $\Delta K \geq 20$ MPa$\sqrt{m}$, the present results were compared with tension specimens of type 309S stainless steel with B = 7.5 mm reported by Thompson [9] because no measurements of da/dN for type 310 stainless steel were available in the literature. This latter steel has a similar but not identical composition to the type 310 stainless steel used in the present work, Table 1. Thompson's results were obtained at values of $\Delta K \geq 25$ MPa$\sqrt{m}$ and gave the exponent of ΔK a value of m = 6.12. These results and those of Shahinian et al [8] for type 300 series stainless steels have been included in Figure 2(b) as dashed lines for comparison with the present data for type 310 stainless steel. Shahinian et al [8] found that types 304, 317, 321 and 348 stainless steels exhibited the same fatigue crack growth behaviour at room temperature. Comparison between these two sets of data and the present data suggests that a similar divergence occurs between the thick and thin specimens for $\Delta K \geq 25$ MPa$\sqrt{m}$, with da/dN for the thin specimens lying below that for the thicker specimens similarly to the type 321 stainless steel, Figure 2(a).

The results of tests on the SEN tension (B = 1 mm) specimens of type 533B pressure vessel steel are compared in Figure 2(c) with published data reported by several groups of investigators [10-12]; the thickness of the latter specimens ranged from B = 6.4 to 25.4 mm. The present SEN tension specimens exhibited crack growth rates which were in reasonable agreement with published data at values of ΔK up to 20 MPa$\sqrt{m}$; the results diverged above this value, with the SEN tension results lying below those for the thicker specimens.

B. Plastic Zone Size

Micro-hardness surveys around crack tips were made in conjunction with the above tests on the SEN tension (B = 1 mm) and SEN bending (B = 10 mm) specimens of type 321 stainless steel. Figure 3 gives typical micro-hardness distributions around the crack tip in a SEN tension specimen at ΔK values of 20, 49 and 103 MPa$\sqrt{m}$. The hardness distributions on the left of the figure are for surveys in front of the crack in the direction of propagation and the distributions on the right, for surveys at right angles to this direction. The curves drawn through the points were obtained by multivariate regression analysis. The micro-hardness values decreased with increasing distance from the crack tip until a plateau was reached. Bathias and Pelloux [13] have shown that this plateau extends until a further decrease in hardness occurs (not determined in the present investigation) which corresponds to the end of the monotonic plastic zone; see insert in Figure 3 which shows a schematic representation of the monotonic and cyclic plastic zones and the corresponding hardness distribution. Figure 3 also shows that the general level of hardness increases with increasing ΔK and that, at a given distance from the crack tip, the hardness values were greater to the side of the crack than in front of it; this latter observation is consistent with the "butterfly" shape of the plastic zone at the crack tip. Similar hardness distributions around crack tips were obtained with the SEN bending (B = 10 mm) specimens and these displayed

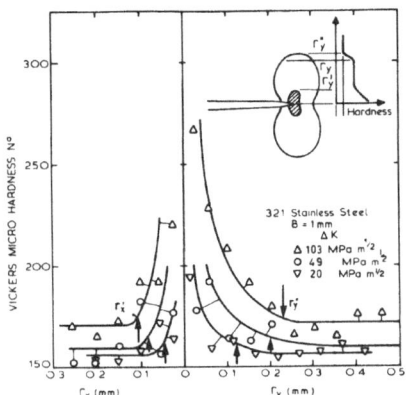

Fig. (3) - Plastic zone size determinations by micro-hardness measurements
on type 321 stainless steel SEN tension (B = 1 mm) specimens for
ΔK = 20, 49 and 103 MPa$\sqrt{m}$. Points to the right of the ordinate
are for hardness measured at distances (r_y) from the crack tip
in the direction at right angles to propagation, and those to
the left are for distances (r_x) from the crack tip in the direction of propagation. r_y' and r_x' are the cyclic plastic zone
sizes. Insert, schematic representation of monotonic and cyclic
plastic zones after Bathias and Pelloux [13]

the same general trends with regard to hardness level and ΔK, hardness and distance from the crack tip, and shape of zone as shown in Figure 3. The extent of
the cyclic plastic zone (r') was taken as the distance at which the micro-hardness
values exceeded the average plateau value by 5 hardness units, i.e., the distance
at which the hardness rose above the plateau scatter. These distances have been
indicated in Figure 3 following Hahn et al [14] by arrows designated r_x' for the
zone size in front of the crack tip in the direction of propagation, and r_y' for
the zone size at right angles to the direction of propagation.

Comparison of the plastic zone sizes in the thick and thin specimens on
the basis of the same ΔK, showed that the zone size in the thicker specimens tended
to be larger than those in thin specimens, Figure 4. However, at lower values of
ΔK, the curves for the thick and thin specimens tended to intersect at $\Delta K \cong 17$ MPa$\sqrt{m}$,
i.e., the value of ΔK at which the crack growth rates diverged in Figure 2(a). Further, r_x' and r_y' for the thick specimens were approximately proportional to $(\Delta K)^2$
whereas the corresponding zone sizes for the thin specimens were less strongly dependent on ΔK and approximately proportional to $(\Delta K)^{\frac{1}{2}}$.

Figure 5 gives the relation between plastic zone size and da/dN. These results indicate that, first, the dependence of r_x' and r_y' on da/dN was not affected

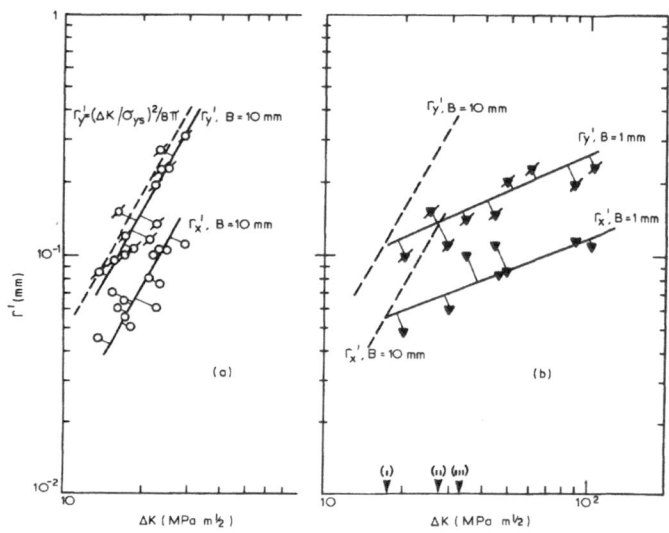

Fig. (4) - The relation between the cyclic plastic zone size (r_x' and r_y')
and ΔK for type 321 stainless steel SEN bending (B = 10 mm)
specimens (a) and SEN tension (B = 1 mm) specimens (b). The
dashed line in (a) represents equation (9), $r_y' = (\Delta K/\sigma_{ys})^2/8\pi$.
Arrows (▼) indicate values of ΔK at which violations of inequalities (6-8) occurred

by thickness; second, da/dN was approximately proportional to (cyclic plastic zone
size)$^{3.5}$, and third $r_y' \cong 2r_x'$. These observations suggest that the fatigue crack
growth rate is closely linked to the size of the plastic zone that can develop in
material of a particular thickness.

FRACTURE OBSERVATIONS

Scanning electron microscopy observations were made on the fracture surface
and the plastic zone at the crack tip of both thick and thin type 321 stainless
steel specimens. The fracture surfaces of both types of specimens were macroscopically planar and showed no evidence of such sizeable shear lips as those observed
by Jack and Price for mild steel [15] and by Schijve for aluminium alloys [16].
At low values of ΔK, the fracture surface of both thick and thin specimens exhibited
ductile facets which were typically ~3x10 μm, i.e., generally smaller than the grain
size, Figures 6(a) and (c). The facets sometimes exhibited fine parallel striation
markings such as those indicated by arrows in Figure 6. At higher values of ΔK, the

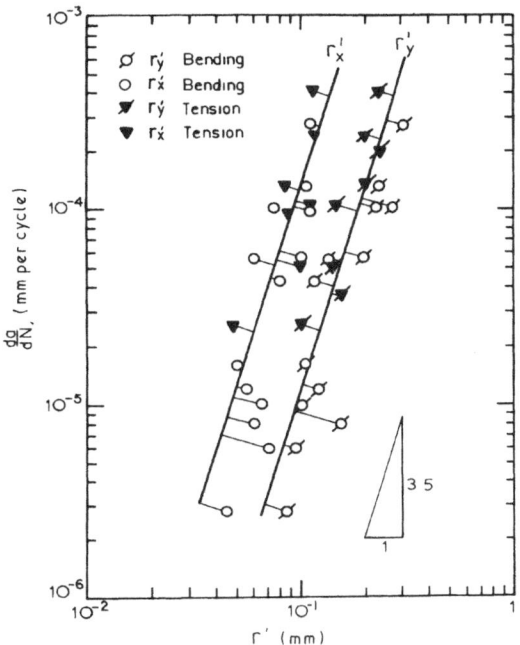

Fig. (5) - The relationship between da/dN and the plastic zone size (r_x' and r_y') in type 321 stainless steel SEN bending (B = 10 mm) specimens, open points, and SEN tension (B = 1 mm) specimens, closed points

ductile facets were progressively replaced by fine and then by coarse striations and secondary cracking, Figures 6(b) and (d). For $\Delta K \geq 30$ MPa$\sqrt{m}$, the striation size was comparable with the crack propagation distance per cycle, whereas below 30 MPa$\sqrt{m}$, the striation size was greater than the propagation distance per cycle. A similar variation in striation size of austenitic stainless steels has been reported by a number of investigators including Bathias and Pelloux [13] and Tomkins [17].

The nature of the deformation within the plastic zone at the crack tip of a SEN tension (B = 1 mm) specimen tested at 20 MPa$\sqrt{m}$ is shown in Figure 7(a). The crack tip zone exhibits a grid of coarse parallel slip lines which have been intensified in the immediate vicinity of the crack tip, i.e., within ~10^{-2} mm or $r_x'/8$. In this region, the grid size is reduced and slip line intensification has occurred along segments of the grid emanating from the crack tip. Figure 7(c)

484

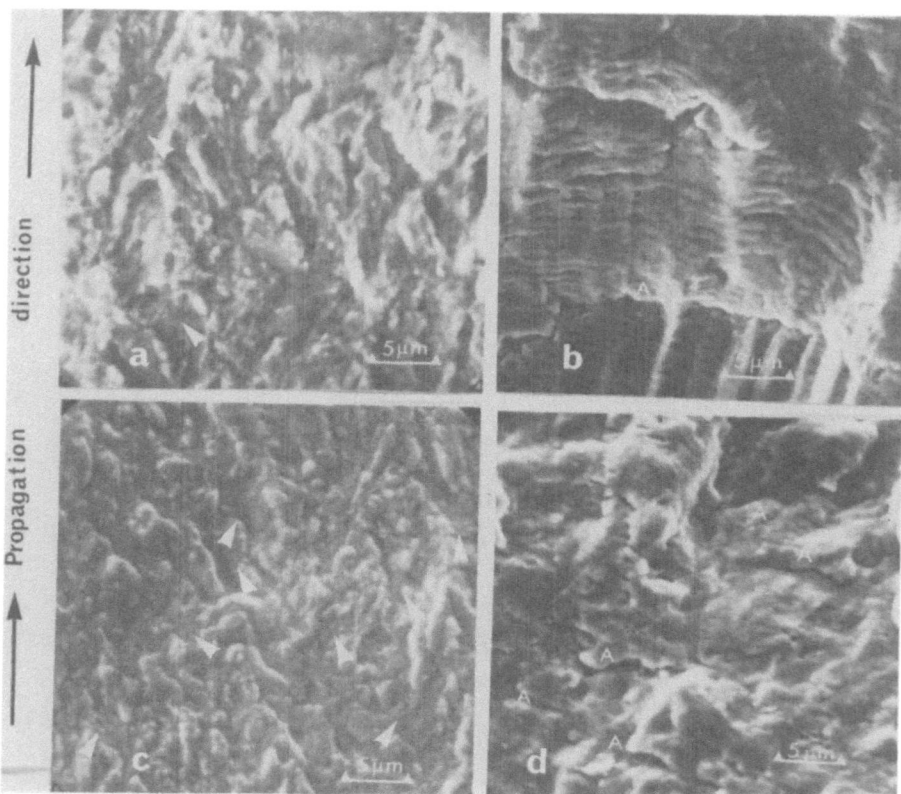

Propagation direction

Fig. (6) - Scanning electron micrographs of the fracture surfaces of type 321
stainless steel thick and thin specimens.
(a) SEN tension (B = 1 mm) specimens, ΔK = 21.5 MPa√m. Area shows
ductile facets. Arrows indicate facets displaying fine striations;
(b) SEN tension (B = 1 mm) specimen, ΔK ≅ 100 MPa√m. Area shows
coarse striations and secondary cracking indicated at A;
(c) SEN bending (B = 10 mm) specimens, ΔK = 17 MPa m̄. Area shows
ductile facets. Arrows indicate facets displaying fine striations;
(d) SEN bending (B = 10 mm) specimens, ΔK = 30 MPa√m. Area displays
striations and secondary cracking indicated at A

shows the same crack as that in Figure 7(a) but further from the crack tip viewed
at an angle of ~45 degrees. The visible fracture surface displays ridge-like features which appear to be associated with the ductile facets shown in Figures 6(a)
and (b).

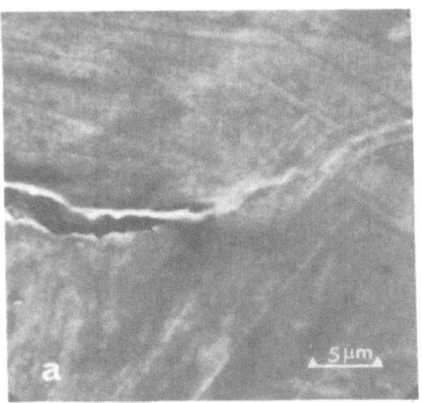

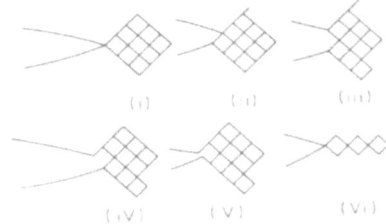

b

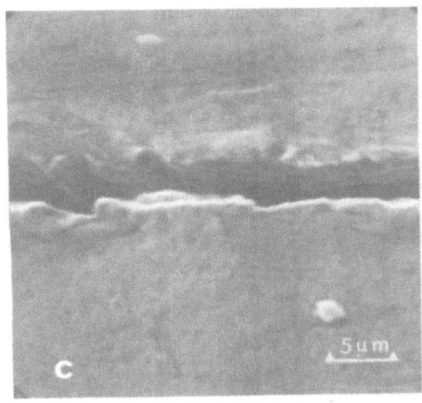

Fig. (7) - (a) Area around crack tip in type 321 stainless steel SEN tension
(B = 1 mm) specimens, ΔK = 20 MPa$\sqrt{m}$. Area shows grid of coarse
slip lines;
(b) Geometric model of fatigue-crack growth by shear sliding (Laird-
Smith-Neumann model). (i) Initial configuration; (ii) Crack opened
in tension with shear sliding on one slip plane above the crack
plane; (iii) Followed by shear sliding on one slip plane below the
crack plane; (iv) Crack unloaded with reversed shear sliding on
slip plane of (ii); (v) Followed by reversed shear sliding on slip
plane of (iii); the crack has advanced; (vi) After a number of repetitions of the mechanism; note that fatigue striations have been
formed. After Weertman [21];
(c) Section of the same crack as shown in (a) but further away from
crack tip. Area observed at ~45 degrees to the surface normal to
give oblique view of fracture surface. This surface shows ridges
and ductile facets

COMPARISON WITH DIFFERENT SIZE CRITERIA

The deviations shown by the da/dN - ΔK curves for the three materials, Figure 2, and by the plastic zone size - ΔK curves, Figure 4, were examined with respect to three criteria for limiting the size for fatigue specimens [2,3]. The first is under consideration for adoption by ASTM and takes the form [2]

$$(W-a) \gtrsim \frac{4}{\pi} \left(K_{max}/\sigma_{ys}\right)^2 \tag{6}$$

The second is similar to equation (6) but makes some allowance for strain hardening and takes the form [2]

$$(W-a) \gtrsim \frac{4}{\pi} \left(K_{max}/\sigma_{flow}\right)^2 \tag{7}$$

The third relates to the onset of gross yielding in the uncracked ligament [3]

$$\sigma_{net} \lesssim 0.8 \; \sigma_{ys} \tag{8}$$

where W is the specimen width, a is the total crack length including the notch, K_{max} is the maximum stress intensity factor in the stress cycle*, σ_{ys} is the gross yield stress, $\sigma_{flow} = (\sigma_{ys} + \sigma_{UTS})/2$ where σ_{UTS} is the tensile strength and σ_{net} is the net section stress. The values of ΔK at which the inequalities (6-8) are violated by the SEN tension (B = 1 mm) specimens are indicated by arrows along the ΔK axis in Figures 2 and 4. The arrow designated (i) relates to inequality (6), (ii) to inequality (7), and (iii) to inequality (8). These arrows indicate that the values of ΔK at which the da/dN - ΔK curves diverge are best predicted by inequality (6) followed by (7) and (8). These results are summarised in Table 3.

TABLE 3 - CRITERIA VIOLATIONS

	Value of ΔK at Violation by SEN Tension Specimens MPa $\sqrt{m}$		
	321 ss	310 ss	533B
Experimental value of ΔK for divergence in growth rate	17	25	20
$(W-a) \geq \frac{4}{\pi} (\Delta K/\sigma_{ys})^2$	17.5	25	24
$(W-a) \geq \frac{4}{\pi} (\Delta K/\sigma_{flow})^2$	27.5	31	29
$\sigma_{net} \leq 0.8 \; \sigma_{ys}$	33	48	48.5

---

*Note, in the present work, $\Delta K = K_{max}$.

DISCUSSION

The present investigation throws light on the inter-relationship between fatigue crack growth rate, cyclic plastic zone size and the thickness of material. The da/dN - ΔK curves for the SEN tension (B = 1 mm) and the SEN bending (B = 10 mm) specimens of type 321 stainless steel showed that a retardation in da/dN of the thin specimens occurred above a particular value of ΔK. Similar behaviour was found for type 310 stainless steel and type 533B pressure vessel steel, Figure 2. At low values of ΔK, the fracture surfaces of both thick and thin specimens of type 321 stainless steel exhibited ductile facets which sometimes displayed fine striations. These features suggest that, at low values of ΔK, fatigue crack growth is highly localised. At higher ΔK, the ductile facets were progressively replaced by fine and coarse striations and secondary cracks which extend over distances comparable with grain size. These changes suggest that fatigue crack growth becomes increasingly less localised with increasing ΔK and that secondary cracking is a consequence of the intense strain hardening that occurs at the crack tip at high ΔK.

The divergencies in the da/dN - ΔK curves were also evident in the cyclic plastic zone size - ΔK curves, Figure 4. For the same ΔK, the cyclic plastic zone sizes, r_x' and r_y', in the thin specimens were smaller than those in the thick specimens for values of ΔK greater than that for divergence in the da/dN - ΔK curves. Both r_x' and r_y' in the thick specimens were approximately proportional to $(\Delta K)^n$, where $n \cong 2$ which is in accord with the results of Bathias and Pelloux [13], Iino [18], and Hahn et al [14], all of whom worked with relatively "thick" specimens, i.e., specimens with B = 9.5, 25.5 and 2.54 to 12.7 mm respectively. The corresponding cyclic plastic zone sizes in the thin specimens exhibited a smaller dependence on ΔK with $n \cong \frac{1}{2}$ which is in agreement with data of Yokobori et al [19] for specimens with B = 2.5 mm. Figure 4 indicates that for both thick and thin specimens, the magnitude of r_y' was about twice r_x' and that for the thick specimens it was close to the value given by [20],

$$r_y' = (\Delta K/\sigma_{ys})^2/8\pi \tag{9}$$

This equation has been plotted as a dashed line in Figure 4. The reason for the dependence of the cyclic plastic zone sizes in the thin specimens on $(\Delta K)^{\frac{1}{2}}$ is not understood at present but may be associated with crack closure effects [21] which appears to be greater at the surface than in the interior [24].

The similarities between the divergence shown by the da/dN - ΔK and the r'-ΔK curves for the thick and thin specimens (Figures 2 and 4) suggested a close link between da/dN and r'. This was confirmed in the plot of da/dN versus r' shown in Figure 5 which indicates that da/dN is approximately proportional to $(r')^p$, where $p \cong 3.5$ and that this relationship does not depend significantly on B. The value of the exponent $p \cong 3.5$ lies between those reported by Iino [18] and Yokobori et al [19] which were $p \cong 2$ and 7.5 respectively. It is also noted that the exponents m, p and n are inter-related, i.e.,

$$da/dN \propto (\Delta K)^m \propto (r')^p, \text{ and } r' \propto (\Delta K)^n$$

giving

$m \cong np$

Examination of the deformation at the crack tip revealed a grid of coarse slip lines in which segments of the grid had been intensified in the immediate vicinity of the crack tip, Figure 7(a). This pattern of deformation appeared similar to that expected from the crack-opening-displacement/shear-sliding mechanism for crack growth elaborated by Weertman [21]. A schematic diagram of the working of this model is given in Figure 7(b). Weertman also pointed out that an analysis of this type of model by Pook and Frost [22] predicted a fatigue crack growth rate proportional to $(\Delta K)^2$,

$$da/dN = \frac{8}{\pi} (\Delta K/E)^2 \tag{10}$$

where E is the elastic modulus. The values of da/dN predicted by equation (10) are in good agreement with the experimental values for the SEN tension (B = 1 mm) specimens of type 321 stainless steel given in Figure 2(a). Equation (10) also indicates that da/dN should not depend on σ_{ys}. This is supported by the similarity between the present results for type 321 and 310 stainless steels (σ_{ys} = 293 and 385 MPa respectively) and those of Shahinian et al [8] who found that σ_{ys} had little effect on the fatigue resistance of type 300 series stainless steels at room temperature.

Comparison between the values of ΔK at which divergences occurred in the da/dN − ΔK curves and the values predicted by the three size criteria given by equations (6)-(8) is presented in Table 3. The table shows that the best agreement was obtained with equation (6). This equation can be combined with equation (9) to give an estimate of the limitation imposed on the size of the cyclic plastic zone in terms of the uncracked ligament, i.e.,

$$\frac{2r'_x}{(W-a)} = \frac{1}{32}$$

This represents ~3% of the uncracked ligament which may be compared with the corresponding value of 25% for the monotonic plastic zone size [23].

CONCLUSIONS

1. The fatigue crack growth rate in SEN tension (B = 1 mm) type 321 stainless steel specimens was up to 10 times slower than that in SEN bending (B = 10 mm) specimens of the same material tested at $\Delta K \geq 17$ MPa√m. Below this value of ΔK, the growth rates were similar.

2. SEN tension (B = 1 mm) specimens of type 310 stainless steel and type 533B pressure vessel steel exhibited similar behaviour to the type 321 stainless steel specimens when results of tests on the former were compared with results reported in the literature for tests on thicker specimens of nominally the same or similar materials.

3. Micro-hardness surveys around crack tips in type 321 stainless steel SEN tension and SEN bending specimens showed that the cyclic plastic zone size was proportional to $(\Delta K)^n$ where $n \cong 2$ for the thick specimens and $\cong \frac{1}{2}$ for the thin speci-

mens. At values of $\Delta K \geq 17$ MPa$\sqrt{m}$, the cyclic plastic zone sizes were smaller for the SEN tension specimens (B = 1 mm) than for the SEN bending specimens (B = 10 mm).

4. The fatigue crack growth rates in the SEN tension and SEN bending specimens of type 321 stainless steel were approximately proportional to (cyclic plastic zone size)$^{3.5}$. This relationship appeared to be independent of specimen thickness.

5. On a macro scale, the fracture surfaces of the SEN tension and SEN bending specimens of type 321 stainless steel were flat, and on a micro scale exhibited ductile facets at low ΔK and fatigue striations and secondary cracks at high ΔK. The changes in the nature of these features suggested that crack growth occurred locally at low ΔK and progressively less locally at higher values of ΔK.

6. The deformation at the crack tip was consistent with the shear-sliding model for fatigue crack growth described by Weertman [21]. The experimental growth rates at $\Delta K \geq 17$ MPa$\sqrt{m}$ for the SEN tension specimens of type 321 stainless steel were in good agreement with the rates predicted by the equation of Pook and Frost [22] based on this model.

7. The divergences in the da/dN - ΔK curves for the thick and thin specimens were best estimated by the criterion $(W-a) \geq \frac{4}{\pi} (\Delta K/\sigma_{ys})^2$. Violations of this inequality were found to occur when the cyclic plastic zone was ~3% of the uncracked ligament.

ACKNOWLEDGEMENTS

The authors wish to thank Mr. K. Veevers for reading the manuscript, Mr. K. Watson and Mr. J. Mellor for assistance with the scanning electron microscopy and the metallography.

REFERENCES

[1] Sadananda, K. and Shahinian, P., "Characterization of materials for service at elevated temperatures", Series MPC-7, G. V. Smith, ed., ASME, pp. 107-127, 1978.

[2] Brose, W. R. and Dowling, N. E., "Size effects on the fatigue crack growth rate of type 304 stainless steel", Elastic Plastic Fracture, ASTM STP 668, J. D. Landes et al, eds., ASTM, pp. 720-735, 1979.

[3] ASTM E 647-78T, "Tentative method for test for constant-load-amplitude fatigue-crack growth rates above 10^{-8} m/cycle, Philadelphia, ASTM 1978.

[4] Hawthorne, J. R. and Watson, H. E., "Strength and notch ductility of selected structural alloys after high fluence, 550F (288C) irradiation", Naval Research Laboratory, Washington, D.C., NRL Report 7813, 1974.

[5] Paris, P. C. and Erdogan, F., Trans. ASME 85, pp. 528-533, 1963.

[6] Brown, W. F. and Srawley, J. E., "Plane strain crack toughness of high strength metallic materials", ASTM STP 410, Philadelphia, Pa., ASTM, pp. 1-65, 1966.

490

[7] Srawley, J. E., Int. J. of Fracture, 12, pp. 475-476, 1976.

[8] Shahinian, P., Smith, H. H, and Watson, H. E., "Fatigue crack growth charac-
     teristics of several austenitic stainless steels at high temperature", ASTM
     STP 520, Philadelphia, Pa., ASTM, pp. 387-400, 1973.

[9] Thompson, A. W., Engineering Fracture Mechanics, 7, pp. 61-68, 1975.

[10] Salivar, G. C. and Creighton, D. L., Engineering Fracture Mechanics, 14, pp.
     337-352, 1981.

[11] James, L. A. and Williams, J. A., J. of Nuclear Materials, 47, pp. 17-22,
     1973.

[12] Stonesifer, F. R., Engineering Fracture Mechanics, 10, pp. 305-314, 1978.

[13] Bathias, C. and Pelloux, R. M., Metallurgical Transactions, 4, pp. 1265-1273,
     1973.

[14] Hahn, G. T., Hoagland, R. G. and Rosenfield, A. R., Metallurgical Transactions,
     3, pp. 1189-1202, 1972.

[15] Jack, A. R. and Price, A. T., Acta Metallurgica, 20, pp. 857-866, 1972.

[16] Schijve, J., Engineering Fracture Mechanics, 14, pp. 467-475, 1981.

[17] Tomkins, B., Metal Science, 14, pp. 408-417, 1980.

[18] Iino, Y., Engineering Fracture Mechanics, 12, pp. 279-299, 1979.

[19] Yokobori, T., Sato, K. and Yamaguchi, Y., Reports of the Institute for Strength
     and Fracture of Materials, 6, pp. 49-67, 1970.

[20] Paris, P. C. and Sih, G. C., Fracture Toughness and its Applications, ASTM,
     pp. 30-76, 1965.

[21] Weertman, J., "Fatigue-crack propagation theories, fatigue and microstruc-
     ture", ASM, pp. 279-306, 1979.

[22] Pook, L. P. and Frost, N. E., Int. J. of Fracture, 9, pp. 53-61, 1973.

[23] Hudak, S. J., Saxena, A., Bucci, R. J. and Malcolm, R. C., "Development of
     standard methods of testing and analysing fatigue crack growth rate data",
     Technical Report AFML-TR-78-40, 1977.

[24] Ogura, K., Ouji, K. and Honda, K., Int. J. of Fracture, 13, pp. 524-527, 1977.

CRACK CLOSURE AND OVERLOAD EFFECTS IN FATIGUE

J. Q. Clayton

Aeronautical Research Laboratories
Melbourne, Australia

ABSTRACT

Observations of crack closure occurring during crack propagation in thin sheet materials under variable amplitude cyclic loading are discussed. The studies show that closure in the region immediately adjacent to the crack tip is important in controlling crack growth. From this basis, the Dugdale model and a simple fracture mechanics analysis is used to investigate the closure stresses produced by an overload during fatigue. The results suggest that this is a promising approach to fatigue crack growth prediction under varying load amplitude.

INTRODUCTION

Under Damage Tolerance Specifications [1,2], modern aircraft structures are assumed to contain flaws of known size (usually established by non-destructive inspection procedures) which propagate under the action of service loads. Reliable estimation of the crack-growth component of the fatigue life is difficult because the loading is of variable amplitude and the complex effects of load history on crack growth are not well understood. The best known and most significant of these effects is the retardation (or temporary decrease) in crack growth rate observed after an intermediate high load (single overload cycle) is applied in a test.

Although the importance of load history effects on crack growth has been recognized for some time, a fundamental understanding of the mechanisms involved has not been obtained. Consequently, the development of a quantitative model has had only limited success. The retardation has been attributed mainly to blunting in the crack tip region [3], the residual stress field ahead of the crack [4] or the residual tensile plastic displacements of the crack faces [5]. Of these possible causes, most attention has been directed towards the suggestion by Elber [5] that plastic deformation in the wake of the crack can cause the crack to be closed, even under an applied tensile stress. The effective stress intensity range ΔK_{eff} is then defined as the range above the crack opening point (K_{OP}),

$$\Delta K_{eff} = K_{max} - K_{OP} \tag{1}$$

$$= U(K_{max} - K_{min})$$

$$= U \Delta K \tag{2}$$

where U is termed the effective stress ratio. Elber measured K_{OP} from the tangent corresponding to the change in slope on a plot of crack-tip displacement versus applied load. The crack-tip displacement was measured by means of a compliance gauge fixed to the specimen at a point just behind the crack tip, thus enabling a stress-strain hysteresis loop to be recorded for material in the crack tip region. The extent to which such measurements of crack closure can be used together with equation (1) to explain the observed effects of mean stress and overloads on crack growth has been the subject of considerable debate [6-8]. Doubt also exists regarding the reliability of the measurement technique [9].

The main aim of the present study was to assess the experimental reliability of the crack-tip displacement gauge as a means of measuring crack closure and to apply the gauge to investigate crack closure in simple overload situations. Attention is focused on the crack propagation rate during the stage immediately following an overload. The results are discussed with the aid of a simple Dugdale model for predicting the stresses causing crack closure.

EXPERIMENTAL DETAILS

The material used for this study was 6061-T6 aluminum alloy (0.25%Zn, 1.0%Mg, 0.25%Cu, 0.7%Fe, 0.6%Si, 0.25%Cr, 0.15%Mn) in sheet form (1.2 mm thick). The material had a 0.2% yield stress of 321 MPa and a tensile strength of 344 MPa. The specimens were of the SEN type (150 mm x 25.4 mm) and were cut with the tensile axis perpendicular to the rolling direction (TL orientation). A 1 mm wide notch (45° tip) was cut in each specimen to a depth of 5 mm and a fatigue crack was grown a further 2.5 mm under constant amplitude loading (all precracking was performed at $\Delta K = 12.0$ MPam$^{\frac{1}{2}}$, stress ratio R=0.

Fatigue crack propagation tests were conducted under load control in an Instron testing machine of 2500 kg load capacity. A cyclic frequency of 0.3 Hz (saw-tooth waveform) was used for most tests; overload cycles were applied under manual operation at ~0.1 Hz.

Strain in the region of the crack tip was measured by mounting a displacement gauge on the specimen surface, Figure 1. The gauge was constructed with a high-compliance aluminum alloy base to which a strain gauge bridge was bonded. The bridge was operated using a Peekel strain gauge amplifier. Linearity in the displacement gauge and freedom from hysteresis were verified in separate cycling tests. Sharp steel pins were used to locate the gauge in scribe marks 0.6 mm apart (± 0.3 mm from the crack) so that the displacement between the crack faces was registered. For all tests, the gauge was positioned 0.5 mm behind the crack tip.

Load was controlled using the Instron testing system load cell. A separate strain gauge bonded to the specimen (between the crack and one grip) and calibrated against the load cell was used to provide the load signal for recording the stress-strain loop. This method was found to provide the best response to dynamic loads.

The crack growth rate during testing was monitored by the AC potential drop method. The equipment employed was not sensitive enough to accurately detect the small changes in crack length occurring at each load level, but provided a useful indication of the general change in crack growth rate. Detailed changes in crack growth rate were obtained by subsequent SEM (scanning electron microscopy) examination of the fracture surfaces (all fracture surfaces were Au sputter-coated).

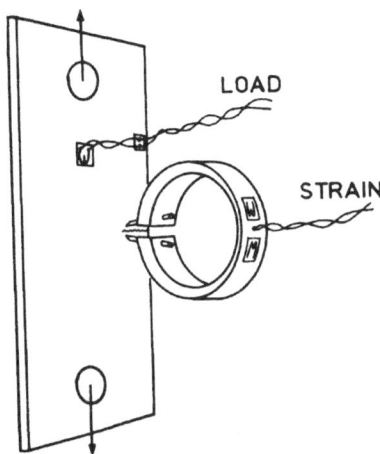

Fig. (1) - Schematic diagram of test specimen showing location
of gauges used for recording stress-strain hysteresis
loop

RESULTS

A. Overload Amplitude Tests

In these tests, the change in measured crack opening stress intensity caused by cycling for a small number of cycles (N=10) at a predetermined amplitude (ΔK = 8.9, 10.7, 12.5, 17.8, 19.6 MPam$^{\frac{1}{2}}$, R=0) was investigated. The maximum load amplitude corresponded to an overload ratio of 1.6. Representative test results are described below.

Figure 2 shows the hysteresis loop (nominal crack-tip strain versus applied stress intensity) recorded after precracking. The maximum stress intensity (8.9 MPam$^{\frac{1}{2}}$) is well below that applied during precracking (12.0 MPam$^{\frac{1}{2}}$). Crack closure is clearly evident from the knee in the hysteresis loop. No change in the loop was detected during subsequent cycling (ΔK = 8.9 MPam$^{\frac{1}{2}}$, N = 300).

The effect on the hysteresis loop of load cycling just above the precracking level (at ΔK = 12.5 MPam$^{\frac{1}{2}}$) is shown in Figure 3. A permanent crack opening (blunting) is evident from the failure of hysteresis loops 1-3 to close on unloading; loops 4-10 close on unloading and are very similar. Figure 3 also shows that the effect of this loading is to produce a small, but measurable, decrease in the crack opening stress intensity.

The hysteresis loop produced by cycling at a relatively high load level (ΔK = 17.8 MPam$^{\frac{1}{2}}$) is shown in Figure 4. It is clear that appreciable crack opening occurs during initial cycling at this amplitude and that the crack opening stress intensity, measured during subsequent low level cycling, decreases markedly.

494

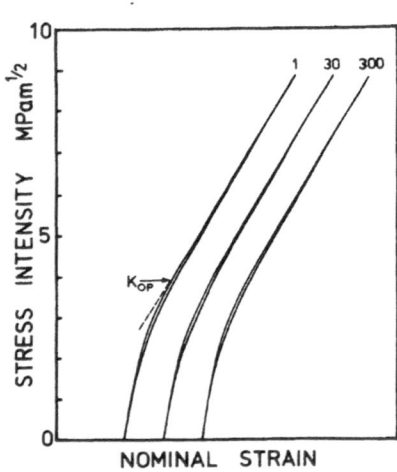

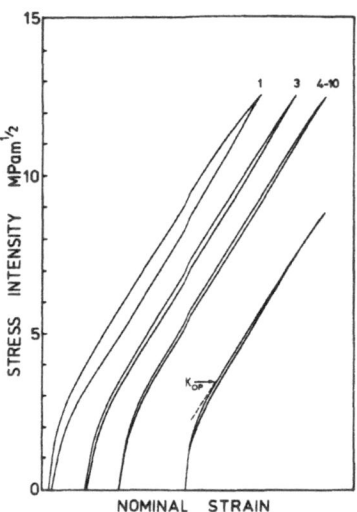

Fig. (2) - Hysteresis loops after pre-cracking, then cycling at $\Delta K = 8.9$ MPam$^{\frac{1}{2}}$ for 1, 30 and 300 cycles. K_{OP} indicates the crack opening stress intensity defined by the tangent method

Fig. (3) - Hysteresis loops during cycling at $\Delta K = 12.5$ MPam$^{\frac{1}{2}}$ for 10 cycles, then cycling at $\Delta K = 8.9$ MPam$^{\frac{1}{2}}$

B. Overload Spectrum Tests

Precracked specimens were subjected to the following types of overload spectra:

(i)   single cycle ($\Delta K = 17.8$ MPam$^{\frac{1}{2}}$ or 19.6 MPam$^{\frac{1}{2}}$, R = 0)

(ii)   multiple cycle, N = 10 ($\Delta K = 17.8$, 19.6 MPam$^{\frac{1}{2}}$, R = 0)

(iii) multiple cycle, N = 300 ($\Delta K = 17.8$, 19.6 MPam$^{\frac{1}{2}}$, R = 0)

(iv)   mean load change ($\Delta K = 8.9$, 9.8 MPam$^{\frac{1}{2}}$, R = 0) with R = 0.5
       (N = 300), then R = 0 or R = 0.15

Hysteresis loops were recorded after each overload spectrum. In most tests, the total number of post-overload cycles applied was 200; these cycles were applied by loading for 100 cycles at R = 0 or 0.15, then 100 cycles at R = 0.5. All post-overload loops for R = 0.15 and R = 0.5 showed no indication of crack closure. Hysteresis loops for the single cycle and ten cycle overload tests were similar; typical results are shown in Figure 5.

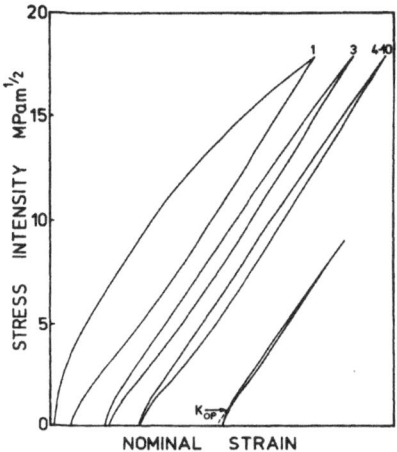

Fig. (4) - Hysteresis loops during cycling at ΔK = 17.8 MPam$^{1/2}$ for 10 cycles, then cycling at ΔK = 8.9 MPam$^{1/2}$

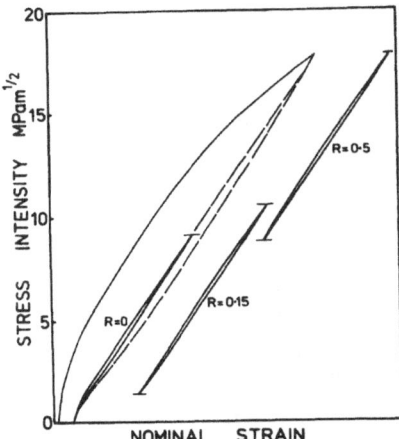

Fig. (5) - Post-overload hysteresis loops formed by cycling at ΔK = 8.9 MPam$^{1/2}$ with R = 0, 0.15 and 0.5

SEM fractographs of representative overload regions are shown in Figures 6 and 7; features corresponding to the load spectra are identified on the fractographs. Crack growth at R = 0 and R = 0.15 was not altered appreciably by either overload spectrum. This is shown most clearly in Figure 7 where the overload spectrum has been preceded by cycling for 100 cycles at R = 0.15.

Hysteresis loops after the 300 cycle overload spectrum and the mean load change stabilized within a few cycles to the basic loop shapes shown in Figure 5 for R = 0.15 and R = 0.5. No change in loop shape was detected over the duration of the tests. It was evident from SEM examination that these overload spectra produced immediate reductions in the crack growth rate, Figures 8 and 9.

DISCUSSION

A. Measured Crack Closure Stress

One of the aims of this study was to investigate the significance of the crack opening stress measured by the displacement gauge technique. The results in Figures 2-4 show that a single overload cycle, or a small number of overload cycles (N $\leq$ 10), when applied to a crack growing under constant amplitude, produces an increase in crack opening. This opening has the effect of lowering the crack opening stress (as defined by the change in slope on the crack-tip strain versus applied stress plot). From equation (1), this would suggest that the initial effect of a substantial overload should be to appreciably accelerate the crack

496

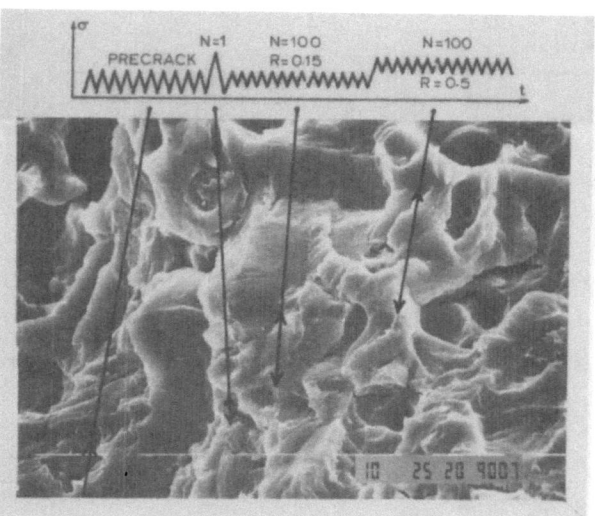

Fig. (6) - SEM fractograph after test with single cycle overload (ΔK = 17.8 MPam$^{\frac{1}{2}}$, R = 0). Arrows mark extremities of crack growth due to components of load spectrum. (In all SEM fractographs, the direction of macroscopic crack growth is from bottom to top of page; bar length is 10 microns)

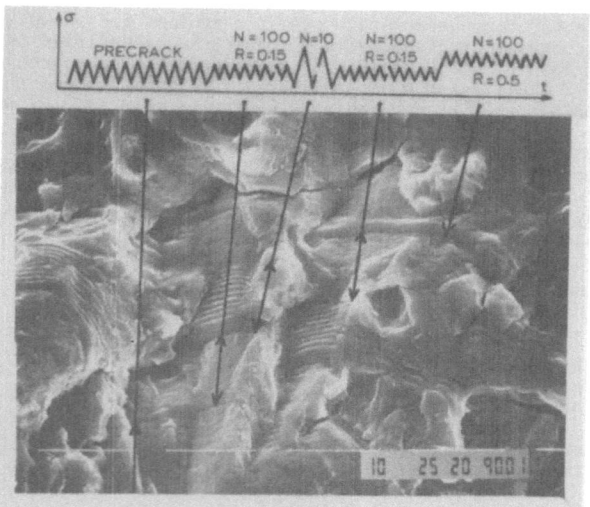

Fig. (7) - SEM fractograph after test with 10 cycle overload (ΔK = 17.8 MPa$^{\frac{1}{2}}$, R = 0)

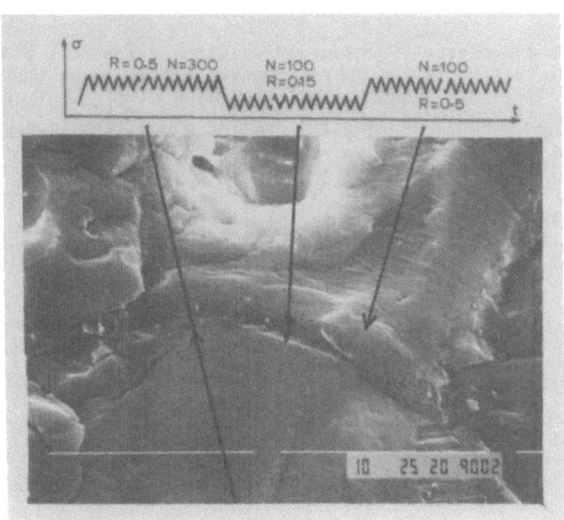

Fig. (8) - SEM fractograph after test with 300 cycle overload
(ΔK = 17.8 MPam½, R = 0)

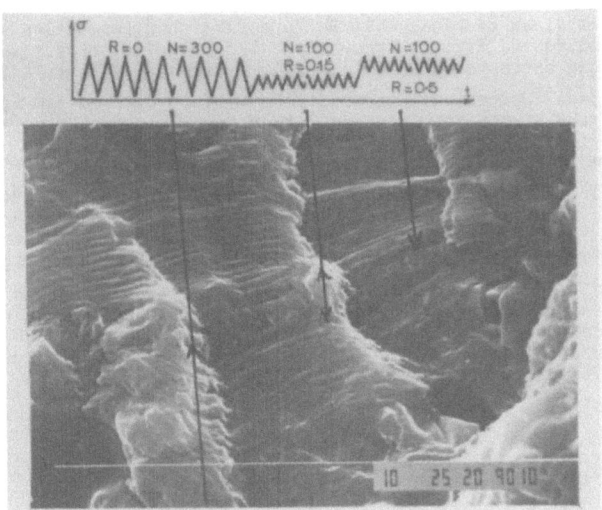

Fig. (9) - SEM fractograph after test with mean load change
(ΔK = 8.9 MPam½, R = 0.5, 0.15, 0.5)

growth rate. However, the present fractographic evidence for 6061-T6 material, Figures 6 and 7, shows that up to ten overloads have little or no immediate effect on the crack growth rate. This suggests that the value of K_{OP} controlling crack growth is not the value measured by the displacement gauge technique. This view is supported by the mean load tests which show that cycling for 300 cycles at R = 0.5 produces no detectable change in the hysteresis loop at R = 0.15, Figure 5. On this basis, K_{OP} would be expected to be below K_{min} and thus the crack growth rate at R = 0.15 should be at a maximum for the applied stress intensity range; fractographic examination shows that, in fact, crack growth is severely retarded, Figure 8. Results for tests with 300 overload cycles show a similar retardation effect.

It is of interest to consider why the local crack growth rate under variable amplitude loading is not correctly predicted by the measured value of K_{OP}.

It seems that this is due to the region of crack closure being located sufficiently close to the crack tip to prevent measurement of the crack closure stress by the displacement gauge technique. The existence of such a closure region close to the crack tip is indicated by the observation that the rate of post-overload crack growth is strongly influenced by the number of overload cycles applied and that the extent to which the crack is grown into the overload-affected plastic zone is therefore important. The two longer overload spectra (i.e., N = 300), which produce appreciable retardation, involve perhaps 100-200 microns crack extension, while the shorter spectra (N = 1, 10), involve just a few microns crack extension. The nature of the stress field within the overload zone is investigated in the following section.

B. Effect of Residual Stress Field

A major effect of an overload will be to establish a residual stress field in the region surrounding the fatigue crack tip. The singular stresses produced by the stress field can be represented by a residual stress intensity K_R. After a simple tensile overload, the residual stress field will be compressive and thus K_R will be a measure of the stress intensity assisting crack closure.

The variation in K_R can be calculated using the Dugdale model [10] if crack growth within the overload zone is treated as a simple separation of material which does not itself contribute to the residual intensity (i.e., alter the residual stress field). Assuming non-hardening material and elastic unloading, the additive σ_{yy} stress distributions for the present Dugdale problem are shown in Figure 10. Attention is directed to a small crack extension Δx; if $\Delta x \ll a$, where a is the crack length, the change in stress intensity due to crack growth can be represented by the change in forces on the crack faces (for simplicity, the forces are applied to a semi-infinite crack as shown in Figure 11). For a set of such forces, corresponding to an arbitrary stress $\sigma_{yy}(s)$ distributed over the region Δx, Figure 11, the stress intensity is given by:

$$K_{(T)} = \int_0^{\Delta x} \sqrt{2/\pi s}\ \sigma_{yy}(s)ds \tag{3}$$

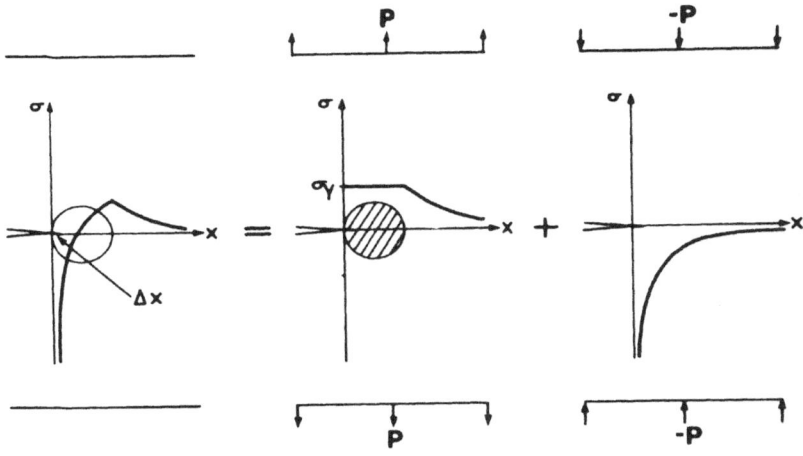

Fig. (10) - Component σ_{yy} stress distributions for elastic unloading

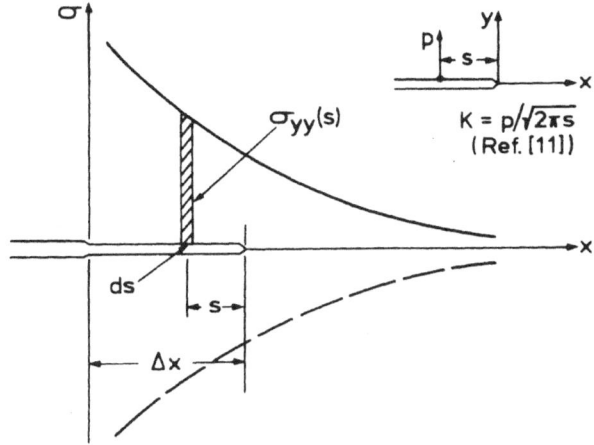

$$K = p/\sqrt{2\pi s}$$
$$(\text{Ref. [11]})$$

Fig. (11) - Semi-infinite crack subject to direct stress over region Δx of crack face

For load P, the overload plastic zone is represented by a thin yield zone of size R along which $\sigma_{yy}(s)$ is equal to the yield stress σ_Y; the stress intensity then becomes

$$K_{(P)} = 2\sqrt{2/\pi}\sigma_Y\sqrt{\Delta x} \tag{4}$$

For load -P, $\sigma_{yy}(s)$ is given by the elastic stress field

$$\sigma_{yy}(s) = -K_1/\sqrt{2\pi(\Delta x-s)} \tag{5}$$

where K_1 is the overload stress intensity. Thus

$$K_{(-P)} = -(K_1/\pi) \int_0^{\Delta x} \frac{ds}{\sqrt{s(\Delta x-s)}} \tag{6}$$

$$= -K_1 \tag{7}$$

If the residual stress intensity is defined in the negative sense

$$K_R = -(K_{(P)} + K_{(-P)}) \tag{8}$$

$$= K_1 - 2\sqrt{2/\pi}\sigma_Y\sqrt{\Delta x} \tag{9}$$

Suppose, next, that unloading from K_1 occurs with reversed plasticity (but without crack closure). Then, for material with a compressive yield stress of $-\sigma_Y$, the y-direction crack-tip stress is

$$\sigma_{yy}(s) = -\sigma_Y \tag{10}$$

The residual stress intensity over the reversed overload plastic zone (of size R') is then

$$K_R = 2\sqrt{2/\pi}\sigma_Y\sqrt{\Delta x} \tag{11}$$

For the region $R'<\Delta x<R$, the y-direction stress is

$$\sigma_{yy}(s) = \sigma_Y - \sigma_Y(4/\pi)\tan^{-1}\sqrt{R'/(\Delta x-R')} \tag{12}$$

It can be shown that K_R is then given by the result for elastic unloading, equation (9). Equations (9) and (11) are shown plotted in non-dimensionalized form in Figure 12.

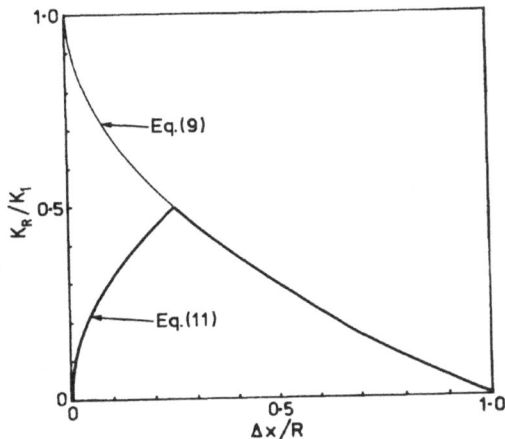

Fig. (12) - Normalized residual stress intensity given by equation (9) (overload zone - elastic unloading) and equation (11) (reversed overload zone), for plastic zone sizes $R = 4R'$ $= (\pi/8)(K_1/\sigma_Y)^2$

The result in Figure 12 represents the stress intensity causing crack closure produced by the overload cycle alone. In a real overload situation, the values will be modified by cyclic plasticity occurring during crack growth and by strain-hardening within the overload plastic zone. Crack closure occurring during the unloading part of the overload cycle will also be important since this will affect the reversed overload plastic zone size and thus the peak value of K_R reached.

Significantly, it can be seen that immediately after the overload cycle, the residual stress intensity is unchanged, but that K_R then increases rapidly to reach a maximum at the boundary of the reversed plastic zone. This is an agreement with the results for the short overload spectra and with the well-known phenomenon of delayed retardation [12]. The effect on K_R of the longer overload spectra (N = 300) is not given by the present theory, but can be visualized by slightly extending the crack into the overload zone; the rapid rise in residual stress intensity is consistent with the severe retardation observed.

The above theory is obviously an incomplete description of variable amplitude loading occurring in practice. However, the approach could provide a basis for the development of a simple quantitative model for crack growth prediction. This will depend on the extent to which complex crack-growth effects such as multiple overloads and underloads can be modelled without explicit consideration of plastic deformation in the wake of the crack.

SUMMARY AND CONCLUSIONS

1. A surface-mounted crack-tip displacement gauge has been employed to investigate crack closure and crack-tip hysteresis behaviour in 6061-T6 material.

2. Following constant amplitude crack growth a reproducible knee was detected in the crack-tip hysteresis loop (nominal strain versus applied stress). The stress given by the tangent above the knee was termed the crack opening stress.

3. Overload cycling (K_{max} = 17.8, 19.6 MPam$^{\frac{1}{2}}$) caused a decrease in crack opening stress; however, no corresponding increase in fatigue crack growth rate was detected in the present material during subsequent low-level cycling (for 100 to 200 cycles). The crack growth rate under these circumstances appears to be determined by a local crack closure condition which is not simply related to the measured crack opening stress.

4. The fatigue crack growth rate for the first 100 or 200 cycles after an overload was found to vary depending on the particular spectrum of overload cycles applied. Extended overload cycling (300 cycles) or a decrease in mean load produced an abrupt retardation in crack growth rate; short overload cycling (10 cycles or less) had relatively little effect.

5. The Dugdale model was used to predict the stress intensity assisting crack closure produced by a single overload cycle. The observed crack growth rates are in qualitative agreement with these predictions.

ACKNOWLEDGEMENTS

The author is grateful to N. T. Goldsmith and Dr. L. R. F. Rose for helpful discussions and suggestions during the course of this work.

REFERENCES

[1] Military Standard, MIL-STD-1530A (11) - Aircraft Structural Integrity Program, Airplane Requirements, USAF, 1975.

[2] Military Specification, MIL-A-83444 - Airplane Damage Tolerance Requirements, USAF, 1974.

[3] Hudson, C. M. and Hardrath, H. F., NASA Tech. Note D-1803, National Aeronautics and Space Administration, 1963.

[4] Wheeler, O. E., J. Basic Eng., Trans. ASME, Series D, 94, (1), p. 181, 1972.

[5] Elber, W., Damage Tolerance in Aircraft Structures, ASTM STP 486, p. 230, 1971.

[6] Lindley, T. C. and Richards, C. E., Mater. Sci. Eng., 14, p. 281, 1974.

[7] Unangst, K. D., Shih, T. T. and Wei, R. P., Eng. Frac. Mech., 9, p. 725, 1977.

[8] Garrett, G. G. and Knott, J. F., Int. J. Fracture, 13, p. 101, 1977.

[9] Haenny, L. and Dickson, J. I., Int. J. Fracture, 16, p. R121, 1980.

[10] Dugdale, D. S., J. Mech. Phys. Solids, 8, p. 100, 1960.

[11] Tada, H., Paris, P. C. and Irwin, G. R., The Stress Analysis of Cracks Handbook, Del Research Corp., Hellertown, Pa., 1973.

[12] Von Euw, E. F. J., Hertzberg, R. W. and Roberts, R., Stress Analysis and Growth of Cracks, ASTM STP 513, p. 230, 1972.

THE INFLUENCE OF CYCLIC IMPACT LOADING AND STRESS RATIO ON FATIGUE CRACK GROWTH RATE IN ALUMINUM ALLOY

R. Murakami and K. Akizono

Department of Precision Mechanics, The University of Tokushima
2-1 Minami-josanjima-cho, Tokushima, 770, Japan

ABSTRACT

Impact fatigue tests were carried out using a rotary type of machine. The frequency and velocity of cyclic impacts and stress ratio on fatigue crack growth rate of aluminum alloy was investigated. The difference of crack growth rate between cyclic impact and slowly-applied load was considerably large. This difference is contributed to periodic impact that is analogous to increase in stress ratio of the more slowly applied cyclic loads. Fractography study is also made.

INTRODUCTION

Repeated impacts of machines and structures are often encountered in practice. Little is understood on this subject because of the transient and complex character of the impact phenomenon [1-5]. Recent developments in fracture mechanics involving the concept of stress intensity factor to crack propagation have provided a means for studying crack growth damage by impact [6].

For a more gradually-applied cyclic load, the conventional crack growth rate da/dN can be connected with the stress intensity factor range ΔK by the relation [7]:

$$da/dN = C(\Delta K)^m \tag{1}$$

where m and C are material constants. The values of m and C have been found to be dependent on the microstructure and fracture mechanism [8-10]. Attempts have been made to explain impact crack growth rate by using the same method as that employed for gradually-applied cyclic loading [11,12], the results are still not clearly understood. In this work, a rotary disk type of machine is used to study the effects of fatigue impact due to material microstructure and temperature. Comparison is made with the standard results on fatigue [10,13]. The fracture surface produced by fatigue impact changed appearance appreciably from that of striation to intergranular cracking. The m value in equation (1) increased with increasing ductile-brittle transition temperature and decreasing test temperature [13]. As the dominant fracture appearance was by striation, impact influenced only the C value. The m value was almost equal to 2, similar to standard fatigue [10]. Such an effect is similar to that obtained by increasing the stress ratio in fatigue tests.

In the present work, an aluminum alloy was used because crack growth would have occurred by striation formation regardless of impact and the stress ratio effect would dominate the crack growth rate. The correlation between fatigue impact and stress ratio change in standard fatigue was investigated by means of fracture mechanics and fractography.

MATERIAL AND EXPERIMENTAL PROCEDURE

The materials selected are aluminum alloys, ZK141F and 2014-TO, having the chemical composition shown in Table 1. All fatigue specimens were machined

TABLE 1 - CHEMICAL COMPOSITION (WT%)

	Cu	Fe	Mn	Mg	Zn	Cr	Al
ZK141F	0.15	0.20	0.37	2.0	4.4	0.10	Bal
2014-TO	4.02	0.25	0.71	0.45	--	--	Bal

from a rolled plate so that the tensile axis is parallel to the direction of rolling. A center cracked specimen is used and has a rectangular cross section of 10 mm (ZK141F) and 5 mm (2014-TO) thickness by 50 mm width with a 6 mm crack length. The mechanical properties of two aluminum alloys are listed in Table 2.

TABLE 2 - MECHANICAL PROPERTIES

	$\sigma_{0.2}$ MPa	UTS MPa	RA %	ε_f	n
ZK141F	151	271	22.1	0.284	0.16
2014-TO	149	239	27.9	0.327	0.13

The Amsler Vibrophore fatigue machine is used to initiate the fatigue crack for a length of 1 mm from each notch tip. The fatigue impact tests were performed using a rotary disk type fatigue impact machine. The details are described elsewhere [13]. For the ZK141F, the fatigue impact test was performed with a constant impact velocity of 5.5 m/sec. The maximum impact tensile stress, σ_{max}, and the stress ratio, R, were 23.5 MPa and -0.4 respectively. For the 2014-TO, the fatigue impact tests were performed for four impact velocities, V = 4.6, 5.1, 5.5 and 6.4 m/sec. σ_{max} was increased with increasing impact velocity as 24.7, 27.2, 28.3 and 32.6 MPa. The R ratio was held constant at -0.44 regardless of the impact velocity. The crack length was measured using a travelling microscope with a resolution of about 10μm. The average crack growth rate was obtained directly from the number of impacts required to increase the crack length by 0.5 mm.

The standard fatigue tests were performed on the ±10 ton Amsler Vibrophore fatigue machine and conducted under a sinusoidal push-pull at about 140 Hz with four R ratios. R = -0.4, 0, 0.3 and 0.6, for ZK141F, and R = 0.5, 0.05, 0.25 and 0.5 for 2014-TO. The crack growth rates measurement method is the same as that for fatigue impact. The stress intensity factor, K, was calculated by the following equation

$$K = \sigma \sqrt{\pi a} \sqrt{\sec(\pi a/W)} \tag{2}$$

for both fatigue impact and standard fatigue. Here, σ is the gross stress, a is the half-crack length and W is the width of specimen.

Two-stage, chromium-shadowed, carbon replicas were obtained from the fatigue fracture surfaces and examined using a transmission electron microscope. The replicas were obtained at the mid-thickness of specimens. The distance from the slip tips to each area examined on the replica was 1 mm in length and 1 mm in width.

RESULTS

Fatigue Crack Growth Rates. Fatigue crack growth rates for both of the aluminum alloys are shown in Figures 1(a) and (b) and compared with the standard fatigue crack growth rates. In standard fatigue, the crack growth rates for a given ΔK level increased with increasing R ratio for a range of positive R ratios. It is indicated that the increase in R from 0 to 0.6 did not affect the m value while the C value increased. The crack growth rate for negative R ratio agreed almost with the result for R = 0.

The standard fatigue crack growth rates for both materials are replotted in Figures 2(a) and (b) against the effective stress intensity, ΔK_{eff}, calculated from

$$\Delta K_{eff} = \begin{cases} (1 - R)^{-1} \Delta K \text{ -- for the ZK141F } (R \geq 0) \\ 0.5 + 0.4R) \Delta K \text{ - for the 2014-TO} \\ 0.5 K_{max} \text{ ---- for both alloys } (R \leq 0) \end{cases} \tag{3}$$

The crack growth rates are approximated by a straight line using ΔK_{eff} regardless of ratios as follows:

$$da/dN = \begin{cases} 4.25 \times 10^{-9} (\Delta K_{eff})^{3.37} \text{ -- for ZK141F} \\ 8.92 \times 10^{-7} (\Delta K_{eff})^{3.5} \text{ -- for 2014-TO} \end{cases} \tag{4}$$

For negative R ratios, da/dN for fatigue impact is much larger than that of standard fatigue. This difference is decreased with increasing R ratio in standard fatigue. That is to say, the m value in fatigue impact is nearly equal to that in standard fatigue while the C value in the former case is almost eight times as large as that of negative R ratio in the latter case. The difference of C value is decreased as R increases. This agreed well with

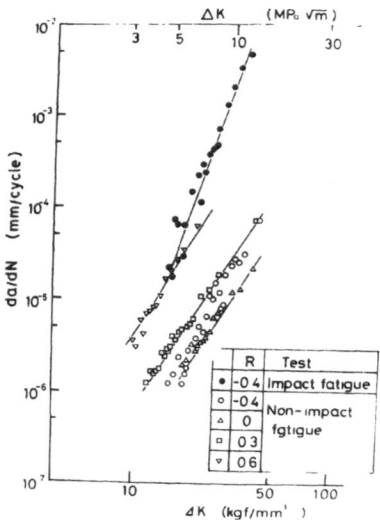

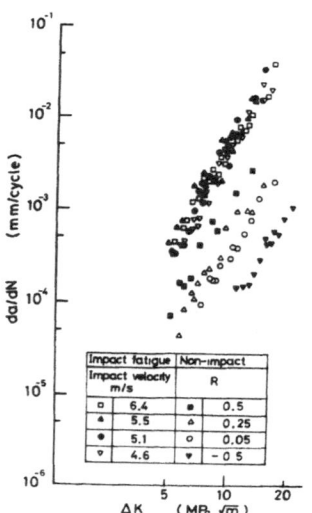

Fig. (1a) - Comparison of impact fatigue crack growth rate with that in non-impact fatigue for the ZK141F

Fig. (1b) - Comparison of impact fatigue crack growth rate with that in non-impact fatigue for the 2014-TO

the results on quenched and tempered Cr-Mo alloy steels [13]. It can therefore be concluded that fatigue impact is equivalent to increasing R in standard fatigue.

Fractography. Fractographic examination showed that the general feature of fracture surfaces varied mainly with the ΔK level regardless of fatigue impact, standard fatigue and R ratio striation dominates. The striation spacing, S, were measured for all specimens tested. For example, in the case of fatigue impact and R = 0.3 in standard fatigue for ZK141F, S can be found in Figure 3 along with the corresponding macroscopic crack growth rate. It is well known that S in standard fatigue is related to ΔK by the equation

$$S = C_s (\Delta K)^{m_s} \qquad (5)$$

It is obvious from Figure 3 that equation (5) can be applied to fatigue impact. An excellent agreement is seen between the crack growth rate measurements for fatigue impact and standard fatigue.

The values of m_s and C_s are listed in Table 3 for ZK141F. The m_s values in

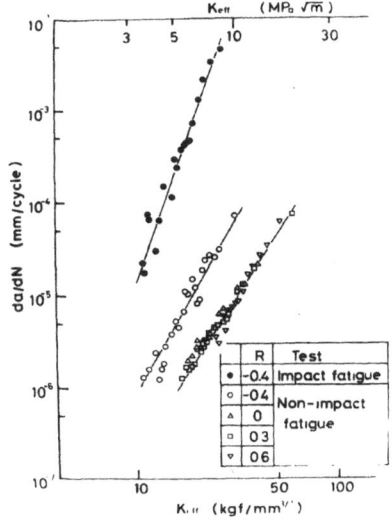

Fig. (2a) - Relation between fatigue
crack growth rate and
effective stress inten-
sity for the ZK141F

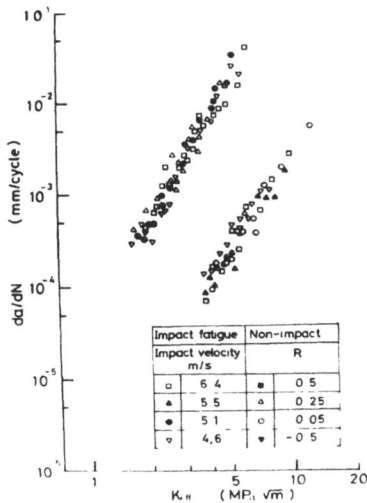

Fig. (2b) - Relation between fatigue
crack growth rate and
effective stress inten-
sity for the 2014-TO

fatigue impact agreed well with those in standard fatigue regardless of the
alloys and R. The C_s values increased as R ratio is increased and when impact
is applied repeatedly.

DISCUSSION

It is obvious from Figure 1 that the fatigue impact crack growth rate is
considerably larger than those of standard fatigue. This difference can be
explained by considering the following three effects; (1) the variation of
mechanical properties with increasing strain rate as impacts are applied, (2)
the interaction between impact stress waves and crack, and (3) the attenuation
of impact stress waves.

The Effect of Strain Rate. In order to assess the effect of strain rate,
fatigue impact tests were performed for four velocities V = 4.6, 5.1, 5.5 and
6.4 m/sec. The strain rate increased about 1.5 times from 2.23 to 3.34 sec^{-1}
as the impact velocity increased. This is about 30 times larger than that in
standard fatigue.

It is well known that the mechanical properties of aluminum alloy are

510

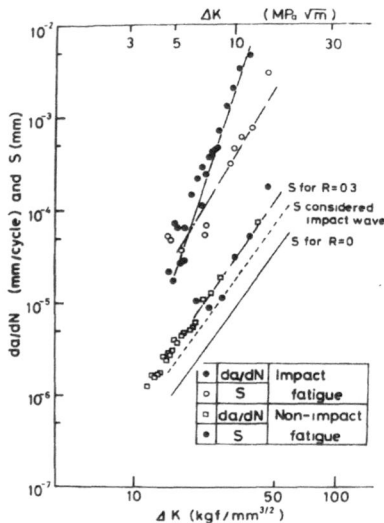

Fig. (3) - Relation between the striation spacing
S, and the macroscopic crack growth rate,
da/dN, in impact fatigue and non-impact
fatigue for the ZK141F

TABLE 3 - THE VALUES OF m_s AND C_s

	Impact fatigue	Non-impact fatigue		
R	-0.4	0	0.3	0.6
m_s	3.35	3.0	3.0	3.0
C_s x 10^{-8}	17.0	0.95	2.78	14.9

generally insensitive to change in strain rate. For example, it was reported
that the yield stress increased only 5% as the strain rate increased 10 times
[16]. The nominal final plastic strain ε_f obtained from the reduction in area
of the fatigue impact specimen, was equal to 0.266, a little smaller than that
obtained under monotonic loading. It can therefore be concluded that the

mechanical properties of this alloy are not influenced very much by impact. Though the crack growth rate tends to increase with increasing strain rate, this change may be explained by influence due to other factors.

Interaction Between Stress Wave and Crack. When repeated impact is applied to the structure, the stress waves propagate through the member and result in complicated stress wave pattern near the crack tip. According to the results in [6,17], the maximum dynamic stress intensity factor K_{dyn}, increased about 1.4 times that of the maximum static stress intensity, K_{max}. This, however, did not always agree with the rectangular wave record shown in Figure 4. The maximum dynamic stress intensity can be expressed as

$$K_{dyn} = \alpha K_{max} \tag{6}$$

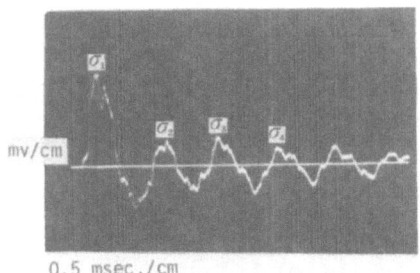

0.5 msec./cm

Fig. (4) - Impact stress wave

where α is the parameter which indicated the effect of impact. Equation (6) can be rewritten using $K_{max} = \Delta K/(1-R)$ in the form

$$K_{dyn} = [\alpha/(1-R_{eq})] \Delta K \tag{7}$$

where the R_{eq} is the equivalent stress ratio defined by the R ratio in standard fatigue.

Effect of Impact Stress Wave. Referring to Figure 4, the interval of impact was 17-24 times the applied time of impact stress. The specimens were suddenly stressed resulting in large strain rates. This is characteristic of fatigue impact. Fracture mechanics can be used to evaluate fatigue impact as carried out in [18]. The crack tip opening displacement CTOD corresponding to the first peak of impact stress is proportional to $K^2/E\sigma_{ys}$. The CTOD for the attenuated portion of the impact stresses is also proportional to $\Delta K^2/2E\sigma_{ys}$ by taking the flow stress at the crack tip to be twice that of cyclic yield stress. Referring to Figure 4, the ratios σ_2/σ_1 and σ_3/σ_1 are respectively

TABLE 4 - VARIATION OF CTOD UNDER IMPACT

		case not considered effect of prior stress cycle		case considered effect of prior stress cycle	
		CTOD	$\frac{(CTOD)_{imp}}{(CTOD)_{non}}$	CTOD	$\frac{(CTOD)_{imp}}{(CTOD)_{non}}$
Impact	0-1	$(K_{oi})^2/E\sigma_{ys}$	2	$\frac{(\Delta K_{e\theta})^2}{2E\sigma_{ys}} + \frac{(K_{oi})^2}{E\sigma_{ys}} - \frac{(K_{osi})^2}{E\sigma_{ys}}$	2.14
	3-4	0	0	0	0
	4-5	$(\Delta K_{45})^2/2E\sigma_{ys}$	0.14	same as left	0.14
	7-8	0	0	0	0
	8-9	$(\Delta K_{e\theta})^2/2E\sigma_{ys}$	0.1	same as left	0.1
	total		2.24		2.38
Non-Impact	0-1	$\frac{(\Delta K_{oi})^2}{2E\sigma_{ys}}$	1	same as left	1
	3-4	0	0	0	0
	4-5	$\frac{(\Delta K_{45})^2}{2E\sigma_{ys}}$	1	same as left	1

0.38 and 0.3. Table 4 gives the comparison of CTOD calculated above values under the impact with those of standard fatigue. The effect of impact wave on crack growth rate was about 2.24 times larger than that in standard fatigue. S_{dyn} can therefore be expressed as

$$S_{dyn} = \begin{cases} C_{dyn} \left(K_{dyn}\right)^{m_s} \\ 2.24\, C_{so} \left(\alpha/1-R_{eq}\right)^{m_s} (\Delta K)^{m_s} \end{cases} \tag{8}$$

where the C_{so} is a material constant that can be equated to C_s at R = 0 in standard fatigue.

It is well known that S in standard fatigue is closely related to $\Delta K/E$ [19]. Refer to the results in [11,12]. Figure 5 gives the relation between S_{dyn} and $\Delta K/E$. S_{dyn} can be approximated by a straight line and expressed by the equation

$$S_{dyn} = 2.75 \times 10^8 \left(K_{max}/E\right)^{3.0} (\sqrt{m}) \tag{9}$$

This is the same as that in standard fatigue. Each cycle of impact involves the formation of ductile striations by alternate blunting and resharpening of crack tip. Since the R ratio in fatigue impact is R = -0.4 for ZK141F, equation (9) can be rewritten using ΔK as

$$S_{dyn} = 1.0 \times 10^8 \left(\Delta K/E\right)^{3.0} \cdot (\sqrt{m}) \tag{10}$$

Comparing equations (8) and (10) for the aluminum alloy with $E = 6.86 \times 10^4$ MPa, it is found that

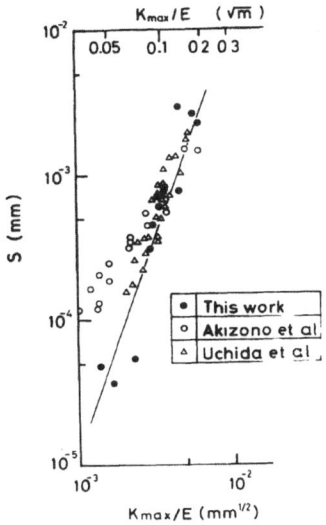

Fig. (5) - Relation of striation spacing in fatigue
impact to K_{max}/E for the two low carbon
steels and one aluminum alloy

$$\alpha/(1-R_{eq}) = 2.44 \tag{11}$$

The quantity R_{eq} in equation (11) defines the equivalent stress ratio for
impact. For example, if K_{dyn} is increased by 40% of the static stress intensity, then $\alpha = 1.4$ and $R_{eq} = 0.43$. The distance of crack extension due to
impact is therefore equivalent to increasing R in standard fatigue from
R = 0 to 0.43.

Nunomura et al [20] have derived the cumulative plastic strain, ε_{pm}, due
to cyclic stress from the monotonic plastic zone result given by Rice and
showed that ε_{pm} is given by

$$\varepsilon_{pm} = \varepsilon_0[4(1+n')(\sigma'_{0.2}{}^2)/(1+n)(\sigma_{0.2}{}^2)(1-R)^2]^{1/(1+n)} \tag{12}$$

where the n and n' are the monotonic and cylcic work hardening factors respectively. The quantities $\sigma_{0.2}$ and $\sigma'_{0.2}$ are the monotonic and cyclic yield stress
respectively while ε_0 is the monotonic yield strain. As the cumulative plastic

514

strain rate to nominal fracture plastic strain ratio $\varepsilon_{pm}/\varepsilon_f$ is increased, the R ratio had a strong influence on the crack growth rate. Figure 6 shows the relation between C_s and $\varepsilon_{pm}/\varepsilon_f$ for the data in Table 2. The resulting relationships follow straight lines. The values of $\sigma'_{0.2}$ and n' correspond to those of monotonic tensile tests.

As mentioned earlier, the mechanical properties of the alloys are not influenced by impact. Assuming that material damaged near the crack tip due to repeated impact is equivalent to that of R = 0.43 in standard fatigue, the cumulative plastic strain of the monotonic plastic zone for fatigue impact is then obtained as ε_{pm} = 0.0232 by making use of the monotonic test data.

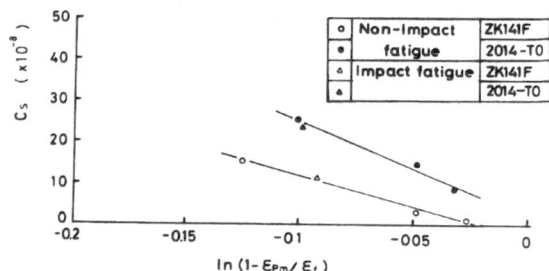

Fig. (6) - Relationship of C_s to $\ln(\varepsilon_{pm}/\varepsilon_f)$ for two aluminum alloys

The C_s value for R = 0.43 becomes 2.14×10^{-9}. These values are plotted in Figure 6 with the results in standard fatigue. The results for fatigue impact and standard fatigue agreed well. The effect of impact can thus be explained by adjusting the R ratio in standard fatigue. Repeated impact tends to increase the cumulative plastic strain at the crack tip. The cumulative damage within the reversed plastic zone is decreased. This accounts for acceleration of crack growth rate in fatigue impact.

CONCLUSION

The fatigue impact tests were carried out using a rotary disk type machine. The influence of repeated impact, velocity change and stress ratio effect on fatigue crack growth in aluminum alloy were investigated by fracture mechanics and fractography. The results obtained can be summarized as follows:

(1) In standard fatigue, the crack growth rate increased with increasing stress ratio and was dependent on the effective stress intensity factor.

(2) The difference of crack growth rate between fatigue impact and standard fatigue is large. It has been found that this difference is due to repeated impact.

(3) The dominant fracture appearance in fatigue impact is striation as in standard fatigue. The striation spacing in fatigue impact can be related to $\Delta K/E$ as given by

$$S = 1.0 \times 10^8 (\Delta K/E)^{3.0} \ (\sqrt{m}) \qquad (13)$$

(4) Three factors must be considered to explain the difference of crack growth rate between fatigue impact and standard fatigue. The impact velocity has little influence on crack growth rate for aluminum alloy. Impact increased only 40% of the maximum stress intensity and resulted in damage equivalent to standard fatigue as R ratio increased to 0.43. Crack growth rate increased 2.24 times in comparison with standard fatigue. The effect of repeated impact is governed by

$$\alpha/(1-R_{eq}) = 2.44 \qquad (14)$$

(5) Comparing the striation spacing in fatigue impact with that in standard fatigue, the effect of repeated impact on fatigue crack growth rate is analogous to increase of stress ratio in standard fatigue.

REFERENCES

[1]   Akizono, K. and Ataki, K., J. Soc. Mater. Sci., 21, p. 660, 1972. (in Japanese)

[2]   Akizono, K. and Murakami, R., Scient. Pap. Fac. Engng. Tokushima Univ., 23, p. 129, 1978. (in Japanese)

[3]   Takemori, M. T., Information of General Electric, No. 81, CRD012, p. 1, 1981.

[4]   Iguchi, H., Tanaka, K. and Taira, S., Fatigue of Engng. Mater. and Struc., 2, p. 165, 1979.

[5]   Johnson, A. A. and Keller, D. J., ibid, 4, p. 279, 1981.

[6]   Aoki, S., Kishimoto, K., Kondo, K. and Sakata, M., Int. J. Frac., 14, p. 59, 1978.

[7]   Paris, P. C. and Erdogan, F., Trans. ASME, Ser. D, 85, p. 528, 1963.

[8]   Kobayashi, H., Murakami, R. and Nakazawa, H., Fract. Mech. and Tech., Edited by Sih, Vol. 1, p. 205, Hong Kong, 1977.

[9]   Rotchie, R. O. and Knott, J. F., Acta Metall., 21, p. 639, 1973.

[10] Murakami, R., Kobayashi, H. and Nakazawa, H., Trans. Jap. Soc. Mech. Engng., 44, p. 1415, 1978. (in Japanese)

[11] Akizono, K. and Murakami, R., J. Jap. Weldng. Soc., 48, p. 971, 1979. (in Japanese)

[12] Uchida, T., Okabe, N., Yano, T., Ishimatsu, M. and Mori, T., J. Soc. Mater. Sci., 29, 1980. (in Japanese)

[13] Murakami, R. and Akizono, K., Fatigue of Engng. Mater. and Struc., 3, p. 357, 1980.

[14] Feddersen, C. E., ASTM STP, 410, p. 77, 1967.

[15] Hertzberg, R. W. and Paris, P. C., Proc. 1st Inter. Conf. Fract., Vol. 1, p. 459, Sendai, Japan, 1966.

[16] Jaul, B., Etude De La Plasticite et Application Aux Metaux, Dunod, 1964.

[17] Itou, S., Engng. Fract. Mech., 16 - 2, p. 247, 1982.

[18] McMillan, J. C. and Pelloux, R. M. N., Engng. Fract. Mech., 2, p. 81, 1970.

[19] Bates, R. C. and Clark, W. C., Trans. ASM., 62, p. 381, 1969.

[20] Nunomura, N. and Fukui, Y., J. Soc. Mater. Sci., 27, p. 1103, 1978. (in Japanese)

THE EFFECT OF CONTACT STRESS INTENSITY FACTORS ON FATIGUE CRACK PROPAGATION

Y. C. Lam and J. F. Williams

Mechanical Engineering Department, Melbourne University
Australia 3052

SUMMARY

Using the technique of Dimensional Analysis the phenomenon of crack closure is modelled using the concept of a contact stress intensity factor K_{con}. For constant amplitude loading, a simple expression, $K_{con}^{max} = g(R)\Delta K$, is obtained without making idealized assumptions concerning crack tip behavior. Further, by assuming that crack closure arises from the interaction of residual plasticity in the wake of the crack and crack tip compressive stresses, the function $g(R)$ is shown to be constant for non-workhardening materials. This implies that any K_{con}^{max} dependent on R must be attributed to the workhardening characteristic of the material. By using the idea of an effective stress intensity factor, the above relationship can be incorporated into the Paris crack growth law. From analysis, it can be deduced that for a workhardening material, as R increases, K_{con}^{max} will decrease and the effective stress intensity factor will increase. This means that the fatigue crack propagation rate will increase with R, in accordance with experimental observations reported in the literature.

THE EFFECT OF CONTACT STRESS INTENSITY FACTORS ON FATIGUE CRACK PROPAGATION

INTRODUCTION

Elber [1,2] has found that crack closure occurs even under cyclic tensile loading. He further uses the crack opening load concept to model this phenomenon. However, by using the same method, Shih and Wei [3] have reported that crack closure alone is not able to fully account for the effect of mean stress. Brown [4], Chanani and May [5] have also found that crack closure alone fails to account for load interaction effects. Further, Macha et al [6] have measured the crack opening load using various methods and found that the so-called "crack opening load" is sensitive to the distance from the crack tip at which the measurement was made.

Thus, an alternative parameter, the contact stress intensity factor, K_{con}, is proposed to model the phenomenon of crack closure. Since this parameter is consistent with the stress intensity range ΔK, in modelling the fatigue crack propagation rates, a simple relationship between K_{con} and ΔK is easily obtainable. The effect of stress ratio, R, on fatigue crack propagation rates can also be obtained easily by investigating its effect on K_{con}.

CRACK CLOSURE AS MODELLED BY THE CONTACT STRESS INTENSITY FACTOR

On unloading, the crack will start to close. A compressive stress system will be set up for that portion of the crack which closes. The stress will not be constant, as depicted in Figure 1. This has been demonstrated experimentally by Williams and Stouffer [7].

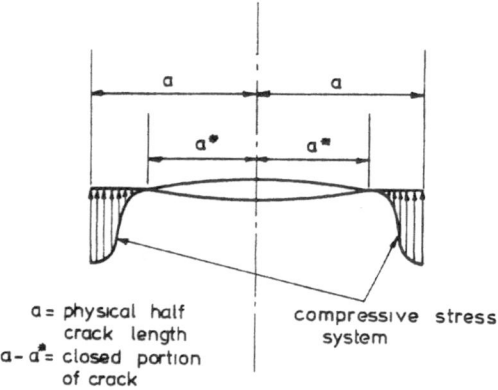

a = physical half crack length

a - a* = closed portion of crack

compressive stress system

Fig. (1) - Compressive stress system between crack surfaces for a crack on unloading

However, provided the compressive stress distribution is known, an equivalent contact stress intensity, K_{con}, may be calculated. Thus, if the compressive stress system between the crack surfaces as shown in Figure 1 is known to be $\sigma_y(x,0)$, then the stress intensity factor may be evaluated by using the expression given by Paris and Sih [8]:

$$K_{con} = 1/(\pi a)^{1/2} \int_{-a}^{a} \sigma_y(x,0)[(a+x)/(a-x)]^{1/2}dx \qquad (1)$$

It should be stressed that K_{con} will depend both on the magnitude and distribution of the compressive stress system.

Having outlined the basic concept of K_{con}, it is now postulated that at any one time during loading or unloading, the actual stress intensity at the crack tip (K_{act}) will depend both on the stress intensity factor caused by the external load (K) and on the current value of K_{con}. That is:

$$K_{act} = K + K_{con} \tag{2}$$

If it is assumed that on loading to maximum external load the crack will be fully open, it follows that K_{con} will be zero and thus:

$$K_{act}^{max} = K_{max} \tag{3}$$

On unloading to the minimum load, K will be at its minimum and K_{con} will be at its maximum, thus:

$$K_{act}^{min} = K_{min} + K_{con}^{max} \tag{4}$$

It should be pointed out that unlike the crack opening load, which, for a given cycle is a single point value, K_{con} is a continuous function during unloading.

Thus, the effective stress intensity range experienced by the material ahead of the crack tip, ΔK_{eff} is equal to:

$$\Delta K_{eff} = K_{act}^{max} - K_{act}^{min}$$

$$= K_{max} - (K_{min} + K_{con}^{max})$$

$$= \Delta K - K_{con}^{max} \tag{5}$$

The effect of K_{con} is therefore to decrease the stress intensity factor range experienced by the material ahead of the crack tip compared to that which would normally be calculated on the basis of the applied external loads only. Thus the Paris crack growth equation:

$$da/dn = C_1 (\Delta K)^{m1} \tag{6}$$

can subsequently be changed into:

$$da/dn = C_2 (\Delta K_{eff})^{m2}$$

$$= C_2 (\Delta K - K_{con}^{max})^{m2} \tag{7}$$

where C_1, $m1$, C_2 and $m2$ are constants in the equations.

THE FUNCTIONAL RELATIONSHIP OF K_{con} WITH EXTERNAL LOADING PARAMETERS UNDER CONSTANT AMPLITUDE LOADING

With either constant stress amplitude or stress intensity amplitude, it may be assumed that after a small number of cycles, a steady or "quasi-equilibrium" state is reached. As a result of this, it will not be necessary to consider any transient effects. The same response will be obtained with the same external excitation. In this case, this will be so when the same crack growth rate is observed for the same applied stress intensity factors, which is satisfied after a small number of cycles under constant amplitude loading.

In order to obtain the functional relationship of K_{con}, it is best approached by means of Dimensional Analysis. Under a "quasi-equilibrium" state with constant amplitude loading and at least in the case of small scale yielding, the stresses and strains in the vicinity of the crack tip can be typified by means of stress intensity factors, which include the conditions of current external loads, crack length and specimen geometry. Since K_{con} arises due to the material behavior around the crack tip region, the relevant parameters are K_{max} and K_{min}.

Thus, the functional relationship of K_{con} may be written as:

$$K_{con}^{max} = f(K_{max}, K_{min}) \tag{8}$$

With three parameters and only one primary dimension, the stress intensity factor, there will be two π products. The dimensionless π products may be chosen as K_{con}^{max}/K_{max}, K_{min}/K_{max}. Thus:

$$K_{con}^{max}/K_{max} = g_1(K_{min}/K_{max}) = g_1(R) \tag{9}$$

In view of $\Delta K = (1-R)K_{max}$, equation (9) can be rewritten as:

$$K_{con}^{max}/\Delta K = g_2(R)$$
$$\tag{10}$$
$$K_{con}^{max} = g_2(R)\Delta K$$

Equation (10) indicates that under constant R, $g_2(R)$ will be a constant. Thus, equation (10) may be rewritten as:

$$K_{con}^{max} = A(\Delta K) \tag{11}$$

where A is a constant for constant R.

Equation (11) can be substituted into equation (7), which gives:

$$da/dn = C_2(\Delta K_{eff})^{m2}$$
$$= C_2(\Delta K - K_{con}^{max})^{m2}$$
$$= C_2(\Delta K - A\Delta K)^{m2}$$
$$= C_2(1-A)^{m2}(\Delta K)^{m2}$$
$$= C_3(\Delta K)^{m2} \tag{12}$$

Comparing this with the original Paris equation, equation (6), it will be seen that equations (6) and (12) are similar, provided that $m1 = m2$ and $C_1 = C_3 = C_2(1-A)^{m2}$. Thus, equations (19) and (20) imply that if the Paris equation applies for one R value, it will also apply to other R values. Only the constant C_3 will be different, because it depends on R.

Using the original Paris equation, for R sensitive materials such as 18/8 austenitic steel, cold rolled mild steel and HS3OWP aluminum alloy (Frost et al [9,10]), a log-log plot of da/dn vs. ΔK appears as a family of straight lines. However, if equation (12) is examined carefully, it will be discovered that K_{con}^{max} and therefore A, are both functions of R. Thus the effect of R on crack propagation rates has already been allowed for in equation (12). This raises the possibility of having a single straight line for all R if a log-log plot of da/dn vs. ΔK_{eff} is attempted. Thus, the equation:

$$da/dn = C_2(\Delta K_{eff})^{m2}$$

will be applicable to all R.

THE VARIATION OF K_{con}^{max} WITH DIFFERENT VALUES OF R.

After establishing that K_{con}^{max} is proportional to ΔK at constant R, attention will now be focused on how K_{con}^{max} will change with different R values, i.e. an examination of the function $g_2(R)$.

It is reasonable to consider that crack closure, or K_{con}, arises from the interference of crack surfaces due to the residual plasticity in the wake of the crack and the residual compressive stress system ahead of the crack tip. To gain a qualitative understanding of the effect of each phenomena, they may be discussed separately, assuming that crack closure does not intervene. The overall effect may then be considered through the intervention of crack closure, which is the result of the interaction between residual plasticity and the compressive stress system around the crack tip region.

RESIDUAL PLASTICITY IN THE WAKE OF THE CRACK

At the crack tip, the plastic deformation may be typified by the crack tip opening displacement (CTOD). For a non-workhardening material, as obtained from Rice [11], for loading to K_{max}:

$$CTOD = K_{max}^2/E\sigma_0 \qquad \text{(for plane stress)}$$

$$= (1-\nu^2)K_{max}^2/2E\sigma_0 \quad \text{(for plane strain)} \qquad (13)$$

where ν is Poisson's Ratio, E is Young's Modulus and σ_0 is the yield stress of the material.

For unloading from K_{max} to K_{min}, assuming that there is no crack closure, ΔCTOD may be obtained by replacing σ_0 by $2\sigma_0$ and K_{max} by ΔK:

$$\Delta CTOD = \Delta K^2/2E\sigma_0 \qquad \text{(for plane stress)}$$

$$= (1-\nu^2)\Delta K^2/4E\sigma_0 \quad \text{(for plane strain)} \qquad (14)$$

In order for plastic deformation to result in crack closure, there must be interference of the crack surfaces. Assuming that crack closure does not occur at this stage, the interference may be visualized as though the surfaces were allowed to pass through each other. This may be termed as the potential or conceptual interference.

In Figure 2a, it is shown that due to plastic deformation at K_{max}, a ligament, length L, outside the plastic zone, will become $(L + \Delta L_1)$ right at the crack tip. ΔL_1 is given by 1/2(CTOD) at K_{max}, since only the upper portion of the crack is being considered. If it is assumed that the crack propagates at K_{max}, then it may be visualized that at K_{max}, the upper and lower ligaments in Figure 1a have already separated. On unloading, if the crack surfaces were allowed to pass through each other, then ΔL_2 of the upper ligament would pass through the centerline of the crack as shown. ΔL_2 will then be given by 1/2ΔCTOD. Thus the total potential interference would be given by 2ΔL_2, i.e. ΔCTOD.

From the above conceptual argument, it is clear that ΔCTOD determines the potential interference in the wake of the crack. Since according to equation (14), ΔCTOD is a function of ΔK alone, it follows that for the same ΔK the potential interference will be the same.

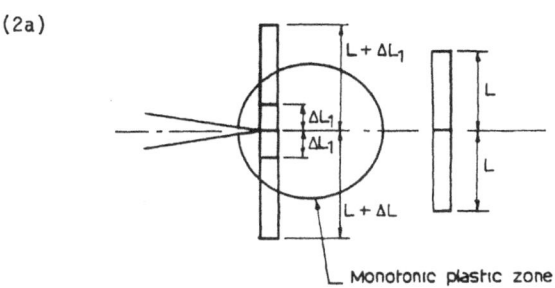

(2a)

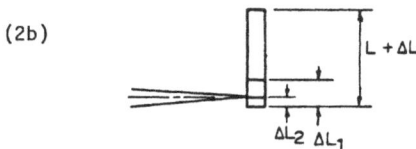

(2b)

Fig. (2) - A demonstration of how the potential
interference arises.

(2a) At maximum load, K - K_{max} where L
increases to L + ΔL, at the crack tip.

(2b) At minimum lead, K = K_{min}, showing
the potential interference ΔL_2 due to
the upper ligament.

Although equations (13) and (14) for CTOD and ΔCTOD are only applicable
for a non-workhardening material, it is reasonable to suspect that for a
workhardening material, there parameters will still be determined by K_{max}
and ΔK respectively but in a different form. Thus, by using the same argument, the potential interference should also be constant for constant ΔK,
irrespective of K_{max} or R.

THE EFFECT OF R ON THE CRACK TIP STRESS

Since it is argued that crack closure originates from potential interference in the wake of the crack and compressive stresses ahead of the crack on
unloading, the stress distribution surrounding the crack on unloading will
be examined.

In order to enable the trend of the function $g_2(R)$ be identified more
easily, the analogy between longitudinal shear and tensile loading (McClintock
[12], McClintock and Irwin [13]) may be utilized, to enable the use of the
simpler solutions for the case of longitudinal shear loading. Thus, for a

material obeying the workhardening stress-strain relationship:

$$\sigma(\varepsilon) = \sigma_0(\varepsilon/\varepsilon_0)^n \tag{15}$$

Rice's [14] longitudinal shear solution in the case of simple tensile loading, within the plastic zone becomes:

$$\sigma_y = \sigma_0[K^2/((n+1)\pi\sigma_0^2 x)]^{n/n+1} \tag{16}$$

$$\varepsilon_y = \varepsilon_0[K^2/((n+1)\pi\sigma_0^2 x)]^{1/n+1} \tag{17}$$

where

σ, ε are stress and strain respectively

σ_0, ε_0 are yield stress and yield strain respectively

σ_y is stress in the y-direction along the crack line at y = 0

x   is the distance ahead of the crack tip

n   is the workhardening exponent

However, these equations are only valid for monotonic loading. For cyclic loading, certain variables in the equations must be modified to take into account the loading and unloading phase, provided certain assumptions are satisfied.

As discussed by Rice [11], if on loading to K_{max}, K is subsequently reduced by ΔK, the same formula will apply, provided that K is replaced by ΔK and σ_0, ε_0 by $2\sigma_0$, $2\varepsilon_0$ respectively. Also, the calculated quantities σ_y, ε_y will in fact be the change in stress and strain $\Delta\sigma_y$, $\Delta\varepsilon_y$ from the stress and strain at maximum load. Thus in order to get the actual stress and strain conditions, the change in stress and strain must be subtracted from those at maximum load.

This simple change in variables does not take into account crack closure. Further, for a workhardening material, the assumptions of (i) kinematic hardening and (ii) that the initial forward and reversed yield stress are the same, must be satisfied. Further, in order for a satisfactory description of the cycle behavior of the material, the workhardening exponent n has to be replaced by n', the cyclic workhardening exponent.

Using the above method, the crack tip stress for a hypothetical, non-workhardening material, with assumed material properties, crack length, loading conditions and the assumption that crack closure does not intervene, are shown in Figure 3.

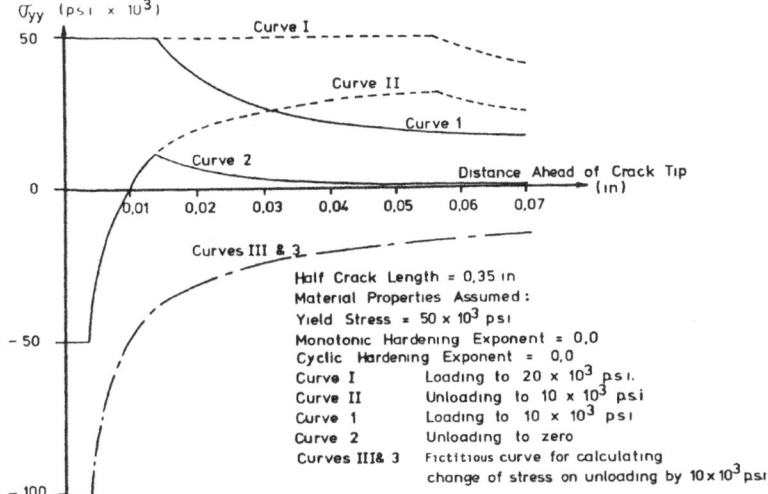

Fig. (3) - σ_{yy} vs distance ahead of crack tip showing the
R - ratio effect for a non-workhardening material

From the stress distribution on unloading, Figure 3, it is clear that
under different R, but with the same ΔK value, the reversed plastic zone
size and more specifically the compressive stress region immediately ahead of
the crack tip will be the same as on unloading. This implies that the compressive stress will only be a function of ΔK. As already determined, the potential interference due to residual plasticity in the wake of the crack is also
a function of ΔK only. Thus, K_{con}^{max} for a non-workhardening material will also
be a function of ΔK alone, since K_{con}^{max} is a function of the potential interference and the compressive stress ahead of the crack.

From the previous dimensional analysis, it is found that, see equation
(10). Thus, for $n' = 0$, i.e. a non-workhardening material, $g_2(R)$ has to be
constant since K_{con}^{max} is independent of R. Therefore, according to equation
(11); A is now independent of R. According to equation (12); this indicates
that the crack growth rate is independent of R for a non-workhardening material.

Since it has been shown that the variation of R has no effect on K_{con}^{max}
and the crack propagation rate for a non-workhardening material, the
hypothesis is now made that any R effect most probably can be explained
by the workhardening behavior of the material.

For workhardening materials, i.e., for n' ≠ 0, a similar analysis can be performed. The crack tip stresses for a hypothetical kinematic workhardening material, with assumed material properties, crack length, loading conditions and the assumption that crack closure does not intervene, are shown in Figure 4.

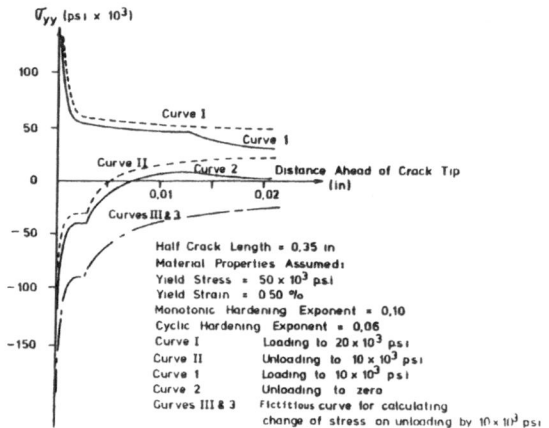

Fig. (4) - σ_{yy} vs. distance ahead of crack tip showing the R - ratio effect for a workhardening material

By concentrating our attention on the compressive stress region immediately ahead of the crack tip, it can be observed upon unloading that the higher the R value the smaller the compressive stress and compressive region (comparing curve II R = 0.5, with curve 2, R = 0). From the previous analysis, it has been determined that the potential will not depend on R, thus any R effect on crack closure will have to be caused by a difference in the compressive stress. Thus, if crack closure is dependent on the compressive stress region ahead of the crack tip, it will be expected that the higher the R value, the smaller the compressive stress and consequently a smaller amount of crack closure. In other words, K_{con}^{max} decreases with an increase in R for the same ΔK.

As R increases, K_{con}^{max} decreases, for the same ΔK. Thus, according to equation (7), this implies that da/dn will increase as R increases for the same ΔK. This is in accordance with experimental observations which indicate that for R sensitive material, an increase in R is generally followed by an increase in da/dn, for example the results given by Frost et al [10].

CONCLUSION

It has been shown that by using the contact stress intensity factor approach in modelling the phenomenon of crack closure, the relationship between crack closure and external loading parameters under constant amplitude loading can easily be explained as a function of the externally applied stress intensity factors.

For loading under constant R, a simple expression $K_{con}^{max} = A\Delta K$ is obtained where A is a constant. By incorporating this into the Paris equation using an effective stress intensity concept, a log-log plot of da/dn vs. ΔK_{eff} will appear as a straight line. Further, by examining the general expression, it is discovered that K_{con}^{max} and thus A, are both functions of R. It is thus argued that the effect of R has automatically been taken into account. Therefore, a log-log plot of da/dn vs ΔK_{eff} should be a single straight line even under different R, in contrast to the original Paris crack growth equation, where for some materials, a family of straight lines results.

The foregoing expressions have been obtained by means of Dimensional Analysis, thus avoiding the necessity of considering the details of how K_{con} arises. These include the effects of the monotonic plastic zone, the residual plasticity in the wake of the crack and the crack tip stresses.

Further, by considering that crack closure arises both from residual plasticity in the wake of the crack and crack tip compressive stress, it is shown that R has no effect on the crack propagation rate for a non-workhardening material. This implies that the R effect is most likely attributable to the workhardening behavior of the material.

REFERENCES

[1]   Elber, W., "Fatigue Crack Closure Under Cyclic Tension", Engineering Fracture Mechanics, Vol. 2, pp 37-45, 1970.

[2]   Elber, W., "The Significance of Fatigue Crack Closure", Damage Tolerance in Aircraft Structures, ASTM STP 486, pp 230-242, 1971.

[3]   Shih, T. T. and Wei, P. P., "A Study of Crack Closure in Fatigue", Engineering Fracture Mechanics, Vol. 6, pp 19-32, 1974.

[4]   Brown, R. D., "Effects of Tensile Overloads on Crack Closure and Crack Propagation Rates in 7050 Aluminum", Engineering Fracture Mechanics, Vol. 10, pp 867-878, 1978.

[5]   Chanani, G. R. and Mays, B. J., "Observation of Crack Closure Behavior After Single Overload Cycles in 7075-T6 Single-Edge-Notched Specimens", Engineering Fracture Mechanics, Vol. 10, pp 65-73, 1979.

[6]   Macha, D. E., Corbly, D. M. and Jones, J. W., "On the Variation of Fatigue-Crack-Opening Load with Measurement Location", Experimental Mechanics, Vol. 19, pp 207-213, 1979.

[7]   Williams, J. F. and Stouffer, D. C., "An Estimate of the Residual Stress Distribution in the Vicinity of a Propagating Fatigue Crack", Engineering Fracture Mechanics, Vol. 11, pp 547-557, 1979.

528

[8]   Paris, P. C. and Sih, G. C., "Stress Analysis of Cracks", Fracture
      Toughness Testing and It's Applications, ASTM STP 381, pp 30-81, 1965.

[9]   Frost, N. E., Marsh, K. J.,and Pook, L. P., "Metal Fatigue", Oxford
      University Press, London, 1974.

[10] Frost, N. E., Pook, L. P. and Denton, K., "A Fracture Mechanics Analysis
      of Fatigue Crack Growth Data for Various Materials", Engineering
      Fracture Mechanics, Vol. 3, pp 109-126, 1971.

[11] Rice, J. R., "Mechanics of Crack Tip Deformation and Extension by
      Fatigue", Fatigue Crack Propagation, ASTM STP 415, pp 247, 1967.

[12] McClintock, F. A., "Discussion of Paper on Fracture Testing of High-
      Strength Sheet Material", Material Research and Standards, pp 277-279,
      April, 1961.

[13] McClintock, F. A. and Irwin, G. R., "Plasticity Aspects of Fracture
      Mechanics", Fracture Toughness Testing and It's Applications, ASTM STP
      381, 1965.

[14] Rice, J. R., "Stresses Due to a Sharp Notch in a Workhardening Elastic-
      Plastic Material Loaded By Longitudinal Shear", Trans. A.S.M.E. Journal
      of Applied Mechanics, June, 1967.

DESIGN OF WELD ATTACHMENTS TO BOILER TUBES UNDER FATIGUE LOADINGS

P. R. Ford and T. Lexmond

State Electricity Commission of Victoria
Richmond, Australia

ABSTRACT

Owing to access problems causing difficulties in producing sound T-butt welds between a load-bearing fin (8 mm thick) and a boiler tube (50.8 mm o.d) pressurized to 18.9 MPa, it was proposed that a double sided fillet weld may give satisfactory performance. This proposal was investigated using a finite element analysis technique, which was also used to determine the weld shape which would move the position of any fatigue failures from the tube to the fin. The direction of fatigue crack growth was predicted using values of stress intensity factor calculated using crack tip elements available in the finite element package. The stress intensity values were calculated from the energy change due to a small extension of the crack tip.

In welds of leg length equal to half the fin thickness the stress distribution and the direction of fatigue crack propagation were similar, since little load was carried by the central portion of the butt weld.

INTRODUCTION

The membrane wall of Loy Yang A Power Station in Victoria consists of integrally extruded finned tube, welded into panels with longitudinal butt welds. In at least one location, around the gas outlet, a small section of the membrane wall is formed using T-butt welds to join the fin to the tube. Access problems were encountered during welding, which resulted in incomplete weld penetration. There was a proposal that a fillet weld would be capable of satisfactory performance in this area. The Research and Development Department was requested to investigate this suggestion, and to try and predict the likely position of a fatigue crack and the direction of crack growth. An important consideration in the design of the tube-fin construction is that any failure should be through the fin rather than through the tube wall. A high water loss from a burst tube can lead to the immediate close down of the boiler. A cracked fin is usually of less importance and can be repaired at the next maintenance overhaul.

It was decided to use a finite element analysis method to investigate this problem, since it enabled one variable at a time to be investigated. Other than the internal tube pressure of 18.9 MPa, no other stresses were known; therefore, three loading conditions on the fin were to be investigated, these being axial tensile, out-of-plane bending and in-plane bending, Figure 1. This paper reports on the findings of the first two of these loading conditions. The stress distri-

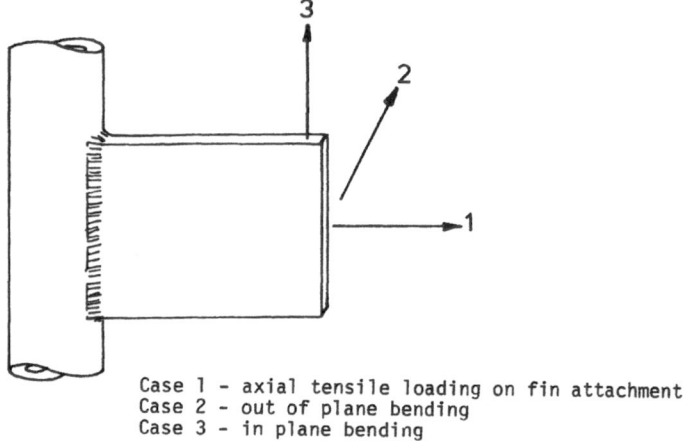

Case 1 - axial tensile loading on fin attachment
Case 2 - out of plane bending
Case 3 - in plane bending

Fig. (1) - Three possible loading conditions on the fin

butions in the butt and fillet welds and the stress intensity factors at the lo-cations of stress concentration, namely the fin and tube-weld toes and the weld-root in the fillet weld were compared. The stress intensity factors were deter-mined using the virtual crack extension method available in BERSAFE [1] and em-ploying the special quadratic isoparametric crack tip elements which have a $\sqrt{r}$ displacement variation from the crack tip. This method considers the change in the strain energy in creating a small extension to the crack tip. Several of these extensions are allowed from the same crack tip in any one computer run. From the energy release rate, the stress intensity factor is calculated for each crack extension. By choosing the direction of crack growth which results in the highest stress intensity value, the direction of crack growth can be predicted.

PROCEDURE

A two-dimensional finite element mesh was produced using three-sided and four-sided elements with six and eight nodes respectively, each node having two trans-lational degrees of freedom. Curved element sides were described by a parabola passing through two corner nodes and a mid side node. Two gauss points per ele-ment were used to calculate the element stiffness matrices.

The basic mesh (BASWELD) represented an 8.0 mm thick fin welded to a 50.8 mm outside diameter, 36.6 mm inside diameter boiler tube. The weld leg-lengths of the basic mesh were equal to half the fin thickness. The meshes used to determine the stress intensity values are seen in Figures 2 to 5. An internal tube pressure of 18.9 MPa was applied to all meshes, together with either an axial tensile pres-sure on the fin end or an out-of-plane bending pressure on the side of the last element in the fin. The basic (BASWELD) mesh was altered to make allowance for

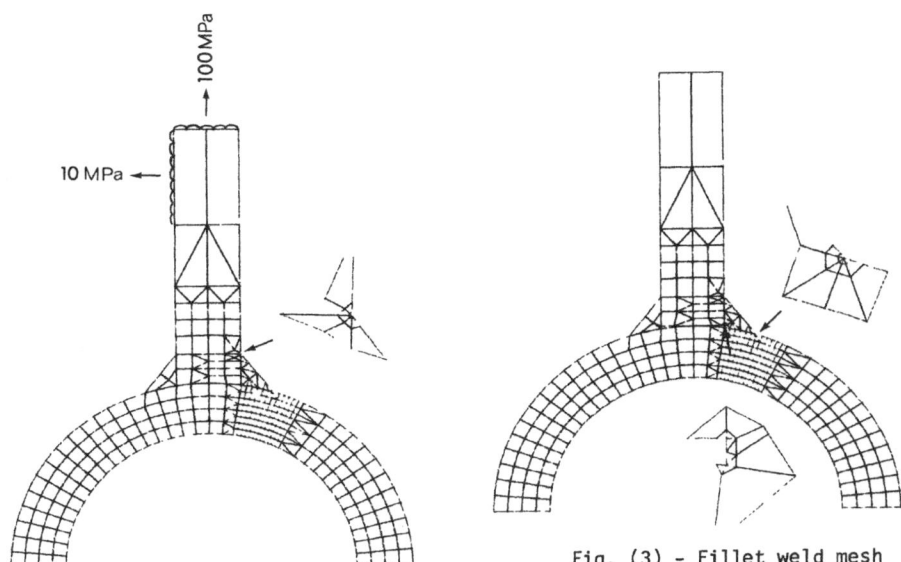

Fig. (2) - Basic mesh for the butt weld
(BASWELD) showing axial ten-
sile and out-of-plane bending loadings

Fig. (3) - Fillet weld mesh
(OPEN WELD)

the effect of varying degrees of weld penetration, tube weld leg-length and weld undercut, such that:

a. fillet welds with zero (OPENWELD) and half weld penetration (HPENWELD):

b. a tube weld leg length twice that of the basic mesh (EXTWELD); and

c. an 0.5 mm deep undercut at the tube weld toe were investigated.

Stress distributions for 100 MPa axial tensile and 10 MPa out-of-plane benders were obtained for the above meshes.

Stress intensity factors (K) for a crack of 0.04 mm situated at either of the weld toes or the weld root, in the case of the fillet welds, were obtained for various stresses on the fin. The direction of a crack propagating from the tube-weld-toe in the basic mesh under axial tensile stresses of 10 and 100 MPa and an out-of-plane bending load of 1 MPa was determined by extending the crack length in the direction of the highest stress intensity factor. Each crack extension re-quired the formation of a new refined mesh around the crack tip.

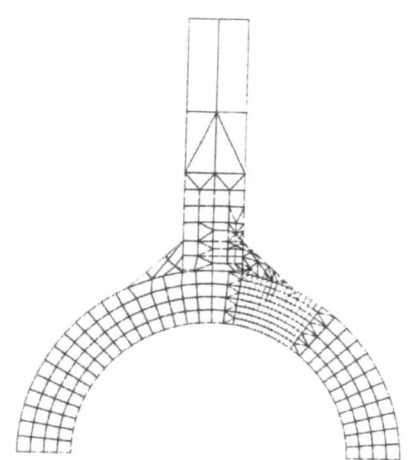

Fig. (5) - Mesh representing undercut
at the tube weld toe of
EXTWELD

Fig. (4) - Fillet weld mesh with extended
tube weld leg length (EXTWELD)

RESULTS

The stress distribution for BASWELD and OPENWELD mesh configurations are seen in Figures 6 to 10. A deformation plot resulting from a 100 MPa axial tensile stress on the fin is shown in Figure 11. Stress concentrations are evident at (a) the weld toes; (b) OPENWELD weld root; and (c) the internal surface of the tube, 90° from the fin location. Stress intensity factors for 0.04 mm cracks occurring at these locations are shown in Figure 12. Extending the length of the tube-weld-leg decreased the stress intensities at the tube toe and weld root and caused a small increase in the fin toe value, Figure 8, under both loading conditions. Under axial tensile loads, the values of stress intensities at these locations in the extended tube leg length weld were similar and the position of highest stress intensity changed from the weld root to the fin toe as the stress intensity value increased. Under out-of-plane bending, the position of highest stress intensity changed from the tube weld toe to the fin weld toe with increasing stress intensity. The presence of undercut at the tube weld toe caused an increase in the stress intensity, Figure 12.

The crack path for a crack initiated at the tube toe was the same for a butt or fillet weld. The direction of the maximum stress intensity factor indicated that for a 10 MPa axial tensile stress or a 1 MPa out-of-plane bending stress on the fin, the crack propagated initially at 143° to the x-axis, Figure 13, before changing to 155° which was the direction normal to the local applied stress. At the higher stress (100 MPa), the direction of the maximum stress intensity factor is less radial being 143° and 145° respectively. However, the maximum stress intensity factor varied only slightly up to 13° either side of the initial growth

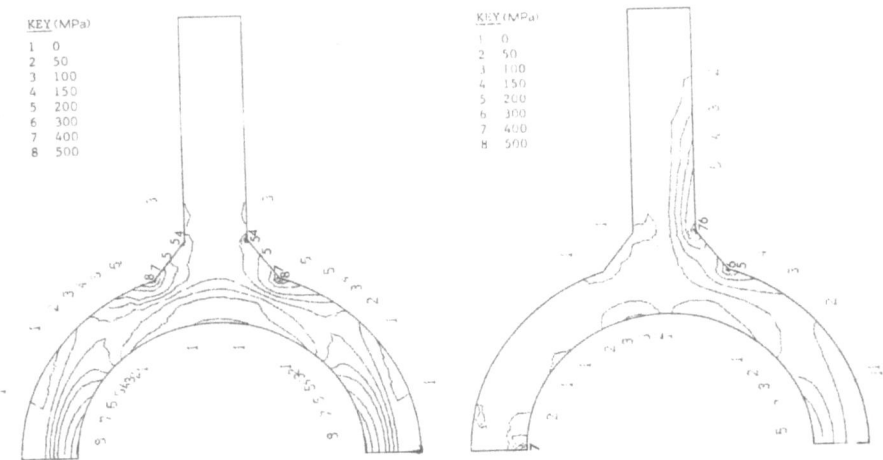

Fig. (6) - Maximum in-plane principal stress for butt weld (BASWELD)
for tube pressure and applied stresses

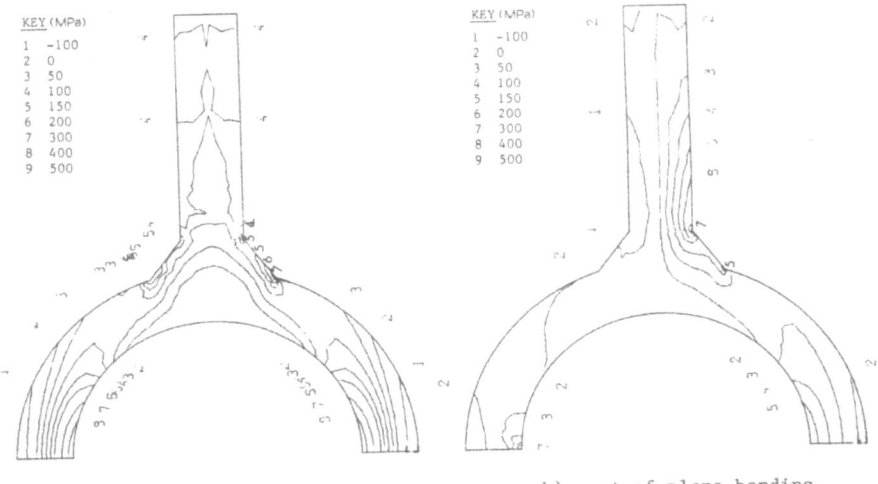

Fig. (7) - Stresses in direction of the fin for butt weld (BASWELD)
for tube pressure and applied stresses

534

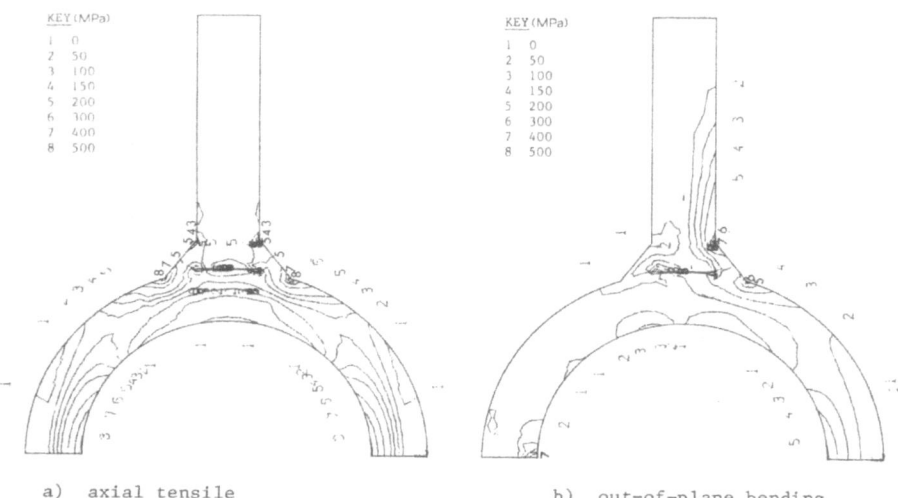

a) axial tensile   b) out-of-plane bending

Fig. (8) - Maximum in-plane principal stress for fillet weld (OPENWELD)
for tube pressure and applied stresses

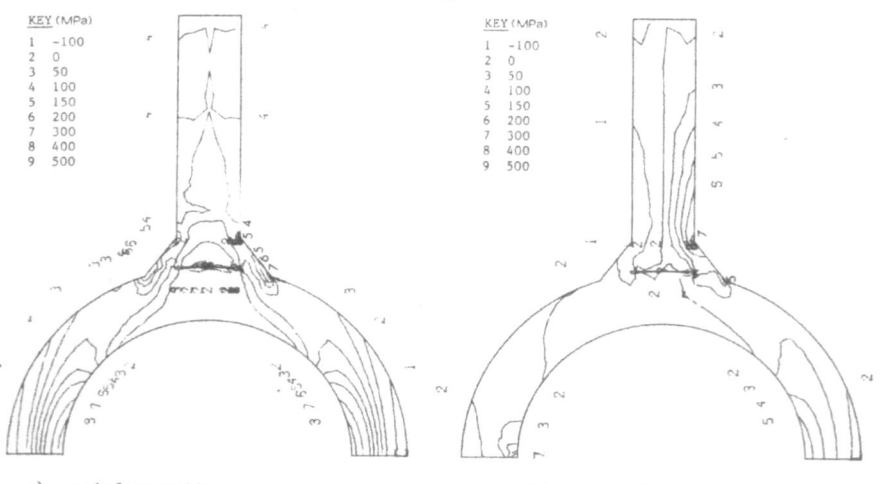

a) axial tensile   b) out-of-plane bending

Fig. (9) - Stresses in direction of the fin for the fillet weld (OPENWELD)
for tube pressure and applied stresses

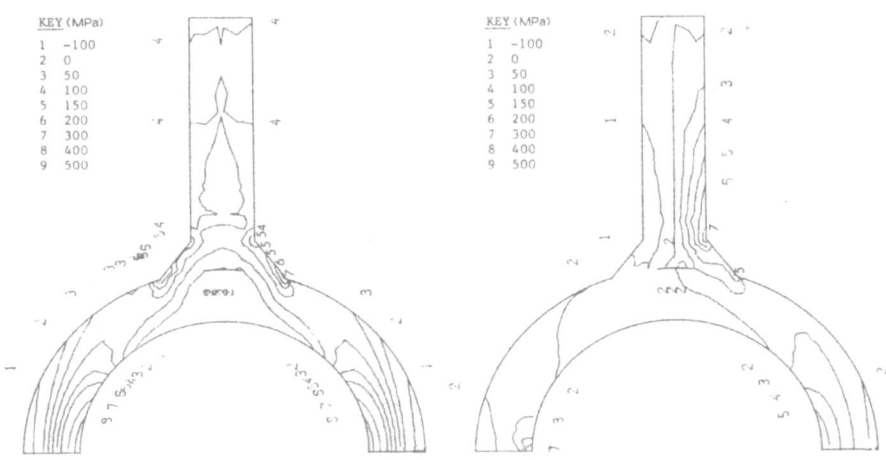

a) axial tensile   b) out-of-plane bending

Fig. (10) - Stresses in direction of the fin for the partial penetration fillet weld (HPENWELD) for tube pressure and applied stresses

Fig. (11) - Deformation plot fillet weld (OPENWELD) with 100 MPa axial tensile pressure on the fin end and tube pressure of 18.9 MPa

536

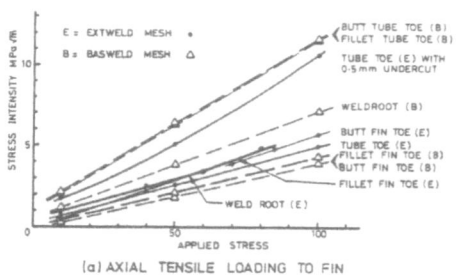

(a) AXIAL TENSILE LOADING TO FIN

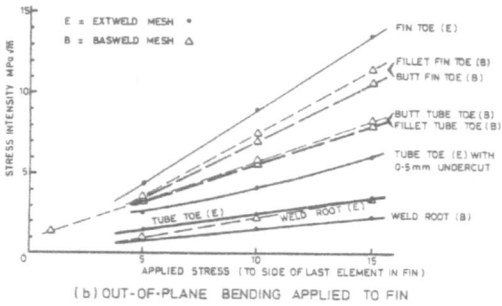

(b) OUT-OF-PLANE BENDING APPLIED TO FIN

Fig. (12) - Stress intensity factor at nominated locations versus applied stress

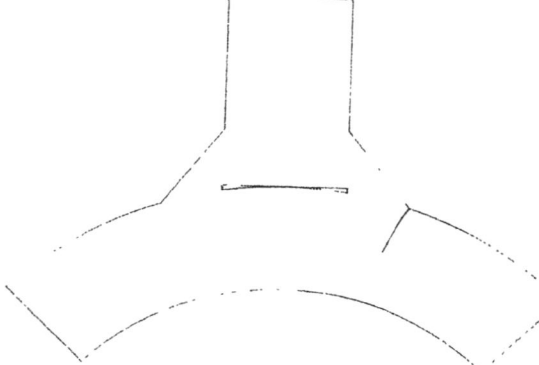

Fig. (13) - Fatigue crack path in a zero penetration fillet weld. Under tube
pressure and 100 MPa axial tensile pressure on fin end

direction given above. Hence, the crack could grow from the tube weld toe over a much wider span.

DISCUSSION

A. Axial Tensile Loads

Under axial tensile loading the butt and fillet welds behaved in a similar manner. This is not surprising since the outer 4 mm of the butt weld carries most of the load to the tube, Figures 6 and 8. The tube itself is deformed so that the center of the butt weld is under a small compressive loading in the fin direction, Figure 7. The central 4 mm of the butt weld carries none of the applied tensile stress and hence the half penetration fillet weld (weld root 2 mm from the fin centre line) behaves the same as the butt weld, Figure 10. Beyond the central 4 mm of the butt weld, the stresses become more tensile and it is in this tensile region that the weld root of the zero penetration fillet weld lies. The surrounding weld metal carries the additional load and a stress concentration is formed at the weld root. The stress intensity factors at the weld toes are similar in both the butt and fillet welds, Figure 12. Since the highest value of stress intensity occurs at the tube weld toe, it is anticipated that it would be the position of crack initiation. In that region, the butt and fillet weld K values are similar resulting in the same fatigue crack propagation rate. Extending the tube weld leg length moved the position of maximum stress intensity to the fin toe.

The fatigue crack path for a crack initiated at the tube weld toe under axial tensile loads or a 1 MPa out-of-plane bending stress was through the tube wall resulting in loss of tube integrity, Figure 13.

Variations in the applied stress had little influence on the crack path. This suggests that the direction of stress flow in the tube beneath the weld did not significantly vary with type of loading on the fin, in the ranges of stress and tube pressure studied.

Whereas the stress intensity at the tube weld toe in the BASWELD mesh was significantly greater than at other locations, the stress intensity factors at the weld toes and root of the EXTWELD mesh were similar and small differences in geometry at these locations may change the position of crack initiation. The presence of a half a millimeter undercut at the tube weld toe caused a significant increase in the stress intensity at that location, Figure 12, which would ensure that the tube toe was the preferred initiation site. The variation in stress intensity at the same locations in the BASWELD and EXTWELD meshes is considered to be due to the reduction in the weld tube contact angle at the tube weld toe and an increase in the weld fin contact angle at the fin weld toe. The stress intensity factor at the weld root would be reduced in the EXTWELD mesh by the increased weld throat distance resulting in lower stresses in that region.

An important effect of the axial tensile load on the fin is to cause ovality in the tube, Figure 11. Thus, the internal surface of the tube beneath the fin is in compression, but in tension at the position 90° from the fin. The tensile stresses at these locations can be large and was 483 MPa for a 100 MPa fin loading. The occurrence of internal defects at these locations in the presence of high axial loads to the fin may therefore lead to premature failure [2].

B. Out-of-Plane Bending Loads

The stress distribution in both fillet and butt welds were similar. The presence of the root land in the fillet weld had little effect on the tensile stress distribution since this area contained only small tensile stresses in the fin direction. The variation in the stress distribution of the butt and fillet welds differed in the area around the weld root. This metal carrying the additional load caused by the presence of the root land. Differences in the stress intensity values at the fillet and butt weld toes were also small and therefore the fillet and butt welds should perform in a similar manner with the probable position of crack initiation being the fin weld toe.

The probability of failure at the weld root of the fillet weld appears remote as the stress intensity at this location is significantly below that at the weld toes. When the tube weld leg length is extended, a fin toe failure is still expected. The presence of the 1 mm undercut at the tube weld toe increases the stress intensity at that location, but not sufficiently to move the location of initiation away from the fin weld toe, Figure 12.

The out-of-plane bending load on the fin caused a stress concentration to occur on the internal surface of the tube 90° from the compressive side of the fin, Figures 6 and 8. There was little effect on the opposite internal surface. This point of stress concentration could become a point of crack initiation in the presence of pre-existing defects.

CONCLUSIONS

1. Under conditions of axial tension and out-of-plane bending applied to the fin attachment, butt and fillet welds were similar with respect to: (a) stress distribution; (b) stress intensity values at the weld toes; (c) crack path for a crack initiating at the tube weld toe.

2. The expected position of crack initiation for the basic mesh geometry for both butt and fillet welds was: (a) at the tube weld toe under axial tension; and (b) at the fin weld toe under out-of-plane bending, except at lower levels of stress intensity.

3. If the leg length of the tube weld was extended (EXTWELD geometry), the expected position of crack initiation under axial tensile stresses on the fin is the tube weld toe. Under out-of-plane bending stresses on the fin stress intensity factors at the weld toes and root were similar and the position of crack initiation would depend upon local variations in weld geometry.

4. The presence of undercut at the weld toes may effect the position of crack initiation.

5. A weld with incomplete penetration should give similar service to a full penetration butt weld for the geometry and loading conditions studied.

ACKNOWLEDGEMENT

The author wishes to thank the State Electricity Commission of Victoria for permission to publish this work.

REFERENCES

[1] Hellen, T. K., "On the method of virtual crack extension", Int. J. Numerical Meth. in Eng., 9, p. 187, 1975.

[2] Lees, D. J. and Siverns, M. J., "Preventing failures in boiler feed pipes carrying saturated steam", I. Mech. E. Conf. on "Tolerance of Flaws in Pressurised Components", London, p. 193, 1978.

SECTION VII
ENVIRONMENTAL EFFECTS

CRITERION FOR FRACTURE OF PLAIN CONCRETE

N. Saeki, N. Takada and Y. Fujita

Department of Civil Engineering, Faculty of Engineering,
Hokkaido University, Sapporo, 060, Japan

ABSTRACT

Concrete is a composite material consisting of aggregate inclusion and cement-paste matrix and the cracks which result from the interface between inclusion and matrix play an important role for the deformation and failure of plain concrete [1]. On the base of the observations of the physical and mechanical properties at the aggregate-paste interface, the model structure of concrete is originated for analyzing the fracture mechanism of plain concrete and the criterion for the fracture is proposed. From an experiment it is ascertained that the criterion is available for determinating the limit surface of the failure of concrete under multi-axial stresses with compression.

INTRODUCTION

Concrete is a popular material used for many structures and it is necessary for designers to have a correct understanding of the behavior of the deformation and failure of plain concrete and then to plan concrete structures safely and less expensively. In order to obtain the elementary information about the behavior of the deformation and failure of plain concrete, a uniaxial compression test is generally used and also available for the investigation of cracking.

The cracking of concrete primarily initiates at the weaker interface between inclusion and matrix at low stress level, namely, the bond cracks at relatively coarse aggregate are formed. Those bond cracks are independently oriented at random and behave in a stable manner due to the crack arrest of aggregate. The further increases in load lead to the onset of continuous cracking, and then the growth of the bond cracks at coarse aggregate and the initiation of the bond cracks of fine aggregate in matrix occur. Those cracks have a decided effect on the failure of concrete. The phenomenological processes of the cracking can be caught by the visual observation of surface cracks, the crack signals by an acoustic emission technique, the characteristics of the stress-strain curve and so on.

Consequently from the behavior of the cracking the model element of concrete for fracture mechanism is assumed to be the three dimensional element with a spherical inclusion in matrix. The criterion for fracture is derived from the application of Mohr-Coulomb's theory. The criterion is practically applied to the ultimate strength envelope of concrete under compression and shear stresses.

FORMATION OF CRACK

It is reported that the properties at the neighborhood of the interface between aggregate and paste are weak and porous due to the affects of bleeding and the increase of water-cement ratio [2,3] and the formation of the thin phase of calcium hydroxide as shown in Figures 1 and 2 by an electron-probe micro analyser [4]. From the physical point of view the interface is situated at the stress concentration due to the incompatibility of the elastic moduli of the aggregate and matrix.

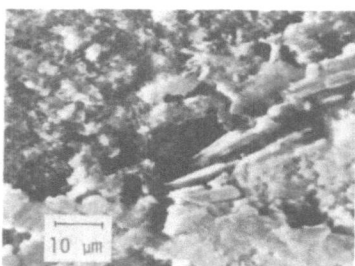

Fig. (1) - Photograph of secondary electron image at aggregate-paste interface (white zone: $Ca(OH)_2$, black zone: aggregate)

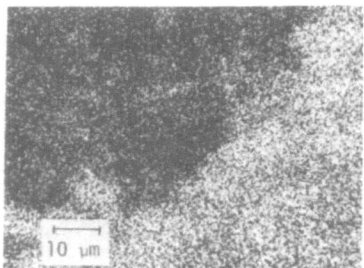

Fig. (2) - Photograph of calcium distribution at same point (white zone: $Ca(OH_2)$)

An acoustic emission technique is used to catch the crack information to three dimensions and noted as the new means of the measurement of the crack initiation and propagation of concrete [5]. The crack signals by AE test have the two remarkable stages until failure as shown in Figure 3. The first

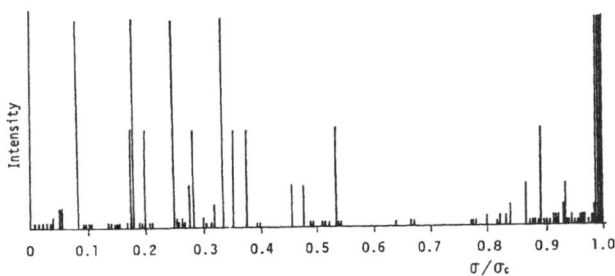

Fig. (3) - Typical crack signals by AE test

stage is about 20-50% of ultimate strength and may correspond to the stress level of the bond crack formation due to coarse aggregate particles. Below this stage the intensity of crack signals is faint and so the stress is proportional to the strain and the behavior of concrete is elastic. The second stage is about 80-90% of ultimate and the continuous cracking may be formed due to the crack initiation of fine aggregate and the propagation at the coarse aggregate interface. The stress level between first and second stage may be in the condition of crack arrest because the behavior of crack signals is gentle as shown in Figure 3.

The optical observation of micro cracking is available to confirm the formation of cracking in two dimensions [6]. The test is carried out by using the model specimens in consideration of the mutual effect of the aggregate particles as shown in Figure 4(a). The diameter of model inclusion

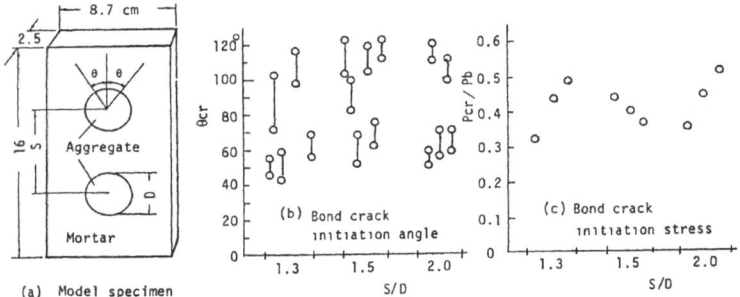

Fig. (4) - Model test of bond crack initiation

(D = 2 cm) is small as compared with the size of specimen, so that the size effect is ignored. It is found that the range of the inclination angle (θcr) of the bond crack initiation is greater than 45° and less than 120° as shown in Figure 4(b). The angles are more and less influenced by each aggregate particle. The ratios of the stress of the crack initiation (Pcr) to the ultimate strength (Pb) vary between 30% and 50% as shown in Figure 4(c). Those behaviors approximately correspond to those of AE test and the estimation by Mohr-Coulomb's theory [7,8,9].

The typical stress-strain curves for concrete under a simple compression are shown in Figure 5. The curves up to about 30-40% of ultimate strength is almost linear, namely, the inclusion and matrix play in one body, when the load is more increased the curves have a tendency to be nonlinear, i.e., the formation of the cracks and creep may be considered. Below about 75~90% of ultimate the behavior of inelasticity is comparatively stable. It may be regarded due to the action of crack arrest of aggregate. From the stress above 75~90%

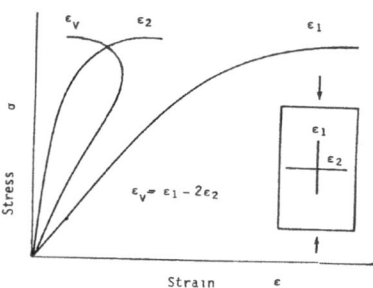

Fig. (5) - Typical stress-strain curves in uni-axial compression

of ultimate the deformation behavior begins to be unstable as the yield point of steel and the volumetric strain of specimen largely varies and concrete structure may be loose. Those behavior are considered to be onset of the continuous cracking. Summarizing the behaviors of cracking mentioned above, the two components of concrete work together elastically in the combined material up to about 30% of ultimate strength. Above this level the local cracking occurs. When load is more increased and above some 80% of ultimate, the continuous cracking occurs due to bridging the bond cracks through the matrix and the sign of the rapid failure arises. The formation of the bond cracks of coarse and fine aggregate has to do with the mechanism for the fracture of concrete.

ANALYSIS OF CRITERION FOR FRACTURE

It is found that the behavior of the bond cracking plays an elementary role for the failure of concrete, so that the structural element with a spherical inclusion in matrix may be assumed as the model structure of concrete as shown in Figure 6, and the analysis of the criterion for fracture is obtained on the base of Mohr-Coulomb's theory.

The normal stresses of σ_1, σ_2 and σ_3 due to the external stresses of P_1, P_2 and P_3 at the interface $(\theta_1, \theta_2, \theta_3)$ of the model element are obtained from Goodier's solution [10] as follows

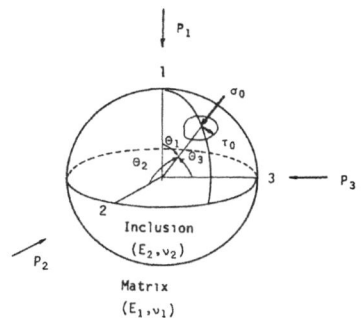

Fig. (6) - Model structure of concrete

$$\sigma_i = P_i(\beta_1 + \beta_2 \cos 2\theta_i) \tag{1}$$

where

$$\beta_1 = \frac{\beta_2}{5} \cdot \frac{2(4-\nu_2-5\nu_1\nu_2) + (1+\nu_1)(7-5\nu_1)\zeta}{(1+\nu_1)[2(1-2\nu_2) + (1+\nu_1)\zeta]} \tag{2}$$

$$\beta_2 = \frac{15}{2} \cdot \frac{(1-\nu_1)(1+\nu_1)\zeta}{(7-5\nu_1)(1+\nu_2) + (8-10\nu_1)(1+\nu_1)\zeta} \tag{3}$$

$$\zeta = \frac{E_2}{E_1} \tag{4}$$

The total normal stress (σ_0) at the interface is as follows:

$$\sigma_0 = \sum_{i=1}^{3} P_i(\beta_1 + \beta_2 \cos 2\theta_i) \tag{5}$$

Similarly the shear stresses τ_1, τ_2 and τ_3 at the interface are obtained

$$\tau_i = P_i\beta_2 \sin 2\theta_i \tag{6}$$

and the total shear stress (τ_0) is expressed

$$\tau_0^2 = \sum_{i=1}^{3} \beta_2^2(P_i^2 \sin^2 2\theta_i - 8\xi_i) \tag{7}$$

where

$$\xi_1 = P_1 P_2 \cos^2\theta_1 \cos^2\theta_2 \tag{8}$$

$$\xi_2 = P_2 P_3 \cos^2\theta_2 \cos^2\theta_3 \tag{9}$$

$$\xi_3 = P_1 P_3 \cos^2\theta_3 \cos^2\theta_1 \tag{10}$$

If the failure occurs at the interface of $\theta_1 = \theta_2 = \theta_3$, the total normal and shear stresses are as follows

$$\sigma_o = \frac{1}{3} \cdot \gamma_1 (P_1 + P_2 + P_3) \tag{11}$$

$$\tau_o^2 = \frac{1}{9} \cdot \gamma_2^2 [(P_1 - P_2)^2 + (P_2 - P_3)^2 + (P_1 - P_3)^2] \tag{12}$$

where

$$\gamma_1 = \frac{3(1-\nu_1)\zeta}{2(1-2\nu_2) + (1+\nu_1)\zeta} \tag{13}$$

$$\gamma_2 = \frac{15(1-\nu_1)(1+\nu_1)\zeta}{(7-5\nu_1)(1+\nu_2) + (8-10\nu_1)(1+\nu_1)\zeta} \tag{14}$$

E_1 and ν_1 are the modulus of elasticity and Poisson's ratio for matrix respectively, likewise E_2 and ν_2 are for an inclusion. The values of ν_1 and ν_2 are calculated by the above equations and shown in Figure 7.

APPLICATION OF THE CRITERION TO FAILURE OF CONCRETE UNDER COMPRESSION AND TORSION

A. Test Procedures

The specimens are made of concrete and mortar. The size of specimens is $\phi 10 \times 40$ cm cylinder and the effective length is 25 cm. The compression load is applied by using an Amsler testing machine and the torque is performed with the two oil jacks fixed by a frame as shown in Figure 8. The frictions due to the rotation and the longitudinal deformation are eliminated by the rollers and the slide plates with teflon sheets and silicon grease respectively. The top and bottom of the specimen are in hinge to avoid an eccentric compression.

B. Test Results

It is assumed that the shear strain (γ) would be linear and in proportion to the radius of a cylinder and the shear stress (τ) would be parabolic to the shear strain. The shear stress (τ) and the applied torque (T)

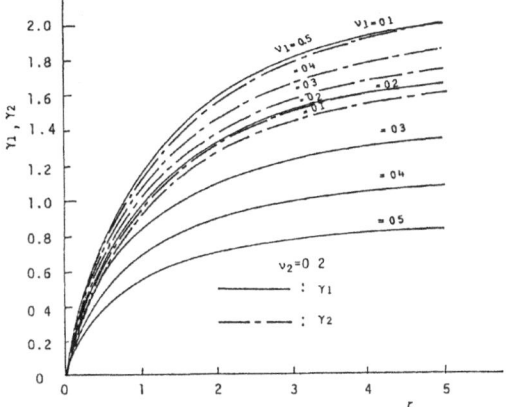

Fig. (7) - Values of γ_1 and γ_2

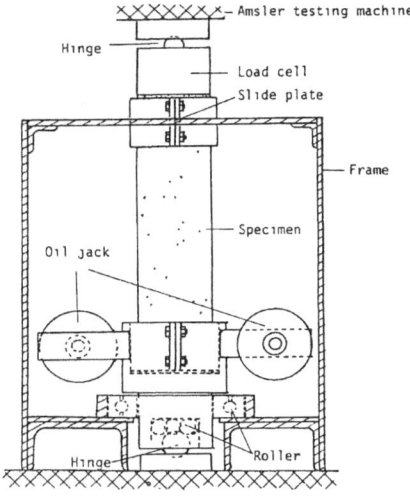

Fig. (8) - Apparatus for loading compression and torsion

550

are defined as follows:

$$\tau = \alpha_1 \gamma^2 + \alpha_2 \gamma \tag{15}$$

$$T = 2\pi R^3 \left(\frac{\alpha_1}{5} \gamma_s^2 + \frac{\alpha_2}{4} \gamma_s \right) \tag{16}$$

where γ_s is shear strain at the surface of a cylinder and R is radius of a cylinder and α_1 and α_2 are estimated from the T-γ_s values by the least squares method. The ultimate shear strains are calculated from equation (16) for the ultimate torque and then substituting the ultimate shear strains in equation (15) the ultimate shear stresses are estimated. The appearances of the specimens after failure under compression and torsion are shown in Figure 9.

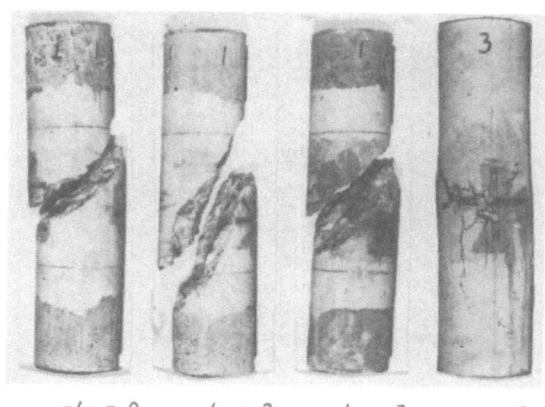

$\sigma/\tau = 0$ $\sigma/\tau = 3$ $\sigma/\tau = 7$ $\tau = 0$

Fig. (9) - Photograph of specimens after failure
under compression and torsion

C. Ultimate Strength Envelopes

The strength of the natural aggregate particles is stronger than that of matrix and the inclusion behaves elastic until failure, so that the modulus of elasticity and Poisson's ratio of the aggregate are maintained constant. On the contrary E_1 and ν_1 of matrix are apparently variable due to the formation of the micro cracks and creep. In consideration of those behaviors it is assumed that γ_1 is constant because the variable range of γ_1 is small below 60% of σ_0/σ_c and γ_2 varies with the ratio of σ_0/σ_c as shown in Figure 10, in which it is supposed that the values of ζ and ν_1 increase reasonably by the ratio of σ_0/σ_c. The values of γ_1 and γ_2 are assumed as follows:

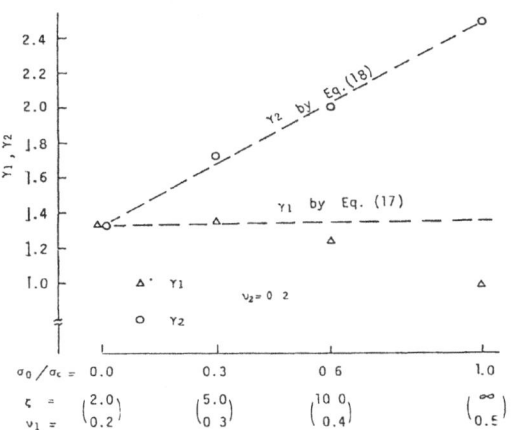

Fig. (10) - Assumed values of γ_1 and γ_2

$$\gamma_1 = \frac{4}{3} \tag{17}$$

$$\gamma_2 = \frac{4}{3} + \frac{7}{6} \cdot \frac{\sigma_0}{\sigma_c} \tag{18}$$

where σ_c is a uniaxial compression strength. The total stresses under compression (σ) and shear stress (τ) are obtained.

$$\bar{\sigma}_0 = \frac{4}{9} \bar{\sigma} \tag{19}$$

$$\bar{\tau}_0 = \frac{4\sqrt{2}}{9} (1 + \frac{7}{18} \bar{\sigma})(\bar{\sigma}^2 + 3\bar{\tau}^2)^{1/2} \tag{20}$$

where

$$\bar{\sigma}_0 = \sigma_0/\sigma_c \ , \ \bar{\tau}_0 = \tau_0/\sigma_c \ , \ \bar{\sigma} = \sigma/\sigma_c \ , \ \bar{\tau} = \tau/\sigma_c \tag{21}$$

The values $\bar{\tau}_0 = \bar{\sigma}_0$ is linear and the relationship is obtained by the least squares method as follows:

$$\bar{\tau}_0 - a\bar{\sigma}_0 + b \tag{22}$$

Substituting the above equation in equations (19) and (20), and then arranging those equations an envelope curve is expressed as follows:

$$\bar{\tau}^2 + \frac{1}{3}\bar{\sigma}^2 = \frac{27}{8}\left(\frac{4a\bar{\sigma} + 9b}{7\bar{\sigma} + 18}\right)^2 \qquad (23)$$

The envelopes illustrated in Figure 11 are calculated by the author's [11] and other researchers' data [12,13,14,15,16,17]. The constants a and b of equation (23) is 1.857 and 0.0774 respectively (case 1). And if equation (22) gets through the point ($\bar{\sigma}$ = 1, $\bar{\tau}$ = 0), the values of a and b are equal to 1.774 and 0.0844 respectively (case 2).

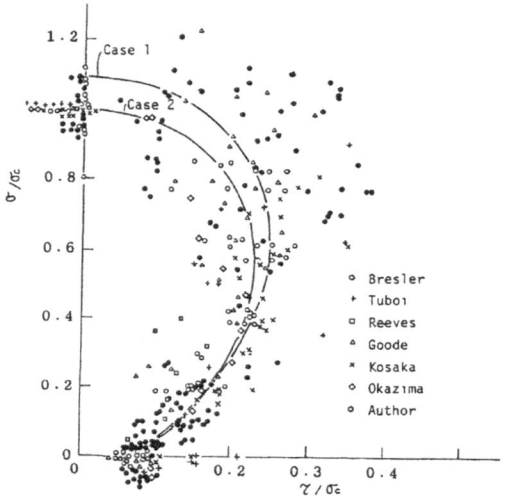

Fig. (11) - Ultimate strength envelopes of concrete under compression and torsion

SUMMARY

The crack behavior of concrete in simple compression until failure has two stages. The first stage is about 30% of ultimate strength, corresponds to the bond crack formation at coarse aggregate and below this state the two components of concrete work together and concrete is elastic body. The second stage is about 80% of the ultimate and the onset of continuous cracking due to the extension of the bond cracks of coarse and fine aggregate in matrix. The behavior of concrete is quasi-elastic. The fracture mechanism of concrete is closely connected with the formation of the bond cracking. On the base of the behavior the criterion for the failure of concrete is analyzed. It is found that the criterion is analogous to the octahedral stress theory [18] and is enable to consider the quasi-elastic behavior due to the physical properties of aggregate and matrix and is available for application of the ultimate strength envelope of plain concrete under compression and shear stresses.

REFERENCES

[1]   Fujita, Y. and Saeki, N., "On crack initiation and propagation of plain concrete", Concrete Journal, Vol. 16, No. 11, p. 1, 1978.

[2]   Suzuki, K. and Mizukami, K., "Bond strength between cement paste and aggregate", CAJ Review, 29, p. 94, 1975.

[3]   Iwasaki, N. and Tomiyama, Y., "Bond and fracture mechanism of inter- face between cement paste and aggregate", CAJ Review, 30, p. 213, 1976.

[4]   Fujita, Y., Saeki, N., Takada, N. and Nara, H., "On properties of cracking of plain concrete", CAJ Review, 31, p. 147, 1977.

[5]   Saeki, N., Takada, N. and Hataya, S., "On studies for cracking and failure of concrete by acoustic emission techniques", CAJ Review, 33, p. 234, 1979.

[6]   Fujita, Y., Saeki, N., Takada, N. and Nara, H., "Formation of bond crack in plain concrete", Proc. of 21st Japan Cong. on Material Research, 21, p. 201, 1978.

[7]   Hsus, T.T.C. and Slate, F. O., "Tensile bond strength between aggregate and cement paste or mortar", J. of ACI, No. 60-4, p. 465, 1963.

[8]   Taylor, M. A. and Broms, B. B., "Shear bond strength between coarse aggregate and cement paste or mortar", J. of ACI, No. 61-52, p. 939, 1964.

[9]   Fujita, Y., Saeki, N., Takada, N. and Nara, H., "Researches on mechan- ism of crack initiation of plain concrete", CAJ Review, 30, p. 204, 1976.

[10] Goodier, J. N., "Concentration of stress around spherical and cylindri- cal inclusion and flaw", APM-55-7, p. 39, 1955.

[11] Saeki, N., Takada, N. and Fujita, Y., "Behavior of deformation and failure of plain concrete under compression and torsion", Proc. JSCE, No. 308, p. 99, 1981.

[12] Bresler, B. and Pister, K. S., "Strength of concrete under combined stresses", J. of ACI, No. 55-20, p. 321, 1958.

[13] Tsuboi, Y. and Suenaga, Y., "Experimental study on failure of plain concrete under combined stresses" Trans. of AIJ, No. 64, p. 25, 1960.

[14] Reeves, J. S., "The strength of concrete under combined direct and shear stresses", Cement and Concrete Association TRA 365, p. 1, 1962.

554

[15] Goode, C. D. and Helmy, M. A., "The strength of concrete under combined shear and direct stress", MCR, Vol. 19, No. 59, p. 105, 1967.

[16] Osaka, Y. and Tanigawa, Y., "Criterion of concrete under compression and torsion in consideration of aggregate", Trans. of AIJ, No. 166, p. 29, 1969.

[17] Okazima, T., "Criterion of concrete under complex stresses (compression-torsion and tension-torsion)", Trans. of AIJ, No. 178, p. 1, 1970.

[18] Nadai, A., "Theory of flow and fracture of solid", McGraw-Hill, 1950.

THERMAL EMBRITTLEMENT AND PURITY IN MAR350 STEEL

J. T. A. Pollock and R. Clissold

CSIRO, Division of Chemical Physics
Lucas Heights, Sutherland, New South Wales 2223, Australia

and

M. Meller, R. Warren, K. Watson and B. Zybenko

Australian Atomic Energy Commission
Lucas Heights, Sutherland, New South Wales 2223, Australia

ABSTRACT

Thermal embrittlement effects in aged MAR350 steels have been studied using K_{IC} fracture toughness measurements. In particular, the effect of time at a sensitizing temperature of 1175°K has been investigated using two steels of differing residual element content. A significant loss of toughness was measured after 300 s. sensitization; this is larger for steel of the highest impurity content. Microstructural and fractography studies are reported which describe the weakening effect of the Ti(CN) precipitation at prior austenite boundaries.

INTRODUCTION

Maraging steels, a class of low carbon, high alloy precipitation hardening steels, are recognized for their combination of considerable strength and fracture toughness. These steels are normally air-cooled from the austenite (γ) regime to produce a tough lathe martensite which is strengthened by precipitation hardening at about 775°K for three hours.

The purity of lower grade maraging steels has a demonstrated importance in the production of very high toughness [1]. By contrast, Carter [2] and Spaeder [3] showed that with higher grades, such as MAR350, fracture toughness and strength of properly heat-treated specimens were not significantly affected by normal variations in the residual element content. Nevertheless, maraging steels may be embrittled if cooled slowly from solution temperatures greater than about 1300°K. Kalish and Rack [4] reported that MAR350 Charpy impact specimens were severely embrittled when held at intermediate temperatures in the range 1070 to 1320°K for four hours. They were unable to use aged specimens as impact energies in this condition were too low for any sensitizing effect to be reliably monitored. Kalish and Rack [4] attributed this embrittlement to the precipitation at prior γ grain boundaries of Ti(CN). Johnson and Stein [5] supported this hypothesis by detecting C and N segregation at the prior boundaries in MAR350 with Auger Spectroscopy. Subsequently, Nes and Thomas [6] studied a relatively high carbon (0.01 wt.%) MAR350 using transmission electron microscopy (TEM) and identified a feather-like precipitate at the prior γ grain boundaries as TiC.

We have investigated the thermal embrittlement of fully age-hardened MAR350 using both fracture toughness measurements and microstructural studies, including optical, transmission and scanning electron microscopy (SEM). In particular, we have examined the effect of purity, employing steels having different levels of residual elements, and determined the role of time at a sensitizing temperature of 1175°K.

EXPERIMENT PROCEDURES

The steels studied were supplied as seamless, hot extruded and annealed tubes of 3 mm wall thickness. Compositions are listed in Table 1.

TABLE 1 - CHEMICAL COMPOSITION OF MAR350 STEELS, (wt.%)

	C	S	P	Si	Mn	Ni	Ti	Mo	Co	Fe
Steel A	0.006	0.008	0.005	0.07	0.04	17.8	1.5	4.2	11.8	Balance
Steel B	0.003	0.003	0.004	0.02	0.02	17.8	1.5	3.7	12.4	Balance

After cutting and then flattening in a press, single edge notch specimens were machined from the slabs according to ASTM specifications [7]. Samples had identical orientation; the long axis was parallel to the extrusion direction and the notch was at right angles to both the tube radius and extrusion direction. Additional material was cold rolled, with an intermediate vacuum anneal at 1275°K to a thickness suitable for tensile strength measurement and microstructural examination. Batches of toughness and sheet specimens were submitted to solution treatment for one hour at 1470°K, rapidly cooled to 1175°K and held for various times before quenching in oil. The procedure was carried out in a furnace and quenching system having a common vacuum of 10^{-3} Pa. Subsequently, age hardening was carried out in a similar vacuum at 755°K for three hours.

Fracture toughness samples were fatigue cracked under cyclic loading before aging, since controlled cracking of aged MAR350 specimens is difficult owing to their high strength. Thus, the toughness value measured, according to ASTM procedures for three point bend testing, is not a true K_{IC} value since the crack was introduced before the heat treatment was completed. Three or more statistically valid K_{IC}^{*} measurements were made at each test treatment condition.

Optical micrographs were obtained from surfaces which had been electropolished and etched with a glacial acetic/perchloric acid solution. Samples for TEM were prepared by thinning to 0.25 mm thick foils using a jet of perchloric acid/methanol solution. Carbon extraction replicas were made of etched surfaces following a standard deposition and electrolyte dissolution technique. After testing, fracture toughness samples were stored in desiccators before SEM examination.

RESULTS

Table 2 summarizes the fracture toughness data measured with age-hardened samples. The overall trend in fracture toughness with time at 1175°K is shown more clearly in Figure 1 together with an average error bar for each steel.

TABLE 2 - AGED MAR350 FRACTURE TOUGHNESS, K_{IC}^*

	Time at 1175°K, s.	Mean K_{IC}^*, MPa$\sqrt{m}$
Steel A	0	46
	1.2×10^2	43
	3×10^2	31
	1.2×10^3	37
	1.2×10^4	35
Steel B	0	45
	1.2×10^2	42
	3×10^2	38
	1.2×10^3	41
	1.2×10^4	37

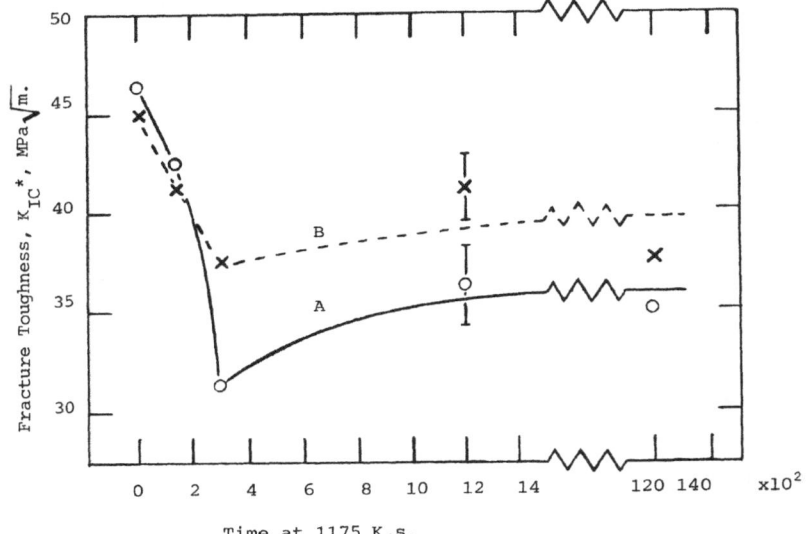

Time at 1175 K,s.

Fig. (1) - Aged MAR350. Fracture toughness, K_{IC}^*, versus time at 1175°K

The data show that before sensitizing, fracture toughness was independent of the variation in residual element content. However, for short times at 1175°K, the degree of embrittlement appeared to be purity dependent. Steel A with a residual element content twice that of Steel B degraded in toughness by about 33% compared with 15% for Steel B. Some recovery of toughness with additional time at 1175°K was recorded for Steel A.

In the unsensitized condition, ultimate tensile strength (UTS) values of about 2.2 GPa were measured for each steel. These values are rather low for MAR350 steel in the full-aged condition and may be attributed to the prior γ grain size of 300 μm following the solution treatment at 1475°K. The values were slightly lowered on sensitizing at 1175°K.

Examination of aged fracture toughness surfaces by SEM showed that before sensitizing, fracture occurred mainly by transgranular cleavage, Figure 2a. In this condition, most cleavage planes extend over several grains. In some areas, microvoid coalescence leading to ductile tearing was observed. With time at 1175°K a gradual change in fracture appearance was noted. Evidence of intercrystalline fracture was noted after two minutes at 1175°K, in the form of prior γ grain boundary separation, Figure 2b. The overall size of the cleavage planes was reduced and, after 20 minutes at 1175°K, the general outline of prior γ grain boundaries could be observed, Figure 2c.

This transition was most noticeable with Steel A; although Steel B did not show the prior γ grain structure as clearly with increasing time at 1175°K, the cleavage plane size decreased and the number of secondary cracks observed increased.

Short times at 1175°K produced subtle though significant changes in the general fracture appearance of aged samples. Kalish and Rack [4] and others [5,6] report that sensitizing for one to four hours caused a dramatic change in the fracture appearance of samples which had only received solution treatment. Several samples of the present study were fractured following short periods at 1175°K and before ageing. Calculated K_{IC} values were invalid. Figure 3 reveals that after two minutes at 1175°K, a transition from dimpled transgranular to severe intercrystalline fracture occurred. In these specimens, it was noticeable that Steel A exhibited considerably more intercrystalline fracture than Steel B after this short sensitizing period. The ratio of intercrystalline to transgranular fracture increased with time at 1175°K. However, as Figure 3d shows, the fracture mode might best be described as quasi-cleavage since the apparently cleaved surfaces are combined with dimpled tearing.

Large precipitates were noted in all fracture surfaces which were identified as sulphides using energy-dispersive X-ray (EDX) analysis. Although related to the large dimples in the unaged samples and the origins of cleavage planes in aged material, these large particles were not involved in the grain boundary weakening.

Optical microscopy revealed the general shape and distribution of the precipitation at 1175°K, Figures 4a-c. Precipitation at the prior γ grain boundaries became more discontinuous at longer times. EDX examination of these precipitates confirmed that they were rich in Ti. In some aged samples, there was clear evidence of grain boundary denudation, Figure 4c. Extraction replicas of sensitized samples, aged and unaged, showed dendritic or feather-like precipitation at the prior γ grain boundaries after 120 s. at 1175°K, Figures 5, a and b. Transmission electron microscopy revealed significant variations in polishing rate which were suggestive of grain boundary segregation, Figure 5c. These precipitation characteristics were not observed in aged or unaged samples which had not been held at 1175°K.

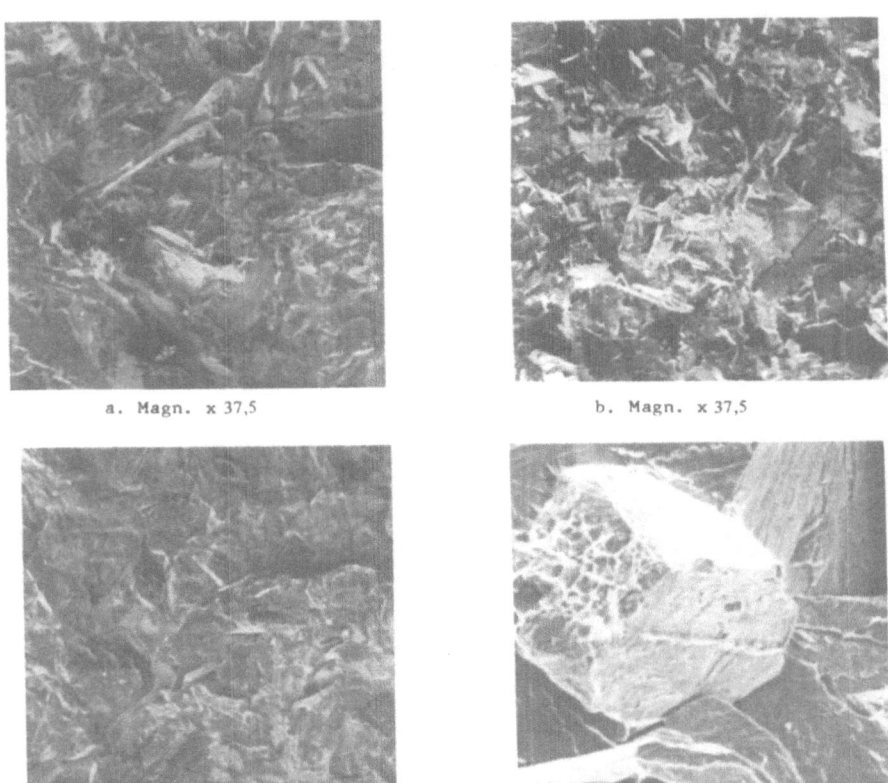

a. Magn. x 37,5   b. Magn. x 37,5

Fig. (2) - Steel A. SEM. (a) Solution treated and aged; transgranular cleavage fracture; (b) Solution treated, sensitized at 1175°K for 120 s. and aged; some signs of intercrystalline fracture; (c) Solution treated, sensitized at 1175°K for 120 s. and aged; clear evidence of prior γ grain boundaries and (d) Solution treated, sensitized at 1175°K for 300 s. and aged; prior γ grain boundary cleavage and areas of void coalescence fracture

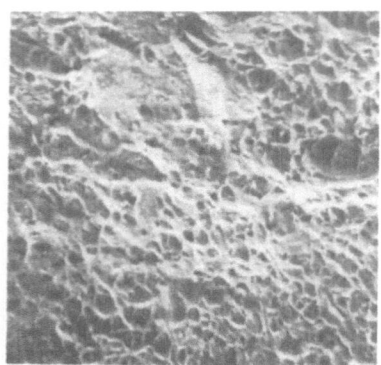

a. Magn. x 37,5

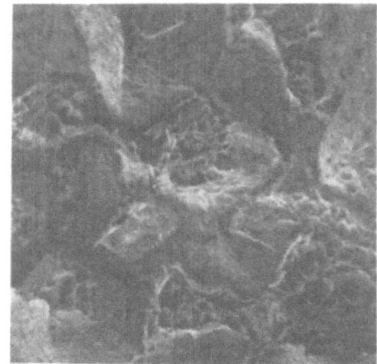

b. Magn. x 37,5

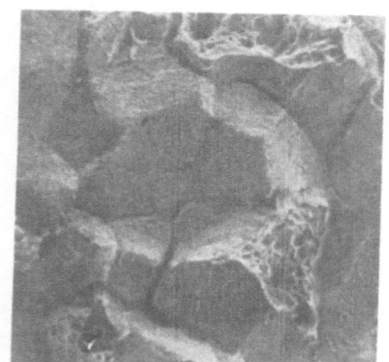

c. Magn. x 37,5

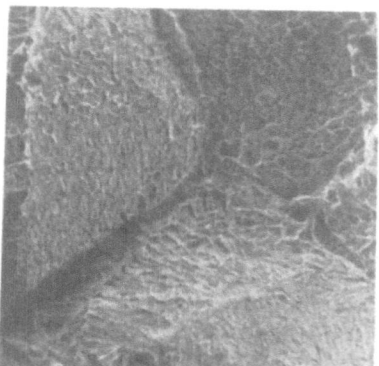

d. Magn. x 187,5

Fig. (3) - Steel B. SEM. (a) Solution treated; ductile transgranular
fracture; (b) Solution treated and sensitized for 300 s. at
1175°K; mixed transgranular fracture; (c) Solution treated
and sensitized for 1200 s. at 1175°K; mainly intercrystalline fracture showing cracking along prior γ grain boundaries,
and (d) as (c) showing quasi-cleavage nature of the fracture
surface

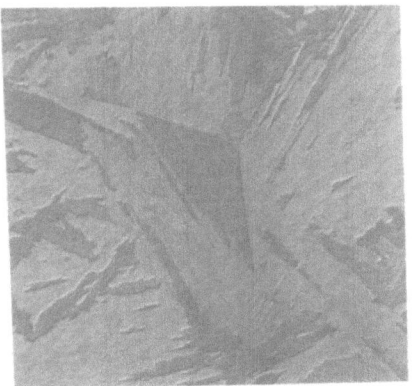

a. Magn. x 375

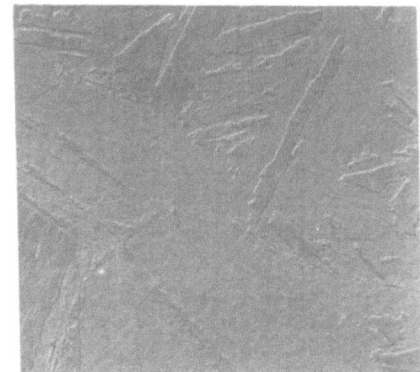

b. Magn. x 375

Fig. (4) - Steel A. Optical. (a) Solution treated; clean prior
grain boundaries; (b) Solution treated, sensitized at
1175°K for 1200 s. and aged: prior γ grain boundary
precipitation, and (c) Solution treated; 12000 s. at
1175°K and aged; denudation and precipitation at prior
γ grain boundaries

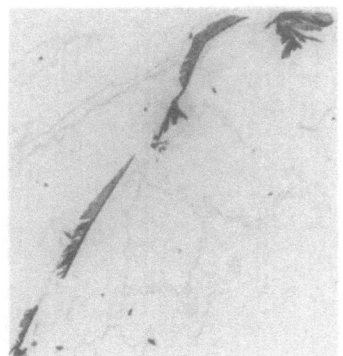

a. Magn. x 1875

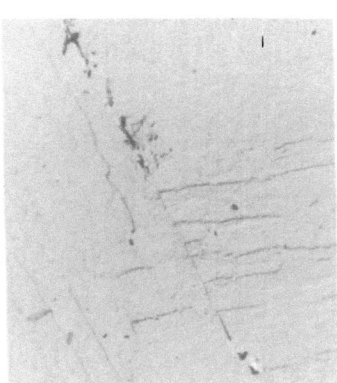

b. Magn. x 1875

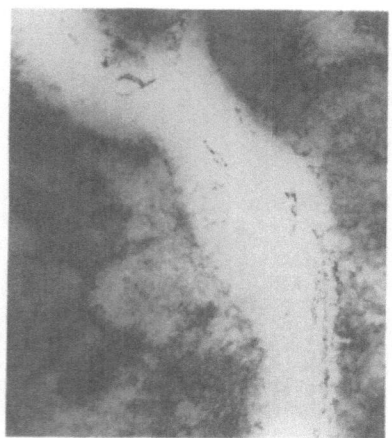

Fig. (5) - TEM. (a) Steel A (extraction replica TEM). Solution treated,
sensitized at 1175°K for 120 s; feathery prior γ grain boundary precipitates; (b) Steel B (extraction replica TEM). Solution treated, sensitized at 1175°K for 1200 s. feathery prior
γ grain boundary precipitation, and (c) Steel A. (TEM). Solution treated 300 s., preferential polishing of grain boundary

DISCUSSION

The data presented in Table 1 and Figure 1 show that if correctly heat treated,
the level of impurity content normally encountered with MAR350 steels has no effect
on fracture toughness in the aged condition. This result is in agreement with an

extrapolation of data reported by Porter [1], Carter [2] and Spaeder [3]. However, the new data show that relatively short times at 1175°K, such as might occur on cooling moderately thick sections from solution treatment temperatures, are sufficient to cause degradation of the fracture toughness level measured on aged samples. Also, the fall in fracture toughness is twice as great for steel of the lowest purity. Since the grain size for each steel is similar, and it is known that grain size is not of primary importance for fracture toughness in maraging steels [2,8], it is clear that impurity content must be the controlling factor. The grain size of the present specimens, 300 μm, is larger than would normally be employed; therefore, the degradation reported here would tend towards a maximum value. Smaller grain sizes would reduce the quantity of precipitation per unit grain boundary area and some reduction in fracture degradation might follow.

In agreement with previous work [4-6], the embrittlement occurring with time at 1175°K may be attributed to the precipitation at prior γ grain boundaries of Ti(CN). To date, we have not carried out a detailed examination and analysis of fracture mode as a function of time at the sensitizing temperature. The results of optical microscopy suggest that the precipitation undergoes a ripening with time which could have caused the observed increase in toughness.

The SEM work indicated that general grain boundary weakening, as reflected in low magnification studies, increases with time. Detailed stereoscopic analysis of the size and size distribution of the dimpling on the quasi-cleavage surface is required. Also, the effect of the ageing process on the sensitizing precipitation needs to be understood. Comparison of the SEM studies of sensitized-aged, Figure 2, and sensitized-unaged, Figure 3, provides clear evidence that the ageing precipitation significantly modifies fracture characteristics. The quasi-cleavage fracture surfaces of the unaged reflect a void coalescence mechanism over the entire grain surface. On ageing, the amount of dimpling is much reduced and where it occurs, smaller.

In conclusion, although much work is required on the mechanism of fracture in sensitized and aged MAR350 steel, it is clear from the present studies that such steels may be embrittled after only a few minutes at sensitizing temperatures, and that the degree of embrittlement is purity dependent.

ACKNOWLEDGEMENTS

We thank P. W. Kelly and C. Ball for helpful discussions during the course of this work.

REFERENCES

[1] Porter, L. F., "The roll of inclusions on mechanical properties in high-strength steels", J. Vac. Sci. Technol., 9, p. 1340, 1972.

[2] Carter, C. S., "The effect of heat treatment on the fracture toughness and subcritical crack growth of a 350-grade maraging steel", Met. Trans., 1, p. 1551, 1970.

[3] Spaeder, G. J., "Impact transition behaviour of high-purity 18 Ni maraging steel", Met. Trans., 1, p. 2011, 1970.

564

[4] Kalish, D. and Rack, H. J., "Thermal embrittlement of 18 Ni (350) maraging steels", Met. Trans., 2, p. 2665, 1971.

[5] Johnson, W. C. and Stein, O. F., "A study of grain boundary segregation in thermally embrittled maraging steel", Met. Trans., 4, p. 549, 1974.

[6] Nes, E. and Thomas, G., "Precipitation of TiC in thermally embrittled maraging steels", Met. Trans., 7A, p. 967, 1976.

[7] ASTM E399-74. "Plane-Strain Fracture Toughness of Metallic Materials".

[8] Carter, C. S., "Fracture toughness and stress corrosion characteristics of a high strength maraging steel", Met. Trans., 2, p. 1621, 1971.

ENVIRONMENTAL EFFECT ON FATIGUE FRACTURE IN HIGH IMPACT POLYSTYRENE

O. F. Yap, Y. W. Mai and B. Cotterell

Department of Mechanical Engineering, The University of Sydney
Sydney, N.S.W. 2006, Australia

ABSTRACT

The fatigue fracture behavior of a high impact polystyrene is investigated
in air and in a series of alcohols. The fatigue data are analyzed using a modified Williams model which gives lower crack tip craze stresses in the liquid environments than in air. The fatigue thresholds are, however, larger in the alcohols due to crack tip blunting. For a given alcohol, the fatigue crack growth
rates and crack tip craze stresses are independent of stress ratios and frequencies.

INTRODUCTION

The analysis of experimental data for fatigue crack propagation in polymers
both in air and in liquid environments [1-3] has been mainly based on the Paris
power law equation [4]. To account for frequency and mean stress effects, empirical parameters have been added to the Paris equation to fit the data. Although
this approach is reasonably successful for the description of fatigue crack growth
in many polymeric materials, it does not have a sound theoretical basis and lacks
any link with established fracture parameters, such as crack opening displacement
and craze stress. Recently, a new model on the fatigue crack growth in polymers
has been presented by Williams [5]. This is based on the Dugdale line plastic
zone model and the assumption of a critical crack opening displacement and a constant cyclically reduced craze stress. The new model seems to give a good description of fatigue crack growth rate, mean stress, frequency and environmental effects
for a range of polymers [5,6]. However, the predicted craze stresses are too high
and modifications to the model have been considered [7,8]. Yap et al [7] have
suggested the following modified fatigue crack growth equation:

$$\frac{da}{dN} = \frac{\pi}{8(1-\alpha)^2 \sigma_c^2} \left| G(R)K^2 - \alpha K_c^2 \right| \tag{1}$$

where $G(R)$ is a function of the stress ratio (R) to be determined either experimentally or theoretically [8] and $G(0) = 1.0$. σ_c is the crack tip stress, α is a
stress reduction factor less than unity and K_c is the critical static stress intensity factor. In order to account for residual compressive stress effects which
give rise to a fatigue threshold (K_t), it has been suggested that in equation (1),

K^2 should be replaced by $(K^2-K_0^2)$ [8]. In our previous work [7], equation (1) is shown to apply well to the fatigue data of a high impact polystyrene (HIPS) in air and give reasonably constant craze stresses independent of mean stress ratios. The present work is a continuation of this previous investigation and it studies the effects of a series of organic solvents on the fatigue fracture behavior of the same HIPS polymer [7,9].

EXPERIMENTAL WORK

All fatigue tests were conducted on single edge notch specimens of dimensions 4.7 x 70 x 210 mm³ in a Shimadzu closed loop servo-pulser testing machine with sinusoidal waveforms at ambient conditions of approximately 22°C and 50% R.H. The liquid environments used in this work were a series of alcohols whose physical properties are given in Table 1. The stress ratio (R) was maintained at zero except for ethanol where both R and frequency were varied. Crack length (a) measurements were obtained periodically with a travelling microscope to an accuracy of 0.01 mm. The fatigue crack growth rates (da/dN) were determined from the a-N plots using a computer program written for this purpose and the stress intensity factor (K) was calculated from the K-equation given by Brown and Srawley [10].

TABLE 1 - PHYSICAL PROPERTIES OF ALCOHOLS USED [11]

Liquid	Surface Tension $\gamma(10^{-5}N/cm)$	Solubility Parameter $(\delta_s)(J/cm^3)^{0.5}$	Molecular Volume $V(cm^3/mole)$
Methanol	22.60	29.67	40.20
Ethanol	22.75	25.99	58.35
n-Propanol	23.78	24.35	74.81
n-Butanol	24.60	23.12	91.80

FATIGUE CRACK GROWTH RESULTS

A. Effects of Alcohol

Figure 1 shows the fatigue crack growth rates (da/dN) plotted against the applied stress intensity factor range (ΔK) on a log-log scale for air and for the series of alcohols used. The range of data available from the liquid environments is limited because at high ΔK the crack tip craze zone is large enough to invalidate the linear elastic fracture mechanics analysis. Neglecting the near threshold low ΔK region the experimental data given seem to fit the Paris power law equation very well, i.e.,

$$\frac{da}{dN} = A\Delta K^m \qquad (2)$$

where A, m are constants to be determined experimentally and are given in Table 2. Clearly for the same ΔK, da/dN is larger in the environments than it is in air. The A and m values given in Table 2 indicate that at a given ΔK, da/dN increases with chain length and surface tension but it decreases with solubility parameter of the solvents, Table 1. This means that the fatigue fracture resistance increases from butanol to methanol and it supports the plasticization theory [12,13] previously used to explain the fracture behavior of polymers in

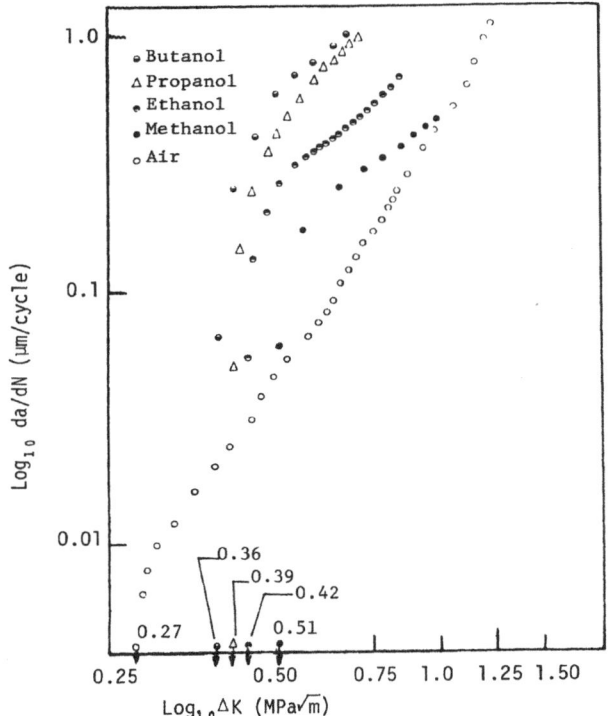

Fig. (1) - Fatigue crack growth rates of HIPS in alcohols at 10 Hz and R = 0

TABLE 2 - VALUES OF A AND m FOR PARIS EQUATION (R = 0, FREQUENCY = 10 Hz), FATIGUE THRESHOLD AND CRAZE STRESS

Environment	A	m	K_t (MPa$\sqrt{m}$)	σ_c (MPa)
Air	0.52	3.32	0.27	20.5
Methanol	0.46	1.69	0.51	17.1
Ethanol	0.89	1.76	0.42	15.2
n-Propanol	1.66	1.74	0.39	13.8
n-Butanol	1.91	1.75	0.36	12.4

organic solvents [13,14]. da/dN decreases with increasing $(\delta_s - \delta_p)$, the solubility parameter difference between the solvent and the polymer, and $\delta_p = 18.62(J/cm^3)^{1/2}$. Figure 1 also shows that the fatigue thresholds K_0 are higher in the alcohols than in air and that they increase directly with $(\delta_s-\delta_p)$ in accordance with the plasticization theory, Table 2. Experimental observations of the crack tip region in the presence of the alcohols reveal that craze bundles are formed at the tip during the incubation period and the tendency of this phenomenon is more marked in methanol than in butanol. We believe that this crack tip blunting mechanism is largely responsible for the larger K_t values in the alcohols. Immediately after a sharp crack has initiated from the craze bundle, the crack propagates with a craze line zone ahead of it and da/dN becomes larger than in air due to the adverse effect of the alcohols.

To examine the effect of the liquid environments on the fatigued craze stress (σ_c) of the polymer, we use Dugdale's model to determine σ_c from direct measurements of the crack tip craze zone by fracturing a previously cyclic loaded specimen in liquid nitrogen. These craze stresses are given in Table 2 and they decrease from methanol to butanol thus confirming the order of aggressiveness of these alcohols.

B. Effect of Stress Ratio and Frequency in Ethanol

The fatigue crack growth rates in ethanol are plotted against ΔK for varying stress ratios (R) and frequencies in Figures 2 and 3 respectively. Fatigue crack growth for R > 0.3 was not studied because it was not possible to constrain K_{max} within the linear elastic range. For the experimental results obtained, there does not seem to be any significant mean stress effect on fatigue crack growth rate, Figure 2. This is in contrast to the behavior of other polymers in aggressive environments [2]. There is also little frequency effect on da/dN, Figure 3, indicating that the environment induced crazes are not very sensitive to rate changes.

ANALYSIS AND DISCUSSION

Although the environmental fatigue data presented here can be successfully analyzed in terms of the Paris power law, equation (2), no information can be obtained with regard to craze stresses and material damage due to fatigue from this simple empirical approach. It is decided therefore to use the modified Williams model, i.e., equation (1), to analyze these experimental data. Figures 4 and 5 plot da/dN versus K^2_{max} in various alcohols and for varying R-ratios and frequencies according to equation (1). As predicted, straight line relationships are obtained. The slope M(R) and the intercept C(R) values are given by [7]:

$$M(R) = \frac{\pi}{8} \frac{G(R)}{(1-\alpha)^2 \sigma_c^2} \tag{3}$$

$$C(R) = \frac{\alpha K_c^2}{G(R)} \tag{4}$$

It should be pointed out here that equation (1) is only valid for fatigue crack growth that is continuous and stable. If the crack growth is discontinuous and unstable, the modified Williams analysis cannot be directly used. For the environmental fatigue crack growth tests conducted in this work, the crack growth

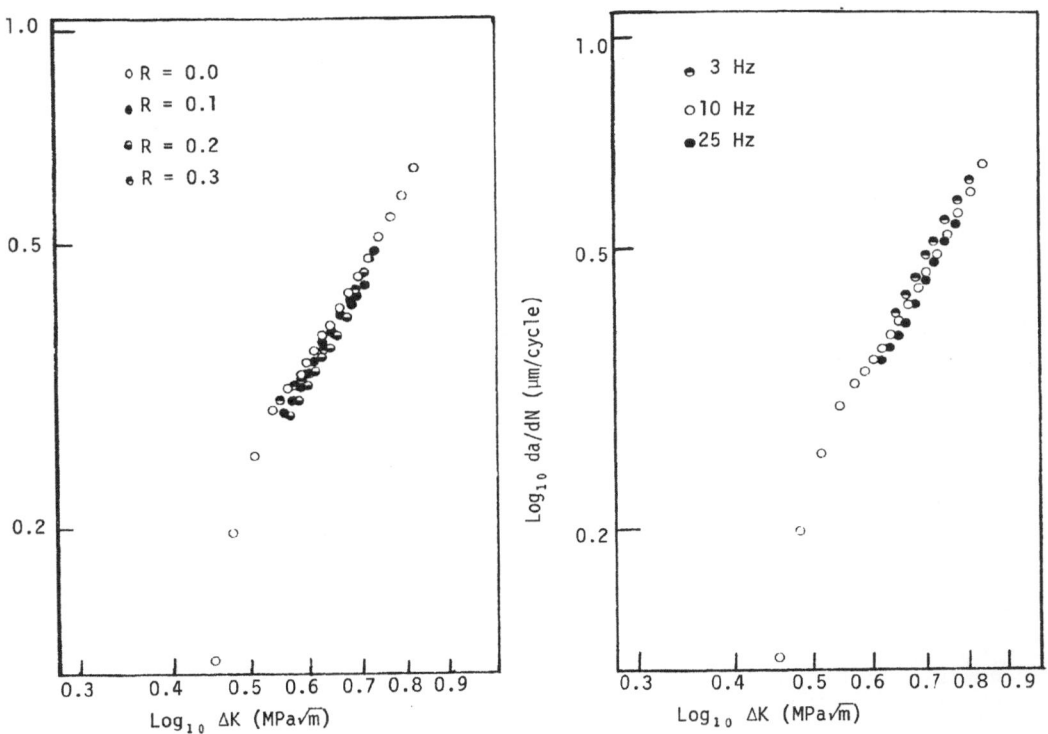

Fig. (2) - Stress ratio effect on
fatigue crack growth in
ethanol

Fig. (3) - Frequency effect on fatigue crack growth in
ethanol

is always continuous and stable. Using equations (3) and (4), the stress reduction factor (α) and the fatigued craze stress ($\alpha\sigma_c$) at the damage zone can be obtained from the results shown in Figures 4 and 5. Table 3 shows that in ethanol,
both α and $\alpha\sigma_c$ are independent of stress ratio and frequency. In particular, it
is noted that the fatigued craze stresses ($\alpha\sigma_c$) are close to those obtained previously by the Dugdale solution and that α is very small implying that there is
a large stress gradient from the damaged to the newly formed undamaged zone at
the craze tip which has a high σ_c value due to large elastic constraints. In a
previous investigation on the environmental fatigue crack growth of polystyrene [6],
one of us has observed that $\alpha\rightarrow0$ so that fatigue crack propagation is really visco-
elastic slow crack growth. In terms of the Paris equation, this gives m→2. It
is suggested there that environment induced crazes are formed at the crack tip

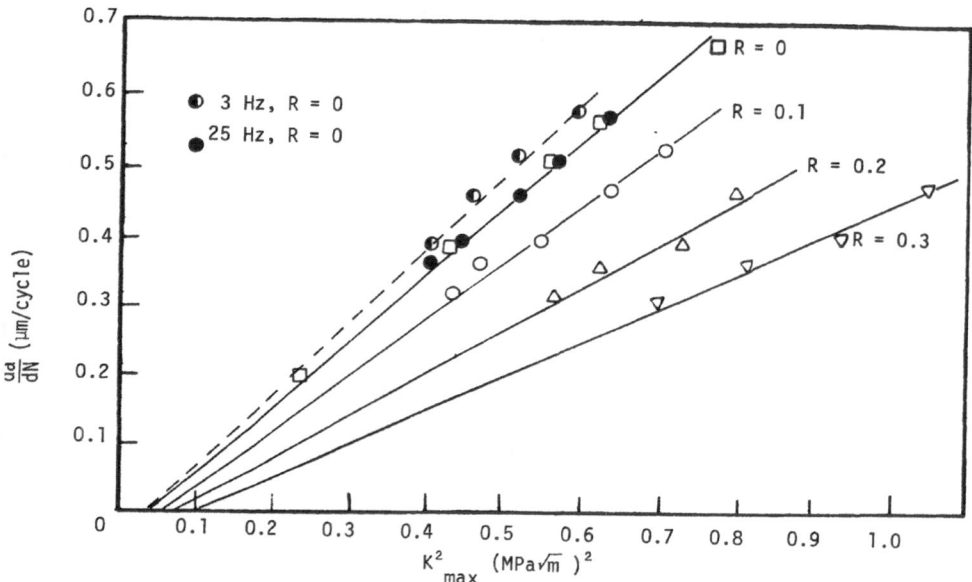

Fig. (4) - Variation of da/dN with K_{max}^2 for HIPS in ethanol at 10 Hz

prior to and not during fatigue crack growth [6]. The present results, however, show that in HIPS, this hypothesis is not valid although on fatigue, the stresses are reduced sharply to the static values indicating the same fracture mechanism. A general conclusion cannot be drawn until further results are obtained for a wide range of other polymers.

If residual compressive stresses are to be considered in equation (1), a term K_o^2 has to be added to the right hand side of equation (4) [8]. Since from Table 3, the product M(R)C(R) is approximately constant, this implies that K_o due to crack closure is close to zero. The fatigue thresholds can be obtained from the intercepts of the straight lines shown in Figure 5 which give $K_t \simeq 0.20$ MPa$\sqrt{m}$ for all alcohols. Agreement with experimental threshold measurements is, however, not very good.

Table 4 shows that the stress reduction factor (α) is independent of the types of alcohol used and is approximately half of that in air. The fatigued craze stresses ($\alpha\sigma_c$) are close to those given in Table 2 and they decrease as the solubility parameter of the solvent decreases thus confirming the plasticization theory for environmental fatigue fracture.

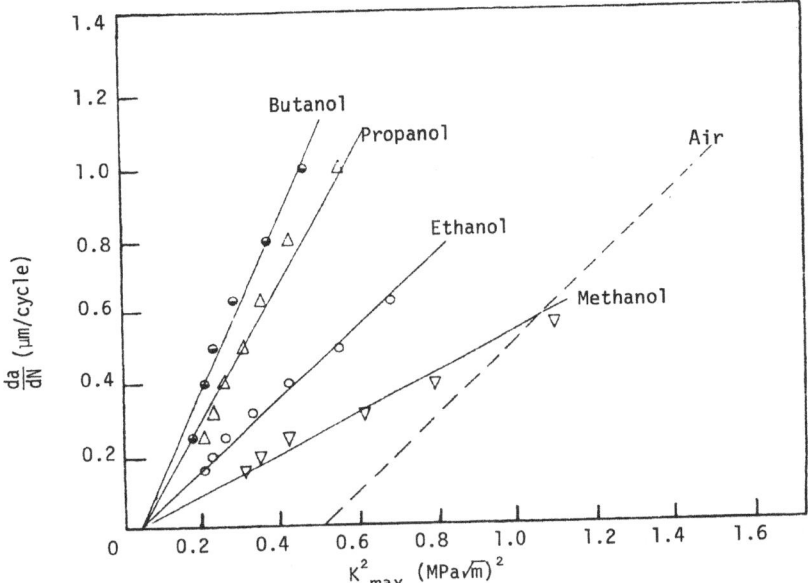

Fig. (5) - A plot of da/dN versus K^2_{max} for HIPS in alcohols at 10 Hz and R=0

TABLE 3 - PARAMETRIC VALUES OF EQUATION (1). (FATIGUE CRACK GROWTH IN ETHANOL K^2_c = 1.55 MPa^{2}m)

Frequency (Hz)	R	M(R) μm(MPa$\sqrt{m}$)$^{-2}$	C(R)	G(R)	α	σ_c (MPa)	$\alpha\sigma_c$ (MPa)
25	0	1.00	0.04	1.00	0.026	650	17
3	0	1.02	0.04	1.00	0.023	650	15
10	0	1.00	0.04	1.00	0.026	650	17
10	0.1	0.80	0.05	0.80	0.026	650	17
10	0.2	0.63	0.07	0.63	0.028	680	19
10	0.3	0.50	0.10	0.50	0.032	650	21

CONCLUSIONS

The effects of the alcohols on the fatigue fracture behavior of a high impact polystyrene polymer are to increase the fatigue crack growth rates and the fatigue thresholds compared to the air results. Stress ratio and frequency have little influence on fatigue crack growth. Using a modified Williams model, fatigued craze stresses can be obtained and they are in good agreement with those values predicted by the Dugdale solution. The proportion of damage by fatigue (α) seems to be independent of environment, frequency and stress ratio.

TABLE 4 - VALUES OF α AND $\alpha\sigma_c$ IN ALCOHOLS. (R=0, FREQUENCY = 10 Hz, K_c^2 = 1.55 MPa^{2}m)

Environment	α	$\alpha\sigma_c$ (MPa)
Air	0.05	49
Methanol	0.026	22
Ethanol	0.026	17
n-Propanol	0.026	12
n-Butanol	0.026	11

REFERENCES

[1] El-Hakeem, H. A. and Culver, L. E., J. Appl. Polym. Sci., Vol. 22, p. 2691, 1978.

[2] El-Hakeem, H. A. and Culver, L. E., Int. J. Fatigue, Vol. 1, p. 133, 1979.

[3] Mai, Y. W., J. Mater. Sci., Vol. 9, p. 1896, 1974.

[4] Hertzberg, R. W. and Manson, J. A., "Fatigue of Engineering Plastics", Academic Press, 1980.

[5] Williams, J. G., J. Mater. Sci., Vol. 12, p. 2525, 1977.

[6] Mai, Y. W. and Williams, J. G., J. Mater. Sci., Vol. 14, p. 1933, 1979.

[7] Yap, O. F., Mai, Y. W. and Cotterell, B., in "Advances in Fracture Research", ICF-5, Vol. 1, Editor: D. Francois, Pergamon Press, pp. 449-456, 1981.

[8] Williams, J. G. and Osorio, A. M. B. A., in "Advances in Fracture Research", ICF-5, Vol. 1, Editor: D. Francois, Pergamon Press, pp. 443-448, 1981.

[9] Yap, O. F., Mai, Y. W. and Cotterell, B., in "Proc. Analytic and Experimental Fracture Mechanics", Editors: G. C. Sih and M. Mirabile, Sijthoff and Noordhoff, pp. 919-930, 1981.

[10] Brown, W. F. and Srawley, J. E., ASTM STP 410, 1966.

[11] McCammond, D. and Hoa, V. S., Polym. Engr. Sci., Vol. 17, p. 869, 1977.

[12] Maxwell, B. and Rahm, L. F., Ind. & Engr. Chem., Vol. 41, p. 1988, 1949.

[13] Kambour, R. P., J. Appl. Polym. Sci., Vol. 11, p. 1879, 1973.

[14] Mai, Y. W., J. Mater. Sci., Vol. 11, p. 303, 1976.

USE OF THERMOMETRIC METHOD OF ANALYSIS FOR EVALUATION OF COLLOIDS FOR 70/30 BRASS IN NITRIC ACID

B. B. Vakil, B. N. Oza and R. S. Sinha

S.V.R. College of Engineering and Technology
Surat-395007, Gujarat, India

INTRODUCTION

Metals and alloys can have spontaneous failure resulting from the combined effects of corrosion and stress. Brass structures are prone to fracture due to combined stress-corrosion effect and the fact that susceptibility to stress corrosion cracking in brass increases with zinc content is well-known. Even general corrosion leads to the reduction in the cross section of metal or alloy structure to the point where it can no longer support applied load. Materials in which stress is not introduced through any external agency, resulting stresses in the already existing microcracks or macrocracks may become high enough to propagate crack either in intergranular or transgranular fashion depending on the condition of material and the type of corrodent. Intercrystalline corrosion penetrating to very shallow pit has been reported to unstressed cartridge brass [1]. According to Graf [2], the alloy at crack tip would be dissolved in the corrodent. Failure of 70/30 brass chain to nitric acid vapour has been reported by Fraser [3]. Corrosion of 70/30 brass in nitric acid can be attributed to mechanical instability arising due to electron concentration. It may be due to pre-existence or formation of a chemically reactive path or paths in the metal that are anodic to the matrix.

Thermometric method have been proved of considerable value and help in studying corrosion behaviour of a number of metals and alloys in various corroding environments [4]. The method is also useful in evaluating the inhibitor efficiency of a number of surface active agents [5,6]. Results obtained by thermometric methods were confirmed by other well-established methods such as weight loss, potential and polarization [5,7] measurements. A. M. Shams Eldin [4] has found that the reaction number RN of metal varies with the composition of corroding solution, both in absence and presence of additives, in such a manner as to allow useful conclusions to be drawn regarding the mechanism of corrosion and inhibition reaction. The variation of the RN with the logarithm of the concentration of the additives in solution is invariably sigmoid in nature. With the inhibitors belonging to same homologous series, the RN-log C curves allow definite conclusions to be drawn regarding the mode of adsorption on the corroding metal. The extent of corrosion and inhibition as determined by the reduction in the RN is the same as that established by weight loss or potentiometric technique [5,7]. In any corroding solution, an additive might cause (i) lowering in RN (inhibitor), (ii) augmentation of RN (accelerator) or (iii) slow effect.

Organic substances of high molecular weight including many colloidal substances appear to form a film over the surface and interfere with the attack whatever its mechanism, partly by hindering the replenishment of corrosive substances and partly in other ways. The first adsorption theory of inhibitive action was proposed by Siverts and Lueg [8]. Kreutzfeld [9] investigated starch, yeast, etc., as inhibitors for steel in sulphuric acid and reported that these substances form a continuous insulating layer on the metal surface. Similar conclusions were reached by Valpert [10] regarding the inhibitive action of gelatine. Evans suggested that many colloidal particles become positively charged in acid solution and hence retardation takes place predominantly in acid solutions. Desai et al studied several colloids as corrosion inhibitor and found that gelatine, acacia, dextrin, agaragar and egg albumin retard the corrosion of α and $\alpha+\beta$ brass in ammonium chloride solutions. Koehler reported the use of the mixture of agaragar and gelatine as inhibitor for corrosion of 70/30 and 60/40 brasses in 0.1N ammonium chloride solution.

In present investigation, an attempt is made to study corrosion behaviour and role of colloids on 70/30 brass in nitric acid environment by thermometric as well as weight loss measurements.

EXPERIMENTAL

For the investigation, brass with the composition Cu-69% Zn-29% pb and Sn less than 0.5% and traces of other impurities were chosen. Test specimen 2.5 cm x 1 cm x 0.2 cm were saw cut from cold rolled and annealed sheet of 0.2 cm thickness. The specimens were cleaned as per the standard procedure by Champion [11]. The volume of the test electrolyte was kept 15 ml. The reaction vessel and procedure for determining corrosion behaviour by thermometric method has been described previously. The initial temperature in all experiments were kept 24.5°C. The nitric acid used was of A.R. grade. The temperature was measured on a calibrated thermometer.

For weight loss measurements, the cleaned specimens of 5.5 cm x 2.5 cm x 0.2 cm were weighed and immersed in 4N HNO$_3$ with and without inhibitors. At the end of the stipulated period, the specimens were removed and cleaned in cleaning solution and weighed. Weight loss is determined in each case and mdd is calculated.

Results are given in Tables 1 and 2 and Figures 1 and 2.

TABLE 1 - INFLUENCES OF NITRIC ACID CONCENTRATION ON REACTION NUMBER
AND TIME REQUIRED TO REACH MAXIMUM TEMPERATURE

Nitric Acid Concentration (N)	RN	ΔT $(T_m - T_i)$	t (Minutes)
0.1	0.003	0.9	300
0.5	0.004	1.2	270
1.0	0.016	4.1	250
2.0	0.033	5.7	190
3.0	0.044	7.1	160
4.0	0.425	17.0	40

TABLE 2 - REACTION NUMBER (RN), (ΔT) AND % INHIBITION

Acid Concentration 4N HNO$_3$				Inhibitor Concentration = 1.5%	
Inhibitor	RN	ΔT ($T_m - T_i$)	t (minutes)	% Inhibition	
				Thermometric Method	Weight Loss Method
-	0.425	17.0	40	-	65.48
Gelatine	0.161	12.9	90	62.11	56.53
Dextrin	0.204	14.3	70	52.00	58.27
Starch	0.198	13.9	70	53.41	50.03
Gum	0.211	16.9	65	50.52	53.42
Agaragar	0.224	15.7	70	47.25	

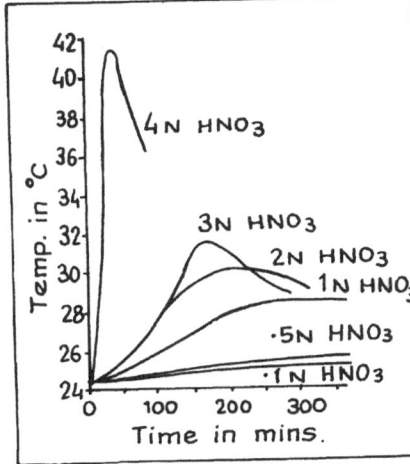

Fig. (1) - Time versus temperature relationship for different concentration of corrodent

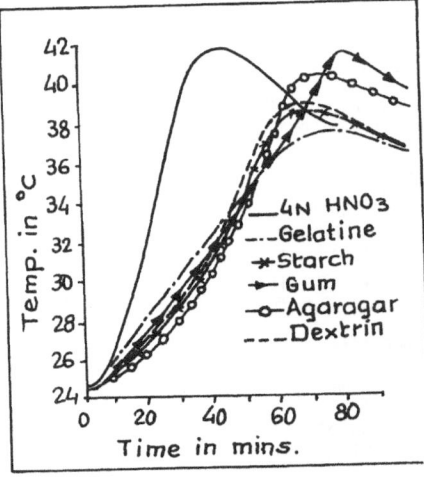

Fig. (2) - Thermometric behaviour of 70/30 brass in nitric acid with and without colloids

DISCUSSION

Percent inhibitor efficiency is calculated using equations

$$i = \frac{w_u - w_i}{w_u} \times 100 \quad \text{(weight loss)} \tag{1}$$

$$i = \frac{RN_F - RN_{Inh}}{RN_F} \times 100 \quad \text{(thermometric)} \tag{2}$$

Reaction Number is calculated as follows:

$$RN = \frac{T_m - T_i}{t} \tag{3}$$

where w_i = weight loss with inhibitor, w_u = weight loss without inhibitor, T_m and T_i are maximum and initial temperature, t = time in minutes to reach T_m, RN_F and RN_{Inh} are reaction number in plain and inhibited electrolytes.

It is clear from Figures 1 and 2 that curves are characterized by an initial period during which the temperature remains constant or varies very slightly. This part of the curves can be described as incubation period in which preimmersion oxide film protects the metal from acid attack or an induction period representing the breakdown of the film and start of attack. It can be noticed that the slope of rising segments of the curve are marked. It is evident from the results presented in Table 1 that increase in acid concentration leads to increase in RN, with rise in T_m value. Higher reaction number indicate the dissolution of brass in the acid solution. Further, it can be said that though there is increase in RN values up to 1N concentration of HNO_3, there is not much effect on the induction period. However, significant effect is observed with higher concentration of HNO_3 on induction period. It is clear from the data presented in Table 2 that with 1.5% inhibitor concentration, the RN can be drastically reduced from 0.425 to 0.160. Gelatine, dextrin, starch, agaragar, gum, etc., can be considered weakly adsorbed inhibitors which cause an increase in time, necessary to reach T_m. However, their inhibitor efficiency can be very well judged by knowing the extent to which RN is affected. The following order can be given for various colloids used in study.

Gelatine > Dextrin > Starch > Agaragar > Gum.

The same order was also reproduced when % inhibition was considered as criteria using weight loss measurements. The narrow difference in % inhibition values on comparison of thermometric data with weight loss measurements established the usefulness of method.

CONCLUSIONS

Evaluation of inhibitor efficiency of various colloids and corrosion behaviour of 70/30 brass in HNO_3 using thermometric method has shown the course of reaction in system. The thermometric results throw light on the following remarkable points during investigations.

1. With lower acid concentration, reaction number RN remains low.

2. Sudden rise in RN is experienced with considerably higher acid concentrations.

3. At lower acid concentrations, the time required to reach maximum temperature was almost constant.

4. With 4N concentration of nitric acid, there is sizeable reduction in the induction period.

5. At 1.5% of inhibitor concentration, there is greater reduction in RN from 0.425 to 0.160 for all the colloids investigated.

6. Gelatine has affected RN maximum thereby leading to highest % inhibition.

ACKNOWLEDGEMENT

The authors express their gratitude to University Grants Commission for granting Junior Research Fellow (B.B.V.) for this work. The authors are thankful to Professor T. P. Sastry, Chemistry Department and college authorities for providing necessary research facility.

REFERENCES

[1] Whitaker, M. E., Metallurgica 39, p. 66, 1948.

[2] Graf, L., Proceedings of the Second International Conference on Metallic Corrosion, Houston, p. 89, 1966.

[3] Fraser, J. P., Metal Protection 2, p. 97, 1963.

[4] Pl. Hosery, A. A., Saleh, R. M. and Shams Eldin, A. M., Corrosion Science 12, p. 897, 1972.

[5] Abdul Waheb, F. M., Saleh, R. M. and El Horsery, A. R., Second International Symposium on Industrial and Oriented Basic Electrochemistry, pp. 6-16, 1-17, 1980.

[6] Abd Kader, J. M. and Shams Eldin, A. M., Corrosion Science 10, p. 551, 1970.

[7] Gouda, V. K., Khedr, M. G. A. and Shams Eldin, A. M., Corrosion Science, 7, p. 221, 1967.

[8] Siverts and Lueg, F., Anorg. Chem., 126, p. 193, 1923.

[9] Kreutzfeld, W., Korrosion and Metallschute 4, p. 104, 1928.

[10] Valpert, G. Z., Phys. Chem. 151, p. 219, 1930.

[11] Champion, F. A., Corrosion Testing Procedure, John Wiley and Sons, New York, p. 188, 1964.

AUGER ANALYSIS OF THE HIGH TEMPERATURE EMBRITTLEMENT OF AN ALPHA BRASS

W. Losch, H. Niehus[*], L. H. de Almeida and S. N. Monteiro

COPPE/UFRJ, C.P. 68505
Rio de Janeiro, CEP. 21.944, RJ, Brazil

ABSTRACT

Auger electron spectroscopy analysis on a leaded high zinc alpha brass specimen, tensile deformed at the ductility minimum temperature, has been performed
to study the possible embrittlement mechanism. It was found that a lead rich
layer is formed at the grain boundaries weakening the atomic bonds and promoting
intergranular fracture.

A dislocation network developed near the grain boundary apparently increases
the efficiency of the lead diffusion causing lower ductility the slower the
strain rate.

INTRODUCTION

From low to moderate temperatures, the alpha brasses present high ductility
associated with other mechanical properties which improve with the Zn content.
For practical purposes, small amounts of Pb are usually added to high zinc alpha
brasses to enhance machinability. Since the maximum solubility of Pb in the
brass matrix is very low, a dispersion of Pb inclusions is formed. At temperatures
around $0.5T_f$ (600K) this class of materials exhibits a minimum in ductility. In
particular, leaded high Zn alpha brass displays ductility minima and become increasingly brittle the lower the strain rate [1].

The phenomenon of high temperature loss in ductility in metals and alloys has
motivated several works including suggestions of theoretical models. One possibility for ductility minima associated with rate dependence would be the effect
of slow strain rate embrittlement, first emphasized by Troiano [2]. In this effect, the cohesive strength of the matrix is lowered by segregation of interstitials at the region near the tip of a dislocation array which then act as a crack
embryo [2]. Fracture could then proceed by the diffusion of more interstitials
to favorable regions such as grain boundaries.

Another model for ductility minima, which could take into account the rate
dependence was that proposed by Rhines and Wray [3]. They attributed the brittle-

---

[*]Permanent address: Institut für Grenzflächenforschung und Vakuumphysik, (IGV),
KFA, Julich, West Germany.

ness of several metals and alloys, including alpha brasses, to grain boundary shearing, GBS, and associated intergranular rupture. At slower strain rates, near the recrystallization temperature, sub-grains are developed and divide the GBS fissures into larger number which are more effective to cause early rupture [3].

Liquid metal embrittlement, LME, due to PB has been proposed [4,5] as a strong possibility for explaining the ductility minima in brasses. This idea is stressed by the fact that the melting temperature of Pb, 600K, corresponds more or less to the temperature of greater reduction in ductility. This type of embrittlement is associated with a ductile to brittle transition due to the reduction of the work of a propagating fracture when liquid Pb is present at the crack tip.

On the other hand, Sanchez-Medina et al [6] disagrees with the LME mechanism [4,5] and, based on pure Cu and leaded Cu mechanical tests results, suggested essentially the same GBS model of Rhines and Wray [3]. In their case [6], the loss of ductility of leaded Cu would be due to the presence of Pb in the form of finely dispersed (solid or liquid) inclusions. They also claimed that their mechanism of embrittlement by inclusion could equally well be used to leaded brasses.

It then appears that an explanation for the ductility minimum in leaded high Zn alpha brass is still open to discussion. Moreover, the strong rate dependence of the phenomenon deserves a more thorough analysis. This was attempted in the present work through an Auger electron spectroscopy investigation.

EXPERIMENTAL PROCEDURE

A specially cast leaded high Zn alpha brass with the following weight composition: Cu - 64,1%, Zn - 35.5%, Pb - 0.125% and Fe - 0.037% was used in this investigation.

The ductility minima were determined with round tensile specimens with 38 mm of gage length and 6 mm of gage diameter. These specimens were annealed at 773K resulting in an equiaxed structure with grain size of 100μm. Tensile tests were conducted in vacuum at temperatures up to 673K and strain rates from 3.5×10^{-5}/s to 3.5×10^{-2}/s.

For the Auger electron spectroscopy analysis, flat 0.2 mm wider tensile specimens, specially spark-cut machined, were deformed to rupture at 623K inside the high vacuum chamber of a Scanning-Auger-Microprobe equipment. These flat specimens were initially rolled and then annealed at 673K resulting in a equiaxed structure with an average grain size of 10μm. Specially designed grips allowed deformation of the specimens at a controlled strain rate which in the present case was 2×10^{-5}/s.

RESULTS

Figure 1 shows the temperature dependence of the elongation for the round specimens tested from 300 to 673K at different strain rates. The loss in ductility starts at 500K and minima are observed from 600 to 640K corresponding to lower uniform elongations the slower the strain rate. Decreasing the strain rate, a tendency for these minima to be shifted to higher temperature is also observed in Figure 1.

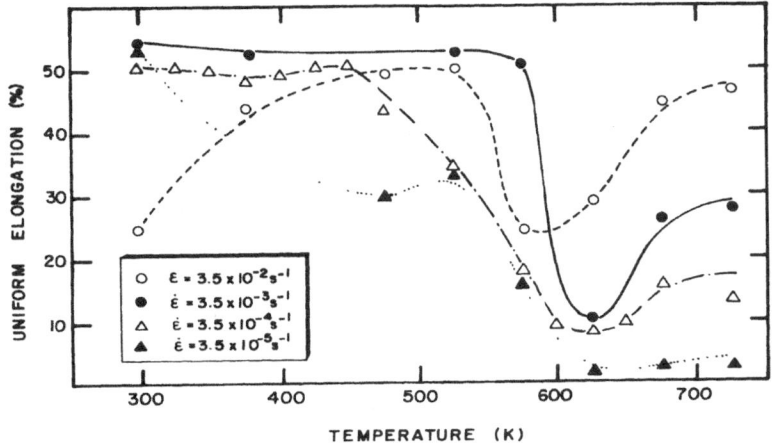

Fig. (1) - The temperature dependence of the uniform elongation,
showing ductility minima around 600K

Figure 2 depicts the fracture surface of round specimens broken at their temperature of minimum in elongation at the strain rates of (a) 3.5 x 10^{-2}/s and
(b) 3.5 x 10^{-5}/s. It should be noted that the dimple covered surface corresponding to the faster strain rate, Figure 2(a), is characteristic of a ductile frac-

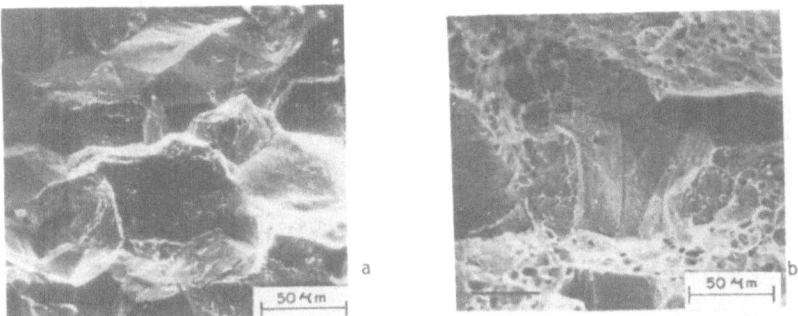

Fig. (2) - Fracture morphology at 623K for the standard round tensile
specimens corresponding to strain rates of (a) 3.5 x 10^{-2}/s,
and (b) 3.5 x 10^{-5}/s

ture and in good agreement with the associated elongation of 25%. On the other
hand, the faceted surface associated with the slower strain rate, Figure 2(b), is
a typical intergranular fracture.

582

A similar fracture surface appearance of that shown in Figure 2(a) was obtained with the flat specimen deformed at 623K inside the Auger chamber, as depicted in Figure 3.

Fig. (3) - Fracture morphology at 623K for the special flat specimens
tensile deformed at a strain rate of 2×10^{-5}/s

The advantage of a Scanning-Auger-Microprobe equipment is that surface elements can be analyzed at local desired spots, lines or areas. The Auger electron spectroscopy analysis at several regions of the flat specimen revealed the existence of Ph together with Zn and Cu. Figure 4 is a typical Auger curve for the regions analyzed. The height of the ever present Pb peak indicates that this element is uniformly spread on the surface with a relatively high concentration.

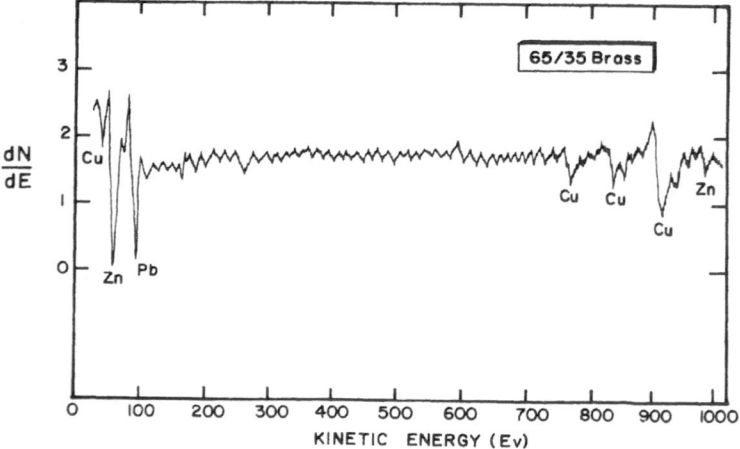

Fig. (4) - Typical Auger electron spectroscopy curve at the fracture surface
of the special flat specimen

Through sputtering technique, the composition gradient form the surface to the bulk of the specimen can also be determined. Figure 5 presents the typical

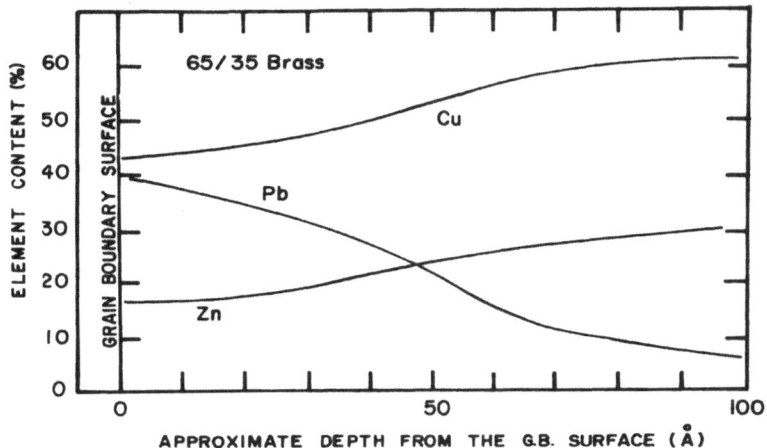

Fig. (5) - Variation of the elements content with depth during sputtering at the fracture surface of the special flat specimen

composition gradient obtained from the intergranular fracture of a flat specimen tested at 623K to a depth of approximately 100 Å. It is important to notice in Figure 5 that the Pb content is very high, 40% at the surface and decreases continuously to the bulk while both Zn and Cu contents increase towards their average value in the alloy.

DISCUSSION

The Auger electron spectroscopy analysis on the fracture surface corresponding to the ductility minimum shows strong evidences of embrittlement in high Zn brass. The spectrogram, in Figure 4, and the concentration curves obtained by sputtering technique, in Figure 5, demonstrate that straining at 623K result in a Pb rich layer at the grain boundary. This is different than the suggestion [4] that Pb would spread under stress forming a liquid Pb film that reduces substantially the surface energy and the related fracture stress.

Moreover, a careful survey of the fracture surface at this condition failed to reveal any inclusion which could support the GBS model [3,6]. It is also worth to mention that the absence of interstitial, Figure 4, at the fracture surface would rule out the possibility of slow strain rate embrittlement as proposed by Troiano [2].

The relatively thick layer with continuously decreasing Pb content form the intergranular fracture surface indicates that a diffusion process was effectively occurring during straining at 623K. This process was probably taking Pb atoms from inclusions inside the grains as well as those existing at the boundaries.

The Pb atoms could not come only from inclusions at the boundaries to justify the amount of Pb in the layer. Considering the average percentage of Pb in the layer and the grain size of alloy, one sees that more than one third of all Pb atoms have diffused to the layer.

This diffusion process is only effective with straining since annealing at 623K followed by tensile deformation at room temperature gives a ductile rupture associated with Pb inclusions. Had the Pb rich layer been formed at the boundaries during annealing, fracture at room temperature would also be intergranular. Similar conclusion was also found by Eborall and Gregory [4].

An attractive explanation for this Pb embrittlement associated with a diffusion process to the boundaries during straining could be found in the Frank-Graf theory [7,8] in connection with Frank's idea [9] on deformation induced LME.

According to this theory [7,8], the dislocation density in alloys is specially high in the vicinity of perturbed regions such as grain boundaries or surfaces. Frank [9] suggested that liquid metal could diffuse, by pipe diffusion through such a grid-like misfit dislocation network. He also predicted [9] a more effective dynamic situation in the case of plastic deformation. During plastic flow, this dislocation network near the grain boundaries operates as sources for new dislocations. These new dislocations and their steps left at the boundaries would act as efficient pipes for Pb diffusion.

Therefore, it is proposed that in alpha brasses, Pb is carried to the grain boundaries mainly by pipe diffusion through dislocations created with the plastic deformation. This diffusion forms a Pb rich layer at the expenses of inclusions existing both in the bulk and at the boundaries.

The lower solubility of Pb in the brass matrix would not permit the formation of a homogeneous mixture with Zn and Cu. On the other hand, the high included (dihedral) angle of contact do not favor the formation of a Pb film [4]. The only way Pb could be dispersed in a relatively thick layer, Figure 5, is by staying at the dislocation lines forming a network with the atoms tending to occupy the more "open" structures, associated with higher lattice distortion, close to the grain boundaries.

The weak Pb atomic bonds at 623K (melting point of Pb is 600K), promote loss of atomic cohesion at regions of higher Pb content associated with the grain boundaries. Fracture stress will probably depend on the amount of Pb which has accumulated at the boundaries. At faster strain rates, diffusion is not effective, Figure 2(a), to pump enough Pb to the grain boundaries and a relatively large plastic deformation has to be reached. In this case, partially ductile fracture occurs. At slower strain rates, Figures 2(b) and 3, Pb can effectively diffuse towards the grain boundaries right at the beginning of plastic flow and grain boundary embrittlement soon occurs.

A word must be said about the tendency to increase ductility above 623K. Increasing the temperature above $0.5T_f$ will promote dynamic recovery of the dislocation network near the grain boundaries and decrease the efficiency of the Pb to form the embrittlement layer.

CONCLUSIONS

The embrittlement of a leaded high Zn alpha brass has been investigated by Auger electron spectroscopy. It was found that:

- Pb is responsible for the ductility minima observed around 600K through the formation of a Pb rich layer at the grain boundaries causing intergranular fracture.

It is proposed that the Pb rich layer be effectively formed by pipe diffusion through a dislocation network developed near the grain boundary following the activation of sources related to misfit dislocations.

At slower strain rates, the Pb is very effective in reducing the ductility. At higher strain rates, larger amounts of plastic deformation occurs before enough Pb has diffuse to the boundaries.

Increasing the temperature, the dislocation network dynamically recovers and the Pb efficiency to promote embrittlement also decreases.

ACKNOWLEDGEMENTS

The authors acknowledge the support to this investigation from the following Brazilian agencies: CNPq, FINEP and CEPG/UFRJ, and the German Volkswagen Foundation.

The author would also like to thank Marise de Albuquerque Lima and Miguel Nehme Saad for their help in practical aspects concerning this investigation.

REFERENCES

[1] Monteiro, S. N., Lima, M. A. and Almeida, L. H., "Conditions for minimum ductility in 65/35 brass as a function of temperature and strain rate", Proceedings of the VI Brazilian Congress on Mechanical Engineering, Rio de Janeiro, pp. 93-101, December 1981.

[2] Troiano, A., Trans. ASM, Vol. 52, pp. 54-80, 1960.

[3] Rhines, F. N. and Wray, P. J., Trans. ASM, Vol. 54, pp. 117-128, 1961.

[4] Eborall, R. and Gregory, P., J. Inst. Metals, Vol. 84, pp. 88-90, 1955.

[5] Roth, M. C., Weatherly, G. C. and Miller, W. A., Canadian Met. Quart. 18, p. 341, 1979.

[6] Sanchez-Medina, M., Sangiorgi, R. and Eustathopoulos, N., Scripta Met., Vol. 15, pp. 737-738, 1981.

[7] Frank, W. and Graf, L., Z. Metallkde, Vol. 66, p. 555, 1975.

[8] Frank, W. and Graf, L., in "Surface Effects in Crystal Plasticity", ed. R. M. Latanision and J. F. Fourie, NATO Advanced Study Institutes Series: Applied Science, No. 17, Noordhoff, Leyden, p. 781, 1977.

[9] Frank, W., Z. Metallkde, Vol. 71, pp. 564-567, 1980.

FATIGUE IN LUBRICANTS ENVIRONMENT

P. Poudou, G. Nouail, J. Ayel

Institut Francais du Pétrole
France

and

G. Pluvinage

Université de Metz
Ile du Saulcy - 57000 Metz, France

ABSTRACT

The lubricant as environment modifies the fatigue behaviour, particularly in fatigue contact. The lubricant influence is studied:

- on four ball machine in hertzian contact. Fatigue life is modified by additives or their concentration;

- on fatigue crack propagation which shows a strong influence of lubricant and temperature effect;

- on crack initiation on which lubricant has a beneficial effect.

Fractography and surface analysis with LAMMA micro-probe analysis indicate that water contained in oil has an effect of stress corrosion on fatigue crack propagation.

INTRODUCTION

Rolling contact fatigue appears on the surface of components such as cams and toppets, gears with cyclic loading. The first step is the formation of little pits and flakes which can develop and finally cause the failure of the machine. Generally, this phenomenon is studied by rolling contact tests conducted on four-ball fatigue machine, full scale ball bearings and thrust bearing rigs, gear rigs, discs machines, cams and toppets rigs, etc. The interpretation of these results is difficult due to the interaction of several parameters, but these tests are convenient to rank lubricants and additives quickly and easily.

The lubricant has two major influences:

- influence on rheological and tribological properties;

- influence on fatigue mechanism.

The lubricant controls the operating conditions of lubricated contacts such as oil-film thickness, friction coefficient, boundary film adherence and surface wear. These lubricant parameters are known to have a strong influence on fatigue pitting by disturbing stress condition and temperature contact. Lubricant influence on fatigue mechanism are not well-known. Galvin and Maylor [1], Kennel [2] and Armstrong et al [3] have studied the influence of the lubricant on the total life of steel specimen in rotating beam fatigue. Polk, Murphy and Rowe [4] are the unique authors studying the lubricant effect on the propagating phase with notched rotating specimens.

Considering that friction leads to introduce tensile stress in the contact stress distribution and the maximum value of these stresses are proportional to the friction coefficient μ ($\sigma_{max} = 2\mu P_{max}$) with P_{max} = maximum hertzian pressure test, and considering that tensile stresses have the major effect on crack propagation, we have studied lubricant influence in fatigue crack process in opening mode.

A particular attention was paid to initiation state and to propagation stage by using specific fatigue tests.

Fatigue Tests with the Four-Ball Machine

In order to show the effect of the lubricant on the rolling fatigue phenomenon, we start testing lubricants with a modified four-ball machine. The modification is such that the three lower balls are allowed to rotate by a change in the design of the ball race. The procedure is derived from the one described in the IP-300-78 Standard. The rotation speed of the upper ball is 1,500 rpm and the applied load is 600 daN. The ball diameter is 12.7 mm. These balls are made of 100 C6 steel. The criterion of the test is the appearance of the pitting on the upper ball. The detection of this pitting is made by an accelerometer fitted on the handle of the ball housing. We use a ball spacer for the three lower balls in order to reduce the vibration level. These results are scattered in the range of 0,1h to 2h and a great number of tests are required. We use to describe time to failure, a probabilistic function given by Weibull [5]:

$$F(t) = 1 - e - (\frac{t-\gamma}{\eta})^{\beta} \tag{1}$$

with $t \geq \gamma$; β shape parameter characterizing shape of the distribution function; η scale parameter which is the mean life for unreliability $Q(t) = \frac{e-1}{e} = 0.632$; γ localizing parameter which represents minimum life duration. After classification of time to failure, unreliability $Q(t_i)$ is calculated with Kimball's method [6]:

$$Q(t_i) = \frac{i}{N+1} \tag{2}$$

where i is rank and N is number of values.

After plotting values on Weibull's graph time to failure data, we obtain the mean life duration L_{50} (50% of non-failure probabilities), and β, η coefficients. The results are reported in terms of relative life which is the ratio of the L_{50} life duration obtained with a given additive to the L_{50} life with basic oil. The additives used are:

- phosphorous compounds: a diphenyl phosphate and a triaryl phosphate;

- sulfur compounds: a sulfurized isobutylene and ditertiododecyl.

For each group of products, we note a strong variation of the relative life with low concentrations. For the concentrations at which these products are commonly used (> 1%), these compounds generally increase life duration. Below these concentrations, the influence is more complex.

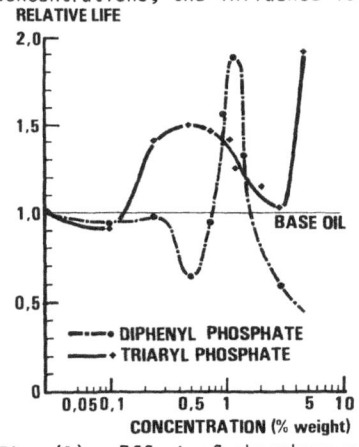

Fig. (1) - Effect of phosphorous compound on the L_{50} mean life

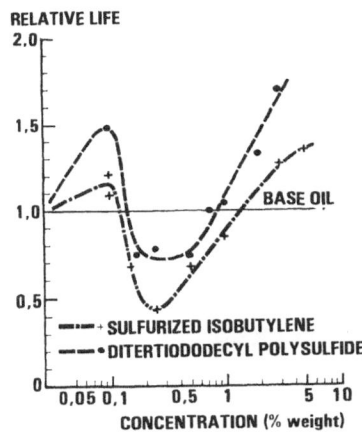

Fig. (2) - Effect of sulfur compound on the L_{50} mean life

Influence of Lubricant on Fatigue Crack Propagation

Crack propagation tests were led in lubricant environment on 35 NCD 16 steel. The chemical composition is the following (in percentage):

C	Mn	Ni	Si	Cr	Mo	P-S
0.3-0.37	0.3-0.6	3.7-4.2	0.1-0.4	1.6-2	0.3-0.5	<0.035%

Tests were performed on CTW 80B 20 samples (20 mm thick) put into an immersion cell filled with a lubricant which is circulated around the specimen. The crack is first initiated in the air, then the lubricant is introduced into the cell and the propagation starts under load frequency of 5 Hz and a load ratio R = 0.1.

590

Crack growth measurement was achieved with an optical method. Results are plotted in a bilogarithmic diagram for the variation of the crack growth rate da/dN versus stress intensity factor range ΔK.

The following parameters were studied (basic oil, additives, temperature, viscosity):

- The effect of basic oil on crack propagation rate as reference; tests were also led in the air and the crack growth rate curve was linear according to Paris law. In presence of 200 NS basic oil, at 80°C, the crack growth curve deviates from linear shape, Figure 3. The maximum deviation is in intermediate stress intensity factor range. At low and high ΔK values, the two curves are quite the same. This generally indicates an environment influence. Residence time of lubricant is reduced at low and high crack growth rates.

- The effect of additives: we have also added three additives in the basic oil:

- ditertiododecyl-polysulphide (S.A.) (3% by weight);

- triaryl phosphate (S.A.) (2% by weight);

- chlorine additive (CA) (3% by weight).

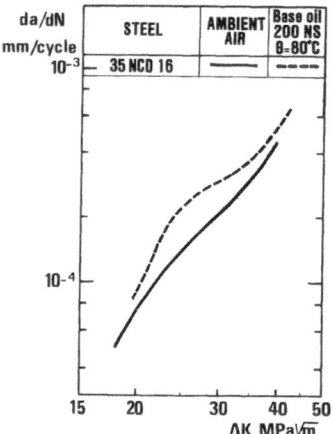

Fig. (3) - Effect of the lubricant on the crack propagation rate

Tests were conducted at 80°C. The results are reported in terms of relative speed versus logarithm of the stress intensity factor. The relative speed is defined as the ratio of the speed measured in a given environment to the speed measured in the air. The results show that the behaviour of lubricants containing these three additives are quite similar; they present a maximum effect in intermediate stress intensity factor range. Ditertiododecyl-polysulphide has a lower influence.

- The effect of temperature on crack growth rate in lubricant environment. The 40°C - 100°C range was investigated with the basic oil. Results are presented in the same manner: relative crack growth rate versus logarithm of stress intensity factor range.

- The effect of viscosity. Tests were conducted with straight mineral oils of different viscosities at 80°C (100 NS, 200 NS, 350 NS, 600 NY). Their viscosities are given in the following table:

Oil	100 NS	200 NS	350 NS	600 NS
Viscosity at 80°C (cst)	6.0	9.8	15.6	21.7

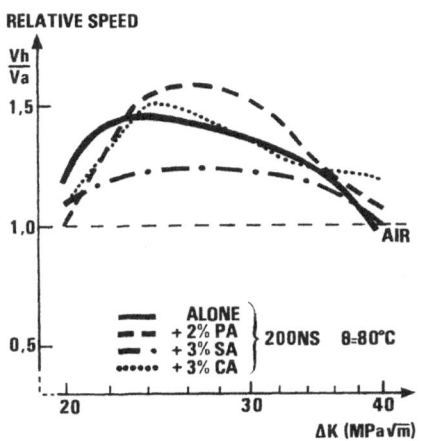

Fig. (4) - Effect of additives on the crack propagation rate

Fig. (5) - Effect of temperature on crack propagation rate

It seems that viscosity is not a governing parameter, Figure 6.

Crack Initiation

Tests for studying crack initiation are conducted on the same CT specimens. Mechanical slots have different acuteness and are more blunt than for crack propagation tests (ρ = 2.5 mm and ρ = 1 mm). The same immersion cell is used. The loading frequency is 10 Hz. The behaviour of the materials towards initiation is expressed by the number of cycles required to initiate fatigue crack. We use for this determination a crack opening displacement gauge which is fixed in the front of the notch. Before any crack develops in the specimen, the signal delivered by the gauge is constant. When fatigue cracks start to develop in the specimen, the signal increases. We determine the number of initiation cycles by

592

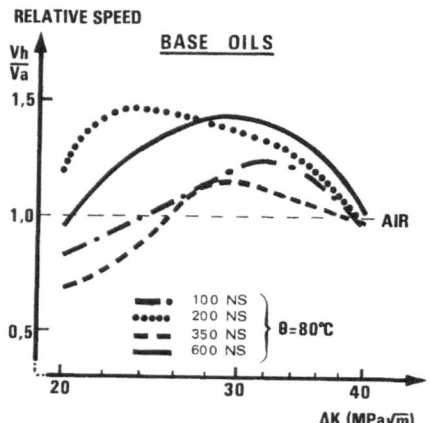

Fig. (6) - Effect of lubricant viscosity on crack propagation rate

the signal changement. A series of tests conducted in ambient air are used as reference. Tests have been led in 200 NS and 200 NS oils plus 3% of a chlorinated wax.

The results, Figure 7, are reported in terms of the number of cycles for fatigue crack initiation Ni, versus the $\Delta K/\sqrt{\rho}$ parameter used by many workers [7]. We find a beneficial effect of lubricant environment slightly higher than the scattering band which we can expect in this kind of test.

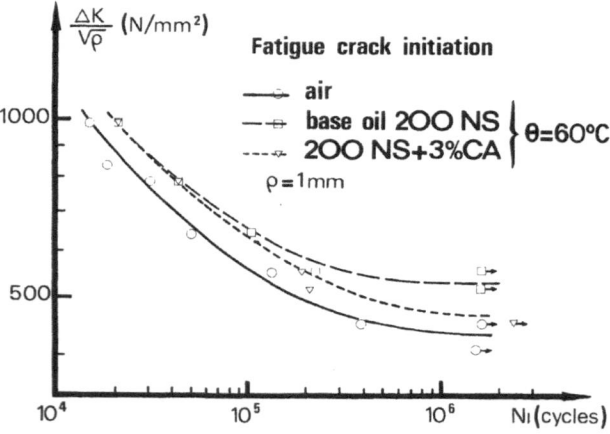

Fig. (7) - Effects of lubricant on the fatigue crack initiation

DISCUSSION

Fatigue tests with the four-ball machine are difficult to interpret because of scattering data. A statistical method is required for plotting data. This test is also a mixture of lubricant and mechanical influences on crack initiation and propagation. The most interest of this test is to select additives and choose the right concentration. The influence of lubricant on crack propagation gives an acceleration of crack propagation with temperature. The maximum effect is for 60°C. An observation of the fracture surface shows a corrosion effect at 40, 60°C. On the contrary, at 80°-100°C, the fracture aspect is similar to that one obtained after testing in the ambient air. The SEM fractographs show intergranular failure areas for the tests conducted at 40°C and 60°C, whereas at 80° and 100°C, we found the same pattern as that one obtained with the tests conducted in the air (ductile tearing fracture mode, Figure 8).

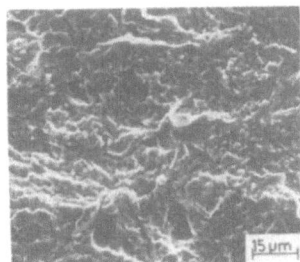 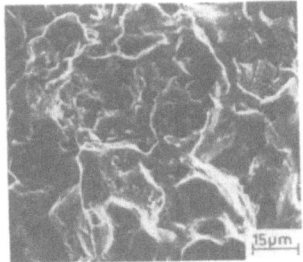

Fig. (8) - SEM micrographs of the fracture surfaces
(a) Test run in ambient air; (b) Test run
in 200 NS oil at 60°C

Surface analysis with LAMMA microprobe analysis (Laser Microprobe Mass Analysis) indicates the presence of iron oxide. We assume that the acceleration of crack propagation is due to a fatigue corrosion mechanism induced by water in oil.

For verification, we make a fatigue crack propagation test in dehydrated oil with molecular screen (20 ppm of water). Crack propagation decreases in comparison with the oil containing 80 ppm of water, but it seems that very few quantities of water in oil are enough to induce stress corrosion fatigue.

The effect of temperature can be explained by a decreasing of viscosity of 200 NS oil (39 cst at 40°C; 6.1 cst at 100°C) and the decreasing of water contents with temperature.

The relative beneficial effect of lubricant on crack initiation can be explained by exclusion of humidity pitting and environment protection. This result is an experimental controversy to literature assumption generally presented. Lubricant influence has not been found on crack initiation but on crack propagation. All tests are conducted in Mode I loading, assuming that lubricant influence, as environment agent, is the same in any kind of loading. Lubricant has also a strong influence on mechanical loading in contact fatigue. Stress distribution is influenced by friction coefficient, tensile stress, and important shearing stress are present in pressure contact. Stress intensity and distribution are modified by the friction coefficient and crack propagated in mixed mode condition (I+II).

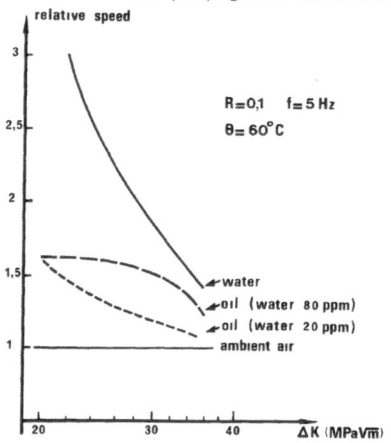

Fig. (9) - Comparison of fatigue crack propagation with water contents

In this condition, verification of the lubricant influences on crack propagation through rupture criteria is required. The modification of crack propagation allowed spalling phenomena on surface.

CONCLUSIONS

Tests with a four-ball fatigue machine are very convenient to test lubricant and their additives in terms of mean fatigue life (L_{50}).

To explain lubricant influence as environment on fatigue, conventional tests are led.

We notice a small beneficial effect on crack initiation.

Fatigue crack propagation is very sensitive to lubricant.

The increasing of crack speed is depending upon the temperature and stress intensity factor range.

It seems that water contents have a stress corrosive effect during each propagation under lubricant environment.

REFERENCES

[1] Galvin, G. and Naylor, H., "Effect of lubricants on the fatigue of steel and other metals", Proc. Inst. Mech. Eng., Vol. 179, Pt. 1, No. 27, pp. 857-875, 1964, 1965.

[2] Kennel, M., Influence de Divers Lubrifiants sur la Résistance à la Fatigue des Métaux, Thesis, University of Paris, 1966.

[3] Armstrong, E. L., Leonardi, S. L., Murphy, W. R. and Wooding, P. S., "Evaluation of water accelerated bearing fatigue in oil lubricated ball-bearings", Lubrication Engineering, Vol. 34, 1, pp. 15-21, 1978.

[4] Polk, C. J., Murphy, W. R. and Rowe, C. N., "Determining fatigue crack propagation rates in lubricating environments through the application of fracture mechanics technique", ASLET Trans., Vol. 18.4, pp. 290-298, 1975.

[5] Weibull, W., "A statistical distribution function of wide applicability", Journal of Applied Mechanics, Vol. 18, pp. 293-297, 1951.

[6] Kimball, B. F., "On the choice of plotting positions on probability paper", J. Am. Stat. Ass., 1960.

[7] Baus, A., Lieurade, H. P., Sanz, G. and Truchon, M., Etude de l'amorcage des fissures de fatigue sur des éprouvettes en acier à très hautes résistance possédant des défauts de formes et de dimensions différentes. Revue de Métallurgie, October 1977.

APPLICATION OF AUGER ELECTRON SPECTROSCOPY TO SURFACE ANALYSIS OF FRACTURES PRODUCED BY HYDROGEN EMBRITTLEMENT

W. R. Broughton[*], P. J. K. Paterson

Royal Melbourne Institute of Technology
Melbourne, Australia

and

W. J. Pollock

Department of Defence Support
Melbourne, Australia

ABSTRACT

Modifications to an Auger Electron Spectrometer were undertaken to permit surface analysis of fractures produced by hydrogen embrittlement. Contamination problems associated with the fracture of steel specimens in hydrogen gas inside the spectrometer were overcome by hydrogen-precharging the specimens and fracturing in vacuum. The technique was used to correlate the stress-corrosion susceptibility of D6ac steel with the degree of impurity segregation at grain boundaries and as a tool for identifying the cause of premature failure of an EN26 steel bolt.

INTRODUCTION

Auger Electron Spectroscopy (AES) has been used extensively to study the detrimental influence of impurity and solute elements on grain-boundary fracture of temper-embrittled steels [1-4]. A number of elements (Bi, S, Sb, Se, Sn, Te, P, As, Ge, Si and Cu) are known to segregate to the grain boundaries and promote temper embrittlement in steels [5,6]. Recent technological developments in high-strength low-alloy steels have significantly increased temper-embrittlement resistance in these alloy systems, although intergranular cracking can often be induced by either stress-corrosion cracking or hydrogen embrittlement. It has been postulated that impurities at the grain boundary act as a catalytic poison for hydrogen recombination, thereby leading to increased concentrations of atomic hydrogen which cause intergranular fracture in the presence of an applied stress [7]. An alternative mechanism has been proposed for stress-corrosion cracking in the case of mild steel. This mechanism suggests that the cracking results from the combined

---

[*] Present address: Aeronautical Research Laboratories, Melbourne, Australia.

action of applied stress and preferential dissolution at the grain-boundary due to the establishment of a galvanic potential between the matrix and the impurity-rich grain boundary [8,9].

Progress in relating grain-boundary chemistry with susceptibility to stress-corrosion cracking and hydrogen embrittlement has been limited by difficulties in obtaining contaminant-free intergranular fracture surfaces for AES analysis. One successful technique involves termination of the stress-corrosion test prior to insertion of the specimen inside the AES chamber with final fracture being induced at a low temperature and 10^{-8} Pa pressure [8]. Small sections of the pre-existing intergranular stress-corrosion crack sometimes extend up to one grain into the virgin steel before cleavage fracture occurs. Although these few exposed grains can be detected using elemental scanning Auger analysis with a 5 μm spot size, the small electron beam diameter reduces the Auger signal strength and signal-to-noise ratio compared with a larger beam diameter (~ 200 μm) which is generally employed in AES analysis. In addition, analysis of only a few grain facets may not produce values representative of the average grain-boundary composition.

In many cases where intergranular stress-corrosion cracking occurs in aqueous environments, the same failure mode can be induced in hydrogen gas or using hydrogen-precharged specimens loaded in air. If hydrogen-assisted intergranular subcritical crack growth can be induced in the AES spectrometer without subsequent contamination of the fracture surface, then average concentration levels of segregated species at the grain boundaries should be obtainable over large areas of the fracture surface using the optimum resolution produced by a large diameter electron beam (200 μm). Results obtained with hydrogen-charged D6ac and EN26 steel specimens are shown to compare favourably with a similar technique where intergranular subcritical crack growth is induced in the Auger spectrometer using uncharged specimens loaded in hydrogen gas.

EXPERIMENTAL

The D6ac steel was commercially produced as 22.8 mm thick cross-rolled plate with the following composition (wt %): 0.45C, 0.75Mn, 0.22Si, 0.004P, 0.005S, 1.10Cr, 0.67Ni, 1.0Mo, 0.09V, balance Fe. Heat-treatment involved austenitization in salt at 930°C for 30 min., quenching into oil at 60°C, cooling to 25°C and tempering in argon at 520°C for either 2, 20 or 200h. Work was also undertaken with material taken from an aircraft-propeller-blade retaining bolt which failed intergranularly during service. The bolt was manufactured from EN26 steel and had the following composition (wt %): 0.4C, 0.59Mn, 0.76Cr, 0.49Mo, 2.55Ni, 0.02P, 0.02S, balance Fe. Heat-treatment of this steel involved austenitization at 830°C for 1h, quenching in oil at 25°C, tempering at 550°C for 1h and quenching in oil at 25°C. The effect of oxygen contamination on the measured AES signals of segregated impurities at grain boundaries was studied using a temper-embrittled steel with the following composition (wt %): 3.5Ni, 1.7Cr, 0.3C, 0.06P, balance Fe. The steel was austenitized in vacuum for 1h at 1000°C, oil quenched, tempered for 1h at 650°C in vacuum and aged at 520°C for 240h in vacuum prior to quenching in oil.

AES specimens of dimensions 23 x 4 x 4 mm were prepared with a 0.5 mm deep notch positioned 7 mm from one end of the specimen. All specimens were metallographically polished and solvent cleaned. Both the D6ac and EN26 steel specimens were hydrogen-precharged for 24h in 0.1M H_2SO_4 containing 50 mg/l sodium arsenite at a cathodic current density of $8mA/cm^2$ prior to notching and mounting in the Auger spectrometer carousel holder. Two methods were used to produce intergranular fracture in D6ac steel:

1. Fatigue pre-cracked AES specimens were inserted in a carousel holder and mounted in a 50 litre AES chamber which was evacuated to 10^{-8} Pa after bakeout using sorption and ion pumps. Specimens were incrementally loaded in 10^5 Pa H_2 gas (99.995% purity) at 25° until subcritical crack propagation occurred. After final fracture, a rotary pump was used to remove hydrogen through a series of liquid nitrogen traps prior to swtiching to the sorption and ion pumps. The total time taken to pump down from 10^5 Pa to 10^{-5} - 10^{-6} Pa and to align exposed fractures in front of the cylindrical mirror analyser for subsequent AES analysis of 5-10 separate sites was 45 min.

2. Hydrogen precharged D6ac and EN26 steel specimens were inserted individually on the carousel inside the AES chamber. A manipulator was used to transfer the specimen into a separate cold stage which was cryogenically cooled to -90°C, after evacuation to a pressure of 5×10^{-2} Pa, to prevent desorption of hydrogen from the precharged steel. When the vacuum pressure reached 10^{-6} Pa, the specimen was transferred back to the fracture device and allowed to warm to 25°C prior to incremental loading of the specimen to induce subcritical intergranular crack growth. The specimen was then broken open by impact loading and AES analysis performed on 5-10 separate sites on the fracture surface. The cold stage was kept at -90°C during both fracture and AES analysis to condense water vapour and other gaseous contaminants.

The temper-embrittled Ni-Cr specimens were cooled to -90°C inside the AES spectrometer and fractured by impact-loading at a vacuum pressure of 10^{-7} - 10^{-8} Pa. AES analysis of the fracture surface was then undertaken at increasing times after fracture and the level of segregated P at the grain boundary determined for increasing levels of oxygen contamination on the fracture surface.

AES analysis was conducted using a Varian Scanning Auger Microprobe with a cylindrical mirror analyser, Figure 1. A 3keV, $10\mu A$ primary electron beam with a spot size of 200 µm diameter was used to irradiate the fracture surface. Conventional AES spectra (EdN(E)/dE versus E) were obtained using a modulation voltage of 5eV peak-to-peak and a frequency of 13kHz. Spectra were recorded at 11eV/s sweep rate using a lock-in-amplifier (LIA) with a sensitivity setting of 30mV and a time constant of 400 ms. Normalized measurements of P, Ni and Mo were recorded as a Peak-Height-Ratio (PHR) relative to 703eV Fe transition, e.g., (P_{120PPH}/ Fe_{703PPH}). The measurement for Cr was taken as negative peak height to background and expressed as a ratio relative to Fe_{703PPH}. Scanning Electron Microscopy (SEM) was used to determine the degree of intergranular fracture.

RESULTS

The isothermally aged 3.5Ni, 1.7Cr, 0.06P alloy steel was highly susceptible to temper embrittlement and fracture was 100% intergranular. AES analysis of the uncontaminated intergranular fracture surface showed a normalized P coverage level

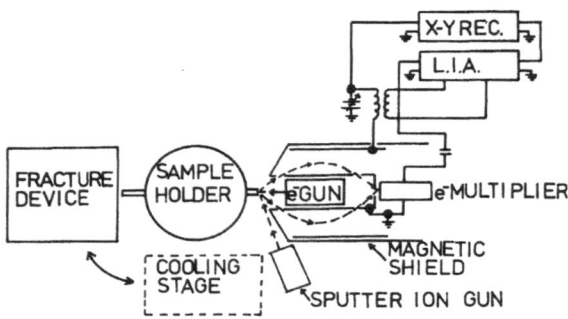

Fig. (1) - Auger Electron Spectrometer

~0.4. Contamination of the fracture surface to an oxygen coverage level ($O_{510PPH}/$ Fe_{703PPH}) ~ 0.4 caused only minimal attenuation of the P signal. The P signal then decreased with increasing amounts of oxygen to reach a value of (P_{120PPH}/Fe_{703}) ~ 0.13 at an oxygen coverage level of (O_{510PPH}/Fe_{703PPH}) ~ 1.0, Figure 2.

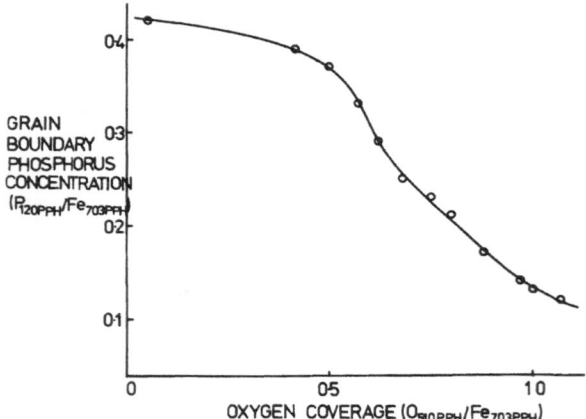

Fig. (2) - Effect of oxygen contamination on the level of grain boundary phosphorus detected on fracture surface of a temper-embrittled Ni-Cr-P steel

Subcritical crack propagation of D6ac steel (2h temper) in 10^5 Pa hydrogen gas produced many localized areas where 100% intergranular fracture occurred. The excessive time delay incurred in re-evacuation and the resulting high background pressure sustained during AES analysis (10^{-5} - 10^{-6} Pa) resulted in oxygen coverage levels (O_{510PPH}/Fe_{703PPH}) ~ 1.5. The results in Figure 2 indicate the impracticality of detecting low levels of phosphorus segregation at this high level of oxygen coverage.

Oxygen coverage levels were reduced to (O_{510PPH}/Fe_{703PPH}) ~ 0.4 when hydrogen-precharged specimens were fractured at 25°C in the AES chamber at 10^{-6} Pa pressure in close proximity to a separate cold-stage designed to condense contaminant gases. The separate cold-stage was found to be necessary since oxygen levels increased to (O_{510PPH}/Fe_{703PPH}) ~ 1.0 when the cold state was removed. It was also found essential to cool the specimen to -90°C during pump-down to retain sufficient hydrogen in the specimen for subcritical crack growth to be induced during subsequent loading at room temperature. In D6ac steel specimens tempered at 520°C for 2h and 20h, substantial regions of 100% intergranular fracture occurred close to the specimen notch and these areas were easily identified for subsequent AES analysis. Subcritical crack growth in D6ac steel specimens tempered for 200h at 520°C produced regions of 100% intergranular fracture that were smaller than the 200 μm diameter electron beam, Figure 3, and AES levels were therefore corrected to compensate for the observed proportion of transgranular ductile fracture. Figure 4 shows a typi-

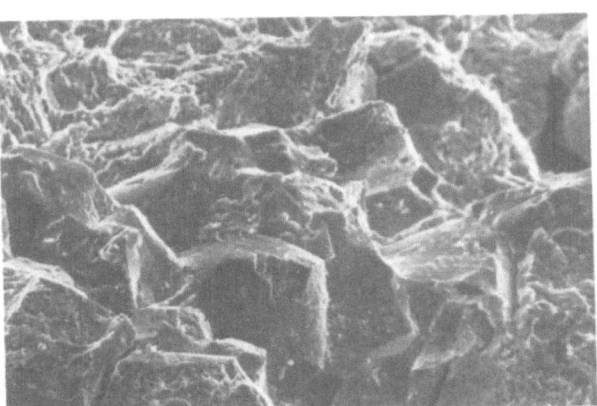

Fig. (3) - Fracture surface of hydrogen-charged D6ac steel specimen tempered for 200h at 520°C and fractured subcritically at 25°C: Magnification 570X

cal AES spectrum taken from a fractured D6ac steel specimen surface. The AES results for D6ac steel in Table 1 are for constant oxygen coverage (O_{510PPH}/Fe_{703PPH}) ~ 0.4 and indicate definite trends in the grain-boundary concentrations of the major segregants (P, Mo and Cr) with tempering time. The P concentration at the grain boundary decreases from 1.7% PHR after tempering for 2h to negligible levels

602

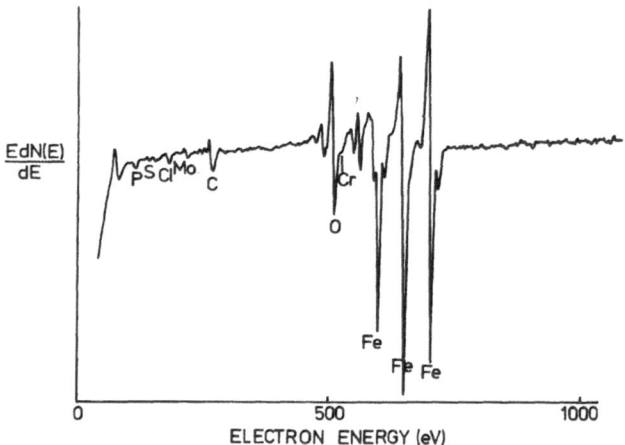

Fig. (4) - Auger spectrum of intergranular fracture surface for hydrogen-charged D6ac steel specimen tempered for 2h at 520°C

TABLE 1 - NORMALIZED GRAIN-BOUNDARY CONCENTRATION (PHR) OF P, Mo AND Cr IN D6ac STEEL AS A FUNCTION OF TEMPERING TIME FOR MINIMUM OXYGEN COVERAGE (O_{510PPH}/Fe_{703PPH} ~ 0.4)

Element	Tempering Time (h)		
	2	20	200
P	0.017	0.013	0.001[*]
Mo	0.037	0.031	0.022[*]
Cr	0.026	0.031	0.043[*]

[*]Values determined after correction for 75% intergranular fracture.

(~0.1%) after tempering for 200h. Similarly, the grain-boundary composition of Mo decreased from 3.7 to 2.2% PHR whereas the level of Cr increased from 2.6 to 4.3% PHR.

The fracture surface of the EN26 steel specimens was ~100% intergranular close to the specimen notch and AES analysis showed an enrichment of P, Mo, Ni and Cr at the grain boundaries, Figure 5. The maximum concentration levels of these elements at the grain boundaries were measured at an oxygen coverage level (O_{510PPH}/Fe_{703PPH}) ~ 0.5 and are shown in Table 2.

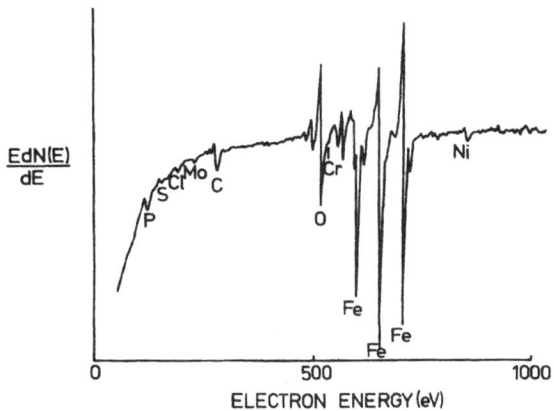

Fig. (5) - Auger spectrum of intergranular fracture surface
for hydrogen-charged EN26 steel specimen

TABLE 2 - NORMALIZED GRAIN-BOUNDARY CONCENTRATION (PHR) OF P, Mo, Ni AND
Cr IN EN26 STEEL FOR MINIMUM OXYGEN COVERAGE (O_{510PPH}/Fe_{703PPH}
~ 0.5)

Element	Concentration
P	0.081
Mo	0.042
Ni	0.038
Cr	0.050

DISCUSSION

AES analysis of intergranular fracture surface produced by subcritical crack
growth of D6ac steel specimens in hydrogen gas shows high levels of oxygen coverage. This contamination is believed to originate from (a) impurities in the hydrogen gas, (b) enhanced desorption of contaminant gases from the walls of the
AES chamber by bombardment of hydrogen gas molecules, and (c) back-streaming of
contaminant gases during evacuation of the hydrogen gas. These problems do not
arise with hydrogen-precharged specimens loaded in vacuum, and contamination is
further minimized using a separate cold-stage to getter condensable contaminants.
The degree of P attenuation produced by adsorption of oxygen on the grain boundaries of the temper-embrittled Ni-Cr steel shows that realistic estimates of seg-
regant levels can be obtained at oxygen coverage levels (O_{510PPH}/Fe_{703PPH}) ~ 0.4.

Optimum conditions might be achieved by inclusion of a secondary vacuum chamber
to permit admission of the specimen to the main AES chamber under vacuum. This

modification would reduce the pump-down time to such an extent that prior cooling of the specimen would not be necessary and the lower attainable pressures would ensure contaminant-free fracture surfaces for subsequent AES analysis.

Previous work has shown that D6ac steel has a banded microstructure in which stress-corrosion cracking occurs preferentially along the solute-rich bands [10]. X-ray microprobe analysis of the banding revealed an excess of Mn, Cr, Mo, S and P in the solute-rich bands with Ni showing no variation in concentration across the banding [10]. Mo and P showed the largest variation in concentration across the banding with bulk concentrations of 1.7 and 0.008 (wt %) respectively in the solute-rich bands. Calculations based on the solubility of P in Fe-Mo-P mixtures predict that the solute-rich bands in D6ac steel are supersaturated in P at 520°C [2,10]. Precipitation of P should therefore occur during tempering, resulting in reduced levels of free segregated P at the grain boundary. These ideas are supported by the present work which show decreasing levels of segregated phosphorus at the grain boundary after long tempering times, Table 1. Recent studies have also shown that the threshold intensity for stress-corrosion cracking (K_{1scc}) of D6ac steel in distilled water at 25°C increases with increase in tempering time at 520°C [11]. This result provides further evidence that susceptibility to stress-corrosion cracking is exacerbated by high levels of free P segregated at grain boundaries, Figure 6.

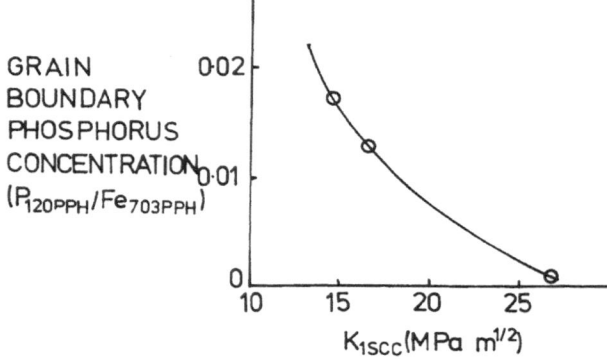

Fig. (6) - Effect of P segregation at grain boundary on K_{1scc} during stress-corrosion cracking of D6ac steel (520°C temper) in distilled water

AES analysis of intergranular fracture surfaces produced by hydrogen embrittlement provides scope for study in a number of areas. In addition to illustrating the deleterious effect of P in promoting hydrogen-assisted intergranular cracking, the present work with EN26 steel demonstrates the value of the technique as a routine tool in the analysis of environment-assisted failure. Since intergranular cracking can be induced in many steels by hydrogen embrittlement, stress-corrosion cracking and liquid metal embrittlement, AES analysis of intergranular fracture surfaces produced by hydrogen embrittlement provides data that may be pertinent to the understanding of the other two processes. AES analysis has been used to investigate the effect of microstructure on solute and impurity segregation to grain boundaries using temper embrittled steels. A possible limitation of this approach involves the narrow range of heat treatments where intergranular fracture can be induced. Similar studies using hydrogen-charged specimens should prove more fruitful since intergranular cracking often occurs over an increased heat-treatment range. AES analysis should also be possible in other alloy systems susceptible to intergranular cracking by hydrogen embrittlement viz. Ni, Ti, Al, Nb, Zr, Mo, V. In some of these alloy systems (e.g., Al, Mg), adoption of slow-strain rate techniques would be required to optimize the degree of intergranular fracture of hydrogen-charged specimens in the AES spectrometer.

CONCLUSIONS

It has been shown that grain boundary chemistry can be determined from AES analysis of intergranular fractures induced by hydrogen embrittlement. Susceptibility of D6ac steel to stress-corrosion cracking was related to the level of free P at the grain boundary. The technique should provide increased scope for studying the effects of heat treatment on impurity and solute segregation to grain boundaries, thereby leading to a better appreciation of the processes causing hydrogen embrittlement, stress-corrosion cracking, liquid metal embrittlement and intergranular corrosion.

REFERENCES

[1] Olefjord, J., Int. Metals Reviews, 4, p. 149, 1978.

[2] Guttmann, M., Surf. Sci., 53, p. 213, 1975.

[3] Mulford, R. A., McMahon, C. J., Pope, D. P. and Feng, H. C., Metall. Trans. 7A, p. 1183, 1976.

[4] Erhart, H. and Grabke, H. J., Met. Sci. J., 15, p. 401, 1981.

[5] Schultz, B. J. and McMahon, C. J., Temper Embrittlement of Alloy Steels, ASTM STP 499, American Society for Testing and Materials, p. 104, 1972.

[6] Seah, M. P., Acta Metall., 28, p. 955, 1980.

[7] Latanision, R. M. and Opperhauser, H., Metall. Trans. 5, p. 483, 1974.

[8] Lea, C., Metal Sci. J., 14, p. 107, 1980.

[9] Hondros, E. D. and Lea, C., Nature, 289, p. 663, 1981.

[10] Pollock, W. J., Ryan, N. E. and Nankivell, J. F., Corros. Sci., 22, p. 215, 1982.

[11] Pollock, W. J., unpublished results.

SECTION VIII
COMPOSITES AND NONMETALS

REPAIR OF BALLISTICALLY IMPACTED CARBON FIBRE REINFORCED (CFR) LAMINATES

J. B. Young and N. Matthews

Cranfield Institute of Technology, Cranfield
Bedford, Mk43 OAL, England

ABSTRACT

This report shows that thin monolithic Carbon Fibre Reinforced Panels impacted with small arms projectiles can be adhesively repaired restoring the damaged panel tensile strength to between 85% and 90% of the original. The compressive buckling strength of the thin CFC panels was found not to be significantly affected by ballistic impacting. The repair is primarily concerned with single-sided repairs utilizing a wet laminating repair patch technique and a precured patch technique. The repair research showed that for the particular impact damage sustained by the CFC panels there existed optimum repair patch physical parameters (i.e. overlap length and ply composition) for strength restorations above 85%. The research highlighted a number of repair techniques which adversely affected strength restorations. These techniques involved the removal of the damage area, use of precured square patches and the use of narrow repair patches.

In addition to the damage panel repair analysis the report also details the construction and the testing of the laminates used in the research and attempts to present theoretical repair analysis which complements the actual repaired panel characteristics.

INTRODUCTION

With the continual rapid increase in aircraft technology, the introduction and rapid development of new materials has occurred. In particular, the use of composite materials has become increasingly popular because they have shown to be ideal where high strength to weight ratios are required. In spite of the advantages of composites it is feared by many experts that the technology and use of composites has advanced too quickly. Unlike metals, which have evolved during many decades of application, composite materials are relatively new and there appears to be few limitations on their use. However their reaction to the many and varied in-service problems have yet to be fully evaluated. These in-service problems include impact resistance, environmental resistance, repairability etc. Of particular interest to the authors was the repairability of the high strength/stiffness composites used in military aircraft.

Because of the limited time available to conduct meaningful research (approximately 4 months) into the relatively large and complex field associated with composites repairs, research was limited to repair of thin monolithic Carbon Fibre Composites (CFC) panels impacted with small arms projectiles. The repairs if possible were to restore the impacted panel strength to a usable degree (e.g. 85% to 90%) and as such could be deemed to be permanent.

COMPOSITION AND CONSTRUCTION OF TEST LAMINATES

A. Material Used

The carbon fibre material used in the construction of the laminates was a Carrfibre high strength bi-directional five-shaft satin woven cloth. This type of material is being increasingly utilized in aircraft applications because of its desirable characteristics i.e., drapability, stable articles of commerce, better production characteristics, etc.

The matrix resin system used for the construction was a two-part liquid epoxy system - Ciba-Geigy XD927. This resin system in addition to the advantages of epoxy systems, namely toughness and good adherence to fibres, exhibited low viscosity which enhanced thorough impregnation and cold curing capability.

B. Production of Laminates

The laminates used for the research were constructed at RAE Farnborough using a vacuum moulding technique. Details of the apparatus and its use are contained in [1]. Laminates of eight plies were selected for construction as it was deemed to represent a typical thickness of thin laminates found in aircraft structures (i.e. skins). Fibre volume ratios of approximately 52% were obtained which are consistent with vacuum moulding technique [2].

MECHANICAL PROPERTIES OF THE LAMINATE AND SELECTED TEST PANELS

The testing and inspection program employed during the research to obtain the various mechanical properties, basically involved:

1) Initial testing on samples of the various laminates to determine respective laminate properties, and

2) Testing and inspection of the selected test panels to determine properties in addition to those obtained for the laminate (i.e. panel buckling and to assess the effect of ballistic impact and the effectiveness of repair).

A. Laminate Properties

The basic mechanical properties of the laminate (i.e. tensile modulus, tensile strength, shear strength, shear modulus, compressive strength etc.) were determined by conducting tests in accordance with recognized procedures. Details of these properties are contained in Table 1.

TABLE 1 - LAMINATE MECHANICAL PROPERTIES

Panel Number	Tensile Strength (MN/m²)	Tensile Modulus (GIN/m²)	Compressive Strength (MN/m²)	Shear Strength (MN/m²)	Shear Modulus (GN/m²)	Poisson's Ratio
1 Warp	660	64	352	97	4.0	.047
Weft	553	60				
2 Warp	682	60	328	91	3.9	.047
3 Warp	660	64	348	105	4.0	.047

B. Test Panel Laminate Properties

    The size of the test panels originally chosen for the research was 250 mm long by 125 mm wide. These dimensions were selected to provide an adequate area to effect repairs after impact. Because of the limited width of available mechanical/hydraulic jaws (i.e., 80 mm max) various methods of alternative testing were employed (Figure 1). These methods proved

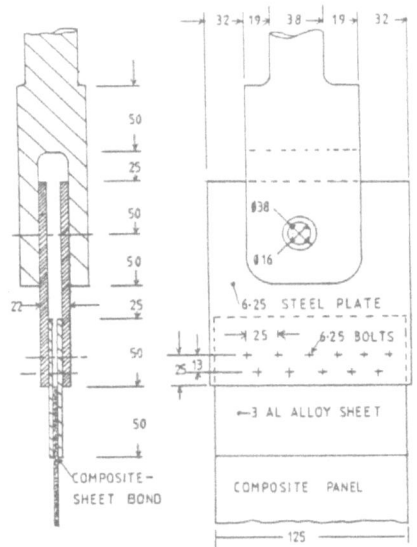

All Dimensions in mm

Fig. (1) - Unsuccessful tensile test set-up

unsuccessful due to high induced peel stresses at the tab ends and the panel width had to be reduced to 80 mm. This reduction in width was initially cause for concern as it was considered that panel edge effects may introduce prohibitive stresses into the repair area. Accordingly, to determine the extent of the edge effects, a panel was strain gauged with the results showing that the edge effects extended into the panel by a distance equivalent to the panel thickness. This result is supported by [3].

The originally selected width presented no problems in compression testing of the panels. The panel was tested to destruction with compressive buckling failure occurring well below ultimate laminate compressive strength (i.e., approximately 40%).

BALLISTIC IMPACTING

All the ballistic damage was achieved with .3 calibre projectiles (i.e., standard NATO 7.62 mm small arm projectile). Although this type of projectile damage would relate to only a small percentage of probably ballistic damage encountered by modern military aircraft, the .3 calibre projectile was selected because:

1) The limited panel width meant that ballistic damage had to be small compared with the width if effective repairs were to be incorporated.

2) The subject cartridges and range facilities were more readily available.

3) Most of the research available on ballistic impacting of fibre panels related to .3 calibre impacting which would enable comparisons to be made between this research and other work.

4) It was anticipated that the results obtained for the .3 calibre impact damage and subsequent repair could be directly applied to larger more realistic panels (i.e., those comparable with aircraft panels) with larger ballistic damage (i.e., damaged caused by 23 mm projectiles).

The ballistic impacting was accomplished by employing two methods of projectile impact. The first method of impacting which was used for the majority of the research was where the projectile was fired directly at the panel with the projectile impacting normally at the centre. The second method involved impacting of the panel with the projectile tumbling.

For both the normal impact and tumbled impact, the test panel was clamped at each end and the firing range was limited to 15 m to ensure that the unstable flight regime of the projectile was not reached (Figure 2). Tumbling of the projectile was employed to assess the significance of the projectile impacting the surface at various angles to incidence.

A. Results of Impact Damage

The impacted panels were examined both visually and ultrasonically. The damage consisted of an area limited to the projectile cross-sectional

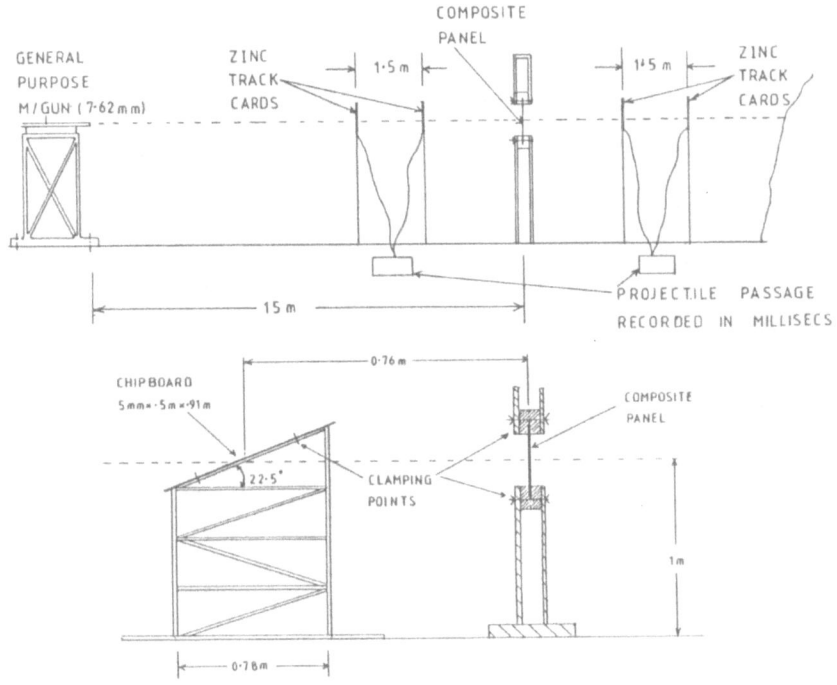

(Additional tumbling apparatus)

Fig. (2) - Ballistic test set-up

area on the entry surface with a slightly larger square/rectangle damage area on the exit surface (Figure 3). Ultrasonic inspection revealed that substrate damage was the same as revealed by visual inspection but slightly larger (i.e., for normal impact: for exit surface - 12 mm visual - 16 mm ultrasonically). Also the observation of little fibre feathering or pull out on the exit surface indicated that a reasonable fibre to matrix bond was present.

All normally impacted panels when tested in tension exhibited an overload failure across the projectile hole with a reduction of panel tensile strength of approximately 50%. The buckling and post-buckling behavior was almost identical to that of an undamaged panel and accordingly for the remaining research only the tensile behavior of the panels was addressed. Residual tensile and compressive buckling strengths for the panels impacted with tumbled projectiles was dependent on the amount of width related damage.

REPAIR OF BALLISTICALLY IMPACTED PANELS

As detailed in the introduction, the aim of the research was to restore

614

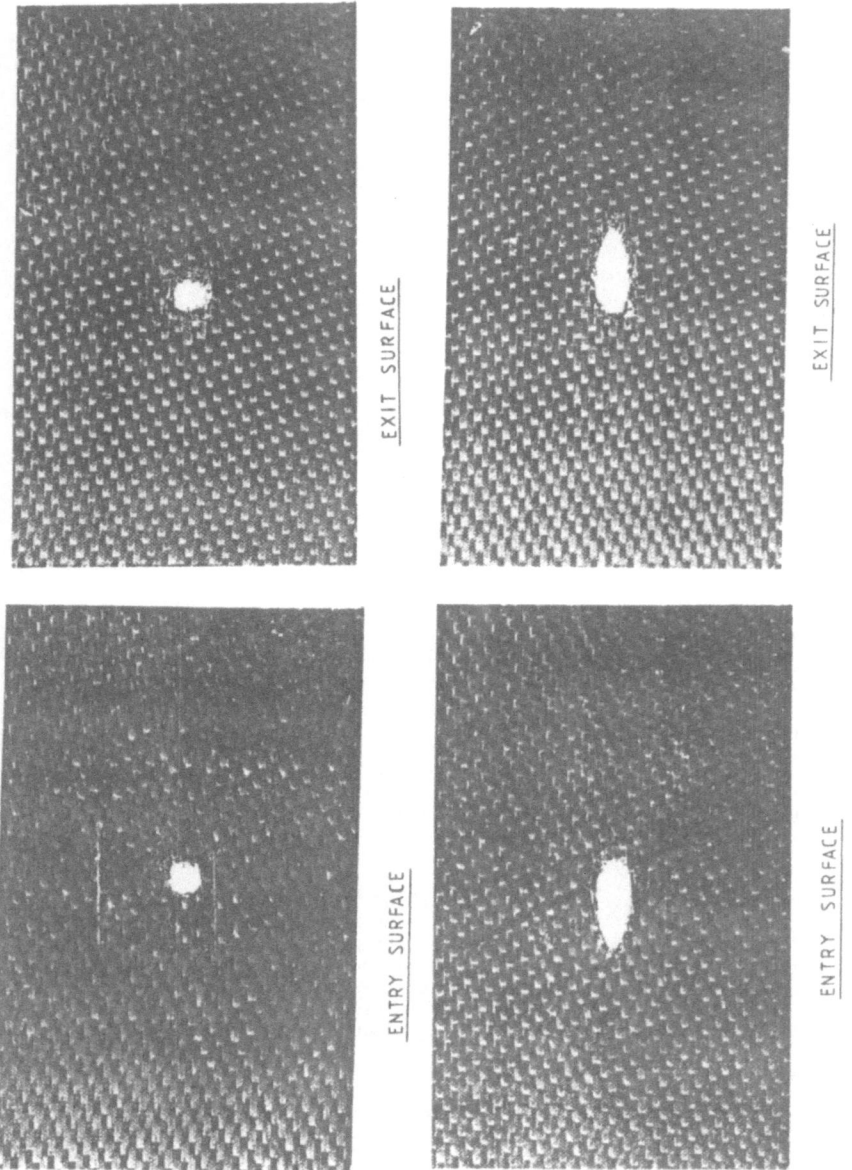

EXIT SURFACE

EXIT SURFACE

ENTRY SURFACE

ENTRY SURFACE

Fig. (3) – Projectile Damage

the strength of the damaged panels to approximately 90% of the original strength.

The repairs were accomplished by basically employing field repair techniques. In conducting the research the relevance to aircraft repairs was always prominent. Accordingly the majority of repairs were restricted to one side. Also, because of the relatively small damage area external repairs were employed. For initial repair work the repairs were accomplished by using the same material as the base adherent, a cold curing epoxy resin and a wet lay-up technique. Once an optimum repair technique had been developed both analytically and experimentally, precured patches were employed.

A. Repair Techniques

Because patches were to be employed on only one side, the eccentricity of the patch to base adherent meant that bending would be induced in the patch and the base laminate resulting in high peel and shear stresses at both the damage edge and patch edge. It was because of the high peel and shear stresses that the first phase of the repair involved the selection of an adhesive which exhibited the best resistance to peel. Specimens were produced to ensure peel failures would occur when the joint was loaded in tension (Figure 4). The choice of available adhesives which would enhance both im-

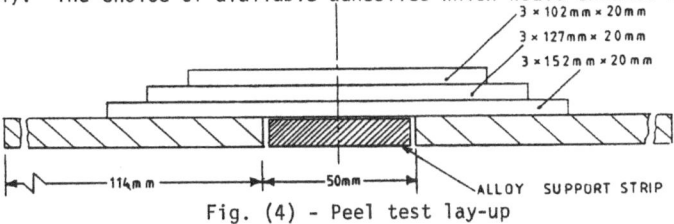

Fig. (4) - Peel test lay-up

pregnation and would cold cure was narrowed initially down to two epoxy systems Ciba-Geigy XD927 and Epikote 828. (Note the epoxy systems were selected in lieu of polyester systems because the epoxy systems enhanced better adhesion to the parent laminate). Subsequent peel tests revealed that the Epikote 828 joints were 60% stronger than the XD927 joints.

Damage area preparation was recognized as one of the most critical operations of the repair procedure. The area was prepared using a fine carborundum cloth followed by a rinse with Acetone. Wet laminated patches were layed-up with a gradual reduction in ply dimensions. This was done to produce a tapered patch effect to enhance peel stress dissipation. Precured patches were later employed in the research work primarily because they were easier to work with and they offered a lighter repair. However, the initial use of precured patch was a square patch which significantly reduced repair effectiveness by approximately 25%. It was felt that this had occurred because of the combination of bending stresses both transverse and longitudinal at the patch corners. Removal of the corners (i.e., 15 mm along each edge) restored panel strength to that obtained by wet laminating patches (i.e., approximately 90%).

616

B. Results of Repair Panels

Two trends were noted from the experimental results with the first relating to repair panel strength versus patch ply thickness. It was noted that after a nominal patch ply thickness was achieved additional plies did not assist in improving the repair strength (Figure 5). The second trend related to repair panel strength versus patch overlap length. Once again when the nominal overlap length was reached further overlap resulted in little strength increase.

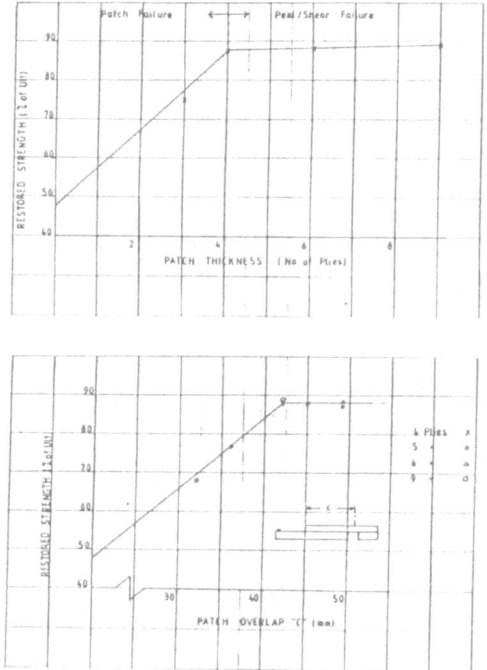

Fig. (5) - Repair strength vs. patch parameters

Other areas which became evident during the research were:

1) Utilization of double side patches resulted in repaired panel strengths of 100% of undamaged strength. This was because bending stresses were alleviated.

2) Removal of damage area reduced panel strength by up to 30%.

THEORETICAL ANALYSIS

In conjunction with the experimental research analytical assessments were conducted. The assessments related to two main areas:

1) Residual strength prediction of impacted panels.

2) Patch parameter prediction (i.e., ply thickness and overlap) for the repairs of impacted panels.

A. Residual Strength Predictions

Experimental data has shown that the application of fracture mechanics to fibre composites is extremely difficult because the failure of composites are dependent on the numerous variable properties of individual composites. These variables include the mechanical properties of the fibres and resin, ply orientation, etc. Because of the difficulties in the definition of the exact failure mechanisms, many models have been proposed for residual strength prediction of damaged composites under tensile loading. These models can be basically divided into two main groups - those which have adopted the basic concepts of Stress Concentration Factors (SCF) and those which have adopted the basic concept of Linear Elastic Fracture Mechanics (LEFM). Numerous papers have been written on the various SCF and LEFM models and these papers adequately describe approaches undertaken in producing the models [4-5]. However, it is worthwhile to note that all the models assume some stress distribution either by a blunting mechanism or by the forming of a damage zone. Accordingly to employ any of the models at least two parameters need to be known. These are the ultimate strength of the undamaged specimen and the size or characteristic of the damage zone. Therefore it must be assumed that the available models are empirical. Fortunately relevant data on size and characteristics of the damage zone for carbon - epoxy laminates was available which permitted the use of available strength prediction models.

The models selected for use were Nuismer and Whitney [11] average stress criteria SCF and LEFM models. These models provided relationship as follows:

$$\sigma_c = \sigma_0(2(1-p_2))/(2-p_2^2-p_2^4 + (K_T-3)(p^6_2-p^8_2))\ldots\ldots(SFC) \tag{1}$$

where

K_T = stress concentration factor

$p_2 = R/(R+a.)$

and

a. is a material property independent of laminate construction and stress distribution

Also

$$\sigma_c = \sigma_0(1+2\ a/a.)^{1/2}\ldots\ldots(LEFM) \tag{2}$$

Application of these models yielded predicted results within 5% of experimental results. (Table 2)

TABLE 2 - RESIDUAL STRENGTH COMPARISIONS

Panel Number	Type of Impact	Predicted % of Ult.	Actual % of Ult.
1,2 & 3	Normal	49	48
2E	Tumbled	44	42
2F	Tumbled	46	43

B. Theoretical Repair Analysis

As detailed in the preceding text experimental results indicated that for a single sided external patch repair, there existed an optimum patch ply overlap and patch ply thickness.

The patch ply thickness is primarily a function of the amount of load being transferred through the patch. However the exact amount of the load through the patch and the load patch is extremely difficult to define. The lack of definition can be attributed to the effect of the patch on the laminate stiffness around the damage. The available models, in the main, adopt a simplistic 2D approach which relates patch loading to the relative linear stiffness of the patch in the damage/repair area (Figure 6):

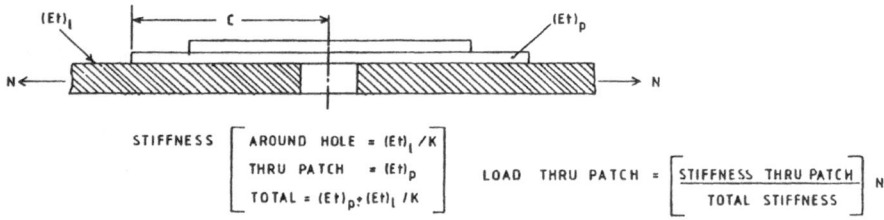

Fig. (6) - Patch repair analysis

i.e.,

$$\text{load through the patch} = \frac{\text{patch stiffness}}{\text{total stiffness}} \times \text{applied load} \qquad (3)$$

Although the linear stiffness of the patch can be readily defined once the patch lay-up is selected, the difficulty arises in defining the laminate stiffness in the damage area. This basically means that the laminate stiffness around the hole (i.e., 3D) is being defined via a 2D model ([7-8]). However utilizing this basic concept the following model for patch load prediction was developed.

$$P_p = \frac{(Et)_p\ P}{(Et)_p + (Et)_L/K} \tag{4}$$

where

$(Et)_p$ is the patch linear stiffness and $(Et)_L/K$ is the laminate stiffness around the hole.

K is the laminate stiffness reduction factor and is the ratio of the ultimate panel load over the net panel load.

P is the applied load.

Having determined the patch thickness, the optimum overlap was obtained by adopting the fundamental analysis of a single lap joint contained in [6]. This analysis, in relationship to peel failure predicts that the maximum peel stress is related to the average laminate stress outside the joint by:

$$(\sigma_p/\sigma_y) = (k/(2)^{1/2}\ (3\ E'c\ t_1/En)^{1/2} \tag{5}$$

where

k is the moment co-efficient of the load path = $2\ M_0/Pt$

E'c is the adhesive peel modulus

t_1 is the adherent thickness

t is the laminate thickness

n is the adhesive thickness

As can be seen from the inclusion of the moment co-efficient in the equation (5) the maximum stresses within and outside the patch area are greatly influenced by the value of the bending moment, M., induced just outside the overlap, by the eccentricity in the load path. This induced moment is a function of the applied load, P, through the adherent and the adherent stiffness can be expressed by the following relationship:

$$M_0 = Pt/2\ (1 + \Sigma\ C + (\Sigma\ C)\ 2/6) \tag{6}$$

where

Σ^2 is the laminate stiffness parameter and is equal to P/D

C is half the patch overlap from the centre of the damage

and

P is the load/width

620

Therefore, once the maximum peel stress has been established, based on the maximum peel strength of the patch adhesive (obtained experimentally), the patch overlap can be predicted.

By employing the above theoretical analysis a ply thickness of 4 layers was predicted for the repair patch. As exhibited in Figure 5, a 4 ply patch was experimentally found to be the optimum patch thickness. Comparison of predicted patch overlap to experimental results are contained in Figure 7.

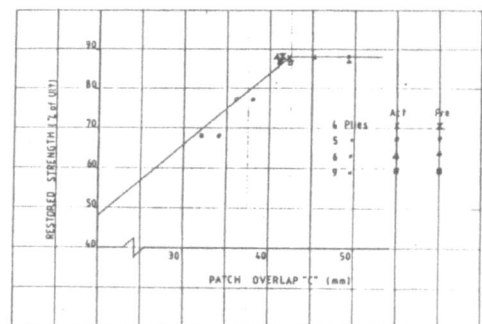

Fig. (7) - Predicted and actual patch overlap

DISCUSSION

Although the limited time available for the research did not permit investigation of areas pertinent to the repairs accomplished (i.e., fatigue, environmental effects, etc.) sufficient time was available to complete the main objective of the research. Before dealing with specific observations it is considered prudent to point out that the term 'strength' used throughout the report refers to tensile strength. Although the normal impacted panels did not reflect changes in compressive characteristics and therefore compressive behavior was neglected for the repairs, this approach could not be taken where thicker panels are used or damage is greater.

Earlier in the report it was stated that the laminates constructed for this research were analogous to those on aircraft. Although this is not entirely correct as most skins contain a substantial amount of ±45% plies, ballistic impact damage for the various laminates are the same [9-10]. Also since the direct strength of the research panels are greater than those containing angled plies, ballistically impacted thin monolithic panels can be repaired using single sided external patches.

With regard to actual repair techniques employed during the research a number of important characteristics were observed. These characteristics were the necessity to ensure proper adhesion between the patch and substrate, the advantage of retaining the damage area and the significant strength degradation when using square patches.

Finally one of the most significant repair characteristics observed was that there appeared to exist optimum patch parameters to achieve a permanent repair (i.e., 90% of undamaged tensile strength). These parameters were patch ply thickness and patch overlap length. By employing various 2D analysis the majority of which is based on joint analysis, the experimental results were paralleled by theoretical analysis (Figure 7). The inherent problems associated with the use of 2D analysis for 3D situations has been discussed in the main part of the text.

CONCLUSION

The research highlighted a number of important finds which can be best summarized as follows:

1) Although it is generally accepted that when carbon laminates are damaged their compressive strength is greatly affected, the research revealed that the compressive buckling strength of the thin monolithic panels was not affected. This revelation is particularly important when one considers that the thin monolithic panels are representative of a large majority of aircraft type panels which have or would be affected by ballistic damage. However, the tensile strength was significantly affected with panels exhibiting only a 20% reduction in cross section area due to ballistic damage suffering degradation of tensile strength in excess of 50%.

2) Single sided external repairs could be utilized to restore impacted panel strength to a usable degree (i.e., 90% of original).

3) There existed optimum patch parameters (i.e., ply thickness and overlap). These became evident during experimental research.

4) These optimum patch parameters can be predicted analytically.

5) Residual tensile strength of impacted panels can also be predicted analytically.

6) Square patches, damage area removal and, poor surface preparation have a significant and detrimental affect on restored strength.

REFERENCES

[1] Rogers, L. F., Kingston-Lee, D. M., and Phillips, L. N., "The development of a lightweight, all-plastic apparatus for vacuum - moulding of composites", Proceedings of Symposium Fabrication Techniques for Advanced Reinforced Plastics, 22/23, IPC Science and Technology Press, pp 27-34, 1980.

[2] DES 8017, "Production aspect of fibre reinforced plastics", Aircraft Design, CIT, 1980.

[3] Agarival, B. D. and Broutman, L. B., "Analysis and performance of fibre composites", John Wiley and Sons, 1980.

[4] Dorey, G., "Relationship between impact resistance and fracture toughness in advanced composite materials", AGARD Conference Proceedings, No. 288, 1981.

[5] Bishop, M. S. and McLaughlin, K. S., "Thickness effects and fracture mechanics in notched carbon fibre composites", RAE, TR79051, 1979.

[6] Hart-Smith, L. J., "Analysis and design of advanced composite bonded joints", NASA CR 2218, 1974.

[7] Murray, J. E., "Shipboard composite repairs using externally bonded patches", McDonnell Douglas Company, Saint Louis, 1980.

[8] Weidemann, J., Griese, H. and Glahn, M., "Stress and strength analysis of reinforced plastic with holes", Consequence of Design, AGARD Conference Proceedings No. 163, pp 7-1 to 7-11, 1975.

[9] Olster, E. F. and Roy, P. A., "Tolerance of advance composite to ballistic damage", Composite Materials, Testing and Design, ASTM STP 546, pp 583-603, 1974.

[10] Avery, J. G. and Porter, T. R., "Comparisons of response of metals and composites for military aircraft applications", Foreign Object Damage to Composites, ASTM STP 568, pp 3-29, 1975.

[11] Whitney, J. M. and Nuismer, R. J., "Stress fracture criteria for laminated composites containing stress concentrations", Journal of Composite Materials, Vol. 8, No. 2, pp 253-265, 1974.

FATIGUE DAMAGE IN GLASS REINFORCED PLASTICS

J. C. Radon and C. R. Wachnicki

Imperial College of Science and Technology
London SW7 2BX, England

ABSTRACT

Low frequency tensile fatigue tests under constant load conditions were carried out on centre-notched biaxial specimens in order to measure the effects of the load biaxiality factor B on a composite material. The fatigue behaviour of chopped strand mat glass fibre reinforced polyester resin in air was also compared to that in a dilute acidic environment. A compliance calibration technique was used to measure the effective crack length so as to extend the applicability of the stress intensity factor to planar composite materials. The Paris law relationship based on the stress intensity factor range was found to be applicable to the results; the value of the exponent was found to decrease slightly with the increase in the biaxiality factor. Also, the magnitude of the exponent was greater by a factor of three than the reported values for metals and polymers when tested in air. However, fatigue tests carried out in an acidic environment showed a reduction in the magnitude of the exponent by at least a factor of two as compared with tests in air. The fracture surfaces viewed were quite different from fracture surfaces in air, the former showing a glassy resin appearance practically devoid of fibre pullout; this is a consequence of macro-crack formation in the resin by the mechanically applied cyclic stress initiating hydrogen ion attack on the exposed glass fibre surfaces.

INTRODUCTION

The introduction of glass reinforced plastics into the structural materials market has made it necessary to study in detail their responses to various types of loading under different environmental conditions. The use of glass fibre-reinforced polyester resins in the manufacture of pipes, storage tanks and pressure vessels for the chemical process industries has highlighted their capability for corrosion resistance and strength. There is concern, however, that the performance of such materials may suffer during and after exposure to corrosive environments especially under dynamic loading conditions in the presence of inherent cracks. A limited amount of work has been done [1] on the influence of acid environments on the stress and strain corrosion of composites, but very little on the combined effects of cyclic loading and environment. This combined influence of stress and environment becomes very important because a further degree of complexity is introduced into the composite material by the large number of microstructural variables already present. Fatigue crack growth studies have shown that the fracture

behaviour will depend on many factors. It may be considered that the bonding
strength of the glass fibre and polyester resin interface in a chemically stable
condition is reduced by the penetration of an aqueous acidic environment, results
in the degradation of the composite material as a whole. The development of fatigue damage in a glass reinforced plastic under the influence of an acidic environment may be accelerated by mechanical attack, by repeated stresses, or by
chemical attack on the glass fibre and interface from the acid, as well as to
their interaction. Moreover, the fatigue behaviour of centre-notched (CN) composite specimens offers an interesting problem in an acidic environment as the interaction of both mechanical and chemical attacks may be more significant. An attempt was therefore made to study the environmental fatigue behaviour of a random
oriented glass fibre-reinforced polyester resin material. The present work is
part of a programme to study corrosive environmental fatigue crack propagation
in a fibre-reinforced plastic under biaxial stress. The framework of the analysis is that of conventional fracture mechanics, so that a relationship is sought
between the crack growth rate, da/dN, and the stress intensity factor range, ΔK.
The optical measurement of crack length, $2a$, under fatigue loading is difficult
in composites because of the development of a damage zone at the crack tip, therefore a compliance technique has been developed to determine an equivalent crack
length.

EXPERIMENTAL PROCEDURE

The cruciform specimens for fatigue testing were individually moulded using
the "hand lay-up" technique, the laminates being impregnated with a powder bound
E-glass fibre in a chopped strand mat reinforcement configuration. To minimize
edge wicking effects, the external surfaces of the laminate were moulded integrally with a C-glass reinforced gel coat. The resin used was an ICI "Atlac" 282-05A
bis-phenol resin catalyzed with 4 gms of benzoyl peroxide (Lucidol CH50) and 0.3 ml
of dimethylanaline accelerator per 100 ml of resin. Twenty-four hours after the
specimen was moulded, it was post-cured for 3 hours at 80°C. The cruciform specimen blanks were then machined to the dimensions presented in Figure 1 which shows
a modified CN biaxial specimen, of 3.5 mm nominal thickness, designed to minimize
interaction between two orthogonally applied loads. This design is the fourth of
a series of such specimens initiated by a previous photoelastic investigation [2]
which showed that this interaction varied from 1 to 3% over a load biaxiality range
of 0<B<2.

Fatigue tests were carried out on a specially built biaxial fatigue rig [3],
a horizontal testing machine of an electro-hydraulic type cycling from zero to a
maximum load amplitude of 30 kN for a cyclic frequency range of 0.1 to 1 Hz. The
biaxial fatigue rig is based on a single double-acting actuator dividing a single
uniaxial force into two orthogonal components of known and controllable magnitude.
The arrangement of linkages and levers provide direct tension stresses for nine
load biaxiality ratios from B = 0 to 4.

To study the effect of various environments, load range and load biaxiality,
fatigue tests were carried out at a constant load range and biaxiality factor B
until fracture in two different environments, namely, air at 20°C and 50% relative
humidity and sulphuric acid of 5% concentration at 20°C. The stress ratio equivalent to K_{min}/K_{max} and the cyclic frequency were kept constant at 0.1 and 0.5 Hz
respectively. For the acidic environment fatigue tests, an environment chamber
was made, composed of rubber "O" ring material and melinex sheet sealed together
with silicone rubber to both sides of the specimen as shown in Figure 1. The di-

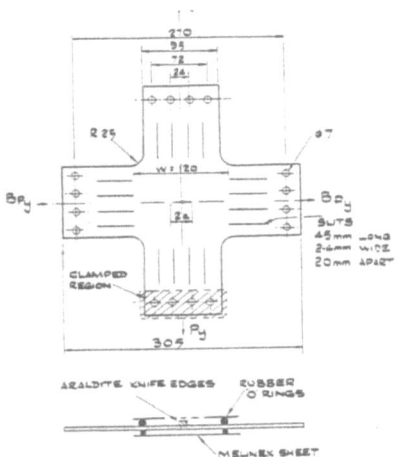

Fig. (1) - Environmental biaxial fatigue specimen geometry with side profile of environmental chamber

lute sulphuric acid was introduced into the chamber using a hypodermic syringe via an aperture in the top section of the environment chamber.

The authors used the compliance method referred to above [4] to determine the effective crack length, 2a, in relation to the biaxial CN specimen compliance. Araldite knife edges were attached to the specimen surface with Araldite adhesive as stress raisers may be formed using a bolt on type of knife edge in the biaxial mode. The highest biaxiality factor that could be used in this study was B = 1, as above B = 1 crack path stability is affected and therefore invalidating the compliance technique. The biaxial specimen was positioned in the test system and the clip gauge associated instrumentation connected and balanced as shown diagrammatically in Figure 2. The experimental determination of compliance is made by extending a slot in the specimen with a fine fret saw by a small increment and for each crack length of 2a measuring the displacement per unit of applied load. Calibration curves for biaxiality ratios B = 0, 0.5 and 1 are shown in Figure 3 where the compliance ϕ is defined as the quotient of the product of crack-opening displacement and specimen thickness divided by a corresponding load increment. The crack length 2a and thus a can be calculated and plotted as a function of the number of cycles N, and therefore the crack growth rate, da/dN can be derived by graphical differentiation.

Fracture mechanics concepts were used in analyzing the crack propagation data obtained from the fatigue tests as the stress intensity factor, K, has been used successfully to describe the behaviour of cracks in composite materials [5]. In the present work, Irwin's tangent formula for K_I [5] was used in a fourth order polynomial form

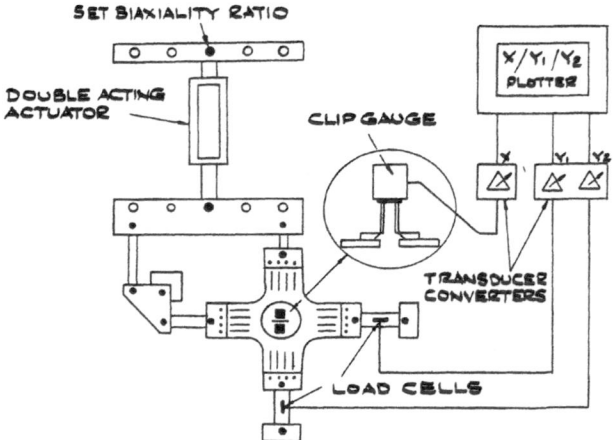

Fig. (2) - Test system used with biaxial fatigue specimen

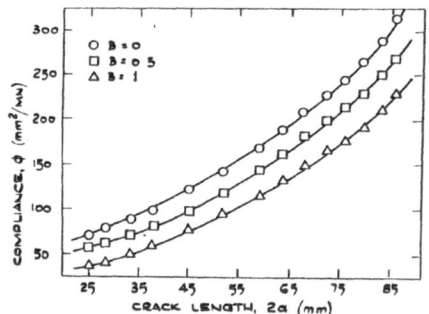

Fig. (3) - Compliance calibration curves for B = 0, 0.5 and 1

$$Y = 1.772 - 0.057 + 1.405\lambda^2 - 1.317\lambda^3 + 3.087\lambda^4 \tag{1}$$

for the solution

$$\Delta K = \Delta\sigma Y a^{1/2} \tag{2}$$

where σ is the gross stress, Y the finite width (CN) correction factor and λ the normalized crack length 2a/W.

RESULTS AND DISCUSSION

It has become conventional to express fatigue crack growth results as a log (da/dN) versus log (ΔK) curve using the Paris relationship [6]

$$\frac{da}{dN} = C(\Delta K)^m \tag{3}$$

where the fatigue cycle is described by ΔK, which is equated to $(K_{max}-K_{min})$. The K_{max} and K_{min} parameters represent opening mode stress intensity factors calculated from the maximum and minimum stress respectively during the fatigue cycle for a stress ratio of $R = K_{min}/K_{max}$. The Paris constants C and m were determined from crack growth data obtained by cycling at a constant load range and thus ensuring that steady state fatigue crack growth rates are recorded free from transient effects. Such relationships between the crack length, a, and the number of cycles, N, are shown in Figures 4 and 5, and are typical of the pattern of accelerating

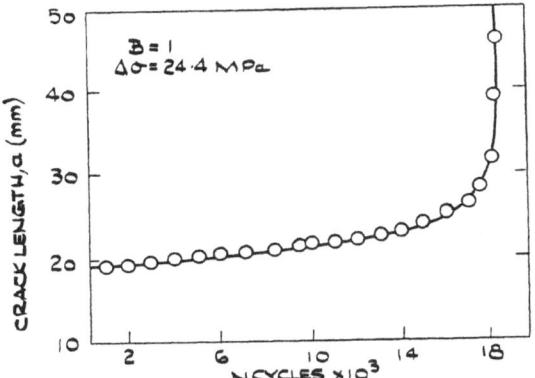

Fig. (4) - Crack length versus number of cycles, test in air

crack growth throughout the test programme. The shape of the curves is characteristic for this type of glass reinforced plastic material irrespective of biaxiality or environment effects and is caused by the rapid acceleration of crack growth leading to instability which defines the failure life rather than gradual crack growth causing a steady increase of the ΔK range, finally reaching the critical crack tip stress intensity factor K_c.

The environmental fatigue crack growth results under biaxial stress are shown in Figures 6 and 7, and analysis of the data shows a unique relationship existing between log (da/dN) and log (ΔK) for the two tests in air and in an acidic environment. Though the scatter is not great for the type of fracture mechanics approach adopted, the complex nature of the fatigue behaviour of the composite material de-

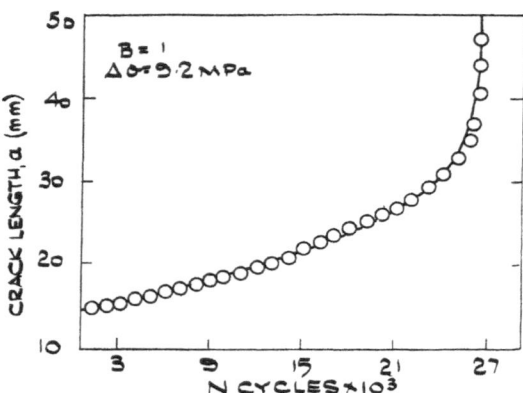

Fig. (5) - Crack length versus number of cycles, test in 5% H_2SO_4

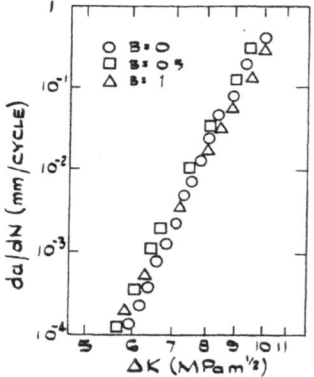

Fig. (6) - The relationship between log (da/dN) and log (ΔK) for tests in air

Fig. (7) - The relationship between log (da/dN) and log (ΔK) H_2SO_4 (5%)

ters one as yet from drawing direct conclusions from the biaxiality effects which are summarized in Table 1.

A. General Features of Fatigue in Air and Acid

The present work is part of a wider investigation of environmental fatigue under biaxial stress of a random oriented glass fibre reinforced polyester resin to determine the effects of material variables, testing parameters and environmental conditions on crack growth propagation. An emphasis on the environmental effects on fatigue behaviour of composites will be made in this paper.

TABLE 1 - PARIS LAW PARAMETER m and C

Environment at room temperature	Cyclic Stress Ranges* $\Delta\sigma$ (MPa)	Biaxiality B	m	C $((mm/cycle)/(MPa\ m^{\frac{1}{2}}))$
Air 20°C	25.2, 20.9	0	15.0	5.62×10^{-15}
Air 20°C	24.7, 21.4	0.5	14.6	1.26×10^{-14}
Air 20°C	24.4, 22.1	1	14.1	2.63×10^{-14}
5% H_2SO_4	12.0, 10.4	0	5.64	3.97×10^{-6}
5% H_2SO_4	12.7	0.5	5.61	4.34×10^{-6}
5% H_2SO_4	11.3, 9.2	1	5.57	4.74×10^{-6}

A number of investigators [7-9] have described the macroscopic appearance of progressive damage and fracture of chopped strand mat/polyester resin laboratory specimens subjected to repeated loading under uniaxial tension in air and other environments. Damage sites of the notched biaxial fatigue specimens have been examined microscopically by scanning electron microscopy and the appearance of the fracture surfaces is very similar to that described previously [7-9]. The fracture surfaces and damage characteristics of the failed biaxial fatigue specimens can be divided into three main categories, the three regions I, II and III are shown in Figures 8 and 9. Figures 8 and 9 show the inherent relationship of log growth rate with number of cycles, the crack growth rates were obtained by

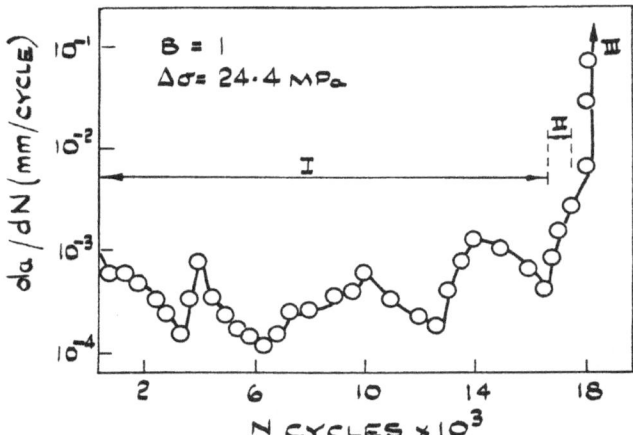

Fig. (8) - Log (crack growth rate) versus log (number of cycles) test in air

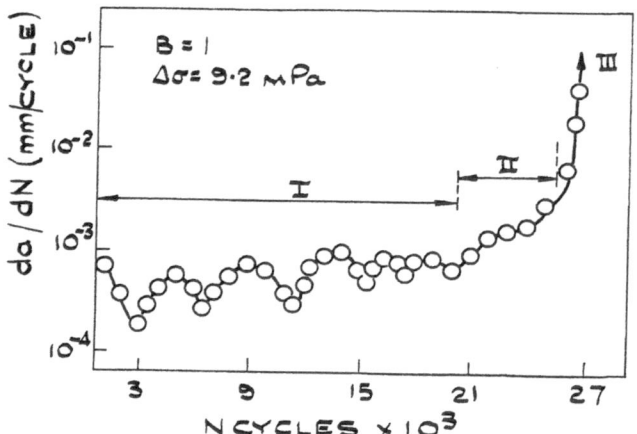

Fig. (9) - Log (crack growth rate) versus log (number of cycles)
test in 5% H_2SO_4

analyzing the raw data of crack length versus number of cycles represented in Figures 4 and 5 respectively. The technique adopted for obtaining the individual crack growth rates was by calculating average growth rates between two adjacent points; the process was continued by moving on to the next point and repeating the sequence until the data ran out.

Region I is associated with slow crack growth, showing the classic "pinning" effect characterized by the slowing down and speeding up of the crack as it travels through local fibre bundles and resin rich areas. Considerable differences in fracture surface topography can be seen in this region, Figure 10 shows a lack of fibre pullout for the acid fatigue test while Figure 11 shows a different topography for the fatigue test in air. The growth rates in Figure 8 for air show considerable scatter, a factor of 8 in the 10^4 to 10^3 mm/cycle growth rate range, the corresponding scatter range in sulphuric acid show a factor of 5 (Figure 9) for the same growth rate range. Incidentally, the scatter range is reduced to a factor of less than 2 when smoothing out techniques are used on the raw a versus N data in producing log (da/dN) versus log (ΔK) curves. The difference in fracture surface topography and crack growth rate characteristics can be explained by the difference in degree and magnitude of various failure mechanisms present in the air and acid fatigue tests. The mechanically applied alternating stress causes the formation of macro-cracks in the resin from previously debonded damage sites allowing access of the aqueous acidic environment to the exposed glass fibres. As the environment makes contact with the exposed glass fibre surfaces, the fibres fail by a mechanism associated with hydrogen ion attack on the glass [1].

Region II is effectively the transition period from slow to unstable crack growth and seems to disappear for the fatigue tests in air compared to the substantial region in acidic fatigue tests. Examination of the fracture surfaces for tests in air reveal a similarly dramatic transition.

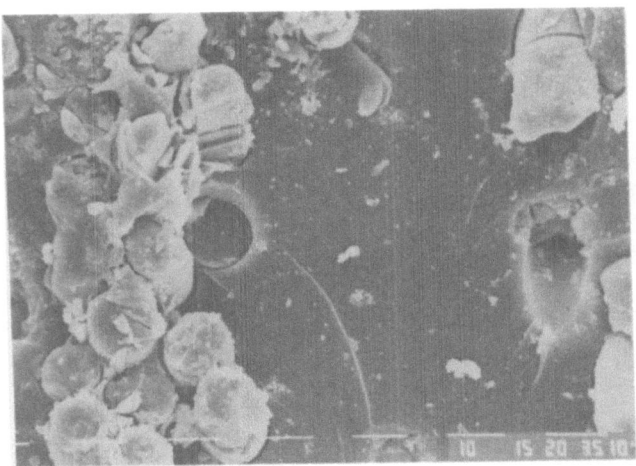

Fig. (10) - SEM of fatigue fracture surface in 5% H_2SO_4, region I

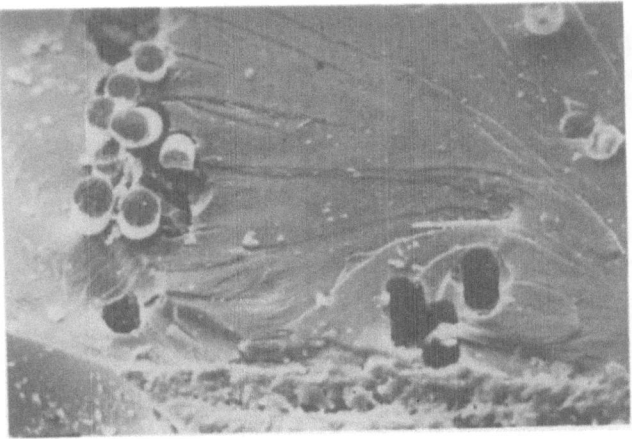

Fig. (11) - SEM of fatigue fracture surface in air, region I

Region I shows significant fibre pullout, Figure 11, up to 200 microns in length, while in region III, substantial fibre pullout giving a bushy appearance, Figure 12, pullout lengths in the order of millimeters. The fracture sur-

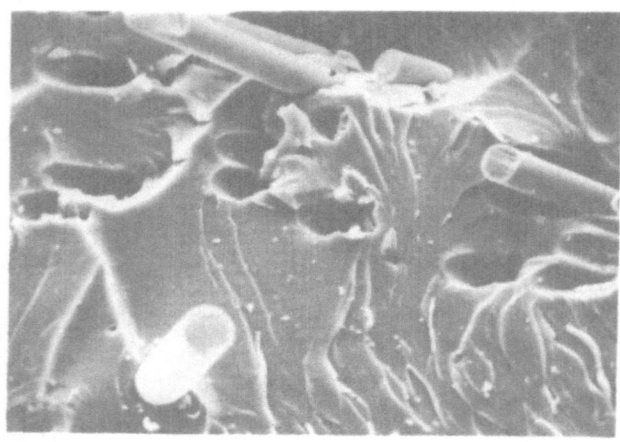

Fig. (12) - SEM of fatigue fracture surface in air, region III

faces in the 5% sulphuric acid environment showed a lack of fibre pullout in region I, Figure 10, while the unstable region III showed some fibre pullout similar
in appearance to region I in air, with pullout lengths of about 400 microns, Figure 13. The fibre pullout in region III for the acid tests was probably the re-

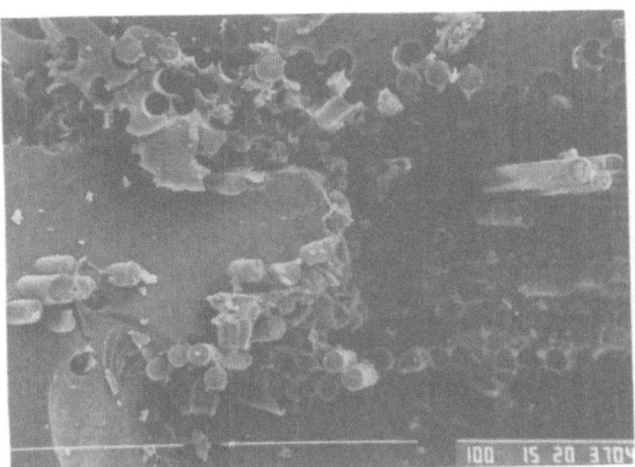

Fig. (13) - SEM of fatigue fracture surface in 5% H_2SO_4 region III

sult of too short a period for the chemical failure mechanism to operate effectively
in the final moments of separation.

Stresses experienced by the specimens tested in acid were much lower than
those conducted in air, i.e., approximately 50% lower, suggesting that the acid
environment weakens considerably the load bearing capacity of the glass fibre reinforcement. The type of instability seen prior to failure in the two different
environments also suggests a unique interaction between the magnitude of the stress
amplitude and the variation in the number and type of failure mechanisms experienced
in composites.

CONCLUSIONS

The present study has demonstrated a satisfactory application of fracture mechanics concepts in characterizing the environmental fatigue crack propagation in
glass fibre reinforced polyester resin composites under biaxial stress with the
use of a compliance technique. The Paris relationship applied to the crack growth
data showed a substantial effect of the acidic environment on fatigue crack growth,
reducing the Paris exponent from 15 in air to 6 in 5% sulphuric acid. It was also
found that the cyclic load range had to be reduced by a factor of 2 to produce comparative crack growth data in acid.

Fatigue damage in the composite specimens is caused mainly by mechanical attack
by repeated stress for tests in air, but in the acidic environment the fatigue dam-

634

age is accelerated by chemical attack to the glass fibres in addition to the alternating stress.

ACKNOWLEDGEMENTS

The authors acknowledge the support received by ICI Ltd., and the assistance of Dr. Prentice who carried out the scanning electron microscopy.

REFERENCES

[1] Roberts, R. C., Design strain and failure mechanism of GRP in chemical environment, Reinf. Plast. Congr. (British Plastics Federation), p. 145, 1978.

[2] Wachnicki, C. R. and Radon, J. C., "Photoelastic study of two biaxial stress fracture specimens", Proc. ECF2, Darmstadt, p. 36, VDI, Dusseldorf, 1979.

[3] Leevers, P. S., Radon, J. C. and Culver, L. E., "Crack growth in plastic panels interbiaxial stress", Polymer 17, p. 627, 1976.

[4] Irwin, G. R. and Kies, J. A., "Critical energy analysis of fracture strength", Welding J. Suppl., Vol. 33, p. 193, 1954.

[5] Owen, M. J. and Rose, R. G., "The fracture toughness and crack propagation properties of polyester resin casts and laminates", J. Phys. D: Appl. Phys., Vol. 6, p. 42, 1973.

[6] Paris, P. C. and Erdogan, F., "A critical analysis of crack propagation laws", Trans. ASME, J. Basic Engng., Vol. 85, p. 528, 1963.

[7] Owen, M. J. and Bishop, P. T., "Crack growth relationships for glass reinforced plastics and their application to design", J. Phys. Appl. Phys., Vol. 7, p. 1214, 1974.

[8] Owen, M. J. and Dukes, R., "Failure of glass reinforced plastics under single and repeated loading", J. Strain Anal., Vol. 2, p. 272, 1967.

[9] Carswell, W. S. and Roberts, R. C., "Environmental fatigue stress failure mechanism for glass fibre mat reinforced polyester", Composites, Vol. II, p. 95, April 1980.

DAMAGE TOLERANCE OF FIBRE COMPOSITE LAMINATES

M. J. Davis and R. Jones

Aeronautical Research Laboratories, Department of Defence Support
Melbourne, Australia

ABSTRACT

This paper presents a detailed review of the application of fracture mechanics to fibre composite materials. Particular attention is paid to the effects of manufacturing defects, low energy impact damage, holes and through the thickness cracks.

INTRODUCTION

The term damage tolerance is used to describe a design philosophy for military aircraft whereby a component is designed such that structural integrity is maintained while a defect of a given size is present in the structure. Modern military aircraft made of metallic materials are designed on this basis, using fracture mechanics to predict the size of a tolerable flaw under the applied loads.

With high performance composites, the field of damage tolerant design is complex, due to the inhomogeneous nature of the material and the failure modes, which differ significantly from those in metals. Composites exhibit near-linear stress-strain characteristics up to failure, while most metals display some ductile deformation. Thus, composites are less tolerant of overload. In fatigue, composites again differ from metals in that metals are sensitive to tension dominated fatigue loading, whereas composites generally exhibit good resistance to tension fatigue. Composites are, however, susceptible to local delaminations which may grow under compression fatigue.

A further complication arises in the manufacturing of composites. Because of the multi-phase nature of the material and the processes used in manufacturing, a substantially higher number of defects may exist in a composite component than would occur in a metallic component. Such defects also need consideration in the damage tolerance assessment of a component.

NATURE OF DEFECTS

A. Manufacturing Defects

Defects in composites due to manufacture are usually of two forms: those produced during the preparation and production of the composite, and those produced during machining, processing and assembly of the completed component. Typical production defects are shown in Table 1. Assembly defects usually arise as

TABLE 1 - TYPICAL PRODUCTION DEFECTS

Problem	Symptom	Effect
under-cure	soft-to-tacky surface	poor load transfer fibre to matrix
over-cure	charred resin	breakdown of resin
uneven cure	areas of uncured/ over-cured resin	thermal stress, defective regions
under-pressurization	high voidage	reduced strength, delaminations
over-pressurization	dry surface	high volume fraction, reduced shear strength
incorrect lay-up	possibly warpage	incorrect strength and mechanical behaviour
incorrect pre-preg	lack of tack, failure of resin flow	voids, low strength
foreign particle contamination	lumps, distortion	delaminations

a result of damage to the final product by such occurrences as scratches, gouges, impact debonding, fibre breakage, incorrect drilling of holes, overtightened fasteners, and so on. In any assessment of the need to repair or reject components, the location and size of the defect, the type of defect, the load spectrum anticipated for the component, and the criticality of the component need to be considered.

B. Impact Damage

The type of damage resulting from impact on composites depends on the energy level involved in tne impact, see Figure 1. High energy impact, such as ballistic damage, results in through penetration with perhaps some minor local delaminations. Lower energy level impact, which does not produce penetration, may result in some local damage on the impact zone, together with delaminations within the structure, and fibre fracture on the back face. Internal delaminations with little if any visible surface damage may result from low energy impact.

Studies [1] on the high energy ballistic damage problem have shown that, for perpendicular impact of boron/epoxy panels, the damage zone size on the entry face is independent of laminate thickness, while the exit face damage zone has been shown to increase linearly with panel thickness. Other work [2] has shown that panel thickness has little effect on residual strength. Also, specimens subjected to tension-tension fatigue after ballistic damage showed little variation in static strength. It was also shown that projectile velocity had no significant effect on residual strength. It was concluded that, while boron/epoxy composites exhibited a substantially larger decrease in static strength when compared to metals, the lower density of boron/epoxy provided a weight saving for structures of comparable damage tolerance.

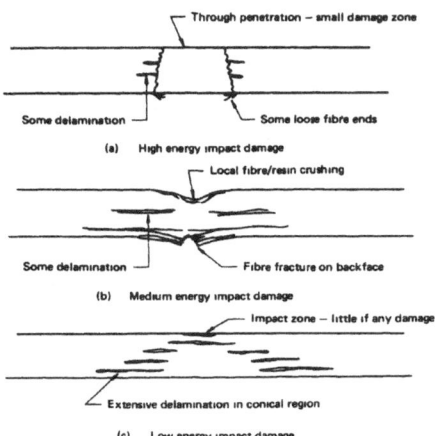

Fig. (1) - Failure modes in laminated composites, resulting from impact at various energy levels

Low energy impact damage is a major problem in practical fibre-composite structural applications. High and medium levels of impact energy cause surface damage which is relatively easily detected, and therefore repairs may be undertaken. Low energy impact can produce damage which is difficult to observe visually, (commonly termed 'barely visible impact damage', BVID). This type of damage is of concern because it may occur at quite low energy levels, and is only detectable using NDI techniques.

The effect of BVID on static and fatigue strengths of graphite/epoxy panels has been studied [3]. It was shown that impact degradation is of more concern in compression loading than tension loading. Specimens tested after impact at a standard energy level were compared with specimens with holes. In tension, the standard energy level impact was equivalent to a 3 mm diameter hole. However, for a similar specimen tested in compression, it was equivalent to a 25 mm diameter hole. Under reversed cycle fatigue ($R = -1.0$), the fatigue performance obtained was lower than for specimens with holes.

Studies have been made [4] of energy levels at which incipient damage to graphite/epoxy composites occur. Monolithic specimens of various thicknesses, and other specimens with honeycomb cores, were subjected to impact under drop weight conditions. Two indenters were used: a blunt spherical indenter and a sharp orthogonal indenter to simulate impact from the corner of a tool box. The specimens were instrumented such that the energy-time history was recorded, giving an accurate measure of the energy required to produce the onset of damage. From the results of these tests, Figure 2, it can be seen that damage to composites oc-

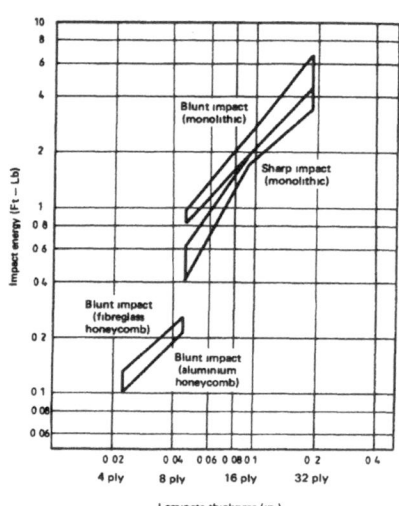

Fig. (2) - Impact energy for incipient damage to graphite/epoxy
laminates from [4]

curs at quite low energy levels, and depends on the thickness of the laminates.
As may be expected, a sharp indenter produces damage at a lower energy level than
a blunt indenter. (It is important to realize that incipient damage may not necessarily imply damage for which repair is required; this will depend on structural
and operational considerations for the component). Honeycomb panels were shown
to be more susceptible to impact damage than monolithic panels of the same thickness. From this and other work [5], it may be concluded that graphite/epoxy laminates experience damage at quite low energy levels. It was shown that the energy
level required to produce damage in graphite/epoxy sandwich panels was half the
energy level for damage in similar panels made from S-glass.

To assess the significance of BVID on graphite/epoxy composites in service, low energy impacts by typical maintenance equipment may be considered.
Graphical representations of energy levels for a typical tool box, spanner (wrench),
and screw-driver are shown in Figure 3, plotted against height of drop. The energy levels for incipient damage for various graphite/epoxy composite laminates
are also shown (dotted lines). From these curves, it is apparent that accidental
tool drops which would hardly cause concern on metal structures may produce damage
in graphite/epoxy structures. For example, a 16-ply monolithic graphite/epoxy
laminate experiences damage from a sharp screw-driver dropped 2 metres, a sharp
edged spanner dropped 1 metre, and a tool box corner dropped 50 mm. Obviously,
upper surfaces of an aircraft are more likely to receive impact from tool drops
than lower surfaces. Since most upper surfaces of an aircraft operate in a compression dominated load regime, and since it has already been shown that impact

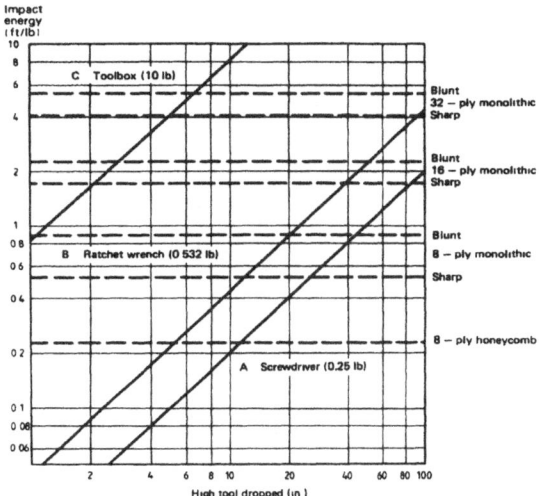

Fig. (3) - Impact energy, dropped tools from [4]

damage is more significant in composite compression members, consideration is required of maintenance procedures and equipment to provide care and protection of composite surfaces during maintenance.

C. Holes and Slots

Because of the anisotropic nature of composite materials, stress concentrations due to the presence of holes and slots may be substantially higher in composites than for an equivalent metallic structure. Also, fibre composites generally exhibit near-linear elastic behaviour to failure. Therefore, the combination of high stress concentrations and the absence of ductile yielding means that composites are relatively intolerant of overloads. The results of experimental evaluation of the effects of holes on fatigue life are discussed in the next section (D), and the procedure for the design of composites with holes is given in Section 3.

D. Fatigue

In general, tension fatigue in composites is not a problem, with fatigue strengths not greatly below static strength. In compression, larger diameter fibres such as boron produce substantially better fatigue strengths than smaller diameter fibres.

The problem of fatigue of composites with holes has been considered in [3] in which experiments were undertaken, involving fatigue regimes of R = -1.0 (tension-compression), R = 0.05 (tension-tension) and R = 10.0 (compression-compression), on graphite/epoxy composites with holes.

Under tension-tension fatigue, specimens achieved 10^6 cycles at maximum loads of up to 90% of ultimate strength, indicating that tension fatigue of composites with holes is not a major concern. For tests involving compression fatigue, failure lives followed the classical fatigue curves produced by metals. Thus, these materials with holes are susceptible to damage under compression fatigue. Further, similar results were obtained for reversed cycle fatigue, despite the fact that the stress excursion for this loading is twice the excursion experienced in compression fatigue. Thus, it was concluded that the compressive stress component of a fatigue spectrum is the dominating parameter in determining the fatigue life of composites with holes.

A further point of interest is the effect of fatigue on the static strength of a composite with holes, Figure 4. Here, the specimens have been subjected to

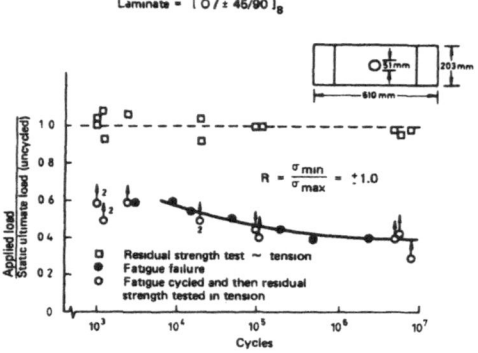

Fig. (4) - Residual strength of fatigue-cycled panels from [3]

a substantial proportion of their demonstrated fatigue life (R = -1.0) prior to tensile testing to failure. From these results, Figure 4, it may be seen that while the specimens were near to failure in fatigue, the tensile strength degradation was minimal. The implications of this with regard to the use of proof testing of structures for recertification after fatigue are significant. An interesting and comprehensive review of compressive fatigue life prediction methodology for composite structures is given in [6].

DAMAGE TOLERANT DESIGN

A. Laminates with Holes

Damage tolerant design for composite materials takes into consideration the types of defects which may by design be present in the structure, such as holes and cut-outs, and those defects which may occur inadvertently during manufacture or service, such as delaminations or cracks.

In the case of holes in composite laminates, several different approaches have been successfully developed. These methods are the progressive ply failure approach, e.g., [7], which may use either the Tsai-Hill or the maximum stress failure criterion for individual plies, the average stress failure criterion and the point stress failure criterion [8]. Of these methods, the latter two seem by far the simplest to use and they will now be described.

Consider a hole of radius R in an infinite orthotropic sheet, Figure 5.

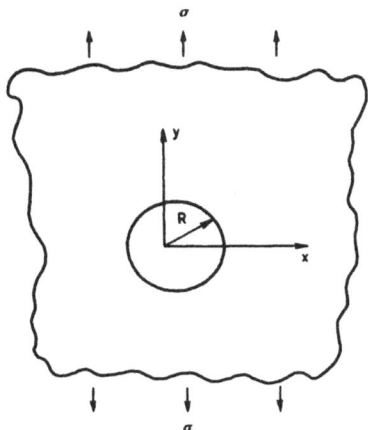

Fig. (5) - Reference axes for a hole in orthotropic panel under uniform tension

If a uniform stress σ is applied parallel to the y axis at infinity, then, as shown in [8], the normal stress σ_y along the x axis in front of the hole can be approximated by

$$\sigma_y(x,o) = \frac{\sigma}{2} \left[2 + \left(\frac{R}{x}\right)^2 + 3\left(\frac{R}{x}\right)^4 - (K_T-3)\left(5\left(\frac{R}{x}\right)^6 - 7\left(\frac{R}{x}\right)^8\right)\right] \tag{1}$$

where K_T is the orthotropic stress concentration factor, which for an infinite sheet is given by

$$K_T = 1 + \left[2\{(A_{11}A_{22})^{\frac{1}{2}} - A_{12} + (A_{11}A_{22} - A_{12}^2)/2A_{66}\}/A_{22}\right]^{\frac{1}{2}}$$

$$= 1 + \left[2((E_x/E_y)^{\frac{1}{2}} - \nu_{xy}) + E_x/G_{xy}\right]^{\frac{1}{2}} \tag{2}$$

Here, the A_{ij} are the stiffness coefficients for the laminate, where the subscript 1 denotes the direction parallel to the applied stress at infinity.

The average stress failure criterion [8] then assumes failure to occur when the average value of σ_y over some fixed length, a_o, ahead of the hole first reaches the unnotched tensile strength of the material. That is, when

$$\frac{1}{a_o} \int_R^{R+a_o} \sigma_y(x,o)dx = \sigma_o \tag{3}$$

Using this criterion with equation (1), gives the ratio of the notched to unnotched strength as

$$\frac{\sigma_N}{\sigma_o} = 2(1-\phi)/[2 - \phi^2 - \phi^4 + (K_T-3)(\phi^6-\phi^8)] \tag{4}$$

where

$$\phi = R/(R+a_o) \tag{5}$$

In practice, the quantity a_o is determined experimentally from strength reduction data.

The point stress criterion assumes that failure occurs when the stress σ_y at some fixed distance, d_o, ahead of the hole first reaches the unnotched tensile stress.

$$\sigma_y(x,o)\Big|_{x=R+d_o} = \sigma_o \tag{6}$$

It was shown in [8] that the point stress and average stress failure criteria are related, and that

$$a_o = 4d_o \tag{7}$$

The accuracy of these methods, in particular the average stress method, can be seen in Figures 6 and 7, where a_o was taken as 3.8 mm. The full lines represent predicted strength using the average stress criterion, while the dotted lines are predicted strengths from the point stress method.

With the use of modern structural analyses programs, it is relatively easy to apply these failure criteria to more complex problems. For example, recent tests [9] were carried out on various 16-ply graphite/epoxy laminates (AS/3501-5) with holes. The laminates were:

1. $(0/\pm45_2/0/\pm45)_s$

2. $(0_2/\pm45/0_2/90/0)_s$

3. $(0/\pm45/90)_{2s}$

The results are shown in Table 2, and are compared to predicted values using the average stress method with $a_o = 2.3$ mm.

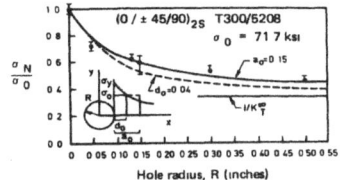

Fig. (6) - Comparison of predicted and experimental failure stresses for circular holes in $(0/\pm45/90)_{2s}$ T300/5208

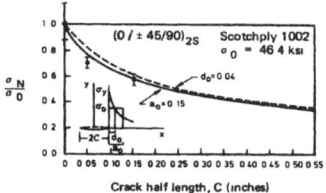

Fig. (7) - Comparison of predicted and experimental failure stresses for centre cracks in $(0/\pm45/90)_{2s}$ Scotchply 1002 from [12]

TABLE 2 - STATIC STRENGTH PREDICTIONS; [9]

Hole Size and Placement	Laminate No.	% of Unnotched Strength	
		Test	Average Stress
2-4.8 mm diameter countersunk cf Figure 8(a)	1	58.9	53.6
"	2	48.1	51.4
"	3	51.8	53.2
2-4.8 mm diameter countersunk cf Figure 8(b)	3	48.7	45.9
2-6.4 mm diameter countersunk cf Figure 8(a)	3	53.1	50.3
1-6.4 mm diameter non countersunk	2	54.0	52.6

As can be seen from the examples given, the average stress criterion provides accurate estimates of the strength reduction due to the presence of holes. This method is widely used in the aerospace industry [10], and has been applied to biaxial stress problems [11], to estimation of strength reduction due to battle damage [12], and to problems in which the stress is compressive [13].

B. Laminates with Cracks

For the problem of through cracks in a fibre composite panel, see Figure 9, there have been a variety of approaches which yield accurate results for

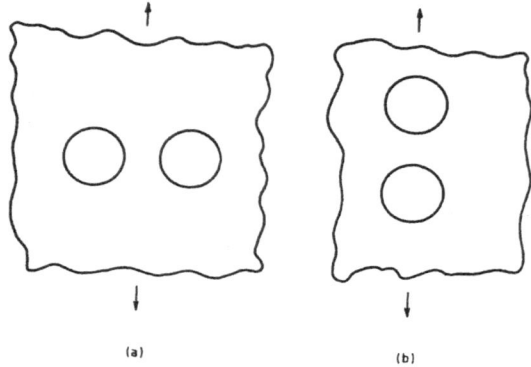

(a)   (b)

Fig. (8) - Location of holes with reference to applied load

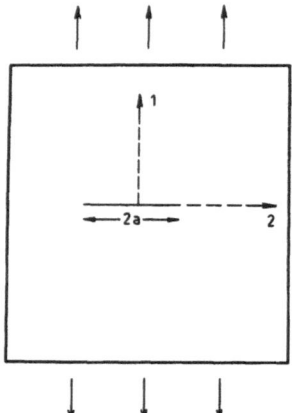

Fig. (9) - Through crack in an orthotropic panel

the reduction in strength due to the presence of a crack. Each is based on fracture mechanics. The most widely used methods are:

1. The strain energy density hypothesis [14],

2. The compliance approach [15],

3. The average and point stress criteria [8].

In the strain energy density hypothesis, the strain energy density factor S is defined as:

$$S = r_0 \frac{dW}{dV} \qquad (8)$$

where (dW/dV) is the strain energy density and r_0 is the radius of a core region surrounding the crack tip. Here,

$$\frac{dW}{dV} = \frac{1}{2} \sigma_{ij} \varepsilon_{ij} \qquad (9)$$

The assumptions made in the strain energy density approach are:

1. Crack growth is directed along a line from the centre of the spherical core (crack tip) to the point on the surface $r = r_0$ with the minimum strain energy density factor S_{min}.

2. Growth along this direction begins when S reaches the maximum critical value S_c which the material will tolerate.

This approach has been useful in the prediction of failure of bonded joints [14] and unidirectional composites [16], as well as for angle ply composites [14]. It has also been used for fatigue crack growth predictions [17]. This method will be discussed in more detail in a following section.

In the compliance method, the strain energy release rate G is calculated from

$$G = \frac{P^2}{2B} \frac{dC}{da} \qquad (10)$$

where P is the applied load, $C = \delta/P$, a is the crack half-length, B is the thickness, and δ is the deflection measured between the loading points. Failure occurs when G exceeds a critical value G_c, which is a material constant. For metallic structures, there is a simple relationship between G and the Mode I and Mode II stress intensity factors K_I and K_{II}. By considering a composite to be a linear elastic orthotropic material with the crack on one plane of symmetry, G is related to K_I and K_{II} by:

$$G = K_I^2((a_{11}/a_{22})/2)^{\frac{1}{2}}((a_{22}/a_{11})^{\frac{1}{2}} + (2a_{12}+a_{66})/2)^{\frac{1}{2}}$$
$$+ K_{II}^2 a_{11}[(a_{22}/a_{11})^{\frac{1}{2}} + (2a_{12}+a_{66})/2a_{11}]^{\frac{1}{2}}/\sqrt{2} \qquad (11)$$

where a_{ij} are the elements of the compliance tensor with the 1 subscript in the direction of the applied load. This procedure has been widely used [18].

For cracks in orthotropic laminates, the stress along the net section beyond the crack tip is given by

$$\sigma_y(x,o) = Y\sigma x/(x^2-a^2)^{\frac{1}{2}} \tag{12}$$

where σ is the applied stress, a is the crack half-length and Y is the finite width correction factor. Making use of this, the average stress failure criterion states that failure occurs when

$$\frac{1}{a_o} \int_a^{a+a_o} \sigma_y dx = \sigma_o \tag{13}$$

which, after using equation (12), gives

$$Y\sigma_N/\sigma_o = ((1-\S)/(1+\S))^{\frac{1}{2}} \tag{14}$$

where

$$\S = a/(a+a_o) \tag{15}$$

and a_o, as previously, is the length of the damage zone in front of the crack. If the point stress criterion is used, this gives

$$\frac{\sigma_N}{\sigma_o} = (1-\S_1^2)^{\frac{1}{2}} \tag{16}$$

where

$$\S_1 = a/(a+d_o)$$

As previously discussed, the two criteria are compatible if $d_o = a_o/4$. These criterion are relatively simple and are widely used. Table 3 shows the experimental tensile strengths and the fracture toughness for various laminate constructions as well as the damage zone sizes, d_o. Table 4 shows similar results, but for cracks oriented at an angle to the load direction. From these, and other works [19,20], it may be seen that the average and point stress methods can accurately predict the reduction in strength due to the presence of cracks. Other interesting works in this field are given in [21,22].

A further extension of the average stress criterion may be considered. For a notched orthotropic laminate, the fracture toughness K_{1c} is given by

$$K_{1c} = \sigma_N(\pi a)^{\frac{1}{2}} \tag{17}$$

TABLE 3 - TENSILE STRENGTH AND FRACTURE TOUGHNESS FROM [20]

Laminate Construction	Tensile Strength MPa	Fracture Toughness $MPa-m^{\frac{1}{2}}$	d_0 (mm)
T300/5208 graphite/epoxy			
$[0/90]_{4s}$	638	42.0	1.01
$[0/\pm45]_{4s}$	542	39.8	1.01
$[0/\pm45/90]_{2s}$	467-494	33.2	-
E-glass/epoxy (Scotch ply 1002)			
$[0/90]_{4s}$	424	30.7	1.01
$[0/\pm45/90]_{2s}$	321	24.3	1.01
Boron/epoxy			
$[0_2/\pm45]_{2s}$	697	62.0	1.27
$[0_3/\pm45/90]_{2s}$	691	59.8	1.27
$[0/\pm45]_{2s}$	608	49.9	1.27
$[0/\pm45/90]_{2s}$	421	38.7	1.27
As/3501 graphite/epoxy			
$[0]_{8s}$	1188	90.7	1.93
$[0/90]_{4s}$	500	24.6	0.38
$[0/\pm60]_{4s}$	452	28.0	0.61
$[0/\pm45/90]_{2s}$	480	32.8	0.72
$[0/\pm36/\pm72]_{2s}$	417	36.2	1.2
$[0/\pm30/\pm60/90]_{2s}$	466	27.2	0.45

where σ_N is the stress applied to the notched panel at failure, and a is the crack half-length. This may be expressed in terms of the unnotched strength, σ_0, by use of equation (14)

$$K_{1c} = \sigma_0(\pi a(1-\S)/(1+\S))^{\frac{1}{2}} \tag{18}$$

TABLE 4 - STATIC TENSILE STRENGTH PREDICTIONS AND TEST RESULTS
FOR SLANT AND NORMAL CRACKS IN Gr/Ep AS/3501-5

Laminate Construction	Crack Orientation to Load (degrees)	% Unnotched Tensile Strength		d_o (mm)
		Test	Average Stress	
$(0_2/\pm45/0_2/90/0)_s$	45	53.4	50.5	0.575
$(0/\pm45/\pm90)_{2s}$	45	51.4	50.7	0.575
$(0_2/\pm45/0_2/90/0)_s$	90	44.1	44.2	0.575
$(0/\pm45/\pm90)_{2s}$	90	47.2	44.3	0.575

It may be shown that as a becomes large, equation (18) approaches an asymptote.

$$K_{1c} \rightarrow \sigma_0(\pi a_0/2)^{\frac{1}{2}} \tag{19}$$

As it would be expected, K_{1c} is independent of crack length for large cracks. However, as $a \rightarrow o$, $K_{1c} \rightarrow o$. Since K_{1c} should remain constant and be independent of crack length, it is necessary to introduce a correction factor C, to provide a modified critical stress intensity K'_{1c}.

$$K'_{1c} = \sigma_N(\pi(a+C_0))^{\frac{1}{2}} \tag{20}$$

This is analogous to the Irwin type correction factor for plastic zone effects in ductile materials. Since fibre composites do not yield, C_0 may represent a damage zone ahead of the crack in which matrix cracking and fibre cracking occur. Use of equation (14) again gives,

$$K'_{1c} = \sigma_0(\pi(a+C_0)(1-\S)/(1+\S))^{\frac{1}{2}} \tag{21}$$

Here, as $a \rightarrow o$, $K_{1c} \rightarrow \sigma_0(\pi C_0)^{\frac{1}{2}}$. By taking $C_0 = a_0/2$, a constant value of K_{1c} is obtained, which is independent of crack length.

$$K'_{1c} = \sigma_0(\pi a_0/2)^{\frac{1}{2}} \tag{22}$$

C. A Unified Approach

The strength predictions using the point stress method and the strain energy density hypothesis may be unified as follows. In the case of uniform tension, $\sigma_{11} = \sigma$ and $\sigma_{ij} = 0$ for all $ij \neq 1$, the point stress criteria gives,

$$\sigma_N/\sigma_0 = (1-\S_1^2)^{\frac{1}{2}} \tag{23}$$

which for $a >> d_0$ reduces to

$$\sigma_N/\sigma_0 = (2d_0/a)^{\frac{1}{2}} \tag{24}$$

Now consider the value of S when $a \to 0$, i.e., $a << r_0$. In this case, $\sigma_{ij} \to \sigma$ for $i=j=1$ and $\sigma_{ij} \to 0$ for $ij \neq 1$ so that at $r = r_0$.

$$S = r_0 \sigma_{11}^2 E_{11}/2 = r_0 \sigma^2/E_{11} \tag{25}$$

The failure load for vanishing small cracks is σ_0, while at failure, S must equal S_c. Hence, at failure

$$S_c = r_0 \sigma_0^2/E_{11} \tag{26}$$

On the other hand, when $a >> r_0$, then from [16]

$$S = \hat{a}_{11} K_1^2 = \hat{a}_{11} I \sigma^2 \pi_a \tag{27}$$

where $\hat{a}_{11}$ depends on material properties and the direction of crack growth and where I is an "orthotropy" factor (I is usually very close to 1). At the failure load $\sigma_N S$ must equal S_c so that

$$\hat{a}_{11} I \sigma_N^2 \pi a = S_c \tag{28}$$

Making use of the previous expression for S_c gives

$$\hat{a}_{11} I \sigma_N^2 \pi a = r_0 \sigma_0^2/E_{11} \tag{29}$$

or

$$\sigma_N^2/\sigma_0^2 = r_0/(E_{11} \hat{a}_{11} \pi a I) \tag{30}$$

Comparing this relationship with that obtained from the point stress failure criteria, we see that the ratio for σ_N/σ_0 will coincide if we have,

$$2d_0 = \frac{r_0}{(\hat{a}_{11} E_{11} I \pi)} \tag{31}$$

For this to be true, we require

$$r_0 = 2d_0 \hat{a}_{11} E_{11} \pi I \tag{32}$$

Consequently, the two failure criteria can be made to coincide if we set r_o = $2d_o I\hat{a}_{11} E_{11} \pi$ and $S_c = r_o \sigma_o^2/E_{11}$. If this is done, the strain energy density hypothesis may be looked on as an extension of the point and average stress criteria, and can readily be used for complex loading configurations.

D. Ballistic Damage

So far, the problems of holes and cracks have been discussed. However, fracture mechanics has also been shown [12,23,24] to be applicable to the estimation of the residual strength of ballistically damaged composite panels. The residual strength of damaged tension panels has been derived [12] from the point stress criterion in the form

$$\sigma/\sigma_o = [(W_s - KW_{ke})/W_s]^{\frac{1}{2}} \tag{33}$$

where σ is the residual strength of the damaged panel, σ_o is the unnotched strength, K is an experimentally determined constant, W_s is the strain energy derived from the stress strain curve for the unnotched material, and W_{ke} is the kinetic energy per unit thickness imported into the structure. This model is applicable for the range of impact velocities which do not result in complete penetration. Where through penetration occurs, equation (4) may be used to estimate the reduction in strength.

The parameter K is dependent on the ratio of panel width to projectile diameter. However, tests [12] have shown that K rapidly asymptotes to a constant value.

E. Delaminations

Delaminations in layered composite materials may occur due to a variety of reasons, such as low energy impact or manufacturing defects. The presence of delaminations is of major concern in the vicinity of bonded joints and in compressively loaded components where the delaminations may grow under fatigue loading by out of plane distortion.

An early study into the growth of delaminations [25] arose from the B-1 composite development program. The effects of delamination size, temperature and moisture on AS/3501-5A graphite/epoxy panels was studied. From these tests, it is clear that compressive strength degrades with increasing size of delamination and with increasing temperature and moisture content. For example, one laminate, (90/0/±45) with a 12.5 mm delamination in a 38.1 mm wide specimen showed a 20% compressive strength reduction in room environment. A similar specimen tested at 132°C, and 1.3% moisture content, with the same size delamination gave a 50% compressive strength reduction. Recent tests [26] on the fatigue growth of a 50.8 mm diameter delamination (due to impact damage) in a wing box, showed that the effect of impact damage on a full scale structure can be estimated from specimen tests.

To date, there have been few analytical studies into delamination growth. A detailed study [27] used the NASTRAN finite element program to determine interlaminar stresses, which were subsequently used to accurately predict delamination growth rates. A similar study [28] into edge delaminations was undertaken, using the strain energy release rate G to predict delamination growth rates. In this work, it appears that the energy release rate remains relatively constant during delamination growth. A detailed three-dimensional analysis [29] of stresses developed during impact has also been undertaken. Sandwich panels with fibre composite faces, and a thick laminate were considered. For the laminate, impact develops large transverse shear stresses in the damage region and it appears that the sub-surface transverse shear stresses are critical. In the case of the sandwich panel, the sub-surface 0° plies would be critical.

Finite element analysis has also been used to investigate the static compressive failure of delaminations [30]. In this study, a detailed three-dimensional finite element analysis of a delaminated graphite/epoxy (viz T300/5208) beam is performed. The orientation of the plies being $(\pm 45/0_2)_{2s}$ and delaminations 12.7 mm, 25.4 mm and 38.1 mm long are considered. The location of the delamination is varied. The maximum strain energy density S at each point along the delamination is computed as is the energy release rate G. The far field failure strain is then estimated on the basis that at failure $G_c = 7.5$ Nm^{-1} and or $S_c = 9.64$ Nm^{-1} as given in [31] and [14] respectively. Both failure criteria give a far field failure strain of approximately 4.5×10^{-3} for a delamination 38.1 mm long between the near surface ± 45 plies and the 0_2 plies. The failure strain for such a delamination is in good agreement with that given in [26] and again confirms the conclusion given in [26] that estimates of the critical flaw size due to delaminations or impact damage may be obtained from specimen tests. Indeed, as can be seen from Table 5, static failure has been found to occur for a variety of laminates and fibre orientation at approximately 4000 microstrain. Furthermore, it has now been clearly shown in [30,25,32] that delamination growth under compressive loading is entirely due to the unsymmetrical nature of the plies above and below the delamination. The compressive forces acting on these plies produce out of plane bending and it is this bending which drives the delamination.

Approximate analytical methods other than the finite element method have also been used [33-35]. These methods are capable of differentiating between laminates whose stacking sequence is susceptible to delamination growth and those laminate stacking sequences more resistant to delamination.

The cause of the final failure of the bulk material is as yet uncertain although a number of hypothesis have been proposed. It is most probable that failure is due to a combination of out of plane bending due to loss of symmetry and a reduction of the net cross section.

DAMAGE TOLERANCE MANAGEMENT

Various papers have shown methods for damage tolerance enhancement of composite materials. It has been shown [36] that surface treatment of graphite fibres to enhance bond strength produces a decrease in Izod impact fracture energy. It is believed that this is due to an alteration in failure mode which for untreated fibres involves more delaminations, which absorb more energy. This work also showed that by use of a plasticiser in the resin system, a higher strain to failure matrix was produced, which resulted in an increase in fracture energy.

TABLE 5 - COMPRESSIVE STRENGTH OF VARIOUS IMPACT DAMAGED LAMINATES.
SPECIMENS LISTED WERE IMPACT DAMAGED, THEN LOADED STATI-
CALLY TO FAILURE. FROM [26]

Layup	Skin Thick- ness mm	Delamination Area mm^2	Compressive Failure Strain μ
AS/3501-6 $(\pm45/0_2/\pm45/0_2/\pm45/0/90)_{2S}$	6.91	1290	3780
AS/3501-6 $(\pm45/90/-45/+22.5/-67.5/-22.5/\pm67.6/ \pm45/+67.5/+22.5/-67.5/-22.5/\pm67.5/ \pm22.5/0_2/\pm22.5)_S$	6.45	1613	4630
AS/3501-6 $(0/\pm45/0_2/\pm45/0)$ Skin on HRP-3/16-5.5 Honeycomb Core	1.07	1161	4270
AS/3501-6 $(\pm45/0*/\pm45/0_2*/90*/0_2*/\pm45*/0_2*/145*/ 0*/+45*/0*/\pm45*/0*/-45*/0*/\pm45*)_S$	12.7	2389	4090
AS/3501-6 $(90*/-45*/\pm67.5/\pm22.5*/0/90*/+22.5*/ 0_2/\pm45*/0*/-22.5*/+45*/+67.5*/+22.5*/ -67.5*/-45*/+22.5_2*/0)_S$	11.1	7097	5460
T300/5208 $(\pm45/0_2/\pm45/0_2/\pm45/0/90)_{2S}$	6.71	1419	4000
T300/5208 $(\pm45/0/90/\pm45/0/90)_{3S}$	6.71	1226	4000

*Double-ply material, all others single ply.

DAMAGE TOLERANCE MANAGEMENT

Various papers have shown methods for damage tolerance enhancement of composite materials. It has been shown [36] that surface treatment of graphite fibres
to enhance bond strength produces a decrease in Izod impact fracture energy. It
is believed that this is due to an alteration in failure mode which for untreated
fibres involves more delaminations, which absorb more energy. This work also
showed that by use of a plasticiser in the resin system, a higher strain to failure matrix was produced, which resulted in an increase in fracture energy.

The effects of laminate stacking sequence on fracture strength have been considered [3], using specimens with the same number of oriented plies, but with differing stacking sequences. The specimens were also slotted. While design procedures would suggest that the laminates would ostensibly have the same unnotched

strength, the fracture strengths of the specimens varied by up to 30%. Stronger panels were those with the outer layers at 0° or 90° to the load direction. Specimens with ±45° fibres on the outer faces produced lower fracture strengths. Also, bunching of 0° layers together was shown to produce higher notched strengths, since extensive delamination was possible between the 0° layers, which absorbs more energy.

A major cause of impact related problems in high performance composites is their low strain to failure. Honeycomb face sheets of graphite epoxy have a strain to failure about 1/3 of that for S-glass face sheets. Various works have evaluated the concept of hybridisation - mixing the fibre types - thus utilizing the advantageous properties of one material to overcome deficiencies of another. Glass fibre face sheets on a graphite core have been compared to an all graphite structure [36]. The higher strain to failure of glass fibres led to substantial increases in threshold energies for impact damage, with the energy level required to produce delamination being increased by a factor of 4.

The more recent availability of aramid fibres has added impetus to studies in hybrid composite systems. Aramid fibres (generally known under their trade name "Kevlar") are cheap, and exhibit a high strain to failure, but also have a substantially higher modulus of elasticity than glass fibres. Therefore, they exhibit a high tensile strength without the stiffness penalty of glass fibre composites. Most work using this fibre has concentrated on the aramid/graphite/epoxy hybrid, since the aramid material provides an increase in impact resistance and a reduced material cost, while the graphite fibres provide compressive strength, which the aramid material lacks. Some studies [37] have shown that aramid/graphite hybridisation lead to a doubling of residual strength after impact compared to all-graphite specimens. For unidirectional layups, the same comparison showed an improvement by a factor of 3.

A further concept for improvement of notch sensitivity is that of specific hybridisation. The incorporation of high strain to failure strips in a structure (softening strips) acts to produce a zone where fracture of the structure is arrested by the high displacement capability of the strips. Similarly, the presence of high modulus strips (hardening strips) provides a restraint to opening of a crack, thus assisting arrest of the crack.

REFERENCES

[1] Avery, J. G. and Porter, T. R., "Comparison of the ballistic response of metals and composites for military aircraft applications", ASTM STP 568, pp. 3-29, 1975.

[2] Suarez, J. A. and Whiteside, J. B., "Comparison of residual strength of composite and metal structures after ballistic damage", ASTM STP 568, pp. 72-91, 1975.

[3] Walter, R. W., Johnson, R. W., June, R. R. and McCarthy, J. E., "Designing for integrity in long-life composite aircraft structures", ASTM STP 636, pp. 228-247, 1977.

[4] NAEC Report 92-136, Naval Air Engineering Centre.

[5] Oplinger, D. W. and Slepetz, J. M., "Impact damage tolerance of graphite epoxy sandwich panels", ASTM STP 568, pp. 30-48, 1975.

654

[6]   Saff, C. R., "Compression fatigue life prediction methodology for composite
      structures-literatures survey", NADC Report No. NADC-78203-60, June 1980.

[7]   Soni Som, R., "Failure analysis of composite laminates with a fastener hole",
      AFWAL Report TR-80-4010, 1980.

[8]   Nuismer, R. J. and Whitney, J. M., "Uniaxial failure of composite laminates
      containing stress concentrations", ASTM STP 593, pp. 117-142, 1975.

[9]   Nuismer, R. J. and Labor, J. D., "Applications of the average stress failure
      criterion: Part 1, Tension", J. Composite Materials, Vol. 12, p. 238, 1978.

[10] Pimm, J. H., "Experimental investigation of composite wing failure", AIAA,
      pp. 320-324, 1978.

[11] Daniel, I. M., "Behaviour of graphite epoxy plates with holes under biaxial
      loading", Experimental Mechanics, pp. 1-8, 1980.

[12] Husman, G. E., Whitney, J. M. and Halpin, J. C., "Residual strength characteri
      sation of laminated composites subjected to impact loading", ASTM STP 568,
      pp. 92-113, 1975.

[13] Nuismer, R. J. and Labour, J. D., "Application of the average stress failure
      criterion: Part 2, Compression", J. Composite Materials, Vol. 13, 1979.

[14] Sih, G. C., Mechanics of Fracture: Linear Response, pp. 155-192 of Numerical
      Methods in Fracture Mechanics, Proceedings of the First International Con-
      ference, Swansea, A. R. Luxmoore and D. R. J. Owen, eds., 1978.

[15] Slepetz, J. M. and Carlson, L., "Fracture of composite compact tension speci-
      mens", ASTM STP 593, pp. 143-162, 1975.

[16] Sih, G. C. and Chen, E. P., "Fracture analysis of unidirectional composites",
      J. Composite Materials, Vol. 7, pp. 230-244, 1973.

[17] Badaliance, R. and Dill, H. D., "Compression fatigue life prediction method-
      ology for composite structures", Vol. II, Technical Proposal, McDonnell Air-
      craft Co., Report No. MDC-A573, 1979.

[18] Barnby, J. T. and Spencer, B., "Crack propagation and compliance calibration
      in fibre-reinforced polymers", J. of Materials Science 11, pp. 78-82, 1976.

[19] Morris, D. M. and Hahn, H. T., "Mixed mode fracture of graphite/epoxy compos-
      ites: fracture strength", J. Composite Materials, Vol. 11, p. 124, 1977.

[20] Carprino, G., Halpin, J. C. and Nicolais, L., "Fracture mechanics in compos-
      ite materials", Composites, October 1979.

[21] Dorey, G., "Damage tolerance in advanced composite materials", RAE Tech.
      Report TR 77172, 1977.

[22] Kanninen, M. F., Rybicki, E. F. and Brinson, H. F., "A critical look at cur-
      rent applications of fracture mechanics to the failure of fibre-reinforced
      composites", Composites, pp. 17-22, 1977.

[23] Olster, E. F. and Roy, P. A., "Tolerance of advanced composites to ballistic damage", ASTM STP 546, pp. 583-603, 1974.

[24] Dorey, G., Sidey, R. and Hutching, J., "Impact properties of carbon fibre/ Kevlar reinforced plastic hybrid composites", RAE Tech. Report 76057, 1976.

[25] Konishi, D. Y. and Johnston, W. R., "Fatigue effects on delaminations and strength degradation in graphite/epoxy laminates", ASTM STP 674, p. 597, 1979.

[26] Gause, L. W., Rosenfeld, M. S. and Vining, R. E., Jr., "Effect of impact damage on the XFV-12 a composite wing box", 25th National SAMPE Symposium and Exhibition, Vol. 25, pp. 679-690, 1980.

[27] Ratwani, M. M. and Kan, H. P., "Compression fatigue analysis of fibre composites", NADC Report 78049-60, 1979.

[28] Rybicki, E. F., Schmueser, D. W. and Fox, J., "An energy release rate approach for stable crack growth in free-edge delamination problem", J. Composite Materials, Vol. 11, p. 470, 1977.

[29] Stanton, E. L. and Crain, L. M., "Interlaminar stress gradients and impact damage", pp. 423-440 of Fibrous Composites in Structural Design, E. M. Lenoe et al, eds., Plenum Press, New York, 1980.

[30] Jones, R. and Callinan, R. J., "Analysis of compression failures in fibre composites", ARL Structures Report, to be published.

[31] Ramkumar, R. L., Kulknarni, S. V. and Pipes, R. B., "Definition and modelling of critical flaws in graphite fiber reinforced epoxy resin matrix composite materials", NADC-76228-30.

[32] Verette, R. M. and Demuts, E., "Effects of manufacturing and in-service defects on composite materials", Proc. Army Symposium on Solid Mechanics, 1976, Composite Materials: The Influence of Mechanics of Failure on Design, AMMRC-MS-76-2, pp. 123-137, 1976.

[33] Rodini, B. T. and Eisenmann, J. R., "An analytical and experimental investigation of edge delamination in composite laminates", pp. 441-458 of Fibrous Composites in Structural Design, see [29].

[34] Pagano, N. J. and Pipes, R. B., "Some observations on the interlaminar strength of composite laminates", Int. J. Mech. Sci., Vol. 15, p. 679, 1973.

[35] Pipes, R. B., "Interlaminar strength of laminated polymeric matrix composites", AFML-TR-7682, 1976.

[36] Bradshaw, J., Dorey, G. and Sidey, R., "Impact resistance of carbon fibre reinforced plastics", RAE Technical Report 72240, 1973.

[37] Dorey, G., "Improved impact tolerance in composite structures", Chapter 9 of of a symposium on the Design and Use of KEVLAR[R] in Aircraft, Geneva, 1980.

PROBABILISTIC ASPECTS OF MECHANICAL DAMAGE IN CONCRETE STRUCTURES

J. Mazars

Ecole Normale Superieure De L'Enseignement Technique
94230 Cachan, France

ABSTRACT

The application of "damage mechanics" to concrete has shown that it was a way to describe the behavior and the rupture, but the problem of the sensibility of the material to the volume effects and the gradient effects was not solved.

The adjunction of a probabilistic theory, presented in this paper, from the works of Weibull, completes its efficiency.

The whole theory has been inserted in a finite element computation program and results of different tests show a good concordance with results of calculation.

INTRODUCTION

The behavior of concrete under mechanical solicitations is the consequence of the existence and development of microcracks [1]. The growth of microcracks leads to the formation of cracks and then to rupture [2].

The evolution of those phenomena is linked to the initial state of the material and in particular to the arrangement of initial flaws (voids, adherence defaults, shrinkage microcracks, etc.).

The distribution of sizes, shapes and orientation of flaws may be considered as a random phenomenon introducing an effect of volume and an effect of loading systems into the strength of concrete structures.

After a brief recall of the principles of damage mechanics application to concrete, we propound in this paper to discuss the probabilistic aspect of this problem.

MECHANICAL DAMAGE OF CONCRETE

The application of mechanical damage has been presented in other papers [3,4]; we intend to recall the principles and the results obtained.

First, we used the concept of effective stress [5] in a coupling damage-elasticity, which is sufficient to describe the main phenomena of strain and rupture of concrete.

A.   Damage in Uniaxial Traction

From the experiment curve $\sigma-\varepsilon$ in uniaxial traction [6], the behavior of concrete is represented by a damage constitutive equation:

$$\sigma = \varepsilon E_0 (1-D)$$

$$\varepsilon \leq \varepsilon_{D_0} \rightarrow D = 0$$

$$\varepsilon > \varepsilon_{D_0} \rightarrow D = 1 - \frac{\varepsilon_{D_0}(1-A)}{\varepsilon_M} - \frac{A}{\exp \cdot [B(\varepsilon_M - \varepsilon_{D_0})]}$$

E_0 is the initial Young modulus

ε_{D_0} is the damage threshold, see Figure 1

A,B are material parameters

ε_M is the maximal strain reached

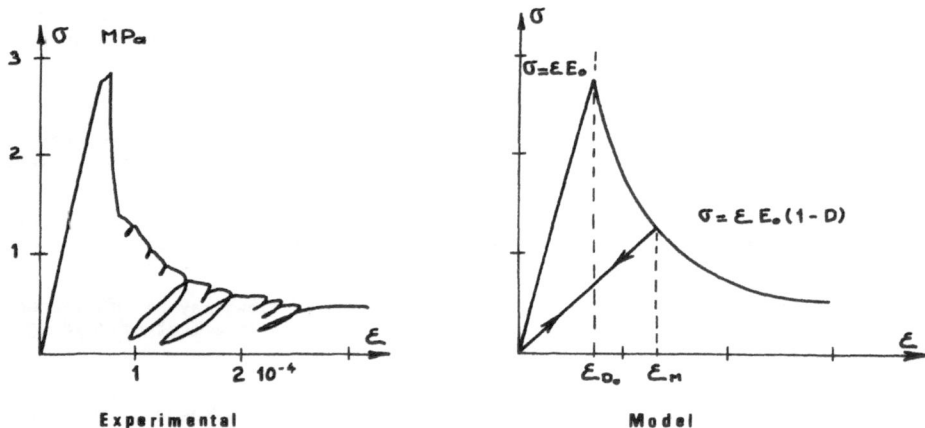

Experimental   Model

Fig. (1) - Behavior of concrete in uniaxial traction

B.   Extension to Tridimensional Problems

In the case of multiaxial solicitation, many tests [7,8] have shown the important part played by tension strains ($\varepsilon > 0$) in the evolution of microcracks, i.e., in damage.

In order to express the state of tension for a particular point, we used the notion of equivalent strain defined by:

$$\tilde{\varepsilon} = \sqrt{<\varepsilon_1>^2 + <\varepsilon_2>^2 + <\varepsilon_3>^2}$$

with

$$<\varepsilon_i> = \varepsilon_i \text{ if } \varepsilon_i > 0$$

$$<\varepsilon_i> = 0 \text{ if } \varepsilon_i \leq 0 \qquad \varepsilon_i: \text{ main strain}$$

As part of the formulation of an isotropic damage (D scalar), the extension of the results from uniaxial traction enabled us to obtain the following constitutive equations:

$$\sigma = (1-D)[2\mu_0 \varepsilon + \lambda_0 \text{ trace}(\varepsilon)]$$

with

$$D = 1 - \frac{\varepsilon_{D_0}(1-A)}{\tilde{\varepsilon}_M} - \frac{A}{\exp \cdot [B(\tilde{\varepsilon}_M - \varepsilon_{D_0})]}$$

σ, ε : stress and strain tensors

μ_0, λ_0: initial Lamé coefficients

$\tilde{\varepsilon}_M$: maximal equivalent strain reached

A numerical calculation using the finite element method, incorporating the evolution of damage by way of variable stiffness has been elaborated.

Such a type of calculation enables us to account for: the local behavior of the material; the global behavior of the structure; the evolution of the damaged zone.

C. Comparison Calculation-Tests

The experimentation realized on specimens made from the same concrete and preserved in the same conditions enabled us to attest the validity of the concepts presented.

Figures 2 and 3 show the results obtained on a three points bending beam and on a notched plate. We can note that:

- the calculation accurately accounts for the local (P-ε) and global behavior (P,δ);

- the evolution of the damaged zone given by calculation enables us to simulate the propagation of a crack such as is shown by the test realized on a plate.

D. Problems Arising from the Identification of the Parameters of the
    Evolution Damage Law

The law of damage evolution includes three parameters:

- ε_{D_o} which is the damage threshold;
- A and B which condition the shape of the curve σ-ε beyond ε_{D_o}.

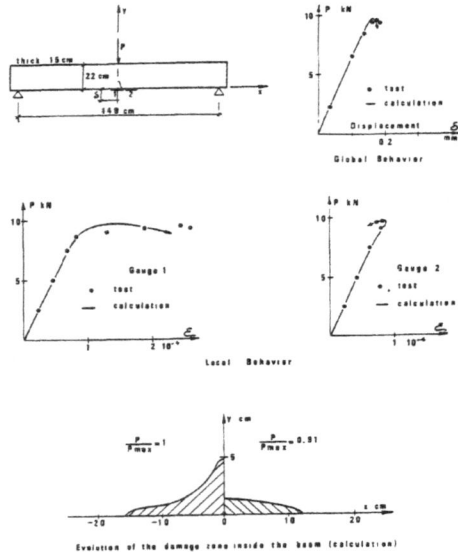

Fig. (2) - Results obtained on a three points bending beam

A series of preliminary tests enabled us to identify parameters A and B
(A = 0.8, B = 2.10^{+4}) but the whole set of tests showed that in order to describe
the various behaviors accurately, it was necessary to adjust the value of ε_{D_o}. The
adjustment is effected so as to have the same maximum load for each test.

Thus we can note:

- for the same kind of tests, a variation of ε_{D_o} of about 10% linked to
  to the dispersion of the results;

- between many kinds of tests, a variation which can exceed 100% and which
  is particularly sensitive to the intensity of volume of the specimens and
  the existence of stress gradients.

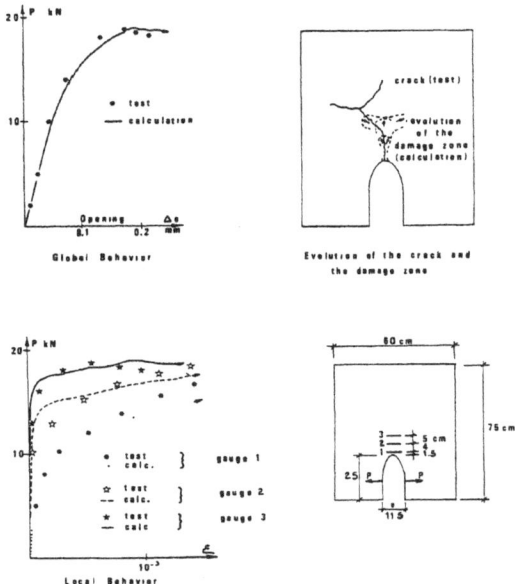

Fig. (3) - Results obtained on a notched plate

Thus, the results shown on Figures 2 and 3 were determined:

- for the three points bending beam with ε_{D_0} = 0.71 10^{-4};

- for the notched plate with ε_{D_0} = 1.31 10^{-4}.

(The initial elastic parameters are E_0 = 3 x 10^4 MPa and ν_0 = 0.2).

The explanation we give for such phenomena links up with the idea according to which the behavior and fracture of concrete depends upon the way the material has been solicited, and goes through a probabilistic analysis of the distribution of flaws inside the material.

PROBABILISTIC ASPECT

As early as 1939, Weibull [9] propounded laws to relate the fracture probability of brittle materials. In this study, the material was supposed to be isotropic and statistically homogeneous, the rupture at the level of a critical flaw leading to complete rupture.

From these considerations, we raised the same kind of hypothesis in connection with damage instead of rupture.

A.   Damage Probability

The notion of equivalent strain $\tilde{\varepsilon}$ can be used to characterize the damage threshold for an elementary volume of material ΔV. We call $\tilde{\varepsilon}_p$ this parameter.

In a volume V constituted by a set of elements ΔV, the distribution of $\tilde{\varepsilon}_D$ can be considered as a random phenomenon; let us call $f(\tilde{\varepsilon}_D)$ the probability density function.

If $\tilde{\varepsilon}_s$ is a solicitation (in terms of equivalent strain) applied to the volume ΔV, the damage probability of ΔV is:

$P_d(\tilde{\varepsilon}_s, \Delta V)$ = probability that $\tilde{\varepsilon}_D \leq \tilde{\varepsilon}_s$

The probability of "non-damage" then becomes

$P_{nd}(\tilde{\varepsilon}_s, \Delta V) = 1 - P_d(\tilde{\varepsilon}_s, \Delta V)$

Let us consider a volume such as $V = n\Delta V$. The hypothesis according to which the "non-damage" probabilities of the different volumes ΔV are independent from each other, enables us to use the multiplicativity rule:

$P_{nd}(\tilde{\varepsilon}_s, n\Delta V) = [P_{nd}(\tilde{\varepsilon}_s, \Delta V)]^n = 1 - [1 - P_d(\tilde{\varepsilon}_s, \Delta V)]^n$

which, when n is large, leads us to write:

$P_d(\tilde{\varepsilon}_s, V) = 1 - \exp[-nP_d(\tilde{\varepsilon}_s, \Delta V)]$

In order to respect the limit conditions,

$P_d(\tilde{\varepsilon}_s, V) = 1$ for $\tilde{\varepsilon}_s \to +\infty$ and $P_d(\tilde{\varepsilon}_s, V) = 0$ for $\tilde{\varepsilon}_s = 0$

We were led to choose:

$nP_d(\tilde{\varepsilon}_s, \Delta V) = k\tilde{\varepsilon}_s^m V$

where: m is the Weibull parameter and k is the constant.

Thus, for a uniform solicitation on a volume V, the damage probability is:

$$P_d(\tilde{\varepsilon}_s, V) = 1 - \exp(-k\tilde{\varepsilon}_s^m V)$$

We can easily show that for a non-uniform solicitation $\tilde{E}_s$ in the volume V

$$P_d(\tilde{E}_s, V) = 1 - \exp(-k \int_V \tilde{\varepsilon}_s^m dv)$$

In any specimen, it is possible to express the local solicitation in any point, $\tilde{\varepsilon}_s$, in terms of the maximum solicitation $\tilde{\varepsilon}_{Ms}$ which appear in the specimen [10]. We can thus deduce:

$$\int_V \tilde{\varepsilon}_s^m dv = \tilde{\varepsilon}_{Ms}^m \int_V g^m dv$$

g is a parameter depending only on the geometry of the specimen and the type of solicitation exerted, but not on the intensity of the solicitation because before damage the material is linear elastic.

Those results enable us to express:

- the density function $f(\tilde{\varepsilon}_D)$ which is obtained by derivation of P_d;

- the mean value of $\tilde{\varepsilon}_o$ which corresponds to the mean value of the maximal local solicitation $\tilde{\varepsilon}_{Ms}$, in a given specimen, when the damage threshold ε_{D_o} is reached; then the expression of ε_{D_o} is:

$$\varepsilon_{D_o} = \frac{\Gamma(1+1/m)}{[k \int_V g^m dv]^{1/m}} \qquad (\Gamma \text{ is the gamma function})$$

The standard deviation of ε_{D_o}

$$s = \varepsilon_{D_o} \sqrt{\frac{\Gamma(1+2/m)}{[\Gamma(1+1/m)]^2} - 1}$$

B. Damage Threshold Calculation

Let us consider two specimens of a same material of respective volumes V_1 and V_2 and solicited in a different way. The damage threshold are expressed:

- for V_1 by $\varepsilon_{D_o,1} = \dfrac{\Gamma(1+1/m)}{[k \int_{V_1} g_1^m dv]^{1/m}}$

- for V_2 by $\varepsilon_{D_o,2} = \dfrac{\Gamma(1+1/m)}{[k \int_{V_2} g_2^m dv]^{1/m}}$

We can deduce:

$$\varepsilon_{D_o,1}^m \int_{V_1} g_1^m dv = \varepsilon_{D_o,2}^m \int_{V_2} g_2^m dv$$

As a conclusion: In an element of volume V subjected to any solicitation, the most probable value for the damage threshold is reached when:

$$\int_V \tilde{\varepsilon}_s^m dv = W_o$$

Like m, W_o is a constant characteristic of the material. That condition also expresses that the damage probability $P_d(\tilde{E}_s,V)$ at the most probable damage threshold, is the same whatever the geometry of the element and solicitation exerted.

C. Application to Concrete

The relation expressing the fact that the most probable damage threshold has been reached can be easily introduced into a finite element calculation.

For a given solicitation expressed by $\tilde{\varepsilon}_{si}$ on the element i of volume ΔV_i, we can calculate for the whole specimen:

$$\sum_{i=1}^{n} \tilde{\varepsilon}_{si}^m \Delta V_i = W$$

When $W = W_o$, the damage threshold is reached and ε_{D_o} corresponds to the value of $\tilde{\varepsilon}_s$ in the most solicited element $\tilde{\varepsilon}_{Ms}$.

The identification of the parameter m and W_o led us to the following values for the concrete tested:

$$m = 6,5 \quad W_o = 3,5 . 10^{-25}$$

They enabled us to find for the standard deviation: $s = 0,180 \varepsilon_{D_o}$.

D. Comparison Calculation-Tests

In the table on Figure 4, we gathered all the results obtained. The most outstanding comparison can be made on the maximum loads given by calculation on the one hand and the average values of that load given by the tests on the other hand, did not differ over 4%.

As far as the standard deviations are concerned, we can note that, the tests give results iftting quite perfectly into the anticipations given by calculations; some reservations have to be made however, because of the number of tests effected in certain categories.

dim : cm				
ε_{D_o}	$9{,}95 \ 10^{-5}$	$7{,}24 \ 10^{-5}$	$6{,}17 \ 10^{-5}$	$14{,}1 \ 10^{-5}$
P_{max} kN	11,7	9,9	14,8	19,6
$s(\varepsilon_{D_o})$	$1{,}79 \ 10^{-5}$	$1{,}31 \ 10^{-5}$	$1{,}12 \ 10^{-5}$	$2{,}55 \ 10^{-5}$
$s^*(P_M)$	1,1	1,3	1,6	2,4
Number of tests	12	3	5	3
$\overline{P}_{max}$ kN	11,9	9,6	15,3	18,9
$s(P_M)$	1	0,5	0,5	9,9

(CALCULATION / EXPERIMENT)

$s^*(P_M)$ is the variation of the maximum load obtained by calculation with a variation $s(\varepsilon_{D_o})$ around the value ε_{D_o} of the damage threshold.

Fig. (4) - Comparison of the results obtained by the statistic analysis with those given by tests

CONCLUSION

Damage mechanics enable us to describe as far as the local and the global behavior of concrete structures.

The volume effects and the gradient effects remained an obstacle for its utilization in estimated calculations. These problems are now solved with the adjunction of a probabilistic method based on the works of Weibull.

The whole results show that from the knowledge of four parameters - two to define the damage evolution (A and B), two to define the damage threshold (m, W_o) and the initial elastic characteristics of the material (E_o, ν_o) can be determinated:

- the mean values of structures limit loads and all the characteristics of the statistical estimation of the results.

The comparison with the experimental results enable us to attest the validity of the hypothesis formulated and allow us to envisage the application of those concepts to reinforced concrete and prestressed concrete structure design.

REFERENCES

[1] Mazars, J., "Evolution of microcracks in concrete: the formation of cracks", Journées d'études "La fissuration des bétons", Paris 1980, Annales de l' I.T.B.T.P. No. 398, October 1981.

[2] Dhir, R. H. and Sangha, M., "Development and propagation of microcracks in plain concrete", Matériaux et Constructions No. 37, pp. 17-23, 1974.

[3] Mazars, J., "Mechanical damage and fracture of concrete structures", Proceedings of ICF5 Cannes, Vol. 4, pp. 1499-1506, April 1981.

[4] Lemaitre, J. and Mazars, J., "Application de la théorie de l'endommagement au comportement non linéaire et à la rupture du béton", Annales de l' I.T.B.T.P. No. 401, January 1982.

[5] Kachanov, L. M., "Time of the rupture process under creep conditions", Tzv. Akad. Nank. SSR. Otd. Teck. Nank. No. 8, pp. 26-31, 1958.

[6] Terrien, M., "Emission acoustique et comportement post-critique d'un béton sollicité en traction", Bulletin de liaison des Ponts et Chaussées, No. 105, pp. 65-72, 1980.

[7] Kupfer, H., Hilsdorf, H. and Ruch, H., "Behavior of concrete under biaxial stresses", J. of Aci., Vol. 60, pp. 209-224, 1969.

[8] Bascoul, A. and Maso, Jc., "Influence de la contrainte intermédiaire sur le comportement mécanique du béton en compression biaxiale", Matériaux et Constructions, No. 42, pp. 411-419, 1977.

[9] Weibull, W., "A statistical theory of the strength of material", Proceedings Roy. Swed. Institute Eng. Res, No. 151, 1939.

[10] Jayatilaka, A. de S., "Fracture of engineering brittle materials", Applied Science Publishers, Ltd., London, 1979.

LIST OF PARTICIPANTS

Ang, K. K.
Department of Civil Engineering
University of New South Wales
P.O. Box 1
Kensington, New South Wales 2033

Atluri, S. N.
Department of Civil Engineering
Georgia Institute of Technology
Atlanta, Georgia 30332 USA

Austin, S.
Engineering Development Establishment
Department of Defence Support
Private Mail Bag 12
P.O. Ascot Vale
Victoria, Australia 3032

Baker, P. W.
Alcan Australia Ltd.
Granville, New South Wales

Bandyopadhyay, S.
Materials Research Laboratories
P.O. Box 50
Ascot Vale
Victoria, Australia 3032

Bathgate, B. G.
Department of Mechanical Engineering
Deakin University
Geelong, Victoria, Australia 3220

Beever, M. A.
Head of School
Engineering Sciences
Bendigo College of Advanced Education
P.O. Box 199
Bendigo, Victoria, Australia 3550

Blicblau, A.
BHP Melbourne Research Laboratories
P.O. Box 264
Clayton, Victoria 3168

Broughton, W. R.
Royal Melbourne Institute of Technology
Latrobe Street
Melbourne, Victoria, Australia 3000

Brown, K. R.
Kaiser Aluminum
P.O. Box 877
Pleasanton, California 94566 USA

Bunyan, P. J.
Department of Mining and Metallurgy
University of Melbourne
Parkville, Victoria, Australia 3052

Callinan, R. J.
Aeronautical Research Laboratories
P.O. Box 4331
Melbourne, Victoria, Australia 3001

Carter, R. A.
Electricity Trust of South Australia
P.O. Box 6
Eastwood, South Australia 5063

Cheong, Y.
Department of Mechanical Engineering
University of Melbourne
Parkville, Victoria, Australia 3052

Chipperfield, C. G.
BHP Melbourne Research Laboratories
P.O. Box 264
Clayton, Victoria, Australia 3168

Clark, G.
Materials Research Laboratories
P.O. Box 50
Ascot Vale
Victoria, Australia 3032

Clayton, J. Q.
Aeronautical Research Laboratories
P.O. Box 4331
Melbourne, Victoria, Australia 3001

Corderoy, D. J.
School of Metallurgy
University of New South Wales
Kensington, New South Wales 2033

Cotterell, B.
Department of Mechanical Engineering
University of Sydney
Sydney, New South Wales 2006

Davis, M. J.
Aeronautical Research Laboratories
P.O. Box 4331
Melbourne, Victoria, Australia 3001

deMorton, M. E.
Materials Research Laboratories
P.O. Box 50
Ascot Vale
Victoria, Australia 3032

Dentry, C. S.
Aeronautical Research Laboratories
P.O. Box 4331
Melbourne, Victoria, Australia 3001

Dixon, B.
Materials Research Laboratories
P.O. Box 50
Ascot Vale
Victoria, Australia 3032

Drew, M.
School of Metallurgy
University of New South Wales
Kensington, New South Wales 2033

Dudley, N.
Hammersly Iron
P.O. Box 21
Dampier, West Australia 6713

Edwards, L.
Department of Metallurgy
Oxford University
Oxford, OX1 3PH
United Kingdom

Ellery, A. R.
Herman Research Laboratories
S.E.C.V.
Richmond, Victoria, Australia 3107

Ferguson, G. W.
University of Auckland
School of Engineering
Auckland, New Zealand

Fletcher, L.
Steel Mains Pty. Ltd.
63 Haigh Street
South Melbourne, Victoria, Australia 3205

Foote, R. M. L.
Department of Mechanical Engineering
University of Sydney
Sydney, New South Wales 2006

Ford, D. G.
Aeronautical Research Laboratories
P.O. Box 4331
Melbourne, Victoria, Australia 3001

Ford, G.
Department of Mechanical Engineering
University of Alberta
Edmonton, Alberta T6G 2G8, Canada

Ford, P. R.
Herman Research Laboratories
S.E.C.V.
Richmond, Victoria, Australia 3121

Goldsmith, N. T.
Aeronautical Research Laboratories
P.O. Box 4331
Melbourne, Victoria, Australia 3001

Graham, D. A.
Aeronautical Research Laboratories
P.O. Box 4331
Melbourne, Victoria, Australia 3001

Gratzer, L. R.
Aeronautical Research Laboratories
P.O. Box 4331
Melbourne, Victoria, Australia 3001

Griffiths, J. R.
Materials Engineering Department
Monash University
Clayton, Victoria, Australia 3168

Grundy, P.
Civil Engineering Department
Monash University
Clayton, Victoria, Australia 3168

Guest, P.
Fluor Australia Pty. Ltd.
GPO Box 1320L
Melbourne, Victoria, Australia 3001

Harris, F.
Aeronautical Research Laboratories
P.O. Box 4331
Melbourne, Victoria, Australia 3001

Ishihara, I.
Ishikawajima Harima Heavy Industry
  Co.
3-5 Mukodai-cho
Tokyo 188, Japan

Isida, M.
Kyushu University
6-10-1 Hakozaki
Fukuoka 812, Japan

Jaffery, D.
Royal Melbourne Institute of Technology
GPO Box 2476V
Melbourne, Victoria, Australia 3001

Jones, R.
Aeronautical Research Laboratories
P.O. Box 4331
Melbourne, Victoria, Australia 3001

Jost, G. S.
Aeronautical Research Laboratories
P.O. Box 4331
Melbourne, Victoria, Australia 3001

Karihaloo, B.
Civil Engineering Department
University of Newcastle
Newcastle, New South Wales 2308

Knott, J. F.
University of Cambridge
Department of Metallurgy
Cambridge, CB2 3QZ
United Kingdom

Kohler, G.
Australian Iron and Steel
P.O. Box 1854
Wollongong, New South Wales 2500

Lam, Y. C.
Department of Mechanical Engineering
University of Melbourne
Parkville, Victoria, Australia 3052

Leicester, R. H.
CSIRO
Graham Road
Highett, Victoria, Australia 3190

Li, Q. J.
Mechanical Engineering Department
University of Sydney
Sydney, New South Wales 2006

Mai, Y.-W.
Department of Mechanical Engineering
University of Sydney
Sydney, New South Wales 2006

Marsh, K. J.
National Engineering Laboratories
East Kilbride
Glasgow, Scotland

Martin, G. G.
Materials Engineering Department
Monash University
Clayton, Victoria, Australia 3168

Mazars, J.
Ecole Normale Superieure de L'ensei
  Griment Technique
61, Avenue du President Wilson
94230 Cachan, France

Moody, P.
Vickers Ruwolt
530 Victoria Street
Richmond, Victoria, Australia 3121

Murakami, R.
Tokushima University
2-1 Minami-josanjima-cho
Tokushima 770, Japan

Murray, M. T.
CSIRO
Box 71
Fitzroy, Victoria, Australia 3065

Murti, V.
Department of Civil Engineering
University of New South Wales
Kensington, New South Wales 2033

Nunomura, S.
Research Laboratory of Precision
Tokyo Institute of Technology
Yokohama 227, Japan

Perger, G.
CSIRO
P.O. Box 71
Fitzroy, Victoria, Australia 3065

Pham, L.
CSIRO
Graham Road
Highett, Victoria, Australia 3190

Pluvinage, G.
Laboratoire de Fiabilite Mecanique
Ile du Saulcy
F-57000 Metz, France

Pollock, J. T. A.
CSIRO
Lucas Heights Research Laboratories
Sutherland, New South Wales 2232

Pollock, W. J.
Aeronautical Research Laboratories
P.O. Box 4331
Melbourne, Victoria, Australia 3001

Radon, J. C.
Imperial College
Exhibition Road
London SW7 2BX, United Kingdom

Raju, K. N.
National Aeronautical Laboratory
Materials Science Division
Bangalore, India

Rao, Y. V. A.
Department of Civil Engineering
University of New South Wales
Kensington, New South Wales 2033

Reid, S. G.
CSIRO
Graham Road
Highett, Victoria, Australia 3190

Revill, G. W.
Aeronautical Research Laboratories
P.O. Box 4331
Melbourne, Victoria, Australia 3001

Ripley, M.
AAEC Research Establishment
Lucas Heights, Sutherland, New South
  Wales 2232

Ritter, J. C.
Materials Research Laboratories
Cordite Avenue
Maribyrnong, Victoria, Australia 3032

Rose, L. R. F.
Aeronautical Research Laboratories
P.O. Box 4331
Melbourne, Victoria, Australia 3001

Ryan, N. E.
Aeronautical Research Laboratories
P.O. Box 4331
Melbourne, Victoria, Australia 3001

Saeki, N.
Hokkaido University
Kit 13, Nishi-8, Kataku, Sapporo 060
Japan

Sih, G. C.
Institute of Fracture and Solid Mechanics
Lehigh University
Bethlehem, Pennsylvania 18015 USA

Smith, C. W.
Virginia Polytechnic Institute and
  State University
Blacksburg, Virginia 24061 USA

Snowden, K.
AAEC Research Establishment
Lucas Heights
Sutherland, New South Wales 2232

Stefoulis, C.
Aeronautical Research Laboratories
P.O. Box 4331
Melbourne, Victoria, Australia 3001

Suezawa, Y.
Nihon University
1-8 Kanda Surugada, Chiyoda-ku
Tokyo 101, Japan

Teh, K. K.
Department of Mechanical Engineering
Melbourne University
Parkville, Victoria, Australia 3052

Teh, S. H.
Department of Civil Engineering
Monash University
Clayton, Victoria, Australia 3168

Turner, C. E.
Imperial College
Department of Mechanical Engineering
Exhibition Road
London SW7 2BX, United Kingdom

Ueda, S.
Osaka City University
3-3-138 Sugimoto-cho
Sumyoshi ku, Osaka 558, Japan

Wade, J. B.
Snowy Mountains Engineering Company
P.O. Box 356
Cooma, New South Wales 2630

Walsh, P. F.
CSIRO
Graham Road
Highett, Victoria, Australia 3190

Wanhill, R. J. H.
National Aerospace Laboratory NLR
P.O. Box 153
8300 Ad Emmeloor
The Netherlands

Ward, I.
Sandvik Australia
Smithfield, New South Wales 2164

Weaver, C.
Materials Research Laboratories
Cordite Avenue
Maribyrnong, Victoria, Australia 3032

Williams, J. F.
Department of Mechanical Engineering
University of Melbourne
Parkville, Victoria, Australia 3052

Woodward, R.
Materials Research Laboratories
P.O. Box 50
Ascot Vale, Victoria, Australia 3032

Yum, Y.-H.
Seoul National University
San 56-1
Sinlim Dong, Kwanak-ku
South Korea

Zhang, D.-Z.
Department of Mechanical Engineering
University of Sydney
Sydney, New South Wales 2006